Bioinformatics with Python Cookbook

Fourth Edition

Solve advanced computational biology problems and build production pipelines with Python and AI tools

Shane Brubaker

<packt>

I0034677

Bioinformatics with Python Cookbook

Fourth Edition

Copyright © 2025 Packt Publishing

All rights reserved. No part of this book may be reproduced, stored in a retrieval system, or transmitted in any form or by any means, without the prior written permission of the publisher, except in the case of brief quotations embedded in critical articles or reviews.

Every effort has been made in the preparation of this book to ensure the accuracy of the information presented. However, the information contained in this book is sold without warranty, either express or implied. Neither the author, nor Packt Publishing or its dealers and distributors, will be held liable for any damages caused or alleged to have been caused directly or indirectly by this book.

Packt Publishing has endeavored to provide trademark information about all of the companies and products mentioned in this book by the appropriate use of capitals. However, Packt Publishing cannot guarantee the accuracy of this information.

Portfolio Director: Sunith Shetty
Relationship Lead: Sanjana Gupta
Project Manager: Shashank Desai
Content Engineer: Tiksha Abhimanyu Lad
Technical Editor: Seemanjay Ameriya
Copy Editor: Safis Editing
Indexer: Tejal Soni
Proofreader: Tiksha Abhimanyu Lad
Production Designer: Deepak Chavan
Growth Lead: Merlyn M Shelley

First published: June 2015
Second edition: November 2018
Third edition: September 2022
Fourth edition: December 2025

Production reference: 1121225

Published by Packt Publishing Ltd.
Grosvenor House
11 St Paul's Square
Birmingham
B3 1RB, UK.

ISBN 978-1-83664-275-6
www.packtpub.com

I dedicate this book to my parents, Dale and Anita, who gave me strength and integrity. I also dedicate this book to my children, Aiden and Preston, who have brought me joy and life.

– Shane Brubaker

Contributors

About the author

Shane Brubaker is a bioinformatics manager living in California. He believes in the power of bioinformatics as an interdisciplinary science to save lives and transform society. Shane has applied bioinformatics in areas ranging from synthetic biology to human health. Over the years, he has taught courses in computer science and biology, co-founded BayBifx, a leading Bay Area bioinformatics networking event, and mentored many bioinformatics professionals. Shane is passionate about training and providing opportunities for the next generation of scientists.

I would like to especially thank the technical reviewers, Betsy Alford and Darryl Leon. Their tireless efforts made the book much stronger. I also wish to thank Jason Chin for writing the foreword. Finally, I wish to thank the entire Packt team for their amazing support and diligent attention to detail.

About the reviewers

Betsy Alford, PhD, is a research scientist in the San Francisco Bay Area. She is fascinated by large-scale cross-species communication and evolution, often focusing on the nexus between biochemistry, molecular biology, and bioinformatics. Betsy has been working in computational biology since 2014, concentrating on genomic data infrastructure and machine learning applications to understand multidimensional biological data.

Darryl Leon has over 15 years of experience in leading product management teams, delivering software product strategies, launching customer-focused products, and partnering with R&D groups to deliver high-quality software, hardware, and cloud-based solutions that support the academic, biopharma, and clinical research markets. He has worked at Agilent, Illumina, Thermo Fisher Scientific, and several small companies. He received his PhD in Biochemistry from U.C. San Diego.

Preface

Bioinformatics is an incredibly transformative field that is drawing more and more interest in this era of computationally driven science. From its early days of developing the first sequence atlases and alignment tools to the excitement of the Human Genome Project, bioinformatics has always been a highly interdisciplinary field. It draws those passionate about using computation to advance science in ways never before achievable.

This book will teach you practical methods for solving real-world problems in bioinformatics. We will cover all the major topics in bioinformatics to give you a solid foundation in each area. The book will be taught using Python, which has become the predominant language in bioinformatics. In each recipe, we'll show you the important Python libraries and programming techniques used to solve relevant problems, and give you a wealth of resources to learn more and grow your skills.

We'll start with a gentle introduction to Python and basic data processing and data science techniques. We will then cover core algorithms in bioinformatics for **genome alignment**, **assembly**, **annotation**, and **variant calling**.

Next, we'll dive into the wealth of public databases available for bioinformatics data and learn how to navigate them. We'll then cover important applications of bioinformatics, such as **phylogenetics**, **metabolic modeling**, and **genome editing**.

We'll give you a solid introduction to **cloud computing** and cover the major **workflow systems** in bioinformatics that are used to orchestrate complex pipelines. Finally, we'll discuss the growing impact of **machine learning** in bioinformatics and show you how scientists are using **large language models** to design new proteins and even entire genomes. We'll finish by talking about exciting frontier topics in bioinformatics, such as **microfluidics**, **single-cell analysis**, and **connectomics**.

This book provides practical examples of using Python libraries in your work. The book takes an AI-first approach to coding. You will learn how to use AI tools to upgrade, debug, and test your code. Throughout the book, AI tips are provided to help you take your learning further. You will be guided to extensive resources on programming, with an emphasis on practical applications in the workplace. You will also be exposed to many scientific resources to inspire meaningful application of the tools to real-world problems.

This fourth edition has been extensively upgraded with modernized code, new libraries, new approaches and applications, and new chapters on the latest topics.

Who this book is for

This book is for software developers and bioinformatics professionals looking to deepen their knowledge and expand their skill set. The book is perfect for scientists and laboratory professionals who want to become interdisciplinary data scientists. The book is also excellent for those who have experience in another language, such as Java, and want to make the bridge into Python programming.

A basic background in programming will be useful for approaching the book. Previous exposure to some scientific training and basic concepts in biology will also be useful. The book is designed with extensive resources along the way, so if you need to brush up on an area as you go, you can pause and take time to deepen your training so that you get the most out of the book.

What this book covers

Chapter 1, Computer Specifications and Python Setup, will show you how to get access to the necessary systems and software, download the code for the book, and set up your environment for success.

Chapter 2, Basics of Data Manipulation, will give you a gentle introduction to Python programming with pandas. You'll learn about DataFrame manipulation using an interesting vaccine dataset.

Chapter 3, Modern Coding Practices and AI-Generated Coding, is an important chapter that gives you a solid grounding in modern coding practices used in the workplace. Here, we introduce you to AI tools for coding that you will use extensively throughout the book.

Chapter 4, Data Science and Graphing, teaches you core concepts in scientific computing using NumPy. You'll use scikit-learn to perform fundamental data science techniques such as principal components analysis and clustering. We'll also give you an early introduction to concepts in machine learning using decision trees. Finally, we'll teach you the fundamentals of graphing and visualization using tools such as Matplotlib and Seaborn, which you'll then use throughout the book to build striking visualizations of your analyses.

Chapter 5, Alignment and Variant Calling, is a critical chapter that introduces traditional bioinformatics algorithms used in next-generation sequencing for quality control, sequence file manipulation, short-read alignment, and variant calling.

Chapter 6, Annotation and Biological Interpretation, dives into the fascinating world of biological function. You'll learn how to annotate variants and entire genomes, and predict the impact of changes in gene and protein structure on an organism.

Chapter 7, Genomes and Genome Assembly, will show you how modern technologies are enabling the construction of complete, highly polished genomes. You'll learn how to access these genomes, assess their quality, and even build them yourself.

Chapter 8, Accessing Public Databases, gives you the tools to navigate the huge wealth of information available on genes, genomes, and protein sequences.

Chapter 9, Protein Structure and Proteomics, covers the fundamentals of protein structures and how to navigate the important file formats used. We'll also introduce proteomics analysis.

Chapter 10, Phylogenetics, explains how to compare sequences across evolutionary time. These techniques are valuable in areas such as intellectual property, disease outbreak analysis, and protein design using comparative genomics.

Chapter 11, Population Genetics, details how scientists can use tools for genome association analysis to track changes in populations over time.

Chapter 12, Metabolic Modeling and Other Applications, goes over several exciting ways that bioinformatics is being used in modern industry. You'll learn how we can model the effects of genetic alterations in organisms to make valuable industrial products. You'll see how powerful RNA technology can be used to alter gene expression. You'll also learn how bioinformatics is being applied to build novel, nutritious foods and improve aquaculture. We'll also see how to discover novel genes from scratch to make powerful drugs and chemicals.

Chapter 13, Genome Editing, explores the exciting world of CRISPR gene editing, which has exploded in recent years. You'll see how to design guide RNAs to edit genomes and resolve the edits using nanopore technology. We'll also look at how to analyze ultra-high-throughput experiments based on genomic barcodes.

Chapter 14, Cloud Basics, will give you a solid grounding in modern cloud computing. You'll set up your own AWS account and learn how to interact with it programmatically. We'll also introduce containers, the fundamental unit of modern bioinformatics workflow-based computing.

Chapter 15, Workflow Systems, introduces you to building structured bioinformatics pipelines using widely adopted workflow engines. You'll learn how to use Galaxy to run analyses through an accessible, web-based platform, and then explore how to create reproducible, scalable workflows with Snakemake and Nextflow. Through these tools, you will see how modern workflow systems streamline complex task orchestration and help you manage data, dependencies, and execution across diverse environments.

Chapter 16, More Workflow Systems, extends your understanding of workflow orchestration by focusing on systems designed for large-scale, cloud-native bioinformatics. You'll learn how to build flexible, maintainable pipelines with Flyte, a Kubernetes-native workflow platform, and then explore how to launch your own orchestration layer in AWS using Step Functions. This chapter shows you how to integrate state machines and event-driven components to run robust, automated pipelines in production-grade cloud settings.

Chapter 17, Deep Learning and LLMs for Nucleic Acid and Protein Design, provides an exciting look at how machine learning is intersecting with bioinformatics. You'll construct a neural network to perform genetic testing analysis. You'll learn how scientists can design novel proteins using language models just like those used in today's popular AI tools. You'll also see how scientists are on the cusp of designing entire genomes from scratch.

Chapter 18, Single-Cell Technology and Imaging, discusses some of the latest exciting fields where bioinformatics is playing a role. You'll see how microfluidics devices can be designed to deliver single-cell precision, revealing new levels of biological insight. You'll learn about the basics of image analysis using machine learning, and see how scientists are now using these tools to map the human brain.

To get the most out of this book

Here are a few things you should possibly know about:

- You should have a basic understanding of a programming language to use this book.

- Take the time to pursue the resources provided in the book if you think you need to brush up on a topic to get the most out of a section.

- The book is best performed on a modern MacBook or macOS computer. However, alternatives are provided if you do not have one.

At the top level of the GitHub repository, you will find a README.md file. This is a Markdown file that can be read with any text editor. This file will contain updates to information and code in the book. There will also be a README.md file within each chapter directory with more detailed information. These files will inform you about important bug fixes and code updates in the recipes.

Download the example code files

The code bundle for the book is hosted on GitHub at https://github.com/PacktPublishing/Bioinformatics-with-Python-Cookbook-Fourth-Edition. We also have other code bundles from our rich catalog of books and videos available at https://github.com/PacktPublishing. Check them out!

Conventions used

There are a number of text conventions used throughout this book.

CodeInText: Indicates code words in text, database table names, folder names, filenames, file extensions, pathnames, dummy URLs, user input, and Twitter handles. For example: "The code for this recipe can be found in Ch05/Ch05-1-qc-data.ipynb."

A block of code is set as follows:

```
import numpy as np
import pandas as pd
vdata = pd.read_csv(
    "2021VAERSDATA.csv.gz",
    encoding="iso-8859-1"
)
vdata.info(memory_usage="deep")
```

Any command-line input or output is written as follows:

```
conda create -n ch04-data-science --clone bioinformatics_base
conda activate ch04-data-science
```

Bold: Indicates a new term, an important word, or words that you see on the screen. For instance, words in menus or dialog boxes appear in the text like this. For example: "Click on Sequence Quality Histograms."

> **Tips or important notes**
> Appear like this.

Sections

In this book, you will find several headings that appear frequently (*Getting ready*, *How to do it...*, *How it works...*, *There's more...*, and *See also*).

To give clear instructions on how to complete a recipe, use these sections as follows:

Getting ready

This section tells you what to expect in the recipe and describes how to set up any software or any preliminary settings required for the recipe.

How to do it...

This section contains the steps required to follow the recipe.

How it works...

This section usually consists of a detailed explanation of what happened in the previous section.

There's more...

This section consists of additional information about the recipe in order to make you more knowledgeable about the recipe.

See also

This section provides helpful links to other useful information for the recipe.

Get in touch

Feedback from our readers is always welcome.

General feedback: If you have questions about any aspect of this book or have any general feedback, please email us at customercare@packt.com and mention the book's title in the subject of your message.

Errata: Although we have taken every care to ensure the accuracy of our content, mistakes do happen. If you have found a mistake in this book, we would be grateful if you reported this to us. Please visit http://www.packt.com/submit-errata, click **Submit Errata**, and fill in the form.

Piracy: If you come across any illegal copies of our works in any form on the internet, we would be grateful if you would provide us with the location address or website name. Please contact us at copyright@packt.com with a link to the material.

If you are interested in becoming an author: If there is a topic that you have expertise in and you are interested in either writing or contributing to a book, please visit http://authors.packt.com/.

Share your thoughts

Once you've read *Bioinformatics with Python Cookbook, Fourth Edition*, we'd love to hear your thoughts! Scan the QR code below to go straight to the Amazon review page for this book and share your feedback.

https://packt.link/r/183664275X

Your review is important to us and the tech community and will help us make sure we're delivering excellent quality content.

Free Benefits with Your Book

This book comes with free benefits to support your learning. Activate them now for instant access (see the "*How to Unlock*" section for instructions).

Here's a quick overview of what you can instantly unlock with your purchase:

PDF and ePub Copies   **Next-Gen Web-Based Reader**

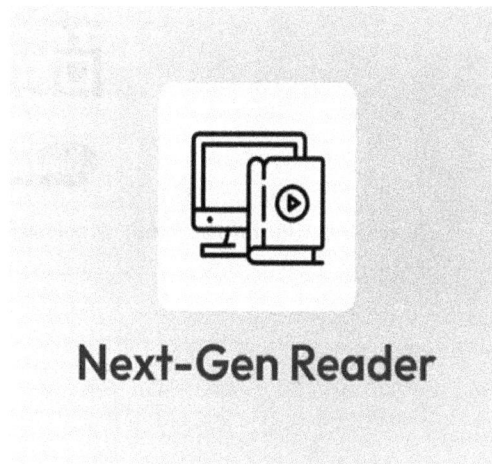

Free PDF and ePub versions

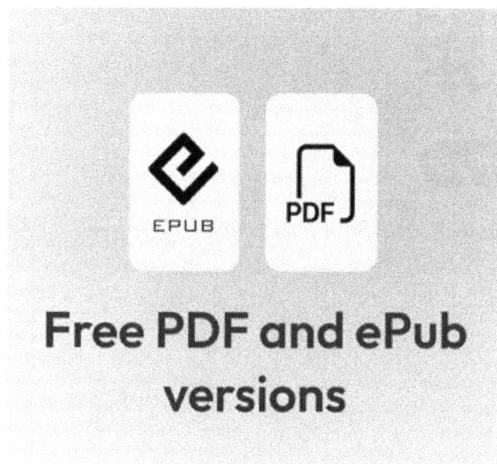

Next-Gen Reader

Access a DRM-free PDF copy of this book to read anywhere, on any device.

Multi-device progress sync: Pick up where you left off, on any device.

Use a DRM-free ePub version with your favorite e-reader.

Highlighting and notetaking: Capture ideas and turn reading into lasting knowledge.

Bookmarking: Save and revisit key sections whenever you need them.

Dark mode: Reduce eye strain by switching to dark or sepia themes

How to Unlock

Scan the QR code (or go to `packtpub.com/unlock`). Search for this book by name, confirm the edition, and then follow the steps on the page.

UNLOCK NOW

Note: Keep your invoice handy. Purchases made directly from Packt don't require one

Table of Contents

3

Modern Coding Practices and AI-Generated Coding   33

4

Data Science and Graphing   55

5

Alignment and Variant Calling   93

6

Annotation and Biological Interpretation   129

9

Protein Structure and Proteomics   211

10

Phylogenetics   239

11

Population Genetics   273

12

Metabolic Modeling and Other Applications   301

13

Genome Editing   353

14

Cloud Basics   395

15

Workflow Systems   423

16

More Workflow Systems   453

17

Deep Learning and LLMs for Nucleic Acid and Protein Design   477

18

Single-Cell Technology and Imaging   523

19

Unlock Your Exclusive Benefits   565

1

Computer Specifications and Python Setup

We will start by installing the basic software that is required for most of this book. This will include the **Python** distribution, some fundamental Python libraries, and our Jupyter Notebook environment. We will also set up our GitHub environment and gain access to the repository for the book. As different users have different requirements, we will cover two different approaches for installing the software. One approach is using the Anaconda Python (http://docs.continuum.io/anaconda/) distribution and the other is via Docker (a server virtualization method based on containers sharing the same operating system kernel; please refer to https://www.docker.com/). This will still install Anaconda for you but inside a container.

If you are using a Windows-based operating system, you are strongly encouraged to consider changing your operating system or using Docker via some of the existing options on Windows. On macOS, you should be able to install most of the software natively, though Docker is also available. Learning using a local distribution (Anaconda or something else) is easier than Docker, but given that package management can be complex in Python, Docker images provide a level of stability.

Most modern data scientists use a Mac due to the ease with which you can interact with a native Linux-style operating system. We recommend using a similar computer for this book. In the *Technical requirements* section, we provide the specifications of the computer and libraries used to develop this book. In most cases, deviations from such a system should work fine with minimal modifications, but if you have trouble, you can try the Docker container. Another alternative could be to use a cloud workstation (some options follow).

In this chapter, we will cover the following recipes:

- Installing the required software with Anaconda
- Installing the required software with Docker
- Introduction to Jupyter Notebook

In this chapter, we will first install some prerequisite software – details of which are given in the *Technical requirements* section. Each recipe will then take you through the software and the steps that are needed to install it. Each chapter and section might have extra requirements on top of these – we will make those clear as the book progresses. An alternative way to start is to use the *Installing the required software with Docker* recipe, after which everything will be taken care of for you via a Docker container.

Technical requirements

It is important to note that, starting around 2020, Apple came out with its own chips instead of using Intel chips – these are referred to as **Apple silicon chips**. This change caused issues with older versions of Python and with some Python libraries. We should not encounter these in this book since we are using newer versions, but it is something to keep in mind if you ever run into issues. This book was developed on an Apple M2 Pro chip, which is an *Apple silicon chip*.

The code in this book was developed and tested on a computer with the following specifications:

- MacBook Pro
- Chip: Apple M2 Pro
- RAM: 16 GB
- OS: Sequoia 15.1
- Python version: 3.12.2

If you do not have a MacBook, there are several options you could explore using **cloud workstations**. These options will give you a handy alternative where you can set up the appropriate environment.

If you have an AWS account, you can set up a macOS EC2 instance to use as a cloud workstation. They offer an M2 Mac Pro option matching the computer this book was developed on.

Here are some additional sites you can check out if you want to use a Mac workstation on the cloud:

- Mac In Cloud: `https://www.macincloud.com/`
- Mac Stadium: `https://www.macstadium.com/vdi`

In the next section, we'll walk through setting up your system and making sure you have Python and some of the basic libraries you need (we'll install more libraries as we go).

Free Benefits with Your Book

Your purchase includes a free PDF copy of this book along with other exclusive benefits. Check the *Free Benefits with Your Book* section in the Preface to unlock them instantly and maximize your learning experience.

Installing the required basic software with Anaconda

Next, we will begin setting up your required software libraries, including Python itself. If you are already using a different Python distribution, you are strongly encouraged to consider Anaconda, as it has become the *de facto* standard for data science and bioinformatics. Also, it is the distribution that will allow you to install software from **bioconda** (`https://bioconda.github.io/`).

Getting ready

Python can be run on top of different environments. For instance, you can use Python inside the **Java Virtual Machine (JVM)** (via **Jython** or with .NET via **IronPython**). However, here, we are not only concerned with Python but also with the complete software ecology around it. Therefore, we will use the standard (**CPython**) implementation, since the JVM and .NET versions exist mostly to interact with the native libraries of these platforms.

For our code, we will be using Python 3.12. If you were starting with Python and bioinformatics, any operating system would work. But here, we are mostly concerned with intermediate to advanced usage, and so we will focus on macOS.

If you are on Windows and do not have easy access to macOS or Linux, don't worry. Modern virtualization software (such as **VirtualBox** and **Docker**) will come to your rescue, which will allow you to install a virtual OS on your operating system. Another option is to use **Windows Subsystem for Linux (WSL2)**, which allows you to run Linux on Windows. For documentation on WSL2, look here:

- `https://learn.microsoft.com/en-us/windows/wsl/install`
- `https://www.windowscentral.com/how-install-wsl2-windows-10`
- `https://docs.docker.com/desktop/features/wsl/`

Another option for you will be to use a cloud workstation (see the *Technical requirements* section).

Bioinformatics and data science are moving at breakneck speed; this is not just hype, it's a reality. When installing software libraries, choosing a version might be tricky. Depending on the code that you have, it might not work with some old versions or perhaps not even work with a newer version. Hopefully, any code that you use will indicate the correct dependencies – though this is not guaranteed. In this book, we will fix the precise versions of all software packages, (or provide you with a minimal version, or specify one in the associated chapter YAML file as appropriate. Check your chapter's README.md file or the Updates section of each notebook for more information.) and we will make sure that the code will work with them. It is quite natural that the code might need tweaking with other package versions.

The software developed for this book is available at `https://github.com/PacktPublishing/Bioinformatics-with-Python-Cookbook-fourth-edition`. To access it, you will need to install Git. First, make sure HomeBrew is installed (`https://brew.sh/`):

```
brew install git
```

You can go to the GitHub page for the book and get the HTTPS link for downloading the source:

```
git clone https://github.com/PacktPublishing/Bioinformatics-with-
Python-Cookbook-Fourth-Edition.git
```

This will download the source code to your computer.

Before you install the Python stack, you will need to install all of the external non-Python software that you will be interoperating with. The list will vary from chapter to chapter, and all chapter-specific packages will be explained in their respective chapters. Most of the software is available via bioconda (`https://bioconda.github.io/`) (also called conda for short) or is `pip` installable (`https://pypi.org/project/pip/`).

Where possible in this book, we will allow you to do everything from your Jupyter notebook, even installing the software. To do this, we will use the `!` command, which allows you to run a command that you would normally run from your Terminal from the notebook instead – for example:

```
! ls
```

This will run the `ls` or list directory command as if it had been run from the Terminal.

In some cases, for more involved installations, you will need to go into the Terminal, but we'll advise you on how to do those steps as we go through the relevant recipes.

You will need to install some development compilers and libraries, all of which are free. On Ubuntu, consider installing the `build-essential` package (`apt-get install build-essential`), and on macOS, consider **Xcode** (`https://developer.apple.com/xcode/`).

We will mention many amazing Python libraries in this book, but here is a brief overview of some of the most important ones:

Name	Application	URL	Purpose
Biopython	All chapters	`https://biopython.org/`	Bioinformatics library
Biotite	Protein Design	`https://www.biotite-python.org/latest/index.html`	MultiTool and Protein Structure
Cython	Big data	`http://cython.org/`	High performance
Dask	Big data	`http://dask.pydata.org`	Parallel processing
DendroPY	Phylogenetics	`https://dendropy.org/`	Phylogenetics
HTSeq	NGS/Genomes	`https://htseq.readthedocs.io`	NGS processing
jupytext	Notebook conversion	`https://jupytext.readthedocs.io/en/latest/`	Convert your notebook to Python text
Keras	Deep Learning	`https://keras.io/`	Higher-level library for ML
Matplotlib	Visualization	`https://matplotlib.org/`	Graphing library

Name	Application	URL	Purpose
NumPy	All chapters	`http://www.numpy.org/`	Array/matrix processing
Numba	Big data	`https://numba.pydata.org/`	High performance
Project Jupyter	All chapters	`https://jupyter.org/`	Interactive computing
PyMol	Proteomics	`https://pymol.org`	Molecular visualization
PyVCF	NGS	`https://pyvcf.readthedocs.io`	VCF processing
Pysam	NGS	`https://github.com/pysam-developers/pysam`	SAM/BAM processing
SciPy	All chapters	`https://www.scipy.org/`	Scientific computing
TensorFlow	Machine learning	`https://www.tensorflow.org/`	Machine learning library
pandas	All chapters	`https://pandas.pydata.org/`	Data processing
scikit-learn	Machine learning	`https://scikit-learn.org`	Machine learning library
seaborn	All chapters	`https://seaborn.pydata.org/`	Statistical chart library

Table 1.1 – Major Python packages that are useful in bioinformatics

We will use `pandas` to process most table data.

How to do it...

To get started, take a look at the following steps:

1. Start by downloading the Anaconda distribution from `https://www.anaconda.com/products/individual`. We will be using version 2024.06, although you will probably be fine with the most recent one. You can accept all the installation's default settings, but you might want to make sure that the `conda` binaries are in your path (do not forget to open a new window so that the path can be updated).

2. As an alternative to downloading from the website, you can use this command:

```
curl -O https://repo.anaconda.com/archive/Anaconda3-2024.06-1-MacOSX-x86_64.sh
```

3. If you have another Python distribution, be careful with PYTHONPATH and existing Python libraries. It's probably better to unset PYTHONPATH. As much as possible, uninstall all other Python versions and installed Python libraries. These steps will help reduce future confusion about which installation of Python you are pointing to.

4. Let's go ahead with the libraries. We will now create a new `conda` environment called `bioinformatics_base` with `biopython=1.84`, as shown in the following command (type it in your Terminal):

    ```
    conda create -n bioinformatics_base python=3.12
    ```

5. Let's activate the environment, as follows:

    ```
    conda activate bioinformatics_base
    ```

6. Let's add the `bioconda` and `conda-forge` channels to our source list:

    ```
    conda config --add channels bioconda
    conda config --add channels conda-forge
    ```

    Note: Conda channels are remote hosting locations that store common packages we may need.

7. Also, install the basic packages:

    ```
    ! conda install biopython==1.84 jupyterlab==4.3.0
    matplotlib==3.9.2 numpy==2.1.0 pandas==2.2.3 scipy==1.14.1
    ```

    As an alternative to the above, you can also set up your conda environment using a file that specifies the packages needed. It is provided as bioinformatics_base.yml. It is a YAML file, which stands for "YAML Ain't Markup Language" (`https://yaml.org/` To use the file run this command:

    ```
    conda env create -f ~/work/CookBook/Ch01/bioinformatics_base.yml
    ```

    This will install the required packages for you.

> **Tip**
>
> We often install the latest version of the package by just typing something like `conda install biopython`, but in this book, we will often do something called "pinning the version." This means we write an explicit version to help with the reproducibility of the code. We won't pin the version in every example throughout the book. In most cases, your code should work fine with the latest version. However, we'll include version pinning where it's necessary. If any version-specific issues arise in the future, notes will be added to the README.md file for each chapter and in the Updates section of the corresponding notebook.

8. Now, let's save our environment so that we can reuse it later to create new environments in other machines or if you need to clean up the base environment:

    ```
    conda list -e > bioinformatics_base.txt
    ```

> **Tip**
>
> On the left side of your Terminal, you will see what Anaconda environment you are in so you can always tell where you are at. For instance, right now, it should say (`bioinformatics_base`).

One thing that can be confusing is that using the `python -V` command in this environment could show an older version. This is because Python 3 is referred to via the `python3` command. To fix this, you want to alias the Python command. Typically, it is easiest to put this in your shell file, which is a file that is always run when you open a Terminal window. In Linux, it was `.bashrc`, but on macOS, you will use the `.zshrc` file (often pronounced *z-shark*).

Solution: Open your `~/.zshrc` file in a text editor

Add the following line to the end of the file:

```
alias python=python3
```

Now save it.

To run it, you can type `source ~/.zshrc`.

Now, when you run `python -V` or `python --version`, you should see that it is 3.12. If you are in a notebook and want to double-check your version, you can run `! python -V` in a cell.

There's more...

If you prefer not to use Anaconda, you will be able to install many of the Python libraries via `pip` using whatever distribution you choose. You can go through the book and keep installing packages in `bioinformatics_base` if you want. But you may, at times, find that you want to create an environment specific to a particular chapter to help isolate any complexity in package installations. Let's look at how to do that real quick:

For example, imagine you want to create an environment for machine learning with `scikit-learn`. You can do the following:

1. First, we need to deactivate our current environment:

    ```
    conda deactivate
    ```

2. Create a clone of the original environment with the following:

    ```
    conda create -n scikit-learn --clone bioinformatics_base
    ```

3. Add `scikit-learn`:

    ```
    conda activate scikit-learn
    conda install scikit-learn
    ```

See Also

*   For more information on Git, see: `https://git-scm.com/docs`
*   For an excellent course on using Git and GitHub, see: `https://www.udemy.com/course/git-and-github-bootcamp/`
*   For a review of the differences between Python 3 and Python 2, see: `https://powerfulpython.com/blog/whats-really-new-in-python-3/`

- For more background on Anaconda, see: `https://www.edureka.co/blog/python-anaconda-tutorial/`

- For a course on Anaconda, see: `https://www.udemy.com/course/anaconda-tutorial/`

Installing the required software with Docker

Docker is the most widely used framework for implementing operating system-level virtualization. This technology allows you to have an independent container: a layer that is lighter than a virtual machine but still allows you to compartmentalize software. This mostly isolates all processes, making it feel like each container is a virtual machine. Containers will be discussed in more detail in *Chapter 14, Cloud Basics*.

Docker works quite well at both extremes of the development spectrum: it's an expedient way to set up the content of this book for learning purposes and could become your platform of choice for deploying your applications in complex environments.

Conda and Docker are key tools to help maintain software compatibility and reproducibility across different systems and libraries. We'll discuss reproducibility more in *Chapter 15, Workflow Systems*.

> **Note**
>
> This recipe is an alternative to the previous recipe. Normally, if you have a Mac and are using it for your Jupyter notebooks, you will not need the Docker container. If you have a Windows machine or cannot get certain code to work in your environment, the Docker container can be useful to provide you with an environment that is set up properly already for you.

Getting ready

The first thing you have to do is install Docker. Go to `https://www.docker.com/`. Install Docker Desktop for your appropriate operating system (remember to check the Apple versus Intel silicon discussion in the *Technical requirements* section if you are using macOS). You'll also need to sign up for a Docker account and record your username and password.

How to do it...

Docker Desktop must be running and you need to be signed in before downloading the Docker file. To get started, follow these steps:

1. Use the following command from your Terminal:

    ```
    docker build -t bio https://github.com/PacktPublishing/
    Bioinformatics-with-Python-Cookbook-Fourth-Edition.
    git#main:docker/main
    ```

> **Tip**
>
> If you are using the digital version of this book, we advise you to type the code yourself or access the code from the book's GitHub repository. Doing so will help you avoid any potential errors related to the copying and pasting of code.
>
> You can find the commands for this section in the chapter's README.md file.

2. Now you are ready to run the container, as follows:

```
docker run -ti -p 9875:9875 -v YOUR_DIRECTORY:/data bio
```

3. Replace YOUR_DIRECTORY with a directory on your operating system. This will be shared between your host operating system and the Docker container. YOUR_DIRECTORY will be seen in the container in /data and vice versa.

    In this case, -p 9875:9875 will expose the container's TCP port 9875 on the host computer port, 9875.

    Especially on Windows (and maybe on macOS), make sure that your directory is actually visible inside the Docker shell environment. If not, check the official Docker documentation on how to expose directories. To access the Docker image while it's running, hover over the Docker Desktop icon. All the files available in the book's GitHub repository will be mirrored in the Docker image.

4. Now you are ready to use the system. Point your browser to http://localhost:9875 and you should get the Jupyter environment.

If this does not work on Windows, check the official Docker documentation (https://docs.docker.com/manuals/) on how to expose ports.

See also

- Docker is the most widely used containerization software and has seen enormous growth in usage in recent times. You can read more about it at https://www.docker.com/.

- A security-minded alternative to Docker is **Red Hat Openshift**, which can be found at https://www.redhat.com/en/technologies/cloud-computing/openshift.

- If you are not able to use Docker, for example, if you do not have the necessary permissions, as will be the case for most compute clusters, then take a look at Singularity at https://www.sylabs.io/singularity/.

- For a good course on Docker, see: https://www.udemy.com/course/docker-and-kubernetes-the-complete-guide/.

Introduction to Jupyter Notebook

All of our work will be developed inside Jupyter Notebook. Jupyter has become the *de facto* standard for writing interactive data analysis scripts. Unfortunately, the default format for Jupyter notebooks is based on JSON. **JSON** is **JavaScript Object Notation** (`https://www.json.org/json-en.html`). This format is difficult to read, difficult to compare, and needs exporting to be fed into a normal Python interpreter. To obviate that problem, we will extend Jupyter with `jupytext` (`https://jupytext.readthedocs.io/`), which allows us to save Jupyter notebooks as normal Python programs. We will start with an overview of Jupyter Notebook, and then look into `jupytext`. Recall that we installed Jupyter Notebook in the first recipe of this chapter, when we installed the `jupyterlab` package using conda.

How to do it...

1. To run Jupyter, on the Terminal, type the following:

   ```
   jupyter notebook
   ```

   This will open the Jupyter browser, and you will see a home page that looks something like this:

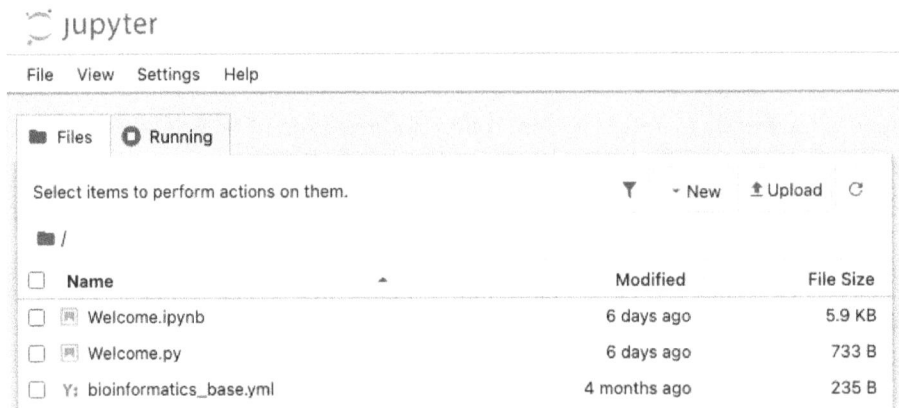

Figure 1.1 – The Jupyter browser home page

This home page gives you an overview of your files, so you can open, rename, and download them, and so on.

2. Let's click on one of the files and open it. We will see something like this:

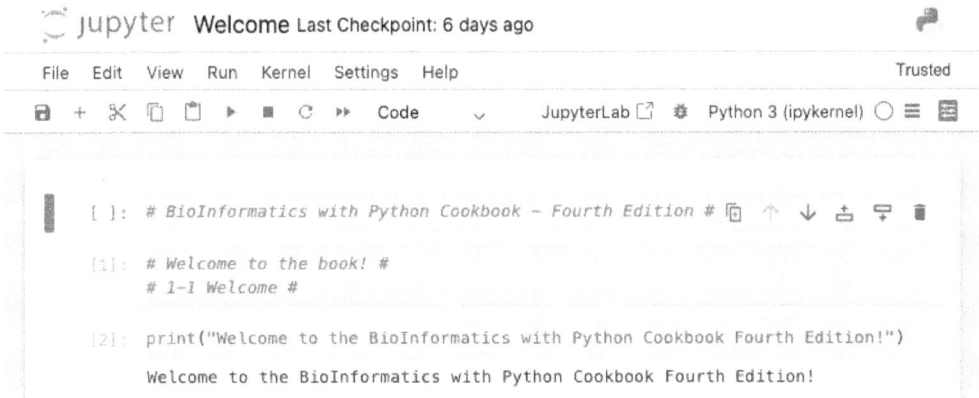

Figure 1.2 – An example of a notebook

Here, we see a menu that allows us to save or download files and perform other actions. Each cell can be executed by clicking the play button. You can also run multiple cells. When you run a cell, you will see its output below.

In some cases, you may need to restart your kernel – use the **Kernel | Restart Kernel...** method.

Jupyter notebook resources

This would be a good time to pause and take some time to learn more about Jupyter notebooks. There are numerous keyboard shortcuts that are worth learning to speed up your development:

Tutorial: https://www.datacamp.com/tutorial/tutorial-jupyter-notebook

Keyboard Shortcuts: https://towardsdatascience.com/jypyter-notebook-shortcuts-bf0101a98330

Now that we have set up our Jupyter Notebook environment, let's take a look at a handy tool called Jupytext.

Jupytext

Sometimes, you will want to convert your notebooks into formats other than `ipynb` – for example, you might want to get them into .py format. For this, we can use `jupytext - https://github.com/mwouts/jupytext`. This handy plugin will allow us to save Jupyter notebooks in formats other than `.ipynb`. Remember to get out of the Jupyter browser first. To do this on a Mac, you would close the Jupyter browser window, then go to the Terminal where you started it. Then, click *Ctrl + Z* to kill the process.

To install `jupytext`, we will run the following:

```
pip install jupytext
```

Now let's start up the Jupyter browser again:

```
jupyter notebook
```

Now we've launched the Jupyter browser again, open the **Welcome** notebook. Go to the **File | Jupytext** menu. Here is what it looks like:

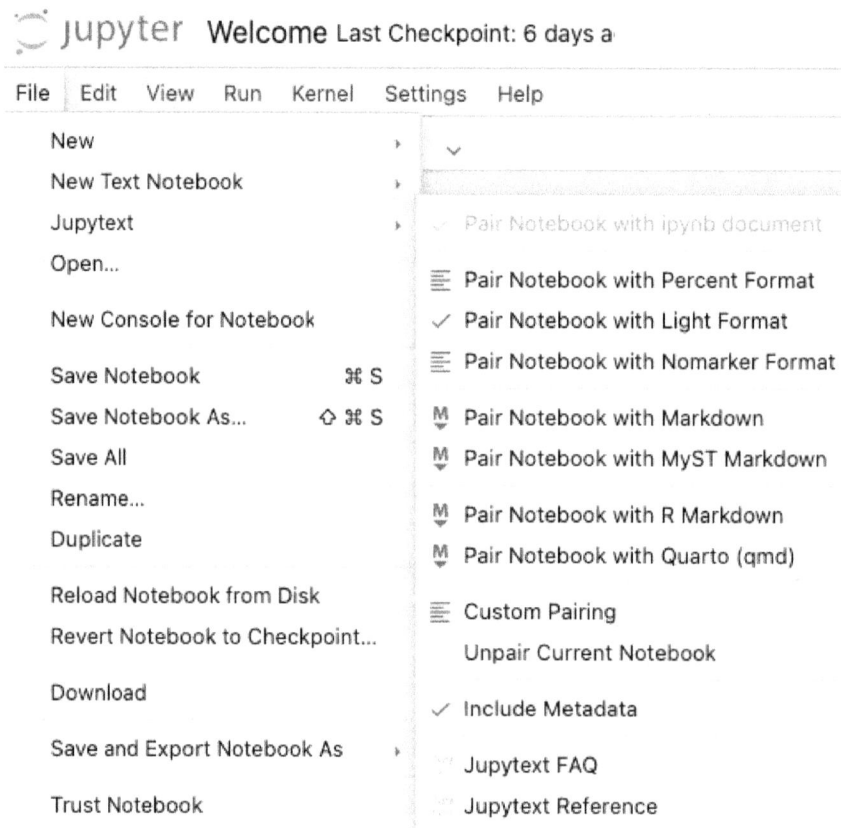

Figure 1.3 – The Jupytext menu within the Jupyter browser

To save your notebook in another format, you **pair** it and choose a format. For instance, if you choose to pair it with the light format, you will get a regular Python (`.py`) formatted file in your current working directory.

Here is what our `Welcome.py` file looks like in our working directory when paired with the `light` format:

```
(base) shanebrubaker@SHANEs-MacBook-Pro Ch01 % cat Welcome.py
# ---
# jupyter:
#   jupytext:
#     formats: ipynb,py:light
#     text_representation:
#       extension: .py
#       format_name: light
#       format_version: '1.5'
#       jupytext_version: 1.17.1
#   kernelspec:
#     display_name: Python 3 (ipykernel)
#     language: python
#     name: python3
# ---

# +
# BioInformatics with Python Cookbook - Fourth Edition #

# +
# Welcome to the book! #
# 1-1 Welcome #
# -

print("Welcome to the BioInformatics with Python Cookbook Fourth Edition!")

# +
# Install packages using Conda
# -

# ! conda install biopython==1.84 jupyterlab==4.3.0 matplotlib==3.9.2 numpy==2.1.0 pandas==2.2.3 scipy==1.14.1

# +
# Install Jupytext
# -

# ! pip install jupytext

# +
## End of Notebook ##
(base) shanebrubaker@SHANEs-MacBook-Pro Ch01 %
```

Figure 1.4 – The Welcome.py notebook in the Light format produced by Jupytext

There are several other popular formats supported by Jupytext. You can read more about them here: `https://jupytext.readthedocs.io/en/latest/index.html`.

> **Warning**
>
> Remember that the recipes in this book are normally meant to be run inside Jupyter notebooks. This means, typically, we will not always use `print` to output content. In a notebook, if you simply put the name of a variable and run it, it will print out the result for you. If you are not using notebooks (e.g. you are writing Python scripts and executing them from the terminal), you may want to add `print` statements to your code. Even within a notebook, you may find it useful at times to add your own `print` statements to inspect variables and debug code.

In addition to the Jupyter browser, there is a more integrated environment called JupyterLab: `https://jupyterlab.readthedocs.io/en/latest/`. It allows you to run Terminals and other widgets inside the same environment as your notebook. To get to it, you can click **View | Open JupyterLab**. You can check it out if you are interested, but it is not necessary to get through the book.

A welcome notebook called `Welcome.ipynb` has been placed in the GitHub repository for this book in the `Ch01` folder. You can use it to test out your notebook environment. This notebook also contains many handy links to help you learn Python and explore bioinformatics!

To recap everything, here are your main options for performing the recipes in this book:

System	Components	Pros	Cons
MacBook Pro Laptop	Anaconda; pip; brew; Jupyter	Best system for compatibility and ease of use	You may not own one
Mac Cloud Workstation or Mac AWS EC2 Instance	Anaconda; pip; brew; Jupyter	Convenient solution; identical to Mac laptop	May incur some costs
Windows Machine + Docker	Docker	Portable solution	Some increased overhead
Windows + VirtualBox or WSL2	Anaconda; pip; brew; Jupyter	Let's you interact with a Linux OS	Some installation or compatibility issues may arise
Linux Machine	Anaconda; pip; brew; Jupyter	Let's you interact with a Linux OS	Some installation or compatibility issues may arise

Table 1.2 – System and OS options for use with this book

See also

- Python Introductory Course: https://www.udemy.com/course/python-for-absolute-beginners-u/

- Intermediate Python Course: https://www.udemy.com/course/100-days-of-code/

- For basic instructions on IPython magics, see: https://ipython.readthedocs.io/en/stable/interactive/magics.html

- A list of third-party extensions for IPython, including magic ones can be found at https://github.com/ipython/ipython/wiki/Extensions-Index

- For a list of JupyterLab widgets, see: https://github.com/search?q=topic%3Ajupyterlab-extension&type=Repositories

- For a deep dive into modern Python 3.12, check out *Modern Python Cookbook* by Steven F. Lott (Packt Publishing): https://www.amazon.com/Modern-Python-Cookbook-updated-techniques/dp/1835466389

Get This Book's PDF Version and Exclusive Extras

UNLOCK NOW

Scan the QR code (or go to `packtpub.com/unlock`). Search for this book by name, confirm the edition, and then follow the steps on the page.

Note: Keep your invoice handy. Purchases made directly from Packt don't require an invoice.

2

Basics of Data Manipulation

In this chapter, we will look at some of the basic aspects of data manipulation in Python. **pandas** is the *de facto* standard for processing tabled data. It is used extensively in bioinformatics and so it is a good core library to become comfortable with.

We will start by looking at the basics of pandas as it provides a high-level library with very broad practical applicability. We will then discuss how to join tables in pandas and perform database-like operations. Finally, we'll briefly touch on some strategies to reduce memory usage in pandas.

In this chapter, we will cover the following recipes:

- Using pandas to process vaccine-adverse events

- Dealing with the pitfalls of joining pandas DataFrames

- Reducing the memory usage of pandas DataFrames

Using pandas to process vaccine-adverse events

We will be introducing pandas with a concrete bioinformatics data analysis example: we will be studying data from the **Vaccine Adverse Event Reporting System** (**VAERS**, `https://vaers.hhs.gov/`). VAERS, which is maintained by the US Department of Health and Human Services, includes a database of vaccine-adverse events going back to 1990.

COVID-19

The COVID-19 pandemic, at its height in 2020-2022, took an estimated 7 million lives globally. The rapid development of COVID vaccines using **messenger RiboNucleic Acid** (**mRNA**) technology represented a technological turning point and stemmed the tide of the pandemic. In this approach, mRNA, which is the code for proteins, is introduced into our cells to teach them to make a harmless piece of the protein in the virus, triggering an immune response. In some cases, vaccines can lead to adverse events themselves, in which side effects may cause problems for some individuals. The VAERS database records such adverse events in a public health database. In this recipe, we will study this database and learn how to process, analyze, and graph the data using Python.

VAERS makes data available in **comma-separated values** (**CSV**) format. The CSV format is quite simple and can even be opened with a simple text editor or a spreadsheet such as Excel. pandas can work very easily with this format.

Getting ready

First, we need to download the data. It is available at `https://vaers.hhs.gov/data/datasets.html`. Please download the ZIP file for 2021: we will be using the 2021 file; do not download a single CSV file only. You will download a file called `2021VAERSData.zip`. After downloading the file, unzip it with this command:

```
unzip 2021VAERSData.zip
```

Then, recompress all the files individually with `gzip -9 *csv` to save disk space. This should leave you with files such as `2021VAERSDATA.csv.gz`, which you will load into your notebook using pandas.

> **Tip on data files and locations**
>
> Our notebooks are typically written expecting the files to be in a `data` folder under the respective chapter folder, that is, `Ch02/data`. If you put your data in another place, just make sure you change the path in your notebook to point there. Remember that if you are working out of your Docker container, you will want to put the data files in whatever directory you have mapped `/data` to (see *Chapter 1*) and then ensure that the path in your notebook is pointing there.

Feel free to have a look at the files with a text editor, or preferably with a tool such as `less` (`zless` for compressed files). You can find documentation for the content of the files at `https://vaers.hhs.gov/docs/VAERSDataUseGuide_en_September2021.pdf`.

We will work with this code in our notebook; the code is provided at the beginning so that you can take care of the necessary processing. If you are using Docker, the base image is enough.

If you are going to write the code in a notebook, remember to activate your environment:

```
conda activate bioinformatics_base
```

And then enter this:

```
jupyter notebook
```

The code can be found in `Ch02/Ch02-1-pandas-basic.ipynb`.

> **Tip**
>
> Remember to start your notebook environment from within the `Ch02` directory. That way, you can see the notebooks, and any paths such as `data/filename` will be relative to that directory.

How to do it...

Follow these steps:

1. Let's start by loading the main data file and gathering the basic statistics:

    ```
    import pandas as pd
    vdata = pd.read_csv(
        "2021VAERSDATA.csv.gz", encoding="iso-8859-1"
    )
    vdata.columns
    vdata.dtypes
    vdata.shape
    ```

    > **Tip**
    >
    > If you get a Dtype error when loading the data, you can ignore it. Each of the commands in the preceding code can be run in an individual cell to see outputs about the data. You can also add the clause low_memory=False when you read in the CSV file to avoid this error.

    We start by loading the data. For most text files, the default encoding of UTF-8 will work. UTF-8 is a Unicode encoding style that can use 1-4 bytes per character. However, this file is in a style called Latin-1, or ISO-8859-1, which uses a single byte per character. So, here we use the encoding parameter of the read_csv() function to specify the type of text encoding. If you don't do this, you will get an error.

2. Next, we print the column names, which start with VAERS_ID, RECVDATE, STATE, AGE_YRS, and so on. They include 35 entries corresponding to each of the columns. Then, we print the types of each column. Here are the first few entries:

    ```
    VAERS_ID            int64
    RECVDATE            object
    STATE               object
    AGE_YRS             float64
    CAGE_YR             float64
    CAGE_MO             float64
    SEX                 object
    ```

    Figure 2.1 – Data types for columns in the dataset

    We next print vdata.shape. By doing this, we get the shape of the data: (753040, 35). This means 753,040 rows and 35 columns. You can use any of the preceding strategies to get the information you need regarding the metadata of the table.

> **Note:**
> The numbers in the output may differ slightly.

To recap, we used the following:

- `vdata.columns`: To get the names of the columns
- `vdata.dtypès`: To get the data types of the columns
- `vdata.shape`: To find out the number of rows and columns we have in the table

3. Now, let's explore the data:

```
vdata.iloc[0]
vdata = vdata.set_index("VAERS_ID")
vdata.loc[916600]
vdata.head(3)
vdata.iloc[:3]
vdata.iloc[:5, 2:4]
```

There are many ways we can look at the data:

- Using the `iloc` function to specify an integer location
- Using the `loc` function to specify a location based on a key
- Using the `head` function to get the first few rows
- Using the `iloc` function to specify particular rows or even slices of rows and columns

We will start by inspecting the first row, based on location. Here is an abridged version:

```
VAERS_ID                    916600
RECVDATE                01/01/2021
STATE                           TX
AGE_YRS                       33.0
CAGE_YR                       33.0
CAGE_MO                        NaN
SEX                              F
```

Figure 2.2 – Inspecting the first row of the dataset

After we index by VAERS_ID, we can use one ID to get a row. We can use `916600` (which is the ID from the preceding record) and get the same result.

Then, we retrieve the first three rows. Notice the two different ways we can do so:

- Using the `head` method
- Using the more general array specification; that is, `iloc[:3]`

4. Finally, we retrieve the first five rows, but only the second and third columns –iloc[:5, 2:4]. Here is the output:

VAERS_ID	AGE_YRS	CAGE_YR
916600	33.0	33.0
916601	73.0	73.0
916602	23.0	23.0
916603	58.0	58.0
916604	47.0	47.0

Figure 2.3 – Restricting the output to certain columns

5. Let's do some basic computations now, namely computing the maximum age in the dataset:

```
vdata["AGE_YRS"].max()
vdata.AGE_YRS.max()
```

The maximum value is 119 years. More importantly than the result, notice the two dialects for accessing AGE_YRS (as a dictionary key and as an object field) for the access columns.

6. Now, let's plot the ages involved:

```
vdata["AGE_YRS"].sort_values().plot(use_index=False)
vdata["AGE_YRS"].plot.hist(bins=20)
```

This generates two separate plots (in the next step, we'll put both plots side by side). We use pandas plotting machinery here, which uses **matplotlib** underneath.

7. Let's have a sneak peek at **matplotlib** here by using it directly:

```
import matplotlib.pyplot as plt
fig, ax = plt.subplots(1, 2, sharey=True)
fig.suptitle("Age of adverse events")
vdata["AGE_YRS"].sort_values().plot(
    use_index=False, ax=ax[0],
    xlabel="Obervation", ylabel="Age"
)
vdata["AGE_YRS"].plot.hist(
    bins=20, orientation="horizontal"
)
```

This includes both figures from the previous steps. Here is the output:

Age of adverse events

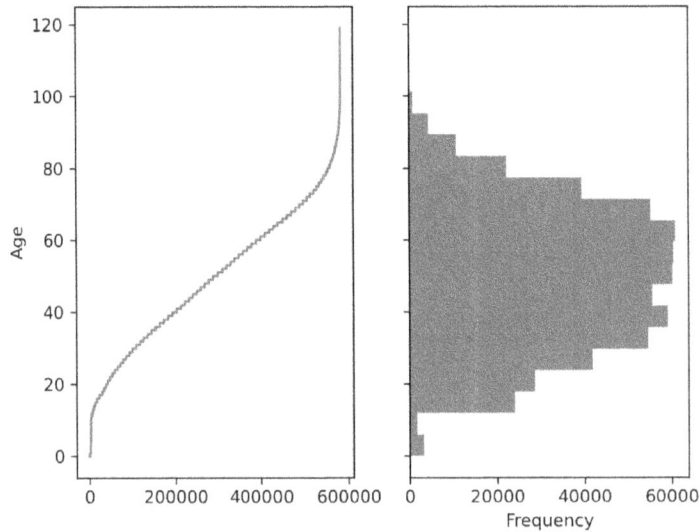

Figure 2.4 – Left – the age for each observation of adverse effect;
right – a histogram showing the distribution of ages

8. We can also take a non-graphical, more analytical approach, such as counting the events per year:

```
vdata["AGE_YRS"].dropna().apply(
    lambda x: int(x)
).value_counts()
```

The output will be as follows:

```
50      12570
65      12410
60      12235
51      12098
59      11955
         ...
115         6
119         4
106         2
113         1
109         1
Name: AGE_YRS, Length: 111, dtype: int64
```

Figure 2.5 – Counting events per year

9. Now, let's see how many people died:

```
vdata.DIED.value_counts(dropna=False)
vdata["is_dead"] = (vdata.DIED == "Y")
```

The output of the count is as follows:

```
NaN      742506
Y         10534
Name: DIED, dtype: int64
```

Figure 2.6 – The number of people who died

Note that the type of DIED is *not* a Boolean. It's more declarative to have a Boolean representation of a Boolean characteristic, so we create is_dead for it.

Tip

Here, we are assuming that NaN is to be interpreted as False. NaN stands for **Not A Number**. In general, we must be careful with the interpretation of NaN. It may mean False or it may simply mean – as in most cases – a lack of data. If that were the case, it should not be converted into False.

---

10. Now, let's associate the individual data about deaths with the type of vaccine involved:

```
dead = vdata[vdata.is_dead]
vax = pd.read_csv(
    "2021VAERSVAX.csv.gz",
    encoding="iso-8859-1"
).set_index("VAERS_ID")
vax.groupby("VAX_TYPE").size().sort_values()
vax19 = vax[vax.VAX_TYPE == "COVID19"]
vax19_dead = dead.join(vax19, lsuffix='_dead', rsuffix='_vax')
```

After we get a DataFrame containing just deaths, we must read the data that contains vaccine information. First, we must do some exploratory analysis of the types of vaccines and their adverse events. Here is the abridged output:

```
             ...
HPV9        1506
FLU4        3342
UNK         7941
VARZOS     11034
COVID19   648723
```

After that, we must choose just the COVID-related vaccines and join them with individual data.

	VAERS_ID	RECVDATE	STATE	AGE_YRS	CAGE_YR	CAGE_MO	SEX	RPT_DATE	SYMPTOM_TEXT	DIED	...	ER_ED_VISIT	ALLERGIES
191	916803	01/01/2021	LA	78.0	78.0	NaN	M	NaN	respitory colase	Y	...	NaN	N?A
491	917117	01/01/2021	AR	82.0	82.0	NaN	M	NaN	After vaccination, patient tested positive for...	Y	...	NaN	None
1254	917790	01/03/2021	AR	90.0	90.0	NaN	F	NaN	At the time of vaccination, there was an outbr...	Y	...	NaN	None
1257	917793	01/03/2021	AR	78.0	78.0	NaN	F	NaN	Prior to the administration of the COVID 19 va...	Y	...	NaN	None

Figure 2.7 – Partial output of the vax19_dead DataFrame

11. Finally, let's see the top 10 COVID vaccine lots that are overrepresented in terms of deaths and how many US states were affected by each lot:

```
baddies = (
    vax19_dead.groupby("VAX_LOT")
    .size()
    .sort_values(ascending=False)
)
for i, (lot,cnt) in enumerate(baddies.items()):
    print(
        lot, cnt, len(
            vax19_dead[
                vax19_dead.VAX_LOT == lot
            ].groupby("STATE")
        )
    )
    if i == 10:
        break
```

The output is as follows:

```
Unknown 221 32
EN6201 157 33
unknown 127 21
EN5318 120 29
EN6200 120 22
EN6198 115 24
EL9261 111 22
EL3248 102 18
EM9810 100 21
EN6202 98 19
039K20A 97 13
```

Figure 2.8 – Top COVID-19 lots

That concludes this recipe!

There's more...

The preceding data about vaccines and lots is not completely correct; we will cover some data analysis pitfalls in the next recipe.

In the *Learning more about matplotlib for Chart Generation* recipe of *Chapter 4, Data Science and Graphing*, we will go into more depth on **matplotlib**, a chart library that provides the backend for pandas plotting. It is a fundamental component of Python's data analysis ecosystem.

See also

The following is some extra information that may be useful:

- Pandas Tutorial: `https://www.w3schools.com/python/pandas/default.asp`
- For a course on pandas, see "Data Analysis with Pandas and Python," Udemy: `https://www.udemy.com/course/data-analysis-with-pandas/`

Dealing with the pitfalls of joining pandas DataFrames

The previous recipe was a whirlwind tour that introduced pandas and exposed most of the features that we will use in this book. While an exhaustive discussion about pandas would require a complete book, in this recipe – and in the next one – we are going to discuss topics that impact data analysis and are seldom discussed in the literature but are very important.

In this recipe, we are going to discuss some pitfalls that deal with relating DataFrames through joins: it turns out that many data analysis errors are introduced by carelessly joining data. We will introduce techniques to reduce such problems here.

Getting ready

We will be using the same data as in the previous recipe, but we will jumble it a bit so that we can discuss typical data analysis pitfalls. Once again, we will be joining the main adverse events table with the vaccination table, but we will randomly sample 90% of the data from each. This mimics, for example, the scenario where you only have incomplete information. This is one of the many examples where joins between tables do not have intuitively obvious results.

Use the following code to prepare our files by randomly sampling 90% of the data:

```
vdata = pd.read_csv(
    "2021VAERSDATA.csv.gz", encoding="iso-8859-1")
vdata.sample(frac=0.9).to_csv(
    "vdata_sample.csv.gz", index=False)
vax = pd.read_csv(
    "2021VAERSVAX.csv.gz", encoding="iso-8859-1")
vax.sample(frac=0.9).to_csv(
    "vax_sample.csv.gz", index=False)
```

Because this code involves random sampling, the results that you will get will be different from the ones reported in this recipe. The code for this recipe can be found in Ch02/Ch02-2-pandas-pitfalls.ipynb.

How to do it...

Follow these steps:

1. Let's start by doing an inner join of the individual and vaccine tables:

    ```
    vdata = pd.read_csv(
        "vdata_sample.csv.gz", low_memory=False)
    vax = pd.read_csv(
        "vax_sample.csv.gz", low_memory=False)
    vdata_with_vax = vdata.join(
        vax.set_index("VAERS_ID"),
        on="VAERS_ID",
        how="inner"
    )
    len(vdata), len(vax), len(vdata_with_vax)
    ```

    The len output for this code is 589,487 for the individual data, 620,361 for the vaccination data, and 558,220 for the join. This suggests that some individual and vaccine data was not captured.

2. Let's find the data that was not captured with the following join:

    ```
    lost_vdata = vdata.loc[
        ~vdata.index.isin(vdata_with_vax.index)
    ]
    lost_vdata
    lost_vax = vax[~vax["VAERS_ID"].isin(vdata.index)]
    lost_vax
    ```

    You will see that 56,524 rows of individual data aren't joined and that there are 62,141 rows of vaccine data.

3. There are other ways to join data. The default way is by performing a left outer join:

```
vdata_with_vax_left = vdata.join(
    vax.set_index("VAERS_ID"),
    on="VAERS_ID"
)
vdata_with_vax_left.groupby(
    "VAERS_ID"
).size().sort_values()
```

A left outer join ensures that all the rows on the left table are always represented. If there are no rows on the right, then all the right columns will be filled with None values.

> **Warning**
>
> There is a caveat that you should be careful with. Remember that the left table – vdata – had one entry per VAERS_ID. When you *left* join, you may end up with a case where the left-hand side is repeated several times. For example, the groupby operation that we did previously shows that VAERS_ID of 962303 has 11 entries. This is correct, but it's not uncommon to have the incorrect expectation that you will still have a single row on the output per row on the left-hand side. This is because the left join returns 1 or more left entries, whereas the inner join above returns 0 or 1 entries, where sometimes, we would like to have precisely 1 entry. Be sure to always test the output for what you want in terms of the number of entries.

4. There is a right join as well. Let's right join COVID vaccines—the left table—with death events—the right table:

```
dead = vdata[vdata.DIED == "Y"]
vax19 = vax[vax.VAX_TYPE == "COVID19"]
vax19_dead = vax19.join(
    dead.set_index("VAERS_ID"),
    on="VAERS_ID", how="right"
)
len(vax19), len(dead), len(vax19_dead)
len(vax19_dead[vax19_dead.VAERS_ID.duplicated()])
len(vax19_dead) - len(dead)
```

As you may expect, a right join will ensure that all the rows on the right table are represented. So, we end up with 583,817 COVID entries, 7,670 dead entries, and a right join of 8,624 entries.

We also check the number of duplicated entries on the joined table and we get 954. If we subtract the length of the dead table from the joined table, we also get, as expected, 954. Make sure you do checks like this when you're using joins.

Example sanity checks for joins:

- Are you using the correct join direction (left, right, inner, outer)?

- Are you missing entries you expected?

- Are row counts what you expected when looking at the size of the respective tables?

- Are you getting duplicates? If this is unexpected, you may want to use a GROUP BY clause and aggregations or calculations on other columns

5. Finally, we are going to revisit the problematic COVID lot calculations since we now understand that we might be overcounting lots:

```
vax19_dead["STATE"] = vax19_dead["STATE"].str.upper()
dead_lot = vax19_dead[
    ["VAERS_ID", "VAX_LOT", "STATE"]
].set_index(["VAERS_ID", "VAX_LOT"])
dead_lot_clean = dead_lot[
    ~dead_lot.index.duplicated()
]
dead_lot_clean = dead_lot_clean.reset_index()
dead_lot_clean[dead_lot_clean.VAERS_ID.isna()]
baddies = dead_lot_clean.groupby(
    "VAX_LOT"
).size().sort_values(ascending=False)

for i, (lot, cnt) in enumerate(baddies.items()):
    print(
        lot, cnt, len(
            dead_lot_clean[
                dead_lot_clean.VAX_LOT == lot
            ].groupby("STATE")
        )
    )
    if i == 10:
        break
```

Note that the strategies that we've used here ensure that we don't get repeats: first, we limit the number of columns to the ones we will be using, then we remove repeated indexes and empty VAERS_ID. This ensures no repetition of the VAERS_ID, VAX_LOT pair, and that no lots are associated with no IDs.

There's more...

There are other types of joins other than left, inner, and right. Most notably, there is the outer join, which ensures all entries from both tables have representation.

Make sure you have tests and assertions for your joins: a very common bug is having the wrong expectations for how joins behave. You should also make sure that there are no empty values on the columns where you are joining, as they can produce a lot of excess tuples.

Reducing the memory usage of pandas DataFrames

When you are dealing with lots of information – for example, when analyzing whole genome sequencing data – memory usage may become a limitation for your analysis. It turns out that naïve pandas is not very efficient from a memory perspective, and we can substantially reduce its consumption. One major reason is that pandas tends to assign data types that are larger than are really needed. For more background on pandas memory usage, see `https://medium.com/@gautamrajotya/how-to-reduce-memory-usage-in-python-pandas-158427a99001`.

In this recipe, we are going to revisit our VAERS data and look at several ways to reduce pandas' memory usage. The impact of these changes can be massive: in many cases, reducing memory consumption may mean the difference between being able to use pandas or requiring a more alternative and complex approach, such as Dask or Spark.

Getting ready

We will be using the data from the first recipe. If you have run it, you are all set; if not, please follow the steps discussed there. You can find this code in `Ch02/Ch02-3-pandas-memory.ipynb`.

How to do it...

Follow these steps:

1. First, let's load the data and inspect the size of the DataFrame:

    ```
    import numpy as np
    import pandas as pd
    vdata = pd.read_csv(
        "2021VAERSDATA.csv.gz",
        encoding="iso-8859-1"
    )
    vdata.info(memory_usage="deep")
    ```

Here is an abridged version of the output:

```
26   HISTORY        374854 non-null   object
27   PRIOR_VAX      36452 non-null    object
28   SPLTTYPE       219962 non-null   object
29   FORM_VERS      753040 non-null   int64
30   TODAYS_DATE    747480 non-null   object
31   BIRTH_DEFECT   459 non-null      object
32   OFC_VISIT      144517 non-null   object
33   ER_ED_VISIT    90288 non-null    object
34   ALLERGIES      298538 non-null   object
dtypes: float64(5), int64(2), object(28)
memory usage: 1.4 GB
```

Figure 2.9 – Memory usage of the initial DataFrame

Here, we have information about the number of rows and the type and non-null values of each row. Finally, we can see that the DataFrame requires a whopping 1.4 GB. Most modern computers have 16 GB of RAM or more, but this means you are already taking up an appreciable fraction of that just with this one data table! In a real, large-scale application, you would need to be very careful about memory efficiency in your code.

2. We can also inspect the size of each column:

```
for name in vdata.columns:
    col_bytes = vdata[name].memory_usage(
        index=False, deep=True)
    col_type = vdata[name].dtype
    print(name, col_type, col_bytes // (1024 ** 2))
```

Here is an abridged version of the output:

```
VAERS_ID int64 5
RECVDATE object 48
STATE object 39
AGE_YRS float64 5
CAGE_YR float64 5
CAGE_MO float64 5
SEX object 41
RPT_DATE object 23
SYMPTOM_TEXT object 496
```

Figure 2.10 – Column sizes

SYMPTOM_TEXT occupies 496 MB, so one-third of our entire table.

3. Now, let's look at the DIED column. Can we find a more efficient representation?

    ```
    vdata.DIED.memory_usage(index=False, deep=True)
    vdata.DIED.fillna(False).astype(bool).memory_usage(
        index=False, deep=True)
    ```

The original column takes 21,181,488 bytes, whereas our compact representation takes 656,986 bytes. That's 32 times less!

4. What about the STATE column? Can we do better?

    ```
    vdata["STATE"] = vdata.STATE.str.upper()
    states = list(vdata["STATE"].unique())
    vdata["encoded_state"] = vdata.STATE.apply(
        lambda state: states.index(state))
    vdata["encoded_state"] = vdata["encoded_state"].astype(np.uint8)
    vdata["STATE"].memory_usage(index=False, deep=True)
    vdata["encoded_state"].memory_usage(
        index=False, deep=True)
    ```

Here, we convert the STATE column, which is text, into encoded_state, which is a number. This number is the position of the state's name in the state list. We use this number to look up the list of states. The original column takes around 36 MB, whereas the encoded column takes 0.6 MB.

As an alternative to this approach, you can look at categorical variables in pandas. I prefer to use them as they have wider applications. Categorical variables take on a certain number of fixed values (e.g. Male, Female, or a list of states) and are stored more efficiently than the corresponding text. This approach can be very intuitive when performing statistics and using data visualization libraries.

5. We can apply most of these optimizations when we *load* the data, so let's prepare for that. But now, we have a chicken-and-egg problem: to be able to know the content of the state table, we have to do a first pass to get the list of states, like so:

    ```
    states = list(pd.read_csv(
        "vdata_sample.csv.gz",
        converters={
            "STATE": lambda state: state.upper()
        },
        usecols=["STATE"]
    )["STATE"].unique())
    ```

We have a converter that simply returns the uppercase version of the state. We only return the STATE column to save memory and processing time. Finally, we get the STATE column from the DataFrame and load it into the states array. In this way, we have created an array of just the states, already pre-processed to save memory.

6. The ultimate optimization is *not* to load the data. Imagine that we don't need `SYMPTOM_TEXT` – that is around one-third of the data. In that case, we can just skip it. Here is the final version:

```
vdata = pd.read_csv(
    "vdata_sample.csv.gz",
    index_col="VAERS_ID",
    converters={
        "DIED": lambda died: died == "Y",
        "STATE": lambda state: states.index(
            state.upper()
        )
    },
    usecols=lambda name: name != "SYMPTOM_TEXT"
)
vdata["STATE"] = vdata["STATE"].astype(np.uint8)
vdata.info(memory_usage="deep")
```

We are now at 714 MB, which is a bit over half of the original. This could be still substantially reduced by applying the methods we used for `STATE` and `DIED` to all other columns.

See also

The following is some extra information that may be useful:

- There is plenty of content available on the web to help you understand pandas. You can start with the main user guide, which is available at `https://pandas.pydata.org/docs/user_guide/index.html`.
- If you need to plot data, do not forget to check the visualization part of the guide since it is especially helpful: `https://pandas.pydata.org/docs/user_guide/visualization.html`.
- If you are willing to use a support library to help with Python processing, check out **Apache Arrow**, which will allow you to have extra memory savings for more memory efficiency: `https://arrow.apache.org/`.
- **Dask and Zarr** allow you to work with larger-than-memory datasets using a pandas-like interface:
  - `https://www.dask.org/`
  - `https://zarr.dev/`

Get This Book's PDF Version and Exclusive Extras

UNLOCK NOW

Scan the QR code (or go to `packtpub.com/unlock`). Search for this book by name, confirm the edition, and then follow the steps on the page.

Note: Keep your invoice handy. Purchases made directly from Packt don't require an invoice.

3

Modern Coding Practices and AI-Generated Coding

In this chapter, we'll learn about key tools that help us write good code and follow solid software engineering practices. First, we'll learn about linting tools that help you automatically format your code and catch a variety of errors. You'll also see how to standardize your code against style guides. Next, we'll learn about using AI for coding, starting with a simple example that manipulates a sequence file. Then we'll try a more complex example and learn how to write code for **alignment** using AI.

We will then discuss the importance of test-writing in your code and how AI can assist you in writing unit tests. Finally, we will go over the importance of **code review** and use a simulated **pull request** process to show you how this practice improves your code.

This chapter will give you a solid grounding in some basic software engineering practices that will serve you well as you go through the book and learn to utilize Python for bioinformatics tasks. You will also get a solid introduction to AI-assisted coding, which can be helpful to you throughout the book for learning and debugging. We'll then provide tips throughout the book about how to use AI tools to take your learning further!

In this chapter, you will learn about the following:

- Using linting tools and style guides to write accurate, well-formed code
- Writing a simple bioinformatics file parser using AI
- Read alignment
- Writing tests with AI-assisted coding
- Code review with AI

Technical requirements

The code for this chapter can be found at `https://github.com/PacktPublishing/Bioinformatics-with-Python-Cookbook-Fourth-Edition/tree/main/Ch03`.

You will want to make a folder for `Ch03` and start your Jupyter environment.

Using linting tools and style guides to write accurate, well-formed code

A variety of tools exist for the modern Python programmer to use to improve their code and ensure consistency of style. Many times, these tools are run automatically when checking code as part of **Continuous Integration/Continuous Deployment (CI/CD)**. Many organizations also adopt style guides that go beyond just "Does the code work?" to specify how code should be written to be understandable and visually clear. **Linting** is a process for analyzing code syntax and style inconsistencies. Pylint is a popular Python tool for this.

One of the most popular style guides to know about for Python is PEP8. This style is enforced by many programs, such as Pylint and PyCodeStyle, which we will use in our Jupyter notebook.

In this recipe, we will learn about a few common tools for style enforcement in Python and learn how to run them both from the terminal and from within our notebooks.

Getting ready

1. First, we'll install `pylint`. From the terminal, type the following:

    ```
    conda install pylint
    ```

    You can check it is working by typing:

    ```
    pylint -h
    ```

2. Next, let's install `flake8`. This tool is fundamental for Python style enforcement. It can be run by itself, but it is also used by `pycodestyle`. We say that `flake8` is a *dependency* of `pycodestyle`, which we will use within our notebooks, so we will need to install that first.

    ```
    pip install pycodestyle pycodestyle_magic
    ```

3. We also install `pycodestyle_magic`, which is a module that enables the `magic` commands for `pycodestyle` in Jupyter.

4. Finally, we will install `black`, which is a comprehensive code formatter for Python:

    ```
    pip install black
    ```

We will begin the recipe by trying out `pylint` on a traditional Python file (`.py` file). This is as opposed to running something on a notebook, which has the `.ipynb` extension. You could use any `.py` file, but for the purpose of this recipe, you can use the `sample.py` file (`Ch03/sample.py`) provided with this recipe.

How to do it ...

Here are the steps to try this recipe:

1. Put the `sample.py` file in your current directory.

2. Now, from the terminal, run the following:

    pylint sample.py

    You will see some output like this:

```
************* Module sample
sample.py:1:0: C0114: Missing module docstring (missing-module-docstring)
sample.py:3:0: C0410: Multiple imports on one line (os, sys) (multiple-imports)
sample.py:6:0: C0116: Missing function or method docstring (missing-function-docstring)
sample.py:6:21: C0103: Argument name "a" doesn't conform to snake_case naming style (invalid-name)
sample.py:6:24: C0103: Argument name "b" doesn't conform to snake_case naming style (invalid-name)
sample.py:13:0: C0115: Missing class docstring (missing-class-docstring)
sample.py:18:4: C0116: Missing function or method docstring (missing-function-docstring)
sample.py:21:4: C0116: Missing function or method docstring (missing-function-docstring)
sample.py:26:0: C0103: Constant name "BADVariableName" doesn't conform to UPPER_CASE naming style (invalid-name)
sample.py:3:0: W0611: Unused import os (unused-import)
sample.py:3:0: W0611: Unused import sys (unused-import)
-------------------------------------------------------------------
Your code has been rated at 3.12/10 (previous run: 5.00/10, -1.88)
```

Figure 3.1 – pylint output for sample.py

This gives you a series of informative messages about potential issues with your code:

*   On *line 1*, we are missing a module `docstring` – a method used for documenting Python code, see `https://www.geeksforgeeks.org/python-docstrings/`.

*   On *line 3*, we see that we have multiple imports on one line; it is typically better form to write each import separately.

*   On *line 6*, we have a missing `docstring` for the function.

*   On *line 6*, we also see variables that do not conform to the case naming style. Typically, you should use snake case with meaningful variable names using lowercase letters joined by underscores – see `https://realpython.com/ref/glossary/snake-case/#:~:text=In%20programming%2C%20snake%20case%20is,in%20an%20identifier%20improves%20readability`.

*   On *lines 13*, *18*, and *21*, we are again missing a docstring.

- On *line 26*, we see a variable that does not conform to naming conventions. Typically, an all-upper-case naming convention is used for constants.

- Finally, we see that we are importing libraries that are unused.

The last thing we see from the `pylint` output is the score. The score is calculated by taking the number 10 and subtracting points based on the number and severity of errors you have. So, the closer the score is to 10, the better. We can see this code got a pretty low score of 3.12 and has room for improvement!

3. Next, let's learn about a handy tool called `pycodestyle`, which is like `pylint` but can be run inside a Jupyter notebook. Start a new notebook.

    I.   First, load `pycodestyle` to use it in your notebook:

    ```
    %load_ext pycodestyle_magic
    ```

    II.   Now you can check out your coding style by including the `%pycodestyle` decorator in a code block and then running it. Let's try out an example.

    III. First, run a cell to load the `magic` extension:

    ```
    %load_ext pycodestyle_magic
    ```

4. Now, in the next cell, run a block of code with the `%%pycodestyle` decorator:

    ```
    %%pycodestyle
    import os, sys
    def example_function(a, b):
        if a > b:
        print("a is greater than b")
        else:
            print("b is greater or equal to a")
    # Long line exceeding 80 characters
    print(
        "This is a really, really, really, really, really, "
        "really, really long line of code."
    )
    example_function(10, 5)
    ```

You will see some output with coding suggestions. It looks like this:

```
%%pycodestyle
import os, sys  # Unused imports and multiple imports on one line
def example_function(a, b):  # Missing docstring
    if a > b:
        print("a is greater than b")  # Improper indentation
    else:
        print("b is greater or equal to a")  # Extra indentation
# Long line exceeding 80 characters
print(
    "This is a really, really, really, really, really, really, really long line of code."
)
example_function(10, 5)  # Function call with no meaningful context

2:10: E401 multiple imports on one line
3:1: E302 expected 2 blank lines, found 0
9:1: E305 expected 2 blank lines after class or function definition, found 0
10:80: E501 line too long (89 > 79 characters)
13:1: W391 blank line at end of file
```

Figure 3.2 – pycodestyle output

This gives us a few warnings similar to what we saw with pylint. You can use this technique throughout the book to help you find issues in your code.

Black is a tool that can be used to automatically format a Python file based on standard conventions. It is often used by programming teams to enforce a universal style.

I.   You can run black from your terminal like this:

```
black sample.py
```

This will automatically reformat your code. You can try it on a file by saving a backup copy of the file first and then running black on the main copy. Next, run diff on the files and see what the differences are.

II.   You can also run black on your Jupyter notebooks. To do that, first we need to install the black for Jupyter dependency. Run this from the terminal:

```
pip install "black[jupyter]"
```

III. Now you can run black on a notebook from your terminal like this:

```
black Ch02-1-pandas-basic.ipynb
```

There's more ...

Many other linting and formatting tools exist for Python. Let's take a quick look at some of them here:

Tool	URL	Description
AutoPep8	`https://github.com/hhatto/autopep8`	This tool will format your code in PEP8 style
Yet Another Python Formatter (YAPF)	`https://github.com/google/yapf`	The Yet Another Python Formatter tool can format in both PEP8 and Google styles
Code Beautifier	`https://codebeautify.org/python-formatter-beautifier`	This tool has an online interface you can paste your code into
MyPy	`https://mypy-lang.org/`	This tool enforces static typing, in which we ensure variables have explicit data types

Table 3.1 – Formatting tools for Python

See also...

- For installation details and an in-depth guide to PyLint, look here: `https://pypi.org/project/pylint/`

- Information on the PEP8 style guide is here: `https://peps.python.org/pep-0008/`

- To learn more about installing and using Flake8 as a tool for style guide enforcement, check here: `https://flake8.pycqa.org/en/latest/`

- A great review of how and why you should use code quality tools can be found here: `https://realpython.com/python-code-quality/`

- For more information on docstrings, see: `https://peps.python.org/pep-0257/`

Writing a simple bioinformatics file parser using AI

Manipulating sequences is one of the most basic functions in bioinformatics. It involves formatting and file type conversion, compression, and manipulation of biological sequences, such as reverse complementation or translation of DNA into protein. We'll be looking at this in *Chapter 5, Recipe 2, Tools for sequence manipulation*. But first, we'll take a quick look at how AI is impacting bioinformatics and coding.

In the past several years, **Large Language Models** (**LLMs**) have taken the AI world by storm. These models use machine learning to model relationships between text. They have become very powerful and can write code and unit tests for you, among other things. In bioinformatics, LLMs are being used to create new proteins and even design entire genomes. For example, ProGen can generate functional proteins of many types (Madani et al, *Large Language Models generate functional protein sequences across diverse families*, Nature Biotechnology, Jan 2023: `https://www.nature.com/articles/s41587-022-01618-2`).

Let's look at an example of how we could use a tool such as ChatGPT to write some code for basic sequence file manipulation.

Getting ready

1. First, you will want to get set up with ChatGPT and familiarize yourself with the interface. You can find it at `https://chatgpt.com/`.

    You can use ChatGPT for free. If you want to get a paid account, it will cost $20/month and will give you access to the latest models and faster response times. For this book, I used the professional version of ChatGPT 4-o.

2. We will learn more about bioinformatics file formats in *Chapter 5, Alignment and Variant Calling*, in the *Tools for sequence manipulation* recipe, but let's briefly review the FASTA format. FASTA is short for "FAST-All" and is a way to represent both DNA and protein sequences. It allows us to easily store and transmit genomic information in a standardized way. You can find more information on it here: `https://www.bioinformatics.nl/tools/crab_fasta.html`.

    Here is what our example file will look like:

    ```
    >Seq-1
    ATTGGGGTAACGGTTGGGCT
    ```

    This format consists of a header that always starts with a greater-than symbol (>) and then a line feed. Next comes one or more lines of sequence data. These can be DNA or protein (amino acid) characters. You can follow this with another header and another sequence. When you have multiple sequences, we typically call this a multi-FASTA file.

Tip

Line break characters can differ between operating systems (e.g., macOS and Windows). This is one of the first things to check if you are having trouble with parsing or opening files.

How to do it...

Here are the steps to try this recipe:

1. Let's go into ChatGPT and type a simple prompt:

    ```
    Write code to parse records from a fasta file
    ```

    You will see that ChatGPT prints out some helpful background information, code for a handy function – `parse_fasta()`, which will read and parse your file, and an example of how to use that function. It then discusses what the code does for you and provides the expected input and output. It even documents the functions using a `docstring` (the text in triple quotes).

    Here's a brief, high-level overview (no step-by-step reasoning): to parse FASTA robustly, read the file **streamingly**, treat any line beginning with **>** as a **new record header**, accumulate subsequent **sequence lines** (removing whitespace), and **emit** the previous record when a new header appears or at end-of-file. Skip blank/comment (;) lines, support multi-line sequences, and (optionally) normalize case or strip gap characters. Handling `.gz` files avoids pre-decompression and keeps memory use low.   Response

    python   Copy code

                                                                          Function
    ```python
    def parse_fasta(path, *, uppercase=True, drop_gaps=False):
        """
        Parse records from a FASTA file (streaming; supports multi-line sequences and .gz).

        Parameters
        ----------
        path : str or os.PathLike
            Path to the FASTA file. Files ending in ".gz" or ".bgz" are read
            transparently via gzip.
        uppercase : bool, optional (default: True)
            If True, convert sequence letters to uppercase.
        drop_gaps : bool, optional (default: False)
            If True, remove '-' characters from sequences.

        Yields
        ------
        dict
            A dictionary per record with keys:
              - "id": first token of the header line (after '>')
              - "description": full header lin ↓ without the leading '>')
    ```

    Figure 3.3 – Example partial output from ChatGPT

> **Note**
>
> At the time of writing, ChatGPT 5 is available and may produce different responses— such as writing a class instead of a function for this example. While the responses might vary, they should still follow the intended logic.

2. Let's try out this code. Paste the code example into your Jupyter notebook. We will use one cell for the `parse_fasta()` function and one for the test example. You can use the handy **Copy code** button at the top right of ChatGPT.

3. We will also need a simple file to play with called `example.fasta`. You can use any text editor (e.g., vi, vim, nano) to make a file like this and copy the example input from the ChatGPT window into the file and save it. Make sure to keep the file in the same directory you are working out of for your Jupyter notebook so it can see the file. You can also use the file that is in the GitHub repository for this course or any other sample FASTA file. In *Chapter 8*, *Accessing Public Databases*, in the, *Accessing Genbank and moving around NCBI databases* recipe, we'll learn how to retrieve files like this programmatically from public data sources.

4. Here is an example of our code being run in the Jupyter notebook:

```
    """
    fasta_dict = {}
    with open(file_path, 'r') as file:
        header = None
        sequence = []
        for line in file:
            line = line.strip()
            if line.startswith(">"):  # Header line
                if header:  # Save the previous sequence
                    fasta_dict[header] = ''.join(sequence)
                header = line[1:]  # Remove ">"
                sequence = []  # Reset sequence list
            else:
                sequence.append(line)
        if header:  # Save the last sequence
            fasta_dict[header] = ''.join(sequence)
    return fasta_dict
```

```
[4]: # Example usage:
fasta_file = "example.fasta"
fasta_records = parse_fasta(fasta_file)
for header, seq in fasta_records.items():
    print(f"Header: {header}")
    print(f"Sequence: {seq}")
```

```
Header: seq1
Sequence: ATCGTACGATCGGATCGTACGATC
Header: seq2
Sequence: CGTAGCTAGCTA
```

Figure 3.4 – Code from ChatGPT used within the Jupyter notebook

5.   After you run the code, you will see the output, which shows the parsed header and sequence information:

```
Header: seq1
Sequence: ATCGTACGATCGGATCGTACGATC
Header: seq2
Sequence: CGTAGCTAGCTA
```

Figure 3.5 – Output of the parse_fasta() function

Let's recap this code:

I.   We define a `parse_fasta()` function using the `def` keyword.

II. We define a dictionary for the results. A dictionary is a set of key/value pairs. In this case, the key will be the sequence name, and the value will be the sequence.

III. We use the `with` keyword to open the file and utilize it.

IV. We use a `for` loop to parse the lines in the file.

V.   We store and return the dictionary of results.

VI. Finally, we utilize the function to set an example file to parse, run the function, store the results, and print them out.

As you can see, ChatGPT is very powerful. Throughout this book, you are encouraged to write and understand your own code. But you should also feel free to turn to ChatGPT to help you write code or debug problems you are having as you go!

There's more...

The field of AI-driven coding is rapidly advancing. In the next few years, it will radically transform how people write and debug code. LLMs are also impacting science, genomics, and bioinformatics in many critical ways. As a bioinformatician, it is worth your time to learn AI-assisted coding tools and stay abreast of the advancements in the field.

Each year, new models that are more powerful or specialized for certain tasks are being built. For instance, ChatGPT now has various model flavors. ChatGPT 4o is the current flagship general model at the time of this writing. 4o-mini is a faster, lighter-weight model. o1-preview and o1-mini have more extensive reasoning, meaning they can be useful when you want to think through a problem more extensively, using the algorithm as a helper. At this time, o1-mini is planned to be released free to users, whereas 01-preview is part of the paid subscription. Within ChatGPT, the DALL-E model is used to provide image generation capabilities, also for free. Canvas, another free tool, is used to assist with writing larger documents or code projects.

Let's see an example of a typical ChatGPT session:

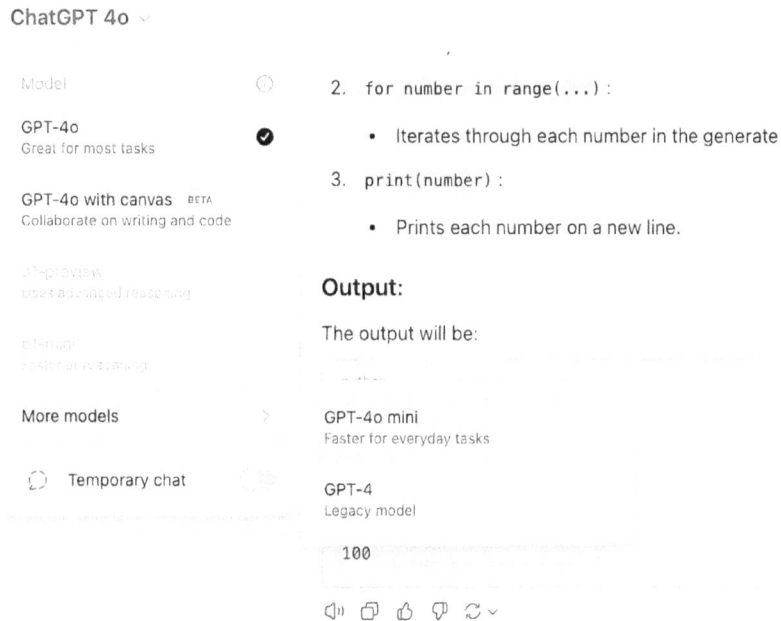

Figure 3.6 – Selecting a model subtype in ChatGPT

At the upper left, you can see the model type selection, currently set to GPT-4o.

There are many other popular sites that offer AI-assisted coding, and more are coming out all the time. As of this writing, some people feel that Claude has a slight edge for coding, while ChatGPT is better at writing text. One thing we can be sure of, though, is that this will keep changing and that significant advances will occur in the next few years in this area. This is an important area for you to keep abreast of.

Here are a few popular sites for AI-assisted coding:

Name	Site	Description
ChatGPT	`https://openai.com/index/chatgpt/`	Popular general-purpose chatbot
Claude	`https://claude.ai`	Excels at editing
GitHub Copilot	`https://github.com/copilot/`	Coding assistant that integrates well with popular Python IDEs
Gemini	`https://gemini.google.com/app`	Google's AI assistant, integrated with its search engine
WatsonX	`https://www.ibm.com/products/watsonx-code-assistant`	IBM's coding assistant

Table 3.2: Popular AI-assistants for coding

See also

- Review on FASTA format: `https://en.wikipedia.org/wiki/FASTA_format`
- Information on docstrings: `https://peps.python.org/pep-0257/`
- Some text editors for Mac: `https://jpolak.org/photo/best-macos-text-editors/`
- Vi Reference Card: `https://www.ks.uiuc.edu/Training/Tutorials/Reference/virefcard.pdf`
- ChatGPT pricing: `https://openai.com/chatgpt/pricing/`

Read alignment

Next, let's look at building a little more sophisticated example using ChatGPT. In this recipe, we will write code to align two reads together, a common bioinformatics task.

A core algorithm in bioinformatics is **alignment**. This is the task of taking two or more sequences and seeing how the sequences can optimally line up with each other. Alignment is used in many bioinformatics tasks – for example, to see how different species may relate to each other (phylogenetics), how proteins may have evolved, or to take many sequencing reads and align them to a reference to do variant calling.

How to do it...

Here are the steps to try this recipe:

1. Go into ChatGPT and get ready to give it the following prompt: `Write an example that aligns two DNA sequences.`

2. This will write example code that uses Biopython's `Bio.Align` module. The code will work; however, we notice that it uses the `pairwise2` algorithm. When you paste this code into your notebook and run it, you will get a message saying that this method will soon be deprecated.

3. We would like to ask ChatGPT to further improve its code and get rid of this warning message. Let's see how we can easily update ChatGPT's previous code with an additional prompt. Note that ChatGPT can remember and refer to the previous work and modify it.

> TIP
>
> ChatGPT may produce different results each time you run a prompt. For consistency, the examples used in this book are included in the accompanying notebooks for reference. In some cases, ChatGPT may choose a different approach. For example, using an alternative algorithm instead of `pairwise2`. The notebook provides both versions so you can compare and understand the differences in output.

4. Here's the next prompt: Change the above code to use PairWiseAligner instead of pairwise2. Now we get code that uses the more modern PairWiseAligner module instead.

5. Paste this code into your notebook and run it.

```
# Perform global alignment
global_alignments = aligner.align(seq1, seq2)

# Display the best global alignment
print("Best global alignment:")
print(global_alignments[0])
print(f"Score: {global_alignments[0].score}")

# Perform local alignment
aligner.mode = 'local'  # Switch to local alignment mode
local_alignments = aligner.align(seq1, seq2)

# Display the best local alignment
print("\nBest local alignment:")
print(local_alignments[0])
print(f"Score: {local_alignments[0].score}")
```

```
Best global alignment:
target            0 ACGT-GCTAGCTAG 13
                  0 ||||-|.|-||||- 14
query             0 ACGTCGAT-GCTA- 12

Score: 6.0

Best local alignment:
target            0 ACGT-GCTAGCTA 12
                  0 ||||-|.|-|||| 13
query             0 ACGTCGAT-GCTA 12

Score: 7.0
```

Figure 3.7 – Results of pairwise alignment code

As you can see, the code worked and printed out both a local and global alignment (more on alignment will be covered in *Chapter 5, Alignment and Variant Calling*).

> **Tip**
>
> Different AI tools, or even future versions of ChatGPT, may give you slightly different results for the prompts we present in the book. In most cases, they will give you something very similar to work with. But if you ever need to refer back to the exact code discussed here, it is available in the notebooks from the GitHub repository.

Let's recap the code:

I.   We define two example DNA sequences (seq1 and seq2).

II.   We initialize the `PairwiseAligner()` class.

III. We set default parameters for match penalties, gap opening penalties, and so on.

IV.   We perform a global alignment of seq1 and seq2.

V.   We print out the alignment and score.

See also

- For more background on sequence alignment, see `https://www.ncbi.nlm.nih.gov/books/NBK464187/`

- For an exciting journey into how AI will transform our society in the coming decades, check out "The Singularity is Nearer" by Ray Kurzweil: `https://www.amazon.com/Singularity-Nearer-Ray-Kurzweil-ebook/dp/B08Y6FYJVY`

Writing tests with AI-assisted coding

In this recipe, we will try out a really useful function of AI-assisted coding: test writing. Tools such as ChatGPT can help you write a variety of tests automatically for your code. This helps ensure completeness and accuracy. We will take the code from read alignment work in the *Read alignment*, recipe and write some tests for it.

Go into ChatGPT and get ready to write a prompt to add testing code for the alignment function that we just wrote in the *Read alignment* recipe.

How to do it...

Here are the steps to try this recipe:

1. We'll begin by writing a prompt to develop testing code for the alignment code we wrote in the previous recipe (make sure it is still available in ChatGPT; if it's not, then run the examples from the *Sequence manipulation* and *Read alignment* recipes first). Here's the prompt: `Write test code for the above.`

> **Note**
>
> We can be very generic with our instructions with ChatGPT. We could have said "write unit tests" or given other instructions, but ChatGPT will understand what is needed, refer to the previous function, and write test code.

2. Now, cut and paste the code into your notebook.

3. When we first write this code, we get an error because it is designed to run on the command line. Let's update this to run it in a Jupyter notebook: `Write the above test code so I can run it in a Jupyter notebook`.

4. Now we can run the code. The following screenshot shows the results of running the code in the notebook:

```
        self.assertAlmostEqual(best_alignment.score, expected_score, places=1)
        self.assertIn(expected_target, str(best_alignment))
        self.assertIn(expected_query, str(best_alignment))

# Run the tests
if __name__ == "__main__":
    unittest.main(argv=[''], exit=False)

FF
======================================================================
FAIL: test_global_alignment (__main__.TestPairwiseAligner.test_global_alignment)
Test global alignment.
----------------------------------------------------------------------
Traceback (most recent call last):
  File "/var/folders/53/kmyyy3057lndfb0bpwx_2pkr0000gn/T/ipykernel_37172/1953800863.py", line 24, in test_global_alignment
    self.assertAlmostEqual(best_alignment.score, expected_score, places=1)
AssertionError: 6.0 != 9.0 within 1 places (3.0 difference)

======================================================================
FAIL: test_local_alignment (__main__.TestPairwiseAligner.test_local_alignment)
Test local alignment.
----------------------------------------------------------------------
Traceback (most recent call last):
  File "/var/folders/53/kmyyy3057lndfb0bpwx_2pkr0000gn/T/ipykernel_37172/1953800863.py", line 39, in test_local_alignment
    self.assertAlmostEqual(best_alignment.score, expected_score, places=1)
AssertionError: 7.0 != 6.0 within 1 places (1.0 difference)

----------------------------------------------------------------------
Ran 2 tests in 0.003s

FAILED (failures=2)
```

Figure 3.8 – Results of the test code

The preceding result looks like an error, but it is the output of the testing code, which, in fact, shows that the tests failed. This is good, this is what we want – when the test results do not match what is expected, we want it to fail so that we can catch errors. ChatGPT tried to guess some of the outputs of the alignment here and wrote some example tests asserting the results, which in this case did not match. If we wanted to, we could run the alignments with the proper scores and adjust this code so that it should pass.

Let's recap this code:

I.   We import the `unittest` library.

II.   We create a class called `TestPairwiseAligner()`.

III. We define the `setUp()` function with initial parameters.

IV.   We define a `test_global_alignment()` function to test the global alignment.

V.   We use `assert` functions to test that the outputs match the expected results.

VI.   We run the tests.

There's more...

This is just a start at showing you how to use ChatGPT to write test code. You could say things such as `Write additional tests` or `Write a test for the gap penalty` to provide additional testing code.

You could even write something such as `What is the coverage over the above test code?`. This will use the `coverage` library to see how thoroughly the provided tests exercise the different elements of the code. In general, you want to aim for something like ninety percent or higher test coverage for your code before you declare it ready.

This can be a great way to help you in your software engineering career by always delivering code that is well-covered by tests from the get-go!

See also

- For a guide to unit testing, look at `https://www.dataquest.io/blog/unit-tests-python/`
- For further discussion of the coverage library, see `https://coverage.readthedocs.io/en/7.6.8/`

Code review with AI

One of the most important jobs of a software engineer is to review the code of other engineers. This is typically done at modern companies to ensure code quality and can be as important and demanding as writing the code itself.

Typically, one or more engineers, often someone more senior, may review your code and make comments to improve the code further, and add more tests, or "nits," which can be just small format improvements. You will be expected to address these comments before the code is ready.

Here is an example of a typical pull request in GitHub, where someone has requested additional changes to the code:

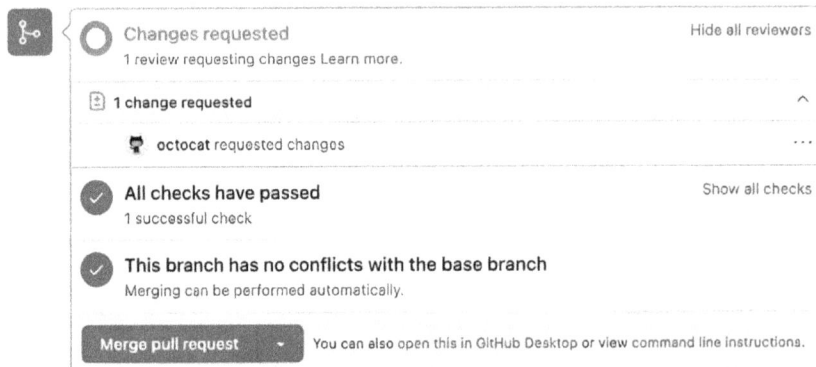

Figure 3.9 – An example of the code review process in GitHub

In this recipe, we will use AI to help us simulate a typical **code review** process and get a feel for how we would use it to improve our code further.

How to do it...

Here are the steps to try this recipe:

1. Let's use ChatGPT to simulate a pull request process. First, grab the `parse_fasta()` function from your `Ch03-2-sequence-manipulation` notebook (also available in the book's GitHub repository). Paste it into ChatGPT. It will automatically summarize the function for you without even being prompted.

2. You will notice that one of the first handy things it does for you is to summarize the `parse_fasta()` function. Here is what the response looks like:

    Your `parse_fasta` function looks **correct** and is **well-structured** — it will successfully parse a standard FASTA file into a dictionary where headers are keys and sequences are values.

    ☑ **Quick points it handles correctly:**

    - Strips whitespace.
    - Detects headers with `>`.
    - Accumulates multi-line sequences.
    - Saves the last sequence after the loop finishes.
    - Clean docstring.

    Figure 3.10 – ChatGPT summarizes the parse_fasta() function

3. Next, let's try and have ChatGPT simulate a pull request process with this prompt: `Review the above code and simulate a pull request review process with me`

    Here are the suggestions we get:

Area	Comment	Suggestion
Error Handling	Right now, if the file doesn't exist or can't be read, the `function` will crash with a Python `FileNotFoundError`.	Wrap the `open()` statement in a `try/except` block to gracefully catch file errors and inform the user.
Blank Line Handling	Some FASTA files might have occasional blank lines between sequences. Your current code would incorrectly treat them as sequence lines.	Add an if not line: continue after stripping whitespace to skip blank lines.
Code Structure (Minor)	Slight duplication in saving the previous sequence inside the loop and after the loop.	This is okay, but you could refactor slightly for elegance (optional).

Area	Comment	Suggestion
Testing	No unit tests provided to verify functionality on various input types.	Provide a few basic test cases using Python's `unittest` or `pytest`.

Table 3.3: Generated code review feedback from ChatGPT simulation

4. Read through the entire response. As you will see, ChatGPT responds with helpful comments and suggestions to improve the code. For example, the first comment exemplifies what should be done if a file does not exist or cannot be opened. It suggests that we consider using a `try..except` block to catch and handle file I/O errors. You can see the details of the suggested improvement in *Figure 3.11*:

```python
try:
    with open(file_path, 'r') as file:
        header = None
        sequence = []
        for line in file:
            line = line.strip()
            if not line:
                continue  # Skip empty lines
            if line.startswith(">"):  # Header line
                if header:  # Save the previous sequence
                    fasta_dict[header] = ''.join(sequence)
                header = line[1:]  # Remove ">"
                sequence = []  # Reset sequence list
            else:
                sequence.append(line)
        if header:  # Save the last sequence
            fasta_dict[header] = ''.join(sequence)
except FileNotFoundError:
    print(f"Error: File '{file_path}' not found.")
return fasta_dict
```

Figure 3.11 – One of the suggested improvements

Several other suggestions are provided that you can review. We can even ask ChatGPT to help us update the code with some of the suggested improvements. Here's a prompt: `Update the parse_fasta function above to include a try except block for catching a missing file on the file open command.`

5. The code is now updated to add the exception handling. This is what the updated code looks like:

```
def parse_fasta(file_path):
    fasta_dict = {}
    try:
        with open(file_path, 'r') as file:
            header = None
            sequence = []
            for line in file:
                line = line.strip()
                if not line:
                    continue
                if line.startswith(">"):
                    if header:
                        fasta_dict[header] = ''.join(
                            sequence)
                    header = line[1:] # Remove ">"
                    sequence = []
                else:
                    sequence.append(line)
            if header:
                fasta_dict[header] = ''.join(sequence)
    except FileNotFoundError:
        print(f"Error: File '{file_path}' not found.")
    return fasta_dict
```

We can see now that our function has been updated to include a `try`/`except` block to handle `FileNotFoundError`.

Of course, in real life, you would often submit the code yourself and work with a human reviewer making comments on the code. But this gives you a very good understanding of the process and how it works in a modern workplace. You can always use ChatGPT to review your code as well!

There's more...

There are many tools and coding assistants that can be used to help make your code better. For example, SonarQube integrates with your CI/CD pipelines and monitors your code quality. Korbit provides AI-powered code review and helps you write pull requests. You will want to take care to look at the terms and conditions of each tool and understand whether your data will be shared if you use the tool (details on the preceding tools can be found in the *See also* section).

You are encouraged to explore this quickly growing space. Future developers will leverage these tools, so you should too!

Agile methodology

Another important thing to learn about on your software engineering journey is Agile methodology and related practices. More and more employers are using these approaches to be effective in their projects. At its core, Agile is about using short sprints to achieve goals and being iterative in feature development. Teams typically work with Product Managers to define requirements and use ticketing systems such as JIRA to track work. These approaches tend to lead to faster, more reliable, and more predictable software delivery. The core elements of Agile sprints, also referred to as Agile Ceremonies, are as follows:

- **Daily Scrum**: This term originally comes from rugby, where players huddle up. It is also called Daily Standup or Standup. It is a quick meeting for the team to check in, discuss what they are doing that day, and most importantly, let the team know about any **blockers** or areas where they cannot make progress.

- **Sprint Planning**: In this meeting, tasks that are ready and on deck are pulled into the next sprint cycle (typically 2 weeks) and assigned to engineers. The team discusses the details of the work and how it may impact others.

- **Sprint Review**: In this ceremony, engineers show their work to stakeholders and discuss whether the features meet the requirements.

- **Sprint Retrospective**: This meeting is focused on the mechanics of the previous sprint. Engineers have a chance to discuss what went wrong and what went well.

- **Backlog Grooming**: This is focused on going into detail on tickets and discussing what needs to be done. Tickets are **pointed** to estimate the amount of time and work each should take.

AI tips

Now that we've seen how powerful AI can be for assisting you in writing code and debugging, we will introduce a great new feature in this edition of the book! In addition to the code you'll be working through yourself, periodically, we'll provide an AI Tip. These callouts will give you some ideas on prompts to use for topics related to that section!

See also

- To learn more about pull requests, read `https://docs.github.com/en/pull-requests/collaborating-with-pull-requests/proposing-changes-to-your-work-with-pull-requests/creating-a-pull-request`

- A handy overview showing the lifecycle of a pull request: `https://devguide.python.org/getting-started/pull-request-lifecycle/`

- For more details on SonarQube: `https://www.sonarsource.com/`

- Korbit is here: `https://www.korbit.ai/`

- Read the Agile Manifesto: `https://agilemanifesto.org/`

- For background on Agile: `https://asana.com/resources/agile-methodology`

- To become a Scrum Master: `https://www.scrum.org/courses/recommended-courses-scrum-masters`

- For a good general background on software design, pick up *A Philosophy of Software Design* by John Ousterhout: `https://www.amazon.com/Philosophy-Software-Design-John-Ousterhout/dp/1732102201`

- *The Missing Readme* by Chris Riccomini and Dmitry Ryaboy (`https://a.co/d/gFqeuni`) provides a great sense of what it is like to be a modern software engineer and how to handle your first workplace experiences

- A Python library for working in JIRA: `https://jira.readthedocs.io/`

- For a great book on working in team environments, check out *Team Topologies* by Skelton and Pais (IT Revolution Press): `https://teamtopologies.com/`

Get This Book's PDF Version and Exclusive Extras

UNLOCK NOW

Scan the QR code (or go to `packtpub.com/unlock`). Search for this book by name, confirm the edition, and then follow the steps on the page.

Note: Keep your invoice handy. Purchases made directly from Packt don't require an invoice.

Data Science and Graphing

In this chapter, we will extend our use of pandas from *Chapter 2* and dive more deeply into data science techniques. Then, we'll go over some important graphing tools.

Data science and bioinformatics are closely related fields, and so many times, bioinformaticians are called upon to get involved in traditional data science tasks. As such, it is important to have a solid grounding in data science techniques. In this chapter, we'll go over **NumPy**, a critical scientific library for Python. We'll also get a quick introduction to some key data science techniques such as **principal components analysis** (**PCA**) and decision trees. In doing so, we'll learn about an amazing toolkit called **scikit-learn**. We'll also develop some basic graphing skills that we'll use throughout the book.

In this chapter, we're going to cover the following recipes:

- Understanding NumPy as the engine behind Python data science and bioinformatics
- Introducing scikit-learn with PCA
- K-means clustering
- Exploring breast cancer traits using decision trees
- Learning more about Matplotlib for chart generation
- Building a UMAP using Seaborn

Technical requirements

The code for this chapter can be found here: `https://github.com/PacktPublishing/Bioinformatics-with-Python-Cookbook-Fourth-Edition/tree/main/Ch04`.

You will want to create a `Ch04` folder and set up your notebooks there. You'll also need to get the data files from `Ch02`.

Remember to activate your `conda` environment before beginning the recipes, like this:

```
conda activate bioinformatics_base
```

Alternatively, if you would like to set up a `conda` environment specific to this chapter, before activating `bioinformatics_base`, run the following:

```
conda create -n ch04-data-science --clone bioinformatics_base
conda activate ch04-data-science
```

You will be able to install the packages for the chapter as you go, or you can use the YAML file provided in the repository:

```
conda env update --file ch04-data-science.yml
```

Understanding NumPy as the engine behind Python data science and bioinformatics

In this recipe, we'll learn about a critical numerical library in Python and explore it using a dataset on adverse responses to vaccines.

Most of your analysis will make use of NumPy, even if you don't use it explicitly. NumPy is an array manipulation library that is behind libraries such as pandas, Matplotlib, BioPython, and scikit-learn, among many others. While much of your bioinformatics work may not require explicit direct use of NumPy, you should be aware of its existence, as it underpins almost everything you do, even if only indirectly via the other libraries.

You should take a moment now to learn more about arrays in Python. Arrays are sets of data that can be one- or multi-dimensional. Here is an example:

```
arr1 = ["a", "b", "c"]
```

The preceding would give you a one-dimensional array with three elements, the letters a, b, and c.

Then, the following would give you the zeroth element of the array, in this case, the letter a:

```
arr1[0]
```

The following gives you the length of the array, which is 3 in this case:

```
len(arr1)
```

Technically, Python does not use arrays; it uses lists, but these can be used as arrays. Before continuing, you should familiarize yourself with this tutorial: `https://www.w3schools.com/python/python_arrays.asp`.

The main way people use arrays in Python, even if indirectly, is via NumPy. When we get into pandas, we will discuss DataFrames, which are essentially structured wrappers around arrays that include additional information such as labels and metadata. So, it is important to understand NumPy arrays first. Read more here: `https://www.geeksforgeeks.org/basics-of-numpy-arrays/`.

In this recipe, we will use the **Vaccine Adverse Event Reporting System** (**VAERS**) data from *Chapter 2* to demonstrate how NumPy is behind many of the core libraries that we use. Our example will extract the number of cases from the five US states with more adverse effects, splitting them into age bins: 0–19, 20–39, up to 100–119 years.

Getting ready

First, grab the data from the *Using pandas to process vaccine-adverse events* (VAERSDATA.csv.gz) recipe and make sure it is available. You can put it under a data subdirectory in your Ch04 folder to put it in the right place relative to your notebook.

How to do it...

1. Let's start by loading the data with pandas and reducing the data so that it's related to the top five US states only:

```
import numpy as np
import pandas as pd
import matplotlib.pyplot as plt
vdata = pd.read_csv(
    "data/2021VAERSDATA.csv.gz",
    encoding="iso-8859-1",
    low_memory=False
)
vdata["STATE"] = vdata["STATE"].str.upper()
top_states = pd.DataFrame({
    "size": vdata.groupby("STATE")
                  .size()
                  .sort_values(ascending=False)
                  .head(5)
}).reset_index()
top_states["rank"] = top_states.index
top_states = top_states.set_index("STATE")
top_vdata = vdata[vdata["STATE"].isin(top_states.index)].copy()
top_vdata["state_code"] = top_vdata["STATE"].apply(
    lambda state: top_states["rank"].at[state]
).astype(np.uint8)
top_vdata = top_vdata[top_vdata["AGE_YRS"].notna()].copy()
top_vdata.loc[:,"AGE_YRS"] = top_vdata["AGE_YRS"].astype(int)
top_states
```

To recap, this code does the following:

I.   Imports our libraries and loads the data

II.   Creates a DataFrame called `top_states`, grouping by the size of the data for each state and selecting the top five

III.   Creates a `rank` column for the states

IV.   Creates a DataFrame called `top_vdata`, and copies over the VAERS data relevant to the top five states only

V.   Cleans up `top_vdata` a little by removing any entries with a blank age

VI.   Prints out the `top_states` DataFrame

This rank will be used later to construct a NumPy matrix. The top states are as follows:

STATE	size	rank
CA	69740	0
FL	41338	1
TX	40929	2
NY	38911	3
PA	25963	4

Figure 4.1 – US states with the largest numbers of adverse effects

2. Now, let's extract the two NumPy arrays that contain age and state data:

```
age_state = top_vdata[["state_code", "AGE_YRS"]]
age_state["state_code"]

state_code_arr = age_state["state_code"].values
type(state_code_arr), state_code_arr.shape, state_code_arr.dtype

age_arr = age_state["AGE_YRS"].values
type(age_arr), age_arr.shape, age_arr.dtype
```

This code does the following:

I.   Gets just the state code and age into a DataFrame called `age_state`

II.   Creates a NumPy array, `state_code_arr`, with just the state code values

III.   Creates a NumPy array, `age_arr`, with the age values

IV.   Prints the data types of the arrays

Note that the data that underlies pandas is NumPy data (the `values` call for both series returns NumPy types). Also, you may recall that pandas has properties such as `.shape` or `.dtype`: these were inspired by NumPy and behave the same.

3. Next, we need to get our data arranged so that we have states with their corresponding age groups. Let's create a NumPy matrix from scratch (a 2D array), where each row is a state and each column represents an age group:

```
age_state_mat = np.zeros((5, 6), dtype=np.uint64)
for row in age_state.itertuples(index=False):
    state_code = int(row.state_code)
    age_bin = min(int(row.AGE_YRS // 20), 5)
    if 0 <= state_code < 5 and 0 <= age_bin < 6:
        age_state_mat[state_code, age_bin] += 1
age_state_mat
```

The array has five rows (one for each state) and six columns (one for each age group). All the cells in the array must have the same type.

We initialize the array with zeros. There are many ways to initialize arrays, but if you have a very large array, initializing it may take a lot of time. Sometimes, depending on your task, it might be OK that the array is empty at the beginning (meaning it was initialized with random trash). In that case, using np.empty will be much faster.

4. We can extract a single row (in our case, the data for a state) very easily. The same applies to a column. Let's take California data and then the 0-19 age group:

```
cal = age_state_mat[0,:]
kids = age_state_mat[:,0]
```

Note the syntax to extract a row or a column. It should be familiar to you, given that pandas copied the syntax from NumPy, and we encountered it in previous recipes.

5. Now, let's compute a new matrix where we have the fraction of cases per age group:

```
def compute_frac(arr_1d):
    return arr_1d / arr_1d.sum()
frac_age_stat_mat = np.apply_along_axis(
    compute_frac, 1, age_state_mat)
```

The last line applies the compute_frac function to all rows. compute_frac takes a single row and returns a new row where all the elements are divided by the total sum.

6. Now, let's create a new matrix that acts as a percentage instead of a fraction – simply because it reads better:

```
perc_age_stat_mat = frac_age_stat_mat * 100
perc_age_stat_mat = perc_age_stat_mat.astype(np.uint8)
perc_age_stat_mat
```

Here, we simply multiply all the elements of the 2D array by 100 to get percentages.

Here is the result:

```
array([[ 8, 27, 33, 26,  4,  0],
       [ 4, 19, 30, 39,  7,  0],
       [ 8, 28, 35, 24,  3,  0],
       [ 7, 27, 33, 27,  4,  0],
       [ 6, 25, 33, 29,  4,  0]], dtype=uint8)
```

Figure 4.2 – A matrix representing the distribution of vaccine-
adverse effects in the five US states with the most cases

7. Finally, let's create a graphical representation of the matrix using Matplotlib (you can check out the resulting figure in the notebook):

```
fig = plt.figure()
ax = fig.add_subplot()
ax.matshow(
    perc_age_stat_mat,
    cmap=plt.get_cmap("Greys")
)
ax.set_yticks(range(5))
ax.set_yticklabels(top_states.index)
ax.set_xticks(range(6))
ax.set_xticklabels([
    "0-19", "20-39", "40-59", "60-79", "80-99", "100-119"
])
fig.savefig("matrix.png")
```

This code does the following:

I.   Initializes a new figure

II. Creates a plot within that figure

III. Displays a grayscale matrix

IV. Sets the check marks and labels for the axes

V.   Saves the figure

You can pass NumPy data structures directly to Matplotlib. Matplotlib, like pandas, is based on NumPy. We'll cover Matplotlib more in the *Learning more about Matplotlib for chart generation* recipe.

In this recipe, we tried to understand how the NumPy library is behind some of the most important data analysis tasks. We started by importing our libraries and loading the data.

Then, we created a DataFrame called `top_states`, grouping by the size of the data for each state and selecting the top five and a `rank` column for the states. We also created a DataFrame called `top_vdata` and copied over the VAERS data relevant to the top five states only. Finally, we cleaned up `top_vdata` a little by removing any entries with a blank age and printed out the `top_states` DataFrame. That was *step 1*. Next, we extracted two NumPy arrays by putting just the state code and age into a DataFrame called `age_state`, followed by creating a NumPy array, `state_code_arr`, with just the state code values, and an `age_arr` NumPy array with the age values. After creating these arrays, we printed their data types. Finally, we created a graphical representation showing the percentage of adverse effects by age group and state for the top five states.

You've now received a basic introduction to NumPy and learned how it serves as the foundation for many key Python libraries. We'll continue building on this knowledge throughout the book. You're encouraged to explore NumPy further using the resources provided here.

See also

The following is some extra information that may be useful:

- NumPy has many more features than the ones we've discussed here. There are plenty of books and tutorials on them. The official documentation is a good place to start: `https://numpy.org/doc/stable/`.
- There are many important issues to discover with NumPy, but probably one of the most important is broadcasting: NumPy's ability to take arrays of different structures and get the operations right. For details, go to `https://numpy.org/doc/stable/user/basics.broadcasting.html`.

Introducing scikit-learn with PCA

In this recipe, we'll use an important data science technique to analyze the key factors in a sample breast cancer dataset.

PCA is a statistical procedure used to find linearly uncorrelated components that explain as much of the variation in a dataset as possible. In this way, it performs **dimensionality reduction**, meaning that we find a simpler or lower-dimensional representation of a more complex, or higher-dimensional dataset, thereby giving us a handle on key features that help explain the data in a powerful way. This step of finding explanatory features is a key first step in machine learning.

In this recipe, we will implement PCA using the scikit-learn library. Scikit-learn is one of the fundamental Python libraries for machine learning. PCA is a form of unsupervised machine learning – meaning we don't provide information about the class of the sample. We will discuss supervised techniques in the other recipes of this chapter; see the *Exploring breast cancer traits using decision trees* recipe.

Getting ready

Before we move any further, here is some background on supervised versus unsupervised learning:

Supervised learning	Unsupervised learning
Input data is labeled with the correct answer	Input data has no labels; we are just trying to find patterns in the data
Outputs predictions for new data	Tries to cluster new data into groups
Methods include linear regression and decision trees	Methods include PCA, K-means, and UMAP
Example: Classifying emails as spam	Example: Finding traits that cluster from tissue samples

Table 4.1 – Supervised vs. unsupervised learning

Let's also learn a little about scikit-learn and the datasets that come with it.

Scikit-learn (`https://scikit-learn.org/stable/`) is a popular machine learning library built on NumPy, SciPy, and Matplotlib. It is also referred to as `sklearn`.

It comes with some handy toy datasets that you can easily bring into your notebook (`https://scikit-learn.org/1.5/datasets/toy_dataset.html`). In this recipe, we will work with the breast cancer dataset. This dataset contains 30 predictive attributes that summarize an image from a breast cancer tissue sample, with two potential outcomes: benign or malignant.

The code for this recipe is in the `Ch04/Ch04-2-PCA.ipynb` notebook. Without any further delay, let's get into it.

How to do it...

1. First, we will import our libraries:

    ```
    from sklearn.datasets import load_breast_cancer
    from sklearn.decomposition import PCA
    from sklearn.preprocessing import StandardScaler
    import matplotlib.pyplot as plt
    import pandas as pd
    ```

    You can see here that we import the breast cancer dataset by using the `sklearn.datasets` package. We also bring in the PCA model from the `sklearn.decomposition` package. This package contains several other variations of matrix decomposition or dimensionality reduction techniques, such as FastICA or truncated singular value decomposition. Next, we bring in `StandardScaler` from the preprocessing package. Most machine learning algorithms expect features to be on a standard scale that looks like a Gaussian, with a mean of 0 and a variance of 1 – this is what we use `StandardScaler` for.

2. Next, we'll load in our dataset:

```
bc_data = load_breast_cancer()
X = bc_data.data
y = bc_data.target
```

This uses a loading function from the `datasets` package to put the data into `bc_data`. We put our features into the X array and our targets (the outcomes) into the y array. If you want, you can print out the type of `bc_data` like this:

```
type(bc_data)
```

You will see that it is an `sklearn` bunch type. Bunch objects are an extension of dictionaries that are used to bundle data and related metadata (`https://scikit-learn.org/stable/modules/generated/sklearn.utils.Bunch.html`).

3. Now, we will normalize the data with `StandardScaler`, as we discussed before:

```
scaler = StandardScaler()
X_scaled = scaler.fit_transform(X)
```

4. Now we have our normalized data in the `X_scaled` array. Next, we'll perform the PCA:

```
bc_pca = PCA(n_components=3)
X_bc_pca = bc_pca.fit_transform(X_scaled)
```

5. We used the `PCA` function of scikit-learn to perform a principal components analysis on the data. We initialize the PCA and then perform the fit on our scaled X data. Note that the number of components is an important parameter that you can play with. In this case, we set it to 3. Three is just a standard number that is easy to think about and visualize in three dimensions – you could pick any number, but the more components you choose, the harder it will be to interpret the meaning of the results. With many components, you can always fit your data, but this can lead to what is called **overfitting**, in which we simply fit our data by being overly specific and providing too much detail in our model. Such models are harder to interpret and will not perform well on future datasets that the model has not yet seen. Let's use the pandas `DataFrame` function to turn our PCA results array into a DataFrame:

```
bc_pca_df = pd.DataFrame(X_bc_pca, columns=[
    'PC1', 'PC2', 'PC3'])
bc_pca_df['label'] = y
```

6. We also set a label for the data based on the result class (benign or malignant). Now, let's plot the results!

```
fig = plt.figure(figsize=(10, 8))
ax = fig.add_subplot(111, projection='3d')
for label, color, marker in zip(
    [0, 1], ['red', 'blue'], ['o', '^']
):
    subset = bc_pca_df[bc_pca_df['label'] == label]
    ax.scatter(
        subset['PC1'], subset['PC2'], subset['PC3'],
        c=color, label=bc_data.target_names[label],
        marker=marker, alpha=0.7
    )
ax.set_title(
    'PCA on Breast Cancer Dataset (3D View: PC1, PC2, PC3)'
)
ax.set_xlabel('Principal Component 1')
ax.set_ylabel('Principal Component 2')
ax.set_zlabel('Principal Component 3')
ax.legend()
plt.show()
```

This code does the following:

I.   Creates a new figure

II.   Adds a 3D plot to the figure

III. Assigns colors to the benign and malignant classes

IV.   Creates subsets of the components

V.   Builds a 3D scatter plot with the principal components as the *X*, *Y*, and *Z* axes

VI.   Sets the titles and axes labels, and shows the plot

Here is a plot of the three principal components:

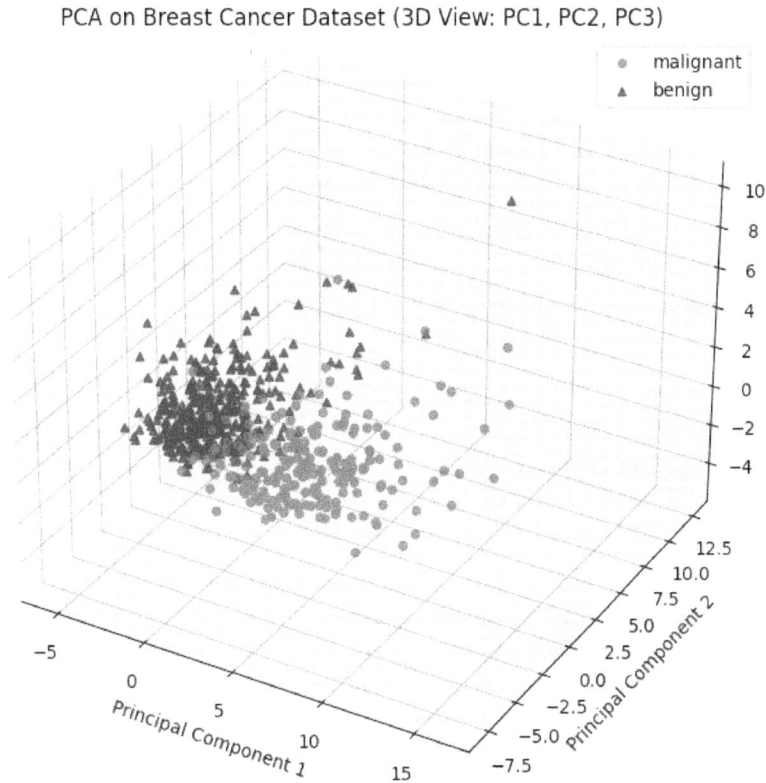

Figure 4.3 – PCA results for breast cancer dataset

We can see the results in *Figure 4.3*. You can definitely notice that **Principal Component 1** explains a lot of the data – as we move along that axis, we can see a correlation with malignant versus benign status pretty clearly.

7. Let's also look at how well the principal components explain the data in a more numeric fashion:

```
explained_variance = bc_pca.explained_variance_ratio_
for i, variance in enumerate(explained_variance, 1):
    print(
        f"Explained variance for PC{i}: {variance:.2f}"
    )
cumulative_variance = explained_variance.sum()
print(
    f"Total variance explained by the first 3 components: "
    f" {cumulative_variance:.2f}")
```

This code uses the `explained_variance_ratio_` parameter of the PCA class to return an array with the percentage variance in the data explained by each component.

It is worth taking a moment here to explain formatting in Python `print()` statements. In the preceding `print` statement, we use something called a Python f-string, which is a formatting convention. The `f` and `"` (double quote) characters start it off. We use curly braces around the string we want to print, and a special formatting code that tells us how to display that variable. In this case, `2f` means to display this as a floating-point number to two decimal places of accuracy.

Take a moment to learn more about Python f-strings here: `https://www.geeksforgeeks.org/formatted-string-literals-f-strings-python/`.

And brush up on your print formatting codes here: `https://www.geeksforgeeks.org/python-output-formatting/`.

Here is the output:

```
Explained variance for PC1: 0.44
Explained variance for PC2: 0.19
Explained variance for PC3: 0.09
Total variance explained by the first 3 components: 0.73
```

We can see that `PC1` is indeed the most powerful and explains 44% of the variance. With three components working together, we can explain 73% of the variance in this dataset. Note here that `PC1` stands for *principal component 1* and is the main set of features that drives most of the variance in the data. For example, it might represent the mean radius or texture score of the tumor, or some combination thereof.

8. Finally, we can graph the cumulative variance explained by the components:

```python
cumulative_variance = explained_variance.cumsum()
plt.figure(figsize=(8, 6))
plt.plot(
    range(1, len(cumulative_variance) + 1),
    cumulative_variance,
    marker='o', linestyle='--'
)
plt.title(
    'Cumulative Explained Variance by Principal Components')
plt.xlabel('Number of Principal Components')
plt.ylabel('Cumulative Explained Variance')
plt.grid(True)
plt.show()
```

This code uses the NumPy cumsum function to add up the sums from the explained variances and then creates a plot with the components on the *X* axis and the cumulative explained variance on the *Y* axis.

The following figure shows the amount of accumulating variance that we can explain as we add more principal components to our model:

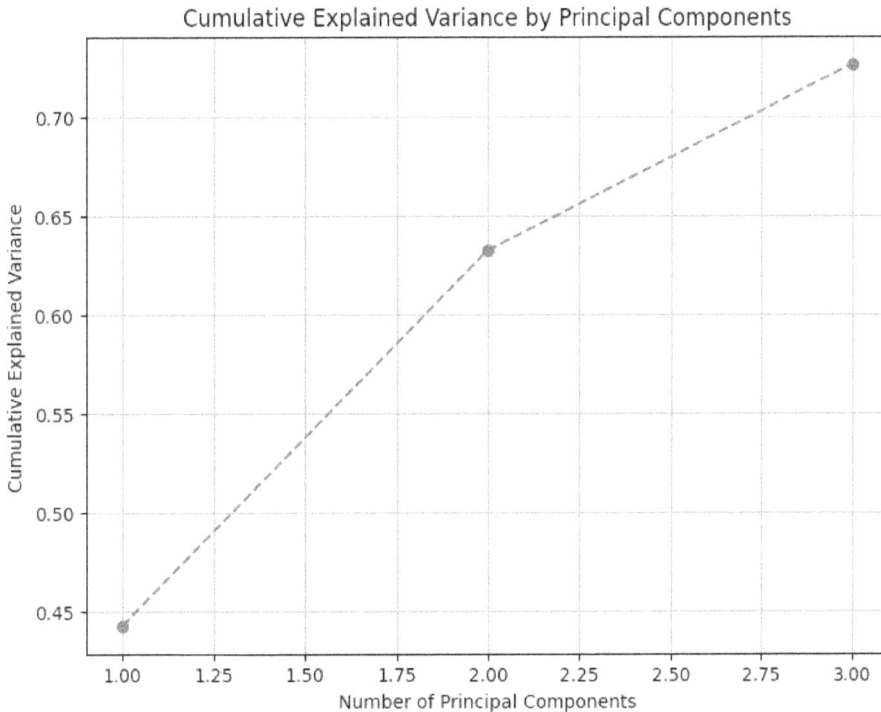

Figure 4.4 – Increasing variance in the dataset is explained as we add more components

As you can see, PCA is a powerful tool for exploring and explaining your data. Scikit-learn is a great toolbox for data science with numerous models and features. In the next recipe, we'll learn about an unsupervised clustering approach called K-means!

There's more...

While we certainly covered some important features, there is a lot more you can do with scikit-learn:

- Preprocess your data with a rich toolbox of functions
- Build a multi-layer Perceptron (a type of neural network) to classify data
- Perform image recognition using the sample datasets
- Explore numerous clustering algorithms such as DBSCAN, agglomerative clustering, spectral clustering, and more

See also

- Check out Human-Learn, a tool that makes scikit-learn easier: `https://koaning.github.io/human-learn/index.html`

- To learn more about scikit-learn, you can read *Machine Learning with PyTorch and SciKit-Learn: Develop machine learning and deep learning models with Python* (`https://a.co/d/gf11T6E` Packt Publishing)

K-means clustering

In this recipe, we'll learn about another data science technique called **clustering** and revisit our breast cancer dataset.

K-means clustering is an example of an unsupervised algorithm. In these types of algorithms, we need a training dataset so that the algorithm is able to learn. After training the algorithm, it will be able to predict a certain outcome for new samples. In our case, we are hoping that we can predict the main classes in the population.

K-means comes from the idea of creating K centers. Points are assigned to centroids based on their Euclidean distance from the center. We then adjust the centers until we have more and more data points falling nearby with minimal distance. In this way, we can attempt to classify our data into approximately K groups.

Getting ready

We will be using the same data as in the previous recipe. The code for this recipe can be found in `Ch04/Ch04-3-k-means.ipynb`.

How to do it...

Here are the steps to try this recipe:

1. First, we'll set up our libraries:

    ```
    from sklearn.datasets import load_breast_cancer
    from sklearn.cluster import KMeans
    from sklearn.preprocessing import StandardScaler
    from sklearn.decomposition import PCA
    import matplotlib.pyplot as plt
    import pandas as pd
    import numpy as np
    ```

We will use the breast cancer dataset from scikit-learn as before and import the KMeans class from the cluster package.

2. Next, we load the breast cancer dataset:

```
data = load_breast_cancer()
X = data.data
y = data.target
```

This loads our features into the X array and labels into y.

3. Next, we'll normalize our data:

```
scaler = StandardScaler()
X_scaled = scaler.fit_transform(X)
```

4. Next, we perform K-means clustering:

```
kmeans = KMeans(n_clusters=2, random_state=42, n_init=10)
clusters = kmeans.fit_predict(X_scaled)
```

We use two clusters. We set the random state explicitly because this helps us have a more reproducible result, but normally, you would not set this. We also use the n_init parameter to re-initialize the clustering 10 times for better results.

5. Next, we'll build a DataFrame to hold the results:

```
bc_kmeans_df = pd.DataFrame(
    X_scaled, columns=data.feature_names)
bc_kmeans_df['Cluster'] = clusters
bc_kmeans_df['True Label'] = y
```

6. Next, we will evaluate the accuracy of our clustering. But first, we need to ensure that we calculate the accuracy using a consistent clustering. K-means cannot be forced to assign specific labels to specific clusters. So, in some cases, you may run the algorithm and get *cluster 0* as benign and *cluster 1* as malignant, and in other runs, you may get *cluster 0* as malignant and *cluster 1* as benign. We are going to write a short function, align_labels(), to ensure consistency:

```
def align_labels(true_labels, cluster_labels):
    new_labels = np.zeros_like(cluster_labels)
    for cluster in np.unique(cluster_labels):
        mask = cluster_labels == cluster
        new_labels[mask] = mode(
            true_labels[mask],
            keepdims=False
        )[0]
    return new_labels
aligned_clusters = align_labels(y, clusters)
```

This function makes sure the cluster assignment matches the actual labels as closely as possible. It initializes new_labels to hold the updated labels. It then loops over the clusters and sets up a mask to hold the samples belonging to the current cluster. It then uses the mode() function to find the most common true label in the cluster. This effectively reassigns the clusters to match the true labels, ensuring consistency across runs.

7. Now, let's look at the accuracy of our clustering:

```
accuracy = accuracy_score(y, aligned_clusters)
print(f"Accuracy of clustering: {accuracy:.2f}")
```

Here is the output:

Accuracy of clustering: 0.91

We see that these two clusters can categorize the samples 91% of the time!

Note that in the preceding print statement, we used the .2f formatting string. This is a Python print formatting string that specifies that we should print a floating-point number to two decimal places of accuracy (two significant digits).

8. Next, we will use PCA again to reduce the dimensionality of the data so that we can graph the clusters along two components, to make it easier to visualize:

```
pca = PCA(n_components=2)
X_pca = pca.fit_transform(X_scaled)
bc_kmeans_df['PC1'] = X_pca[:, 0]
bc_kmeans_df['PC2'] = X_pca[:, 1]
```

This assigns groupings in our clustering results to one principal component or the other.

9. Now, we can plot the results:

```
plt.figure(figsize=(8, 6))
for cluster, color, marker in zip(
    [0, 1], ['blue', 'red'], ['^', 'o']
):
    subset = bc_kmeans_df[bc_kmeans_df['Cluster'] == cluster]
    plt.scatter(
        subset['PC1'], subset['PC2'], c=color,
        label=f'Cluster {cluster}', marker=marker, alpha=0.7
    )

plt.title('K-Means Clustering on Breast Cancer Dataset')
plt.xlabel('Principal Component 1')
plt.ylabel('Principal Component 2')
plt.legend()
plt.grid()
plt.show()
```

Here are the results:

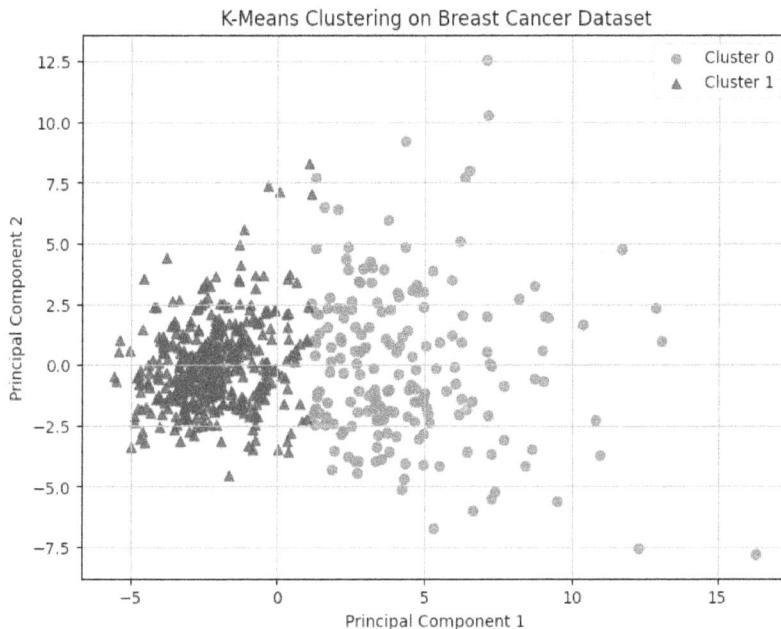

Figure 4.5 – Results of K-means clustering

As you can see, K-means clustering can be a powerful way to find patterns in your data.

> **AI tip**
>
> To make our preceding display simpler, we only chose two of our three main principal components, but we can extend our code even further to handle the 3D case!
>
> - **Prepare**: First, cut and paste the entire notebook code for the third recipe. Alternatively, you can try uploading your entire notebook. ChatGPT can upload your notebook, alter it, and provide a new notebook for you to download.
>
> - **Prompt**: Update the preceding to provide a 3D visualization as well as the 2D plot.
>
> - **Prompt**: Update the preceding notebook (or code) to use PCA and animate the clusters.
>
> You should see the updated code to perform PCA dimension reduction with three components and then plot and visualize the results. If you used the preceding approach to upload your notebook, you will also get a fully functional notebook to download. You will see that a 3D toolkit is imported called `mpl_toolkits.mplot3d`. To animate the clusters, the code will use the IPython HTML and Matplotlib animation modules. You will be able to view a rotating 3D plot of the principal components!
>
> The example result of this AI tip is available in the GitHub repository as `Ch04-3-k-means-PCA-animated.ipynb`.

There's more...

K-means is just one type of clustering algorithm included in scikit-learn. For instance, **density-based spatial clustering of applications with noise (DBSCAN)** can identify clusters of any shape and hence can be more versatile than K-means. For example, the DBSCAN-CellX software (Kuchenhoff et al., see the following reference) can be used to classify cell populations.

See also

- Go deeper by reading *Deep-learning based clustering approaches for bioinformatics*, Karim et al., Briefings in BioInformatics, 2021: https://academic.oup.com/bib/article/22/1/393/5721075

- Kuchenhoff et al., *Extended methods for spatial cell classification with DBSCAN-CellX*, Nature Scientific Reports, November 2023: https://www.nature.com/articles/s41598-023-45190-4

Exploring breast cancer traits using decision trees

Next, we will discuss exploratory analysis based on decision trees. **Decision trees** are a set of rules that classify our data – they may sound simple at first, but they can be very powerful. The big advantage of decision trees is that they will give us the rules that constructed the decision tree, providing some understanding of what is going on with our data.

Getting ready

We'll use the `sklearn` breast cancer dataset as before. The code for this recipe can be found in `Ch04/Ch04-4-decision-trees.ipynb`.

How to do it...

Here are the steps to try this recipe:

1. First, we'll import our libraries:

```
from sklearn.datasets import load_breast_cancer
from sklearn.model_selection import train_test_split
from sklearn.tree import DecisionTreeClassifier, plot_tree
from sklearn.metrics import (
    accuracy_score,
    confusion_matrix,
    classification_report,
    precision_score,
```

```
        recall_score,
        f1_score
)
import matplotlib.pyplot as plt
import seaborn as sns
import numpy as np
```

We brought in some additional libraries this time, including the scikit-learn `model_selection` package. This package contains several useful tools for things such as splitting up your data into test and training sets or performing **cross-validation**.

We also import the `DecisionTreeClassifier` model and several metrics.

2. Now, we will load our breast cancer dataset:

```
data = load_breast_cancer()
X, y = data.data, data.target
```

Now, we have our features and labels in the X and y arrays, respectively.

3. Next, let's split our data into training and test sets. We will use the `train_test_split()` function from the `model_selection` package for this:

```
X_train, X_test, y_train, y_test = train_test_split(
    X, y, test_size=0.2, random_state=42)
```

4. The next step is to actually train the decision tree classifier:

```
dt_classifier = DecisionTreeClassifier(
    random_state=42,
    max_depth=5,
    criterion='gini' )
dt_classifier.fit(X_train, y_train)
```

This initializes the classifier. Again, we set `random_state` *explicitly* to help with reproducibility. We also limit the tree depth to help avoid overfitting, a common problem in machine learning. If you are allowed to keep making more and more precise trees, it would be easy to fit this dataset as much as we'd like; however, it would then become very poor at predicting any future dataset, which is what we want it to do. We use the default model called `gini`, which is for the **Gini importance**, a measure of how much each feature we add improves the model. We then use the `fit` function of the classifier to fit the data.

5. To test the classifier, we now make predictions:

```
y_pred = dt_classifier.predict(X_test)
```

We use our test dataset from the earlier splitting to make predictions.

6. Now, let's examine the performance of our model:

```
print("Decision Tree Performance Metrics:")
print("-" * 30)
print(f"Accuracy: {accuracy_score(y_test, y_pred):.4f}")
print(f"Precision: {precision_score(y_test, y_pred):.4f}")
print(f"Recall: {recall_score(y_test, y_pred):.4f}")
print(f"F1 Score: {f1_score(y_test, y_pred):.4f}")
```

We see the accuracy, precision, recall, and F1 score of the model.

7. We can also print a detailed classification report using the following:

```
print("\nDetailed Classification Report:")
print(classification_report(y_test, y_pred,
target_names=data.target_names))
```

8. And we can build a confusion matrix with the following code:

```
plt.figure(figsize=(8, 6))
cm = confusion_matrix(y_test, y_pred)
sns.heatmap(
    cm, annot=True, fmt='d', cmap='Blues',
    xticklabels=data.target_names,
    yticklabels=data.target_names
)
plt.title('Confusion Matrix for Decision Tree')
plt.xlabel('Predicted Label')
plt.ylabel('True Label')
plt.tight_layout()
plt.show()
```

Here's the confusion matrix:

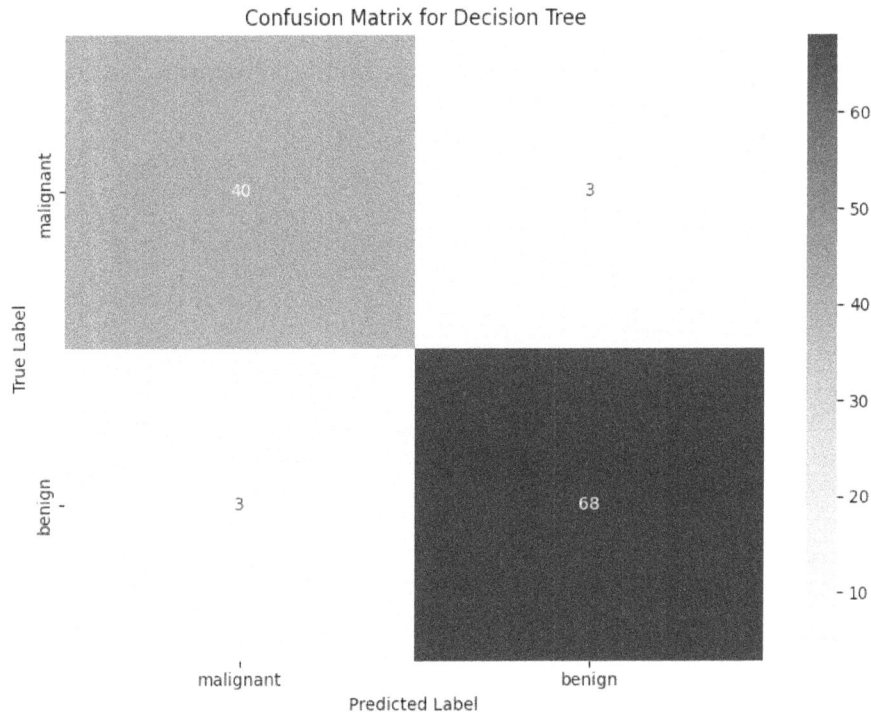

Figure 4.6 – Confusion matrix

A confusion matrix shows the true positives in the upper-left quarter and true negatives in the lower-right quarter. In the lower-left quarter, we have the false positives, and in the upper-right quarter, we have the false negatives.

9. Now, let's build another visualization that shows the importance of various features:

```
plt.figure(figsize=(10, 6))
feature_importance = dt_classifier.feature_importances_
sorted_idx = np.argsort(feature_importance)
pos = np.arange(sorted_idx.shape[0]) + .5

plt.barh(
    pos, feature_importance[sorted_idx], align='center')
plt.yticks(
    pos, [data.feature_names[i] for i in sorted_idx])
plt.xlabel('Feature Importance')
plt.title('Decision Tree Feature Importance')
plt.tight_layout()
plt.show()
```

This code uses the `feature_importances_` parameter from the classifier and builds a bar chart:

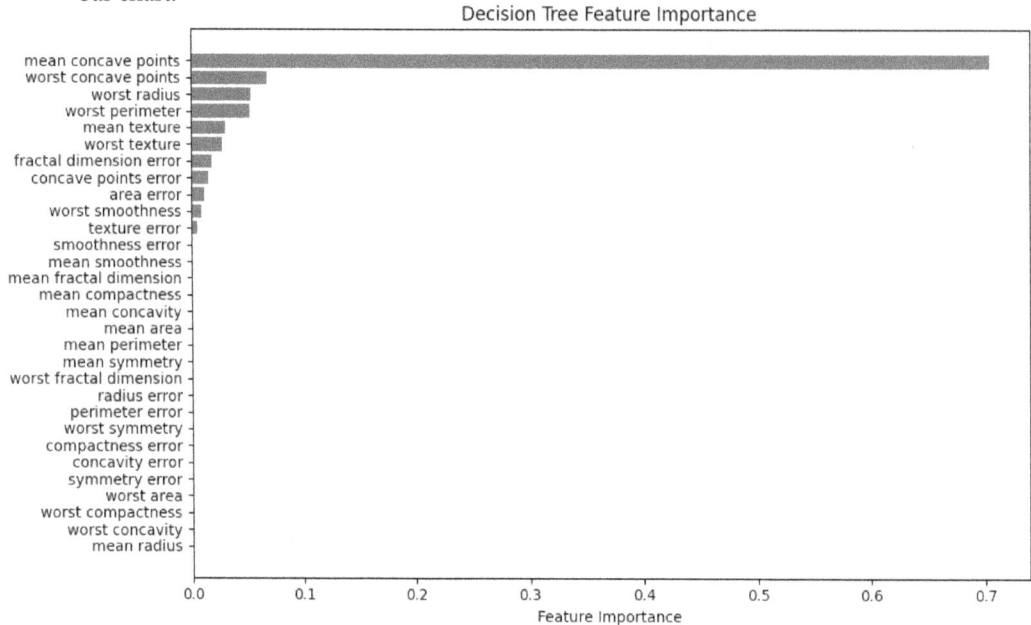

Figure 4.7 – Feature importances in the decision tree

We can see that certain aspects of the tissue sample, such as `mean concave points`, lend the most explainability to our model.

10. This code uses the `plot_tree` function to show the features of the decision tree:

```
plt.figure(figsize=(20,10))
plot_tree(
    dt_classifier,
    feature_names=data.feature_names,
    class_names=data.target_names,
    filled=True,
    rounded=True
)
plt.title('Decision Tree Classifier')
plt.show()
```

Now, let's take a look at our actual decision tree:

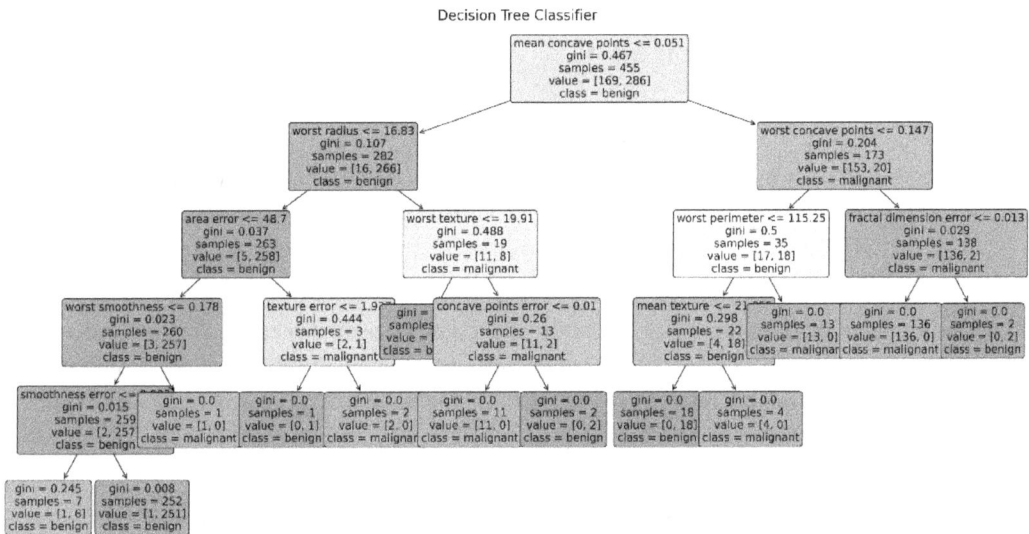

Figure 4.8 – Decision tree

This shows us an outline of our decision tree. You can see some interesting things about how it is making decisions. For instance, if the `mean concave points` feature is less than a certain number, we go down the left side of the tree; otherwise, we go to the right. This continues until we make a decision, or classification, of the sample as malignant or benign.

Already, you can see how useful this is. As we mentioned in the introduction, one nice thing about decision trees is that a human can interpret their results. This is part of a trend called **explainable AI**, in which machine learning models are meant to be understood and interpreted in ways that could be meaningful to a human. A more complex model, such as a neural network, might be very hard to visualize and explain.

11. Finally, let's try a cross-validation:

```
from sklearn.model_selection import cross_val_score
cv_scores = cross_val_score(dt_classifier, X, y, cv=5)

print("\nCross-Validation Scores:")
print(f"Mean CV Score: {cv_scores.mean():.4f}")
print(f"Standard Deviation: {cv_scores.std():.4f}")
```

This method trains the model on subsets of the data and tests it iteratively to provide a more meaningful test of model accuracy – as opposed to just training and testing on the same dataset, which can lead to overfitting and poor predictive accuracy.

We can see that our model has an overall mean CV score of ~91%. This is the model's average score over multiple tests in the cross-validation, and a measure of how well it generalized to new data. Our model is pretty good!

See also

- Check out **iTree**, which allows you to interactively build a decision tree on a website (Sokolowski et al., *ITree: a user-driven tool for interactive decision-making with classification trees*, BioInformatics, April 2024): `https://academic.oup.com/bioinformatics/article/40/5/btae273/7651198`

- Learn about random forests, an extension of decision trees, in Su et al., *Colon cancer diagnosis and staging classification based on machine learning and bioinformatics analysis*, Computers in Biology and Medicine, June 2022: `https://www.sciencedirect.com/science/article/abs/pii/S0010482522002013`

Learning more about Matplotlib for chart generation

In this recipe, we will dive deeper into a core charting library in Python. We'll revisit our vaccine data and make some beautiful charts!

Matplotlib is the most common Python library for generating charts. There are more modern alternatives, such as Bokeh, which is web-centered, but the advantage of Matplotlib is not only that it is the most widely available and widely documented chart library, but also, in the computational biology world, we want a chart library that is both web- and paper-centric. This is because many of our charts will be submitted to scientific journals, which are equally concerned with both formats. Matplotlib can handle this for us.

Many of the examples in this recipe could also be done directly with pandas (hence, indirectly with Matplotlib), but the point here is to exercise Matplotlib.

Once again, we are going to use VAERS data to plot some information about the DataFrame's metadata and summarize the epidemiological data.

Getting ready

Again, we will be using the data from the first recipe, and the code can be found in `/Ch04/Ch04-5-matplotlib.ipynb`.

How to do it...

Here are the steps to try this recipe:

1. The first thing that we will do is plot the fraction of nulls per column:

```python
import numpy as np
import pandas as pd
import matplotlib as mpl
import matplotlib.pyplot as plt
vdata = pd.read_csv(
    "2021VAERSDATA.csv.gz", encoding="iso-8859-1",
    usecols=lambda name: name != "SYMPTOM_TEXT"
)
num_rows = len(vdata)
perc_nan = {}
for col_name in vdata.columns:
    num_nans = len(vdata[col_name][vdata[col_name].isna()])
    perc_nan[col_name] = 100 * num_nans / num_rows
labels = perc_nan.keys()
bar_values = list(perc_nan.values())
x_positions = np.arange(len(labels))
```

The `labels` variable contains the column names that we are analyzing, `bar_values` is the fraction of null values, and `x_positions` is the location of the bars on the bar chart that we are going to plot next.

2. Here is the code for the first version of the bar plot:

```python
fig = plt.figure()
fig.suptitle("Fraction of empty values per column")
ax = fig.add_subplot()
ax.bar(x_positions, bar_values)
ax.set_ylabel("Percent of empty values")
ax.set_ylabel("Column")
ax.set_xticks(x_positions)
ax.set_xticklabels(labels)
ax.legend()
fig.savefig("naive_chart.png")
```

We start by creating a figure object with a title. The figure will have a subplot that will contain the bar chart. We also set several labels and only used defaults. Here is the result:

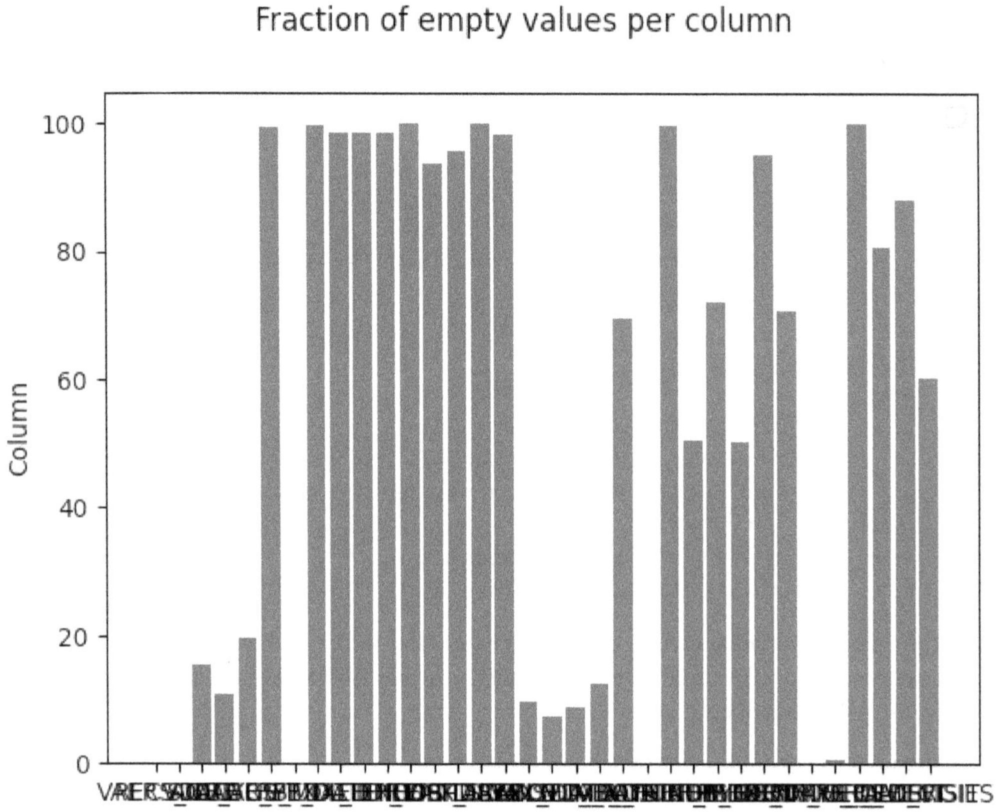

Figure 4.9 – Our first chart attempt, just using the defaults

3. Surely we can do better. Let's format the chart more substantially:

```
fig = plt.figure(
    figsize=(16, 9), tight_layout=True, dpi=600)
fig.suptitle(
    "Fraction of empty values per column", fontsize="48")
ax = fig.add_subplot()
b1 = ax.bar(x_positions, bar_values)
ax.set_ylabel(
    "Percent of empty values", fontsize="xx-large")
ax.set_xticks(x_positions)
```

```
ax.set_xticklabels(labels, rotation=45, ha="right")
ax.set_ylim(0, 100)
ax.set_xlim(-0.5, len(labels))
for i, x in enumerate(x_positions):
    ax.text(
        x, 2, "%.1f" % bar_values[i], rotation=90,
        va="bottom", ha="center",
        backgroundcolor="white"
    )
fig.text(0.2, 0.01, "Column", fontsize="xx-large")
fig.savefig("cleaner_chart.png")
```

The first thing that we do is set up a bigger figure for Matplotlib to provide a tighter layout. We rotate the *x* axis tick labels 45 degrees so that they fit better. We also put the values on the bars. Finally, we do not have a standard *x* axis label as it would be on top of the tick labels. Instead, we write the text explicitly. Note that the coordinate system of the figure can be completely different from the coordinate system of the subplot – for example, compare the coordinates of `ax.text` and `fig.text`. Here is the result:

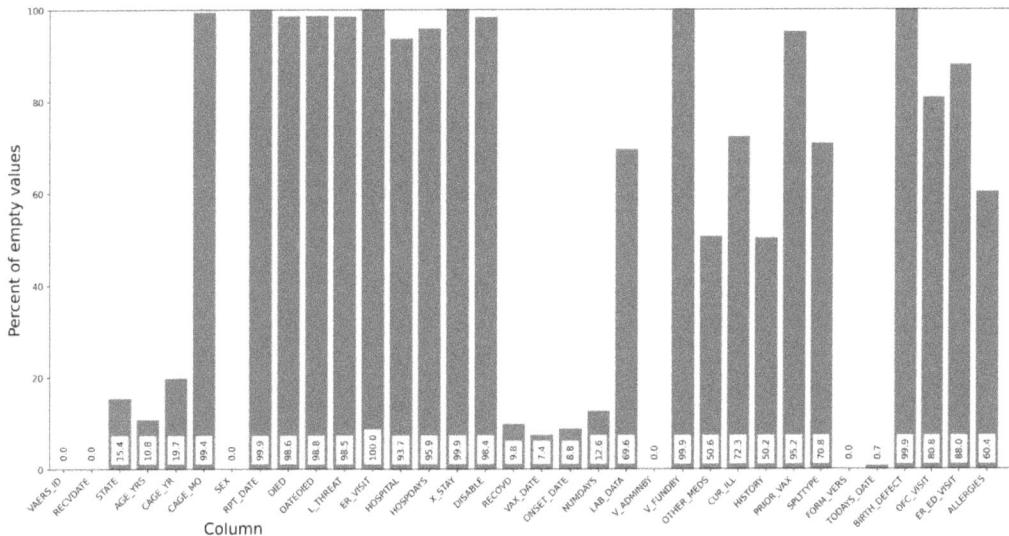

Figure 4.10 – Our second chart attempt, while taking care of the layout

4. Now, we are going to do some summary analysis of our data based on four plots on a single figure. We will chart the vaccines involved in deaths, the days between administration and death, the deaths over time, and the sex of people who have died for the top 10 states in terms of their quantity (remember to get the `2021VAERSVAX.csv.gz` file from Ch02):

```
dead = vdata[vdata.DIED == "Y"]
vax = pd.read_csv(
    "2021VAERSVAX.csv.gz",
    encoding="iso-8859-1"
).set_index("VAERS_ID")
vax_dead = dead.join(vax, on="VAERS_ID", how="inner",
                     lsuffix='_data', rsuffix='_vax')
dead_counts = vax_dead["VAX_TYPE"].value_counts()
large_values = dead_counts[dead_counts >= 10]
other_sum = dead_counts[dead_counts < 10].sum()
large_values = pd.concat(
    [large_values, pd.Series({"OTHER": other_sum})])
```

We've now set up some of our key datasets. Let's calculate our time distance DataFrames next:

```
distance_df = vax_dead[
    vax_dead.DATEDIED.notna() & vax_dead.VAX_DATE.notna()]
distance_df["DATEDIED"] = pd.to_datetime(
    distance_df["DATEDIED"])
distance_df["VAX_DATE"] = pd.to_datetime(
    distance_df["VAX_DATE"])
distance_df = distance_df[
    distance_df.DATEDIED >= "2021"]
distance_df = distance_df[
    distance_df.VAX_DATE >= "2021"]
distance_df = distance_df[
    distance_df.DATEDIED >= distance_df.VAX_DATE]

time_distances = (
    distance_df["DATEDIED"] -
    distance_df["VAX_ DATE"]
)
time_distances_d = (
    time_distances.astype(int) / (10**9 * 60 * 60 * 24)
)
```

Now, we will calculate the total deaths and find the states with the highest number of deaths:

```
date_died = pd.to_datetime(
    vax_dead[vax_dead.DATEDIED.notna()]["DATEDIED"])
date_died = date_died[date_died >= "2021"]
date_died_counts = date_died.value_counts().sort_index()
cum_deaths = date_died_counts.cumsum()
state_dead = vax_dead[
    vax_dead["STATE"].notna()
][["STATE", "SEX"]]
```

```
top_states = sorted(
    state_dead["STATE"].value_counts().head(10).index)
top_state_dead = (
    state_dead[state_dead["STATE"].isin(top_states)]
    .groupby(["STATE", "SEX"]).size()
)#.reset_index()
top_state_dead.loc["MN", "U"] = 0  # XXXX
top_state_dead = top_state_dead.sort_index().reset_index()
top_state_females = top_state_dead[
    top_state_dead.SEX == "F"][0]
top_state_males = top_state_dead[
    top_state_dead.SEX == "M"][0]
top_state_unk = top_state_dead[
    top_state_dead.SEX == "U"][0]
```

The preceding code is strictly pandas-based and was made in preparation for the plotting activity.

5. The following code plots all the information simultaneously. We are going to have four subplots organized in 2 x 2 format. In this way, we can see all four plots in a convenient format that fits on one page:

```
fig, (
    (vax_cnt, time_dist),
    (death_time, state_reps)
) = plt.subplots(
    2, 2,
    figsize=(16, 9), tight_layout=True
)
vax_cnt.set_title("Vaccines involved in deaths")
wedges, texts = vax_cnt.pie(large_values)
vax_cnt.legend(wedges, large_values.index, loc="lower left")

time_dist.hist(time_distances_d, bins=50)
time_dist.set_title("Days between vaccine administration and
death")
time_dist.set_xlabel("Days")
time_dist.set_ylabel("Observations")

death_time.plot(date_died_counts.index, date_died_counts, ".")
death_time.set_title("Deaths over time")
death_time.set_ylabel("Daily deaths")
death_time.set_xlabel("Date")
```

```
tw = death_time.twinx()
tw.plot(cum_deaths.index, cum_deaths)
tw.set_ylabel("Cummulative deaths")

state_reps.set_title("Deaths per state stratified by sex")
state_reps.bar(top_states, top_state_females, label="Females")
state_reps.bar(
    top_states, top_state_males, label="Males",
    bottom=top_state_females)
state_reps.bar(
    top_states, top_state_unk, label="Unknown",
    bottom=top_state_females.values + top_state_males.values
)
state_reps.legend()
state_reps.set_xlabel("State")
state_reps.set_ylabel("Deaths")
fig.savefig("summary.png")
```

We start by creating a figure with 2 x 2 subplots. The subplots function returns, along with the figure object, four axes objects that we can use to create our charts. Note that the legend is positioned in the pie chart; we have used a twin axis on the time distance plot, and we have a way to compute stacked bars on the death per state chart. Here is the result:

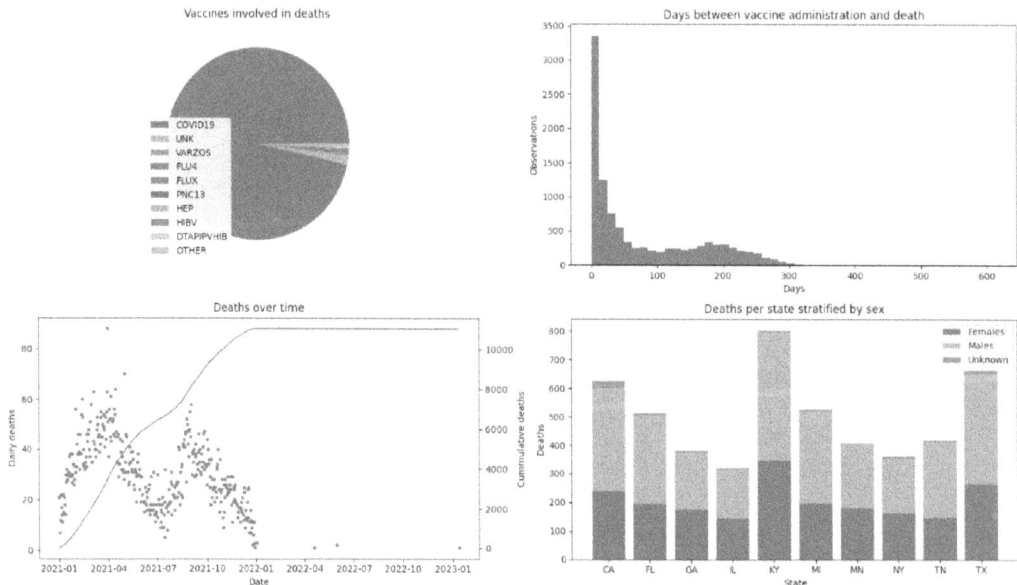

Figure 4.11 – Four combined charts summarizing the vaccine data

There's more...

Matplotlib has two interfaces you can use – an older interface, designed to be similar to MATLAB, and a more powerful **object-oriented (OO)** interface. Try as much as possible to avoid mixing the two. Using the OO interface is probably more future-proof. The MATLAB-like interface is below the `matplotlib.pyplot` module. To make things confusing, the entry points for the OO interface are in that module – that is, `matplotlib.pyplot.figure` and `matplotlib.pyplot.subplots`.

The following is some extra information that may be useful.

The documentation for Matplotlib is really, really good. For example, there's a gallery of visual samples with links to the code for generating each sample. This can be found at `https://matplotlib.org/stable/index.html`. The API documentation is generally very complete.

Another way to improve the look of Matplotlib charts is to use the Seaborn library. Seaborn's main purpose is to add statistical visualization artifacts, but as a side effect, when imported, it changes the defaults of Matplotlib to something more palatable. We will be using Seaborn throughout this book; check out the plots provided in the next recipe.

Building a UMAP using Seaborn

In this recipe, we'll learn about a newer and very visually appealing clustering algorithm called UMAP, using our breast cancer dataset!

UMAP stands for **Uniform Manifold Approximation and Projection**. It is useful to understand the structure of higher-dimensional data, even when it is non-linear. UMAP essentially takes a high-dimensional space and represents it using the most equivalent lower-dimensional graph it can find. It is also fast and efficient.

Getting ready

The code for this recipe can be found in `Ch04/Ch04-6-seaborn.ipynb`.

You will need to install the `seaborn` and `umap-learn` packages if you don't already have them. You can do this from the terminal by typing the following:

```
pip install seaborn
pip install umap-learn
```

Or you can install these from the notebook like this:

```
! pip install umap-learn
! pip install seaborn
! pip install ipywidgets
```

This installs the umap package. We also install seaborn here, although you may already have it installed from previous exercises. When trying this code, I got an error on displaying a progress bar, so I included the installation of ipywidgets to remove this error.

How to do it...

Here are the steps to try this recipe:

1. First, we will import our libraries:

```
import numpy as np
import matplotlib.pyplot as plt
import seaborn as sns
from sklearn.datasets import load_breast_cancer
from sklearn.preprocessing import StandardScaler
import umap
```

2. We will use the breast cancer dataset as before and bring in the umap module and seaborn for plotting. Let's load our test data:

```
data = load_breast_cancer()
X, y = data.data, data.target
```

3. Now, we will normalize our data as we have before:

```
scaler = StandardScaler()
X_scaled = scaler.fit_transform(X)
```

4. Now, we will initialize umap:

```
umap_reducer = umap.UMAP(
    n_neighbors=15,
    min_dist=0.1,
    n_components=2,
    random_state=42,
    n_jobs=1
)
X_umap = umap_reducer.fit_transform(X_scaled)
```

This code initializes umap. It sets the number of neighbors, n_neighbors, which controls local versus global structure in the clustering. We use the min_dist setting to control the compactness of clusters. We also set n_components to 2, which gives us a *2D* plot. This can be set to 3 for a *3D* view. Finally, we set the random state to 42 for easier reproducibility. We pass in the scaled features as the input.

5. Now, we can visualize our UMAP!

```
plt.figure(figsize=(10, 8))
for i in [0, 1]:
    mask = y == i
    plt.scatter(
        X_umap[mask, 0],
        X_umap[mask, 1],
        label=data.target_names[i],
        alpha=0.7,
        edgecolors='black',
        linewidth=0.5
    )
plt.title(
    'UMAP Visualization of Breast Cancer Dataset',
    fontsize=16
)
plt.xlabel('UMAP Dimension 1', fontsize=12)
plt.ylabel('UMAP Dimension 2', fontsize=12)
plt.legend()
plt.grid(True, linestyle='--', alpha=0.7)
plt.tight_layout()
plt.show()
```

This initializes a scatter plot and then iterates over the two classes of data, setting a mask to control which class we will process as we make the subplots. We set the color based on the target class and other visual parameters.

Next, we set a plot title and label the X and Y axes. We add a grid, a legend, and adjust the layout, and then show the plot:

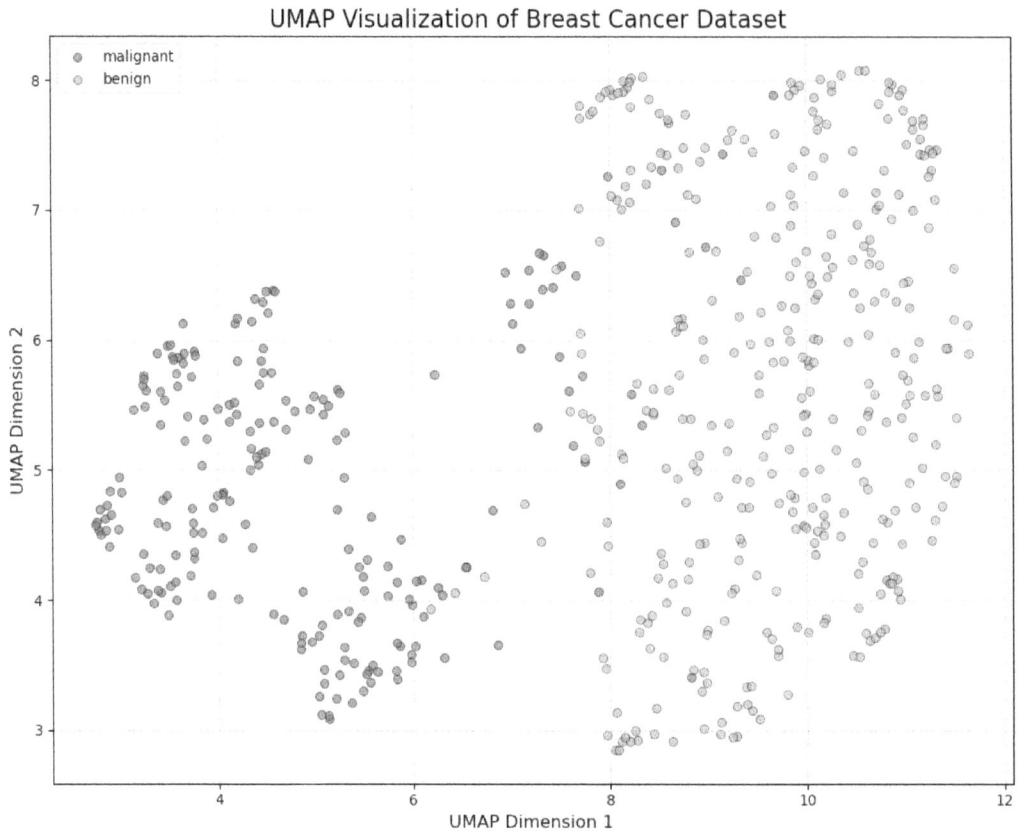

Figure 4.12 – UMAP visualization for the breast cancer dataset

As you can see, we actually get a fairly good clustering, although some points do seem to cross into the other cluster.

6. Now, let's see whether we can improve upon the clustering by trying different parameters:

```python
def plot_umap_parameter_comparison():
    fig, axs = plt.subplots(2, 2, figsize=(16, 16))
    neighbors_values = [5, 15, 30, 50]
    for i, n_neighbors in enumerate(neighbors_values):
        row = i // 2
        col = i % 2
        umap_reducer = umap.UMAP(
            n_neighbors=n_neighbors,
            min_dist=0.1,
            n_components=2,
            random_state=42,
            n_jobs=1
        )
        X_umap = umap_reducer.fit_transform(X_scaled)
        axs[row, col].scatter(
            X_umap[:, 0],
            X_umap[:, 1],
            c=y,
            cmap='viridis',
            alpha=0.7,
            edgecolors='black',
            linewidth=0.5
        )
        axs[row, col].set_title(
            f'UMAP (n_neighbors = {n_neighbors})')
        axs[row, col].set_xlabel('UMAP Dimension 1')
        axs[row, col].set_ylabel('UMAP Dimension 2')
    plt.tight_layout()
    plt.show()
```

We define a function to loop over a list of `neighbors` values. For each value, we initialize a UMAP and fit the data as before. We then show the plots. Here is what we get:

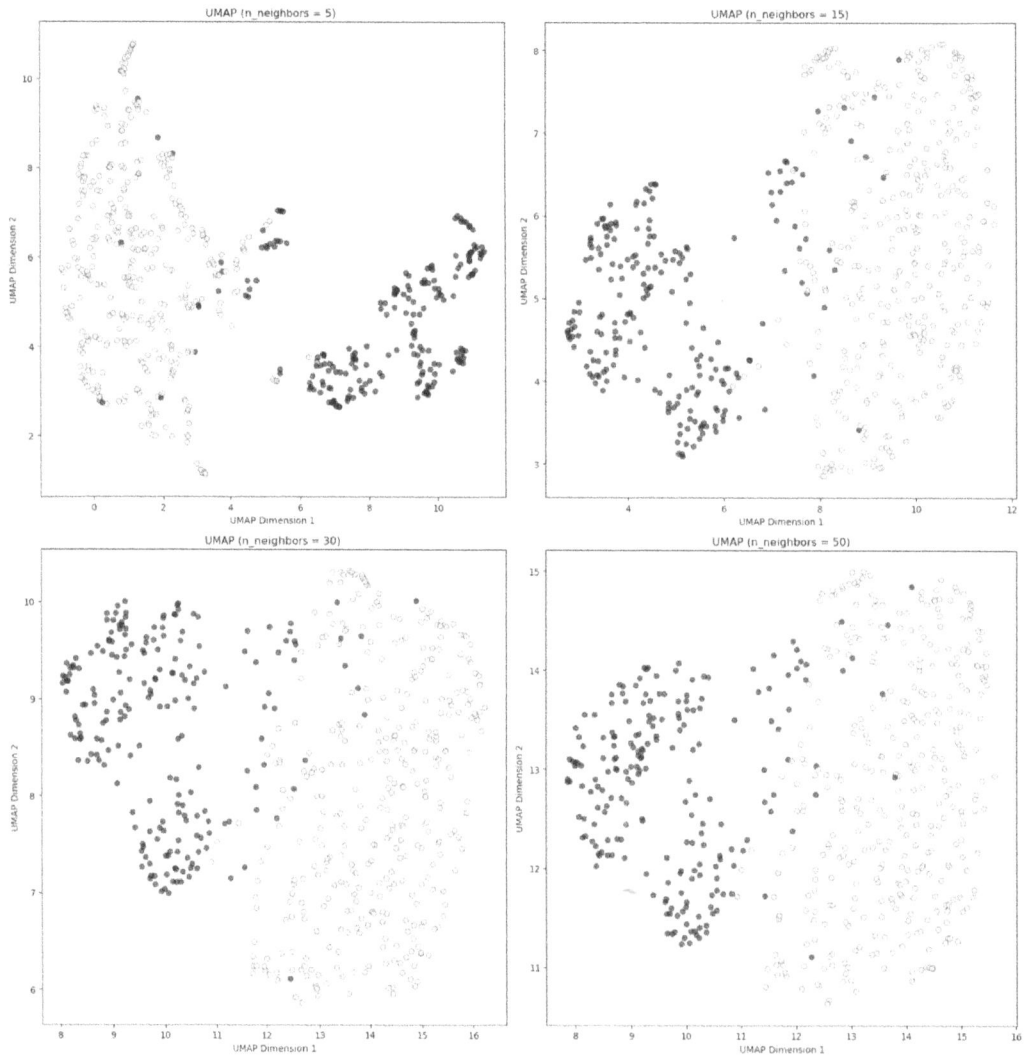

Figure 4.13 – Comparison of neighbor parameters

We can see that different parameters give us different results. Having `neighbors` at 5 seems to give the clearest separation. In this way, we can tune our parameters and find the best approach.

There's more ...

UMAP is very similar to a popular technique called **t-distributed Stochastic Neighbor Embedding (t-SNE)**. It is also available in scikit-learn. You will see it used extensively in bioinformatics literature (e.g., Kobak & Berens, *The art of using t-SNE for single-cell transcriptomics*, Nature Communications, November 2019). For a good comparison of UMAP and t-SNE, see `https://pair-code.github.io/understanding-umap/`.

When you are finished, clean up your environment by running the following:

```
conda deactivate
```

See also

- To learn more about the math behind UMAP, visit `https://umap-learn.readthedocs.io/en/latest/how_umap_works.html`

- This paper shows how UMAP can be used to systematically analyze the transcript profiles of gene deletion mutants of yeast: Dorritty et al., *Dimensionality reduction by UMAP to visualize physical and genetic interactions*, Nature Communications, March 2020 (`https://www.nature.com/articles/s41467-020-15351-4`)

- More information on Seaborn can be found here: `https://seaborn.pydata.org/`

Get This Book's PDF Version and Exclusive Extras

UNLOCK NOW

Scan the QR code (or go to `packtpub.com/unlock`). Search for this book by name, confirm the edition, and then follow the steps on the page.

Note: Keep your invoice handy. Purchases made directly from Packt don't require an invoice.

Alignment and Variant Calling

In this chapter, we will cover some of the core aspects of bioinformatics work. We will deeply explore how to examine the quality of sequencing data for different technologies. We will also go deeper into various tools for manipulating relevant file formats and briefly cover some recent advances in data compression for bioinformatics files. Then, we'll discuss how sequencing data is aligned to reference sequences and review the importance of different aspects of sequencing algorithms. Finally, we'll get into variant calling and see how the information resulting from an alignment can be used to find critical biological insights about people and organisms.

In this chapter, we're going to cover the following main topics:

- Quality control for sequencing data
- Tools for sequence manipulation
- Sequence alignment with BWA
- Variant calling with FreeBayes

Technical requirements

You will find the code for this chapter at `https://github.com/PacktPublishing/Bioinformatics-with-Python-Cookbook-Fourth-Edition/tree/main/Ch05`.

Create a GitHub folder named Ch05 for your notebooks. Inside it, make a subdirectory called `data` for our practice files.

Remember to activate your `conda` environment before beginning the recipes, like this:

```
conda activate bioinformatics_base
```

Or, if you would like to set up a `conda` environment specifically for this chapter, before activating `bioinformatics_base`, run the following:

```
conda create -n ch05-alignment --clone bioinformatics_base
conda activate ch05-alignment
```

You will be able to install the packages for the chapter as you go, or you can use the YAML file provided in the repository:

```
conda env update --file ch05-alignment.yml
```

Quality control for sequencing data

The discovery of DNA and the ability to read, or sequence, it has revolutionized the world. The first such technology was Sanger sequencing, which uses fluorescent dyes sorted by a capillary. This is also called first-generation sequencing. But it was the advent of **Next-Generation Sequencing (NGS)**, the second generation, that allowed dramatic advances, such as routine human genome sequencing. The most popular example of this is the Illumina sequencing technology (`https://www.illumina.com`), which uses short reads approximately 100 to 300 **base pairs** long to read out DNA sequences.

> **Base pairs**
>
> **Base pairs (bp)** are the pairs of complementary nucleotide bases (A goes with T, C goes with G) that are located on the two complementary strands of DNA. We normally don't repeat both sides of the two strands; we read one strand and assume that the complementary bases are on the other side. So, for example, if we read out the sequence ACGT from one strand, we would say we have "read four base pairs". A **kb** is a **kilobase**, or 1,000 bp. See `https://www.genome.gov/genetics-glossary/Base-Pair` for more information.

More recently, **third-generation** technology has matured in which much longer reads can be generated, often up to thousands of bp long.

These technologies are used by companies such as PacBio and Oxford Nanopore. These long-read technologies often suffer from high error rates, but are excellent at resolving long-range structural variations in genomes. Each type of technology requires different approaches to quality control and data interpretation.

Here is an overview of the current sequencing technologies:

Technology	Read Types	Website	Description
Illumina	Short reads (~100-300 bp)	`https://www.illumina.com/`	Excels on volume/cost
Pacific Biosciences	Long reads (10-25 kb)	`https://www.pacb.com/`	Great at structural variation; high error rate
Oxford Nanopore	Long reads (10-300 kb)	`https://nanoporetech.com/`	Excels at structural variation; adaptive sampling mode
Element Biosciences	Short reads (75-300 bp)	`https://www.elementbiosciences.com/`	High accuracy

Technology	Read Types	Website	Description
Ultima Genomics	Short reads (~300 bp)	`https://www.ultimagenomics.com/`	No flow cell; extremely low cost
BioNano Genomics	Very long-range structure	`https://bionano.com/`	Optical genome mapping for long-range analysis

Table 5.1 – Sequencing technologies

In this recipe, we'll learn how to perform the first step in sequence analysis, basic quality control. We'll see how to interpret quality control outputs and make decisions on whether the data quality is sufficient to move forward to downstream analysis.

Getting ready

Before we begin, make sure you have installed `wget`:

```
brew install wget
```

Now, we will get some FASTQ data to practice with (we'll go over the FASTQ format in the *Tools for sequence manipulation* recipe). Run the following commands in your terminal (or you can run them from your notebook using the `!` command syntax):

```
wget ftp://ftp.sra.ebi.ac.uk/vol1/fastq/SRR390/SRR390728/SRR390728_1.
fastq.gz
wget ftp://ftp.sra.ebi.ac.uk/vol1/fastq/SRR390/SRR390728/SRR390728_2.
fastq.gz
mv SRR390728_1.fastq.gz data/
mv SRR390728_2.fastq.gz data/
```

We will also need to install FastQC and MultiQC with these commands:

```
conda install -c bioconda fastqc
conda install -c bioconda multiqc
```

The code for this recipe can be found in `Ch05/Ch05-1-qc-data.ipynb`.

How to do it...

Here are the steps to try this recipe:

1. First, we'll import the modules we need:

    ```
    import os
    import subprocess
    ```

The `os` module provides operating system functions. The `subprocess` module is used to run system (terminal) commands.

2. Next, we'll define a function called `run_fastqc()` to execute FastQC (`https://github.com/s-andrews/FastQC`). FastQC is a tool that analyzes many basic sequence qualities, such as the quality of each base, the content of the sequences, and their length distribution. This helps lab scientists and bioinformaticians to determine the overall quality of the data they generate. FastQC can be run directly from the command line - have a look at `https://olvtools.com/en/documents/fastqc`. However, here we will run it using a Python function so that it can be incorporated into our notebook code. Here is the code for this function:

```python
def run_fastqc(input_dir, output_dir):
    os.makedirs(output_dir, exist_ok=True)
    fastq_files = [
        f for f in os.listdir(input_dir)
        if f.endswith((".fastq", ".fastq.gz"))
    ]
    if not fastq_files:
        print(
            "Could not find any FASTQ files"
            "in the input directory."
        )
        return
    print("Running FastQC...")
    fastqc_command = (
        ["fastqc", "-o", output_dir] +
        [os.path.join(input_dir, f)
         for f in fastq_files]
    )
    subprocess.run(fastqc_command)
    print("FastQC analysis Completed.")
```

This code does the following:

- Defines the `run_fastqc()` function with parameters for the input directory, which will contain our FASTQ files, and an output directory for us to place the FastQC reports.

- Uses the Python `os` module to create the `output` directory using the `makedirs()` function.

- Builds a `fastq_files` array by looping over the files in the input directory using the `os.listdir()` function. We use the `endswith()` function on the file to restrict it to filenames ending with the `.fastq` extension.

- Prints a warning if we cannot find any FASTQ files.

- Constructs a `fastqc_command` array that contains the appropriate command for each file.

- Uses the `subprocess` module to run the FastQC commands.

3. Next, we'll define a `run_multiqc()` function to aggregate our FastQC reports. Here it is:

```
def run_multiqc(input_dir, output_dir):
    os.makedirs(output_dir, exist_ok=True)
    print("Running MultiQC...")
    multiqc_command = [
        "multiqc", input_dir, "-o",
        output_dir
    ]
    subprocess.run(multiqc_command)
    print("Finished...MultiQC report(s) generated.")
```

This code creates a directory for the MultiQC output files. It then constructs a `multiqc_command` to run MultiQC against the input directory and place the output in the output subdirectory.

4. Now, we'll define our `main()` function and run it:

```
def main():
    input_dir = "./data"
    fastqc_output_dir = "fastqc_output"
    multiqc_output_dir = "multiqc_output"
    run_fastqc(input_dir, fastqc_output_dir)
    run_multiqc(fastqc_output_dir, multiqc_output_dir)
    print(
        f"MultiQC report saved in: "
        f"{os.path.abspath(multiqc_output_dir)}"
    )
if __name__ == "__main__":
    main()
```

This code does the following:

- Points our input directory at the `data` subdirectory.

- Defines the FastQC and MultiQC output directories.

- Calls our `run_fastqc()` function to process the FASTQ files.

- Calls our `run_multiqc()` function to aggregate the reports.

- Prints out the path to the report.

- Uses the special `dunder` method to define the `main()` function. The term **dunder** is short for **double underscore** and defines special methods. Here, we use it to make sure the `main()` function is only called when we are executing this script directly and not when importing it into another module. Executing this code block will run the code and process our FASTQ files.

5. Now, we can review our MultiQC report and examine the quality of our data:

    ```
    ! open multiqc_output/multiqc_report.html
    ```

This last command will open the MultiQC report in your browser. This is what it looks like:

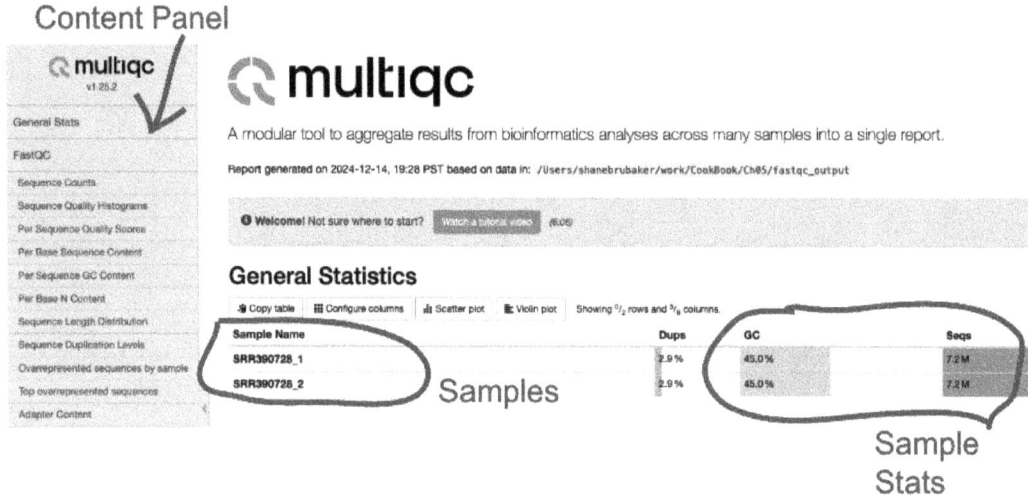

Figure 5.1 – MultiQC main page

In this screenshot, we can see that MultiQC has aggregated our samples into a single report. In the **General Statistics** section, we see the sample names, the percentage of duplicate sequences, GC content (the percentage of G and C bases in the genome), and the total number of reads. Each of our samples has about 7.2 million reads.

6. On the left side, we can see a content panel that will take us to various sections of the report. Let's take a look at some of the sections. Click on **Sequence Quality Histograms**:

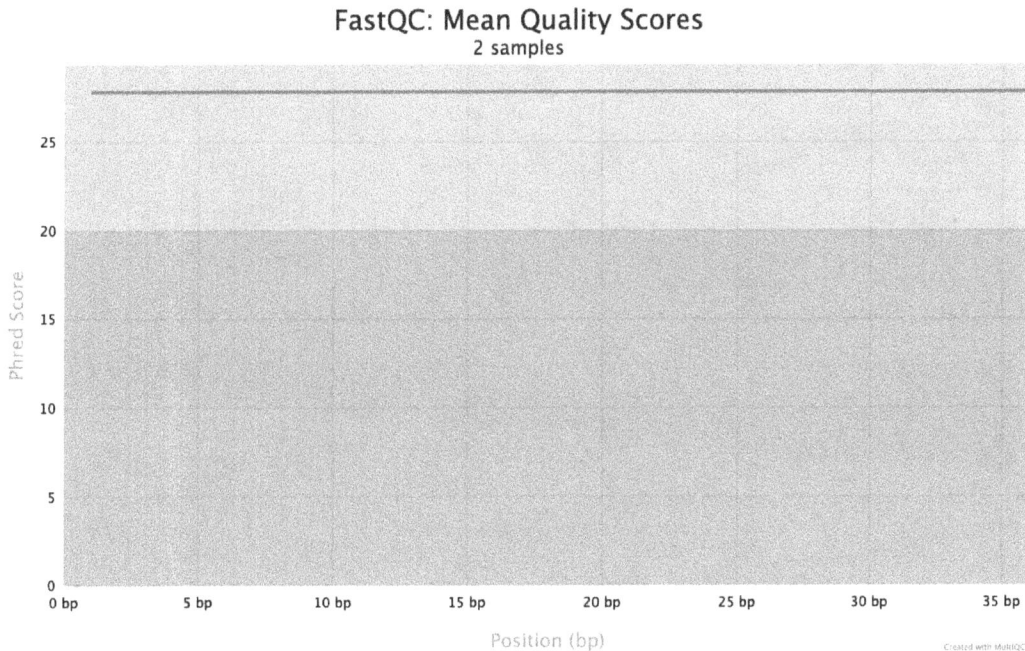

Figure 5.2 – Sequence Quality Histograms

We see the read length on the *x* axis and the quality score or **Phred Score** on the *y* axis. It is important to take a moment to understand the Phred score, which is also called the **Q score** or **Q value**:

Phred scores

The Phred score is a measure of the accuracy of a base call in a sequence. It is calculated as the negative log of the probability of the base being wrong: $Q = -10 * log10(P)$, where P is the probability of an error.

So, for example, if we say *these bases all exceed Q30*, what we mean is that the chance of any base being wrong is less than one in one thousand.

We can see that the sequences from our test data are 35 bp in length and that every base has a nice uniform Q score around 28 (our two samples are right on top of each other). They are in the green section of the graph because they fall within the acceptable quality range.

7.   Here's an example where read quality falls off towards the end of the read:

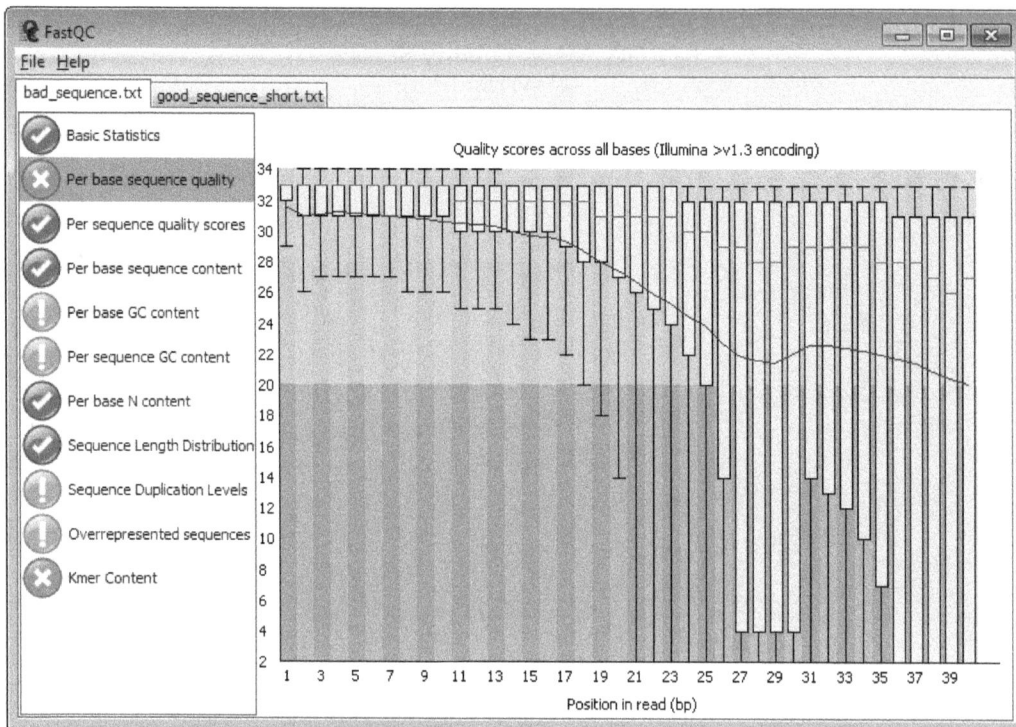

Figure 5.3 – FastQC sequence quality graph (source: FastQC website)

This is an example from FastQC of a read where the average quality does not stay the same and falls off toward the end of the read. The bars represent histograms of the range of qualities that we see at that position in the read. The line that goes from left to right represents the average, so you can see there is a downward trend as we move forward in the read. This is very common for many sequencing technologies. In severe cases it might represent bad data, and you would simply discard the data. It is also quite common to use **read trimming** tools to simply trim off part of the end of the read to yield higher-quality data for use in downstream analysis. We'll cover read trimming in more depth in the *Tools for sequence manipulation* recipe.

Let's next look at another graph, **Per Sequence GC Content**:

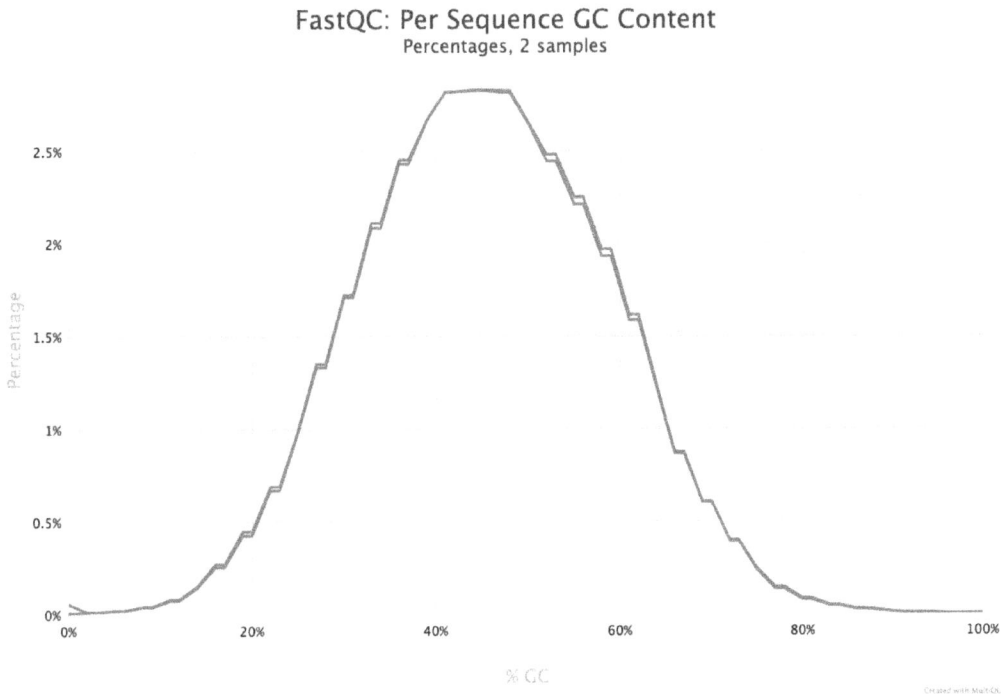

Figure 5.4 – Per Sequence GC Content

This plot shows the distribution of GC content in the reads. We should typically expect a normal distribution with a peak around the average GC content of the organism we are sequencing. For example, the average GC content of the human genome is forty-one percent. So, if we were trying to sequence a human and saw a different average at the peak here, we would suspect something is wrong. If you see an unusual shoulder or double-peak effect, it could indicate contamination, suggesting that another organism (such as a bacterium that has a different GC content) has gotten into your sequencing preparation.

You can investigate the MultiQC output further and look at some of the other available reports. These include things such as the number of Ns (places where the sequencer could not call a base) and the level of **adapter content**, which can indicate that there is an overrepresentation of the adapters used to make the sequencing library as opposed to the actual DNA you want to sequence.

It is important to understand sequencing quality control and its implications. You will want to work with lab personnel to manually inspect entire sequencing runs for quality and develop automated tools in your pipelines to help control for and correct quality issues.

There's more...

In this recipe, we only aggregated results from the FastQC tool, but MultiQC can also recognize other tools and integrate their results. It can use output from SAMtools, picard, GATK, SnpEff, and many others. You can use this to aggregate many types of reports together in a single handy HTML report.

Another important topic we should cover is the use of **paired reads**. When we sequence DNA, we first extract DNA from our sample and then shear it into fragments. In **single-end** read mode, we read the DNA from various parts of each fragment. In paired-end mode, we read from each end of the fragment inward. This means we know these two read pairs go with each other, and we know the approximate distance between them. This increases the range of information we have, and this information can be used by aligners and assemblers. In RNA sequencing, this can be especially useful to get reads that span the junction between exons that are spliced together:

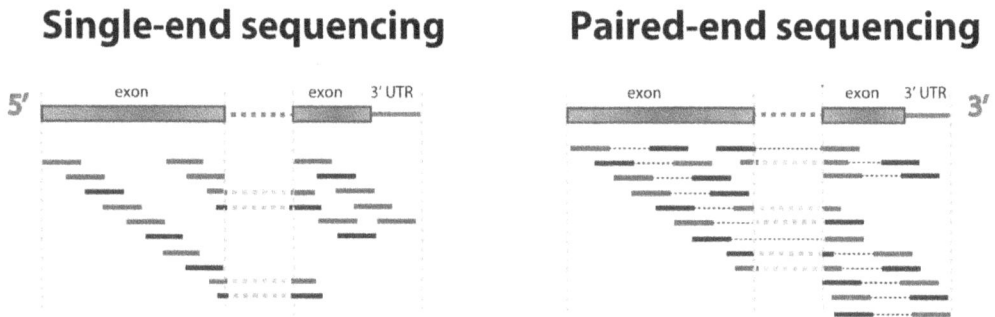

Figure 5.5 – Single-end versus paired-end sequencing (source: https://statomics.github.io/SGA/sequencing_intro.html)

We should also discuss quality control of long reads. Long reads have traditionally had higher error rates than shorter reads. However, short reads tend to have errors that are **systematic** – meaning they occur in the same locations in reads – and long reads tend to have **random** errors. The random error profile of long reads actually turns out to be very useful because the more coverage we have of a certain base, the less likely it is that we will see the same random error. In this way, we can overcome the error rate of long reads simply by increasing coverage. Indeed, long reads have ushered in an era where we can achieve chromosome-scale assemblies at Q70 or greater quality. This means genomes are assembled into entire chromosomes instead of thousands of contigs, and the quality of each base is such that there is only a one in ten to the seventh power chance that it is incorrect!

For a great discussion of this, see Taylor et al., *Beyond the Human Genome Project: The Age of Complete Human Genome Sequences and Pangenome References*, Annual Reviews, Vol 25, 2024.

> **Contig**
>
> A **contig** is a **contiguous region** of a genome. Typically, when overlapping reads can be assembled into one long region of a genome that goes together, it is referred to as a contig. If a genome were completely assembled, the contigs would represent entire chromosomes.

There are several tools that specialize in quality control of long reads. LongQC provides quality control for both PacBio and Nanopore technologies (Fukusawa et al., *LongQC: A Quality Control Tool for Third Generation Sequencing Long Read Data*, G3, Feb 2020). PycoQC is a tool focused on Oxford Nanopore data (`https://a-slide.github.io/pycoQC/`).

See also

- You can find the documentation on FastQC here: `https://www.bioinformatics.babraham.ac.uk/projects/fastqc/`

- More information on MultiQC can be found at `https://seqera.io/multiqc/`, along with a video tutorial: `https://www.youtube.com/watch?v=qPbIlO_KWN0`

- More information on the Phred quality score can be found here: `https://gatk.broadinstitute.org/hc/en-us/articles/360035531872-Phred-scaled-quality-scores`

- For a good description of single-end versus paired-end sequencing, visit `https://www.illumina.com/science/technology/next-generation-sequencing/plan-experiments/paired-end-vs-single-read.html`

- If you want to explore the science further, the paper by Merker et al. describes how long-read sequencing allowed them to find disease-causing variants in patients that were missed by short-read technology: Long-read genome sequencing identifies causal structural variation in a Mendelian disease, Genetics in Medicine, January 2018, `https://pubmed.ncbi.nlm.nih.gov/28640241/`

Tools for sequence manipulation

In this recipe, we will learn about fundamental tasks for sequencing manipulation that are part of the core toolkit of any bioinformatician. We'll cover basic operations such as reverse complementation and translation, as well as more advanced topics such as read trimming.

We look at BioPython, which is an important library to learn about. It includes tools for storing and manipulating sequence information, performing alignments, running **Basic Local Alignment Sequence Tool** (**BLAST**), manipulating protein structures, and much more.

Before we go any further, let's get familiar with some of the core bioinformatics file formats:

Format	Description	Reference	Comments
FAST-All (FASTA)	Plain text DNA or protein sequences	`https://www.ncbi.nlm.nih.gov/genbank/fastaformat/`	Great for storing human-readable sequences
FASTA with Quality score (FASTQ)	Like FASTA but includes quality scores as an additional line	`https://learn.gencore.bio.nyu.edu/ngs-file-formats/fastq-format/`	Most common output of sequencers
Sequence Alignment Map (SAM)	SAM format specification	`https://samtools.github.io/hts-specs/SAMv1.pdf`	Human-readable alignment output
Binary Alignment Map (BAM)	Binary version of SAM file	`https://davetang.github.io/learning_bam_file/`	Takes up less storage space
Compressed BAM (CRAM)	Compressed reference-oriented alignment map	`https://samtools.github.io/hts-specs/CRAMv3.pdf`	Up to 40% less storage than BAM
Original Read Archive (ORA)	New FASTQ format from Illumina with improved compression, used in DRAGEN	`https://support-docs.illumina.com/SW/DRAGEN_v38/Content/SW/DRAGEN/ORA_Compression_fDG_swHS.htm`	Up to 5x smaller than FASTQ files

Table 5.2 – Common bioinformatics file formats

Note that some of the file formats listed here do not use strict abbreviations per se, but rather just come from historical terms in bioinformatics. The **FASTA** format, which we discussed briefly in *Chapter 3, Modern Coding Practices and AI-Generated Coding*, is a human-readable format with the ability to store one or more DNA or protein sequences. The term came from a *fast* algorithm that could store *all* types of sequences, meaning nucleotide and protein.

The FASTQ format adds sequencer quality information to the FASTA format. For each base, we have a quality score on the next line. Here is an example section:

```
@SRR000001.2 EM7LVYS02GCAPL length=115
TATATTTTCCTTCTTAGATTCCACGGCAGCCCTGTGAGTTAACAATCAACTCTGTTTCAAAGCTGAGGACACTGAGGCTCTAAGAGGTTAAATTATTGACCCAGATCACAAGAAT
+SRR000001.2 EM7LVYS02GCAPL length=115
<=;<GC7*C<B:=C=<=<D=C<<<C==<<<FB1<=<==<D=D==C<=<C=<=<<<FA/<FB0<====B;===<===B:====B<;<B;C=FB1C<<C<9=FB2;==;=<=D==B;<
```

Figure 5.6 – Example section from a FASTQ file

We see an @ symbol first – this line gives us a read identifier. Next, we see the actual sequence of the read. Then there is a separator, which is just a + sign. Finally, we have the Phred quality scores. These codes translate into a score – the codes can be found here: `https://en.wikipedia.org/wiki/Phred_quality_score`. For example, E means the Phred quality score was 69.

SAM files are used for sequence alignments and are human-readable. They have a header section with a variety of information about the alignment. These headers can vary and are often a source of confusion when trying to compare similar alignments. We then have a series of lines representing each read alignment to a position on the reference. An important concept included here is the CIGAR string. **CIGAR** stands for **Concise Idiosyncratic Gapped Alignment Report**. It consists of a series of positions and codes for those positions indicating whether they match, are in a gap, are not included in a reference sequence, and so on. It helps us represent the exact details of how a query sequence matches the reference. You can find more information on CIGAR strings here: `https://www.drive5.com/usearch/manual/cigar.html`. You may also want to check out the `cigar` library: `https://github.com/brentp/cigar`.

BAM files are compressed binary versions of SAM files. They are compressed in the BGZF format. BAM files take up less space and can be indexed for faster access. BAM files use a compression level setting that allows you to trade off CPU time during compression for additional storage savings. For more details, see `https://medium.com/@acarroll.dna/looking-at-trade-offs-in-compression-levels-for-genomics-tools-eec2834e8b94`.

CRAM files introduce a reference file into the BAM file to increase efficiency and compression levels. With the new CRAM 3.1 standard, these files can be 50-70% smaller than a corresponding BAM file. When using CRAM, you want to make sure you track carefully the reference file used with it (although it does not strictly require a reference). CRAM files may introduce increased speed in your pipelines in some cases but may also increase times slightly due to the decompression needed, depending on the situation.

You should read Bonfield, *CRAM 3.1: advances in the CRAM file format*, BioInformatics, March 2022, `https://academic.oup.com/bioinformatics/article/38/6/1497/6499262`, for more information on the CRAM format.

HTSLib is a popular C-based library for accessing bioinformatics files and includes support for CRAM: `https://github.com/samtools/htslib`.

There are many other important bioinformatics file formats that we'll continue to cover as we go through the book. These are the key formats for primary (sequence-level) and secondary (alignment-level) analysis.

Getting ready

First, let's get some sample data to work with. We will get a sample FASTA file and a **Coding DNA Sequence (CDS)** file:

```
! wget -O sample.fasta "https://www.ncbi.nlm.nih.gov/sviewer/viewer.
cgi?db=nuccore&id=NM_001200.1&report=fasta"
! mv sample.fasta data/


! wget -O cds_sequence.fasta "https://www.ncbi.nlm.nih.gov/sviewer/
viewer.cgi?db=nuccore&id=NM_000518.5&report=fasta_cds_na&retmode=text"
! mv cds_sequence.fasta data/
```

In this recipe, we will also learn about the **SRA Toolkit**, a powerful library for retrieving sequence data from public databases:

1. First, we will download it (The following commands are executed from the terminal):

   ```
   curl --output sratoolkit.tar.gz https://ftp-trace.ncbi.nlm.nih.
   gov/sra/sdk/current/sratoolkit.current-mac64.tar.gz
   ```

2. Next, we will unzip and untar it:

   ```
   gunzip sratoolkit.tar.gz
   tar -xvf sratoolkit.tar
   ```

3. We also need to add it to our path:

   ```
   export PATH=$PATH:~/Software/sratoolkit.3.1.1-mac-x86_64/bin
   ```

> **Note**
>
> In the preceding path, `sratoolkit.3.1.1-mac-x86_64` refers to the version of the SRA Toolkit that was current at the time of writing this book, which was 3.1.1. In your case, you should replace it with the current version. Do the same in the upcoming instructions.

4. You will want to add this to your `.zshrc` file The following commands add the path to your `.zshrc` file and test that the SRA Toolkit is working by running the `fasterq-dump` help command.:

   ```
   echo 'export PATH=$PATH:~/Software/sratoolkit.3.1.1-mac-x86_64/
   bin' >> ~/.zshrc
   source ~/.zshrc
   fasterq-dump -h
   ```

5. Now, let's use it to download a sample FASTQ file. Move back over to your Ch05/data directory and type the following:

   ```
   fasterq-dump SRR000001
   ```

This will use the `fasterq-dump` SRA tool to download the FASTQ file with the `SR000001` accession number. This will download three files. The ones ending in _1 and _2 represent the paired reads (left and right, respectively). The third one is an **interleaved** file in which the paired reads are alternated to produce a single file.

For more information on installing SRA Toolkit, visit `https://github.com/ncbi/sra-tools/wiki/02.-Installing-SRA-Toolkit`.

For this recipe, you will also need to install the `biopython` Python modules:

```
conda install -c conda-forge biopython
```

The code for this chapter can be found in `Ch05/Ch05-2-sequence-manipulation.ipynb`

How to do it...

Here are the steps to try this recipe:

1. Let's start by looking at how we can use BioPython to manipulate sequences. First, we will import our modules:

   ```
   from Bio.Seq import Seq
   from Bio import SeqIO
   ```

2. This brings in the BioPython `Seq` class. In BioPython, sequences are typically stored in `Seq` objects. The `Seq` object has a number of handy methods for working with sequences. We also bring in `SeqIO`, which is the standard input/output system for BioPython. It will help us read (or write) files in common bioinformatics formats.

3. Next, we'll load our sample FASTA files. We'll want to end up with `Seq` objects for each of our sequences:

   ```
   fasta_file1 = "data/sample.fasta"
   fasta_file2 = "data/cds_sequence.fasta"
   sample_sequence = []
   cds_sequence = []
   def read_fasta(file_path, seq_list):
       with open(file_path, "r") as handle:
           for record in SeqIO.parse(handle, "fasta"):
               seq_list.append(str(record.seq))
   read_fasta(fasta_file1, sample_sequence)
   read_fasta(fasta_file2, cds_sequence)
   sample_seq_str = " ".join(sample_sequence)
   cds_seq_str = " ".join(cds_sequence)
   dna_seq = Seq(sample_seq_str)
   cds_seq = Seq(cds_seq_str)
   ```

To recap, this code does the following:

- Defines the location of our two FASTA files

- Defines a `list` to hold each sequence

- Creates a `read_fasta()` function, which takes a path to a file and a list as input, then uses the SeqIO `parse` method to read in the file and append the sequence contents to the list

- Uses the `read_fasta()` function to bring in our sample DNA and CDS sequences

- Turns the resulting sequence lists into strings by using the `"  ".join()` method (this tells Python to join the elements of the list into a string)

- Finally, creates BioPython `Seq` objects for each of the two sequence strings

4. Now, let's use our `Seq` objects to alter our sequences! First, we will `Complement` the DNA sequence:

```
print("Complement:", dna_seq.complement())
```

5. DNA is double-stranded, so if one strand has ACTG, the other strand will contain TGAC. Complementation to determine the opposite strand is a common sequence operation. For more information on the structure of DNA, please review this Khan Academy tutorial: `https://www.khanacademy.org/test-prep/mcat/biomolecules/dna/a/dna-structure-and-function`.

6. When we are reading the forward DNA strand, or the 5' to 3' direction, we often say we are going in the **sense** direction. Note that the tick mark here is read as the word *prime*. This terminology comes from the numbering of atoms on the sugars in the nucleic acid, which gives us a sense of direction when reading DNA. The 5' sugar is on one side of the molecule and the 3' sugar ends up on the other side of the molecule. When we go in the opposite direction, 3' to 5', or the **anti-sense** direction, we will be reading the complementary strand backward, so ACTG will become CAGT. This is what the reverse complement operation does:

```
print("Reverse Complement:", dna_seq.reverse_complement())
```

7. The **central dogma** says that DNA is transcribed into RNA and then translated into protein (amino acids). In this step, we will use the seq object's `transcribe()` method to convert the CDS into RNA (the **CDS** is the **Coding DNA Sequence**, meaning the part that can be translated into protein). The main difference here is that Uracil (U) is used instead of Thymine (T). You should make sure you understand the basic concepts of transcription, tRNA, and translation. This page contains a good overview.

`https://en.wikipedia.org/wiki/Central_dogma_of_molecular_biology`

Here is the code to transcribe the CDS to RNA:

```
print("Transcription (DNA to RNA):", cds_seq.transcribe())
```

8. Finally, we translate the CDS sequence to get the amino acid or protein sequence. The amino acids fold into a 3-dimensional structure to make a protein or enzyme (an enzyme is just a protein that has catalytic or chemical activity, whereas some proteins can be structural). These are the workhorses of the cell that make life possible!

```
print("Protein Translation (DNA to Protein):",
      cds_seq.translate(to_stop=True))
```

Here, we see some of the output from our code:

```
Reverse Complement: TATATATATATTTATGTATTTAATTTTGCTGTACTAGCGACACCCACAACCCTCCACAACCATGTCCTGATAGTTCTTTAATACAACCTTTTCATTCTCGTCAAGGTACAGCATCGAGATA
GCACTGAGTTCTGTCGGGACACAGCATGCCTTAGGAATCTTAGAGTTAACAGAGTTGACCAACGTCTGAACAATGGCATGATTAGTGGAGTTCAGATGATCAGCCAGAGGAAAAGGGCATTCTCCGTGGCAGTAAAAGGCG
TGATACCCCGGGGGAGCCACAATCCAGTCATTCCACCCCACGTCACTGAAGTCCACGTACAAAGGGTGTCTCTTACAGCTGGACTTAAGGCGTTTCCGCTGTTTGTGTTTGGCTTGACGTTTTTCTCTTTTGTGGAGAGGA
TGCCCTTTTCCATCATGGCCAAAAGTTACTAGCAATGGCCTTATCTGTGACCAGCTGTGTTCATCTTGGTGCAAAGACCTGCTTATCCTAACATGTCTCTTGGAGACACCTTGTTTCTCCTCCAAGTGGGCCACTTCCACC
ACGAATCCATGGTTGGCGTGTCCCTGTGCAGTCCACCGCATCACAGCGGGGGTGACATCAAAACTTTCCCACCTGCTTGCATTCTGATTCACCAACCTGGTGTCCAAAAGTCTGGTCACGGGGAATTTCGAGTTGGCTGTT
GCAGGTTTTATGATTTCATAAATATTAATTCGGTGATGGAAACTGCTATTGTTTCCTAAAGCATCTTGCATCTGTTCTCGGAAAACCTGAAGCTCTGCTGAGGTGATAAACTCCTCCGTGGGGATAGAACTTAAATTAAAG
TGCCTGCGATACAGGTCTAGCATGTAGGGGGGCACCACGGCGTCCCTGCTGGGGGTGGGTCTCTGTTTCAGGCCGAACATGCTGAGCAGCCGCAACTCGAACTCGCTCAGGACCTCGTCAGAGGGCTGGGATGAGGGGCGG
CCCGACGACGCCGCCGCGAACTTCCTGCGGCCCAGCTCCGGAACGAGGCCAGCCGCGCCGCCGCCCAGGAGGACCTGGGGAAGCAGCAACGCTAGAAGACAGCGGGTCCCGGCCACCATGGTCGACCTTTAGGAGACCGCAGTC
CGTCTAAGAAGCACGCGGGGACACGTCCATTGAAAGAGCGTCCACATGGAAAAACTCTGGTCAAAGGACCTGGCGCAAGGACCGAATGTCCGTTCCTTTTCTTTGCCTCCTTCTCCCGGGTGCCGGCCGCGCAGTCTC
TCTTTTCACGCTGGGAACAGCGTCTCAGTGTCGGGCAAGGCCGAGGAGTGGAGGGGCGGTGGGGGCTCGGAGATGGCGAAGCAGGCTCCGCTGGGGCAAAGGCACCGGCGCAAAGTGGGGTGCGCAAGTTATTCTCCCTGCA
AGTTCAAGAAGTCCCC
Transcription (DNA to RNA): AUGGUGCAUCUGACUCCUGAGGAGAAGUCUGCCGUUACUGCCCUGUGGGGCAAGGUGAACGUGGAUGAAGUUGGUGGUGAGGCCUGGGCAGGCUGCUGGUGGUGCUACCCUUG
GACCCAGAGGGUUCUUUGAGUCCUUUGGGGAUCUGUCCACUCCUGAUGCUGUUAUGGGCAACCCUAAGGUGAAGGCUCAUGGCAAGAAAGUGCUCGGUGCCUUUAGUGAUGGCCUGGCUCACCUGGACAACCUCAAGGGCAC
CUUUGCCACACUGAGUGAGCUGCACUGUGACAAGCUGCACGGUGGAUCCUGAGAACUUCAGGCUCCUGGGCAACGUGCUGGUCUGUGUGCUGGCCCAUCACUUUGGCAAAGAAUUCACCCCACCAGUGCAGGCUGCCUAUCA
GAAAGUGGUGGCCUGGUGUGGCUAAUGCCCUGGCCCACAAGUAUCACUAA
Protein Translation (DNA to Protein): MVHLTPEEKSAVTALWGKVNVDEVGGEALGRLLVVYPWTQRFFESFGDLSTPDAVMGNPKVKAHGKKVLGAFSDGLAHLDNLKGTFATLSELHCDKLHVDPEN
FRLLGNVLVCVLAHHFGKEFTPPVQAAYQKVVAGVANALAHKYH
```

Figure 5.7 – Part of the output from our sequence manipulation calls

In this screenshot, you can see the reverse complement, transcription, and translation outputs.

9. Now, let's look at some basic quality control steps, such as quality-based read trimming and adapter removal. As we learned in the *Quality control for sequencing data* recipe, it can be important to remove low-quality bases from our data. It is common to perform **3' end trimming**, in which we remove bases with a quality below a certain threshold from our reads before further processing.

   Another common task is **adapter trimming**. When DNA is prepared for sequencing, short sequences called adapters are added to bind to sequencing primers that amplify the DNA, and to provide **index sequences**, which allow **sample demultiplexing**. The indexes allow scientists to put several samples on a single **sequencing lane** and then deconvolute (demultiplex) the reads associated with each sample later, thereby providing significant cost savings.

   Let's begin by importing the libraries we'll need for read trimming:

```
from Bio import SeqIO
from Bio.SeqRecord import SeqRecord
from Bio.Seq import Seq
```

We used the BioPython SeqIO and SeqRecord libraries, and the Seq class in the preceding code.

10. Next, we'll define a function for trimming bases based on quality:

```
def trim_low_quality_bases(record, quality_threshold):
    qualities = record.letter_annotations[
        "phred_quality"]
    trimmed_index = len(qualities)
    for i in range(
        len(qualities) - 1, -1, -1
    ):  # Iterate over qualities
        if qualities[i] >= quality_threshold:
            break
        trimmed_index = i

    trimmed_seq = record.seq[:trimmed_index]
    trimmed_qual = qualities[:trimmed_index]
    trimmed_record = SeqRecord(
        Seq(str(trimmed_seq)),
        id=record.id,
        description=record.description,
        letter_annotations={
            "phred_quality": trimmed_qual}
    )
    return trimmed_record
```

This code performs the following functions:

- It takes in a `SeqRecord` object resulting from the parsing of our FASTQ file. This record contains the sequence as well as some annotations of its properties.

- The `letter_annotations` property of the record contains per-letter (per-base) annotations – in this case, the sequence quality values. The code reads this into a `qualities` array.

- Next, it initializes `trimmed_index` – we start at the end by default (assuming no trimming).

- We next create a `for` loop to iterate backward, starting from the 3' end of the sequence. As soon as we meet a base falling below our threshold, we will update our `trimmed_index` and stop there. This is where the read will be trimmed.

- We then trim the sequence and quality string, respectively, based on the index position.

- Finally, we create a new `SeqRecord` object using the updated sequence and quality strings and return it.

11. Next, we'll set up a function to do adapter trimming:

```
def remove_adapter(record, adapter_seq):
    seq_str = str(record.seq)
    adapter_position = seq_str.find(adapter_seq)
    if adapter_position != -1:
```

```
        trimmed_seq = record.seq[:adapter_position]
        trimmed_qual = record.letter_annotations[
            "phred_quality"
        ][:adapter_position]
        record = SeqRecord(
            Seq(str(trimmed_seq)),
            id=record.id,
            description=record.description,
            letter_annotations={
                "phred_quality": trimmed_qual
            }
        )
    return record
```

This code does the following:

- It reads in a SeqRecord object as well as the desired sequence for the adapter.

- It converts the sequence to a string and uses the find() method to locate the adapter within the sequence.

- If the adapter is found, it takes the sequence from the startup to the adapter position. It also takes the corresponding quality scores.

- It then creates a new SeqRecord object based on the trimmed sequence and qualities and returns it.

12. For our last function, let's process a FASTQ file and call our other two functions to do the trimming:

```
def process_fastq(
    input_fastq, output_fastq,
    quality_threshold=20, adapter_seq=None
):
    with open(input_fastq, "r") as input_handle, \
         open(output_fastq, "w") as output_handle:
        for record in SeqIO.parse(input_handle, "fastq"):
            record = trim_low_quality_bases(
                record, quality_threshold)
            if adapter_seq:
                record = remove_adapter(
                    record, adapter_seq)
            if len(record.seq) > 0:
                SeqIO.write(
                    record, output_handle,
                    "fastq"
                )
    print(f"Processing complete. "
        f"Trimmed reads saved to {output_ fastq}")
```

Note that in our function header, we define the inputs and set some defaults. Our default quality threshold is 20. By default, we assume there is no adapter sequence.

This code defines a `process_fastq()` function that does the following:

- Opens a FASTQ file and an output file to which we will write our trimmed sequence.

- Parses each sequence record in the file using the `SeqIO.parse()` method and calls our quality trimming function. Note that the default Q value is set to 20 in the function definition but can be overridden when you call the function.

- Calls the adapter removal function if an adapter sequence has been defined.

- Assuming any sequence remains after trimming, it uses the `SeqIO.write()` method to write out the trimmed record to the output FASTQ file.

13. Finally, let's try out our functions:

```
input_fastq = "data/SRR000001_1.fastq"
output_fastq = "data/processed_reads.fastq"
quality_threshold = 30
adapter_sequence = "AGATCGGAAGAGC"
process_fastq(
    input_fastq, output_fastq,
    quality_threshold, adapter_sequence
)
```

This code defines our input FASTQ file using the one we downloaded in the *Getting ready* section. Our output will go into a file called `processed_reads.fastq`. We set our Q score cutoff at 30 (a common setting) and provide an adapter sequence. Normally, you would get this adapter sequence from your lab personnel based on the method they are using.

We call our `process_fastq()` function, which will trim the reads and output the processed file. You can check the resulting file sizes, and you will see that some of the sequence has been removed!

There's more...

We have had a basic introduction to some of the core sequence file formats used in bioinformatics and some tools for manipulating those files. This could include translating sequences or processing files to improve their quality before they go into further analysis in a pipeline.

Today's modern bioinformatics files can be huge in size, depending on the organism and sequencer you are using. The latest Illumina NovaSeqX can produce up to 8 TB of data per run (see https://www.illumina.com/systems/sequencing-platforms/novaseq-x-plus/specifications.html). As such, storage and data transfer costs can be quite significant. It is therefore important to stay on top of modern advances in compression and related tools. For example, the CRAM format is now gaining in popularity for alignments, and the ORA format has been introduced by Illumina to provide significant compression of FASTQ data.

Advances in the field of compression are being made all the time. The Genozip program (Lan et al., *Genozip: a universal extensible genomic data compressor*, BioInformatics, August 2021, `https://academic.oup.com/bioinformatics/article/37/16/2225/6135077`) is able to achieve significant compression ratios even over already-compressed file types such as fastq.gz and CRAM. It includes some nice features such as co-compression of FASTQ and BAM data.

It is worthwhile to stay on top of the latest advances in bioinformatics file formats and compression, as well as to understand how to interrogate your storage costs using modern tools such as AWS Storage Lens (`https://aws.amazon.com/s3/storage-lens/`). This will put you in a position to save your organization a lot of money and optimize your bioinformatics pipelines! This will only increase in importance in the coming years as modern sequencing advances are not slowing down. In the near future, we can expect to see sequencing costs come down even further and the enablement of cost-effective long-read genome sequencing. This will mean we can routinely assemble entire human genomes in a highly accurate way with costs that are in line with current short-read approaches. These advancements will transform genetic testing, cancer research, and many other areas, and lead to huge impacts on human health and well-being.

See also

- For a deeper background on BioPython's Seq class, visit `https://biopython.org/wiki/Seq`

- Details on the SeqRecord object: `https://biopython.org/wiki/SeqRecord`

- To learn about the **central dogma**, visit `https://www.yourgenome.org/theme/how-is-dna-turned-into-protein-the-central-dogma-of-molecular-biology/`

- SciKit-Bio is another good library to check out, which provides sequence manipulation among other tools: `https://scikit.bio/index.html`

- ISeq is a modern library for fetching sequencing data from public databases, including SRA: `https://academic.oup.com/bioinformatics/article/40/11/btae641/7840256`

- To learn more about adapter trimming, visit `https://knowledge.illumina.com/software/general/software-general-reference_material-list/000002905`

- A good read on practical CRAM usage can be found at `https://www.ga4gh.org/news_item/guest-post-seven-myths-about-cram-the-community-standard-for-genomic-data-compression/`

Sequence alignment with BWA

Next, we'll dive into a core algorithm at the heart of bioinformatics: alignment.

Alignment is the process of matching two sequences together so that they share the maximum amount of common sequence, or aligning many reads to a reference sequence. An aligner is any software that specializes in lining up sequencing reads to longer sequences, or in some cases, sets of longer sequences to each other.

One of the first core aligners was **BWA**, the **Burrows-Wheeler Aligner** (to read more about these aligners, look at the *See also* section of this recipe).

Another important aligner that has been developed recently is the DRAGEN aligner from Illumina. DragMap is an open source version of the DRAGEN aligner.

SAMtools (`https://www.htslib.org/`) is another important program that we'll be using in this recipe (Li et al., *The Sequence Alignment/Map Format and SAMtools*, Bioinformatics, June 2009). SAMtools helps us read, write, index, and view SAM and BAM files. It is part of a suite of tools that includes BCFtools for `vcf` parsing and HTSlib, a C library for reading sequence data.

Getting ready

For this recipe, we use the Escherichia coli (E. coli) reference genome. E. coli is a small bacterium with a genome size of ~4.6 MB (4.6 million bases).

We'll keep our genome in `data/ecoli_genome` and our sequencing reads in `data/ecoli_reads`.

You will need to download the genome sequence (you can run this from the notebook, or if you want to do this in your terminal, just remove the `!` character from in front of each command):

```
! mkdir -p data/ecoli_genome
! wget -O data/ecoli_genome/ecoli_reference.fasta.gz
"https://ftp.ncbi.nlm.nih.gov/genomes/all/GCF/000/005/845/
GCF_000005845.2_ASM584v2/GCF_000005845.2_ASM584v2_genomic.fna.gz"

! gunzip data/ecoli_genome/ecoli_reference.fasta.gz
```

Now, let's get some reads we can align to the reference (this step may take some time):

```
! fasterq-dump --split-files --outdir ./ecoli_reads SRR31783077
! mv ecoli_reads data/
```

Note that these are **paired-end** reads, so we will end up with two files ending in _1 and _2.

We will also need to install bwa, samtools, and pysam:

```
! brew install bwa
! brew install samtools
! pip install pysam
```

We'll also be using matplotlib in this exercise, which you used in *Chapter 4*:

```
! pip install matplotlib
```

The code for this chapter can be found in Ch05/Ch05-3-alignment.ipynb.

How to do it...

Here are the steps to try this recipe:

1. Our first task will involve aligning reads to a reference genome (in this case, E. coli). Let's import our libraries:

   ```
   import subprocess
   import os
   ```

   We will use the subprocess and os libraries to run BWA and define a function to index our reference genome.

2. When aligning reads to a reference genome, we must **index** the reference first. Indexing a genome is an important first step for alignment. Indexing preprocesses the genome, providing a guide to where sequences originate from, making alignment faster:

   ```
   def index_reference_genome(reference_fasta):
       print("Indexing the reference genome with BWA...")
       cmd = ["bwa", "index", reference_fasta]
       subprocess.run(cmd, check=True)
       print("Reference genome indexing complete.\n")
   ```

   The code defines a function called index_reference_genome() that takes in a reference file in FASTA format and then constructs a command to run the BWA index on the file. It then uses subprocess.run() to perform the indexing.

3. Next, we'll define a function to perform the alignment:

```python
def align_fastq_to_reference(
    reference_fasta, fastq_file1,
    fastq_file2, output_sam, threads=4
):
    print("Performing alignment with BWA-MEM...")
    cmd = [
        "bwa", "mem",
        "-t", str(threads),
        reference_fasta,
        fastq_file1,
        fastq_file2,
    ]
    with open(output_sam, "w") as out:
        subprocess.run(cmd, stdout=out, check=True)
    print(f"Alignment complete. "
          f"SAM file saved to: {output_ sam}\n")
```

This function takes in a reference FASTA file, a set of paired reads, and a SAM file for output. It then constructs a BWA command with the reference file, FASTQ files, and the number of threads. The number of threads is the number of parallel compute cores to use – in this case, we will default to 4. This is just a default – you can set it to another number based on the number of cores your machine supports. Note that we use the mem version of BWA. This will produce a SAM file.

4. Next, we will create a function to sort the SAM file and convert it to a BAM file:

```python
def convert_sam_to_sorted_bam(
    sam_file, bam_file, threads=4
):
    print("Converting SAM to sorted BAM using Samtools...")
    cmd_sort = [
        "samtools", "sort", "-@",
        str(threads), "-o", bam_file, sam_file
    ]
    subprocess.run(cmd_sort, check=True)
    cmd_index = ["samtools", "index", bam_file]
    subprocess.run(cmd_index, check=True)
    print(f"Sorted BAM file saved to: {bam_file}\n")
```

This function takes the SAM file as input and runs samtools sort with the default number of threads, 4. It then runs samtools index and turns the SAM file into a BAM file.

5. Finally, let's set up our main code to run these functions:

```python
def main():
    reference_fasta = "data/ecoli_genome/ecoli_reference.fasta"
    fastq_file1 = "data/ecoli_reads/SRR31783077_1.fastq"
    fastq_file2 = "data/ecoli_reads/SRR31783077_2.fastq"
    output_sam = "data/output/aligned_reads.sam"
    output_bam = "data/output/aligned_reads_sorted.bam"
    os.makedirs("data/output", exist_ok=True)

    try:
        index_reference_genome(reference_fasta)
        align_fastq_to_reference(
            reference_fasta, fastq_file1,
            fastq_file2, output_sam, threads=4
        )
        convert_sam_to_sorted_bam(
            output_sam, output_bam, threads=4
        )
    except subprocess.CalledProcessError as e:
        print(f"Error occurred during execution: {e}")
    except Exception as e:
        print(f"Unexpected error: {e}")

if __name__ == "__main__":
    main()
```

This code does the following:

- Specifies the location of the genome reference file.

- Specifies the location of the paired-end reads.

- Defines locations for the output SAM and BAM files and makes a directory to hold them.

- Indexes the reference genome.

- Runs BWA to align the reads.

- Sorts the sam file and creates a bam file.

- Note that we use a Try..Except block for the operation. If there is an error, we use the subprocess.CalledProcessError property to report the error.

Now, we have a BAM file containing our alignment!

6. We can also get some basic statistics from our BAM file using `samtools`. We will import our libraries and define a `run_command()` function. These should look familiar by now:

```python
import subprocess
def run_command(cmd):
    try:
        result = subprocess.run(
            cmd, stdout=subprocess.PIPE,
            stderr=subprocess.PIPE,
            text=True, check=True
        )
        return result.stdout
    except subprocess.CalledProcessError as e:
        print(f"Error executing command: {' '.join(cmd)}")
        print(e.stderr)
        raise
```

7. Next, we'll write a function to get statistics from a BAM file:

```python
def get_bam_statistics(bam_file):
    print(f"Getting basic statistics for BAM file: {bam_file}")
    stats_output = run_command([
        "samtools", "stats", bam_file])
    stats = {}
    for line in stats_output.splitlines():
        if line.startswith("#"):
            continue
        parts = line.split("\t")
        if len(parts) > 1:
            stats[parts[0].strip()] = parts[1].strip()
    return stats
```

This uses `samtools stats` to get information about a BAM file.

8. Finally, let's define our `main()` function, which will run our statistics against our sorted BAM file:

```python
def main():
    bam_file = "data/output/aligned_reads_sorted.bam"
    try:
        stats = get_bam_statistics(bam_file)
        print("\nBAM File Statistics:")
        for key, value in stats.items():
            print(f"{key}: {value}")
    except Exception as e:
        print(f"Error: {e}")
```

```
if __name__ == "__main__":
    main()
```

This will give us some output that looks like this:

```
BAM File Statistics:
CHK: 7fb89a1e
SN: percentage of properly paired reads (%):
FFQ: 151
LFQ: 151
GCF: 86.68
GCL: 97.49
GCC: 151
GCT: 151
FBC: 151
FTC: 32085562
LBC: 151
LTC: 32056282
IS: 887
RL: 151
FRL: 151
LRL: 151
MAPQ: 60
ID: 66
IC: 148
COV: [479-479]
GCD: 55.0
```

Figure 5.8 – Output statistics for a BAM file

This shows various statistics on the BAM file. For example, COV gives us the coverage (depth) distribution within the region. GCF gives us information on GC content. These statistics are commonly used by bioinformaticians to get information on average coverage or other aspects of a BAM file, to make further decisions in processing, or to feed back quality information to lab scientists. For a full list of the statistics provided by SAMtools, visit `https://www.htslib.org/doc/samtools-stats.html`.

9. Let's look at some ways to visualize the BAM file. First, we will import our libraries:

    ```
    import pysam
    import matplotlib.pyplot as plt
    ```

We introduce **pysam**, a handy Python module for interfacing with SAM and BAM files (`https://pysam.readthedocs.io/en/v0.16.0.1/api.html`). PySam can also run SAMtools commands.

10. Next, we define a function that uses matplotlib to visualize a region in our BAM file:

```python
def visualize_bam_coverage(
    bam_file, region, output_file=None
):
    contig, positions = region.split(":")
    start, end = map(int, positions.split("-"))
    bam = pysam.AlignmentFile(bam_file, "rb")
    coverage = [0] * (end - start)
    for pileup_column in bam.pileup(
        contig, start, end
    ):
        pos = pileup_column.reference_pos
        if start <= pos < end:
            coverage[pos - start] = pileup_column.nsegments
    bam.close()
    plt.figure(figsize=(10, 5))
    plt.plot(
        range(start, end), coverage,
        label="Coverage"
    )
    plt.xlabel("Position")
    plt.ylabel("Read Depth")
    plt.title(f"Coverage Plot for {region}")
    plt.legend()
    if output_file:
        plt.savefig(output_file)
        print(f"Coverage plot saved to: {output_file}")
    else:
        plt.show()
```

To recap, this code will do the following:

I.   Define a function called `visualize_bam_coverage()` that will take in an input BAM file, a region, and an optional output file.

II.   Determine the contig and position range by splitting out based on the colon. This means parsing the string by splitting everything from the colon to the left into one substring, and the right side from the colon onward into another substring. This gives us the two variables we need, "contig" and "positions". In this example, we are going to use NC_000913.3:1000-1500 as the contig and range. Note that often the words chromosome and contig are used interchangeably for these ranges. When an organism can only be assembled into small pieces, we call this a contig. When we can assemble things into entire chromosomes or place contigs within the context of a known physical chromosome, we may use notation such as Chr1:10000-15000. E. coli is a bacterium with a single chromosome, whose accession number is NC_000913.3, and so you see the preceding notation.

III.   Use the Python map () function to extract the start and end positions of the range as integers. Here, map () simply applies the int () function to each part of the range as split out on the "-" character In other words, we first split the left side of the string before the. dash, which becomes the start position, and convert it to an integer, and then do the same on the right side of the dash, converting the end position to an integer.

IV.   Use pysam to open the BAM file in binary read mode (rb).

V.   Initialize a coverage list with zeroes. The pileup property of the BAM represents how many reads have "piled up" at that position, giving us the reference position and the depth of reads covering the position (nsegments). It then updates the coverage list with the depth at each position.

VI.   Close our BAM file.

VII.   Initialize a plot using matplotlib. We set the title and other parameters for the plot. We will plot the position along the reference as the X axis and the read depth as the Y axis.

VIII.   Finally, it will save or display the plot.

11.   Now, let's use our charting function on our data:

```
bam_file = "data/output/aligned_reads_sorted.bam"
region = "NC_000913.3:1000-1500"
output_file = "coverage_plot.png"
visualize_bam_coverage(bam_file, region, output_file)
```

We get a nice chart that looks like this:

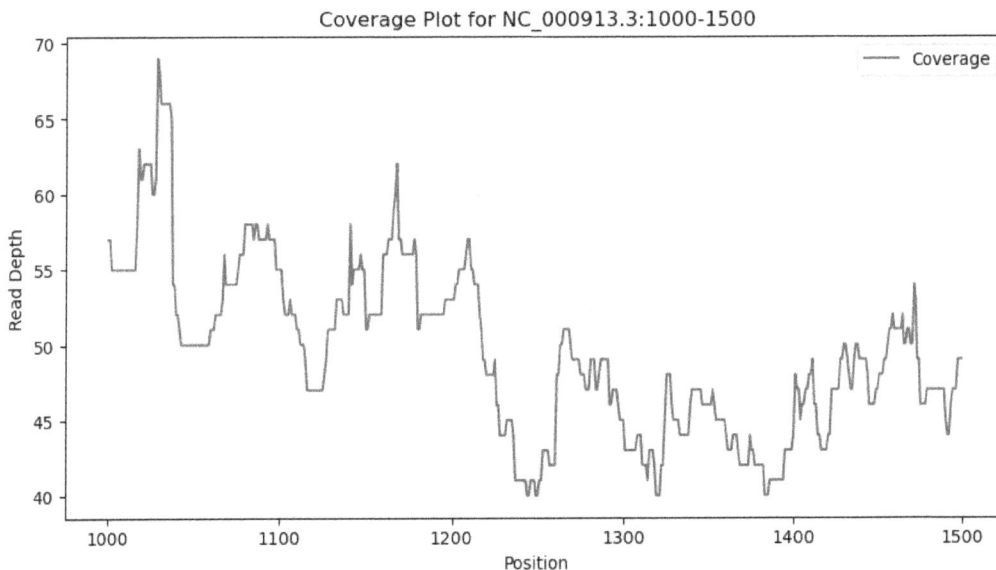

Figure 5.9 – Coverage plot from a BAM file

We can see the coverage as it varies over the range we provided. This type of chart can be very helpful to inspect variations in coverage that may come from lab processing or sequencer issues.

As you can see, there are many downstream tools for analysis that you can run once you have your completed alignment in BAM format.

There's more...

Sequence alignment is an important area for any bioinformatics professional to understand deeply. Alignment quality forms the basis of variant calling and subsequent analysis of the data and as such has a critical impact on downstream interpretation.

Long-read methods have their own alignment tools. For instance, pbmm2 (`https://github.com/PacificBiosciences/pbmm2`) is used for PacBio data. Long-read alignments can have powerful impacts on variant calling and improve accuracy significantly. For example, the **All of Us** initiative seeks to sequence the DNA of over one million Americans to gain expanded information on ethnic diversity and improve healthcare. They use long-read sequencing and alignment to study the impact of these technologies on improving variant calling and provide comparisons with the DRAGEN aligner for short reads from Illumina technology (Mahmoud et al., *Utility of long-read sequencing for All of Us*, Nature Communications, Jan 2024, `https://www.nature.com/articles/s41467-024-44804-3`).

Continued advancements in sequencing and alignment will greatly improve our understanding of human variation and lead to a revolution in medicine. These tools are also being actively applied in plant genomics to improve crops and provide new foods and medicines.

See also

- BWA is described in Li and Durbin, *Fast and accurate short read alignment with Burrows-Wheeler transform*, Bioinformatics, May 2009, `https://academic.oup.com/bioinformatics/article/25/14/1754/225615`

- DRAGEN is discussed in the paper by Behera et al., *Comprehensive genome analysis and variant detection at scale using DRAGEN*, Nature, October 2024, `https://www.nature.com/articles/s41587-024-02382-1`

- DRAGMAP can be found here: `https://github.com/Illumina/DRAGMAP`

- For more information on SAMTools, read Danecek et al., *Twelve years of SAMTools and BCFTools*, GigaScience, February 2021, `https://academic.oup.com/gigascience/article/10/2/giab008/6137722`

- Another cool tool for BAM visualization is SamBamViz: `https://github.com/niemasd/SamBamViz`

- This article discusses aligners for nanopore data: Helal et al., *Benchmarking long-read aligners and SV callers for structural variation detection in Oxford nanopore sequencing data*, Scientific Reports, March 2024, `https://www.nature.com/articles/s41598-024-56604-2`

Variant calling with FreeBayes

In this recipe, we will see how to use an alignment file to call variants, or alterations in a genome. We will look at FreeBayes (`https://github.com/freebayes/freebayes`), which is a popular variant caller.

In variant calling, we look at the pile-up of reads in the alignment and try to determine whether there are any variations from the reference sequence.

Let's take a look at an example:

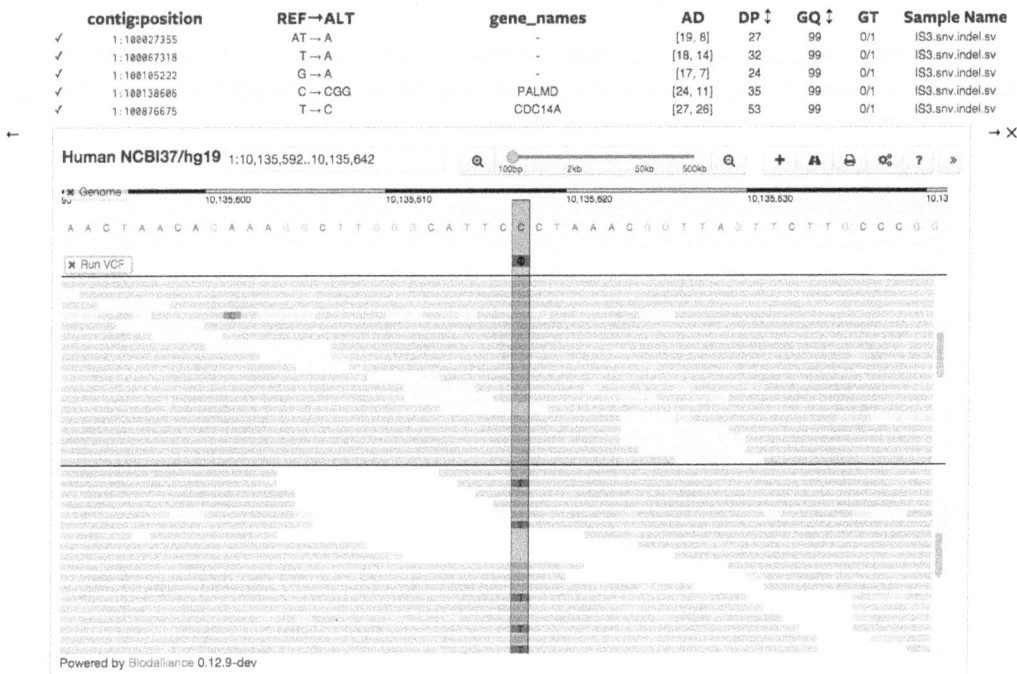

	contig:position	REF→ALT	gene_names	AD	DP ↕	GQ ↕	GT	Sample Name
✓	1:100027355	AT → A	-	[19, 8]	27	99	0/1	IS3.snv.indel.sv
✓	1:100067318	T → A	-	[18, 14]	32	99	0/1	IS3.snv.indel.sv
✓	1:100105222	G → A	-	[17, 7]	24	99	0/1	IS3.snv.indel.sv
✓	1:100138606	C → CGG	PALMD	[24, 11]	35	99	0/1	IS3.snv.indel.sv
✓	1:100876675	T → C	CDC14A	[27, 26]	53	99	0/1	IS3.snv.indel.sv

Figure 5.10 – Variant Calling example (source: `https://www.hammerlab.org/2015/01/23/faster-pileup-loading-with-bai-indices/`)

The image above shows sequencing reads "piled up" against a reference genome (top). Reads containing a variant are highlighted to indicate the position of the variation. In some reads, the base pair remains a G, matching the reference sequence, while in many others it is a C. Because the majority of reads show a C, we make a consensus call that this position in the genome now contains a C. This process lies at the core of variant calling.

Getting ready

First, we will install FreeBayes using the following code. You may want to add it to your PATH:

```
! brew install freebayes
```

The code for this chapter can be found in `Ch05/Ch05-4-variant-calling.ipynb`.

How to do it...

Here are the steps to try this recipe:

1. First, we will import our libraries:

    ```
    import subprocess
    import os
    ```

2. Next, we define our `run_command()` function:

    ```
    def run_command(cmd):
        print(f"Running: {' '.join(cmd)}")
        subprocess.run(cmd, check=True)
    ```

3. Now we define a function to index our reference genome:

    ```
    def index_reference(reference_fasta):
        print("Indexing the reference genome...")
        run_command(["samtools", "faidx", reference_fasta])
        print("Reference indexing complete.\n")
    ```

    This code will use the `samtools faidx` command to index the reference file in preparation for running FreeBayes on it.

4. We define a function to sort and index our BAM file:

    ```
    def sort_and_index_bam(input_bam, output_sorted_bam):
        print("Sorting and indexing the BAM file...")
        run_command([
            "samtools", "sort", "-o",
            output_sorted_bam, input_bam
        ])
        run_command([
            "samtools", "index", output_sorted_bam
        ])
        print(f"Sorted BAM file: {output_sorted_bam}\n")
    ```

5. Now, we define a function to call variants with FreeBayes:

```python
def call_variants_with_freebayes(
    reference_fasta, input_bam, output_vcf
):
    print("Calling variants with FreeBayes...")
    cmd = [
        "freebayes",
        "-f", reference_fasta,
        input_bam
    ]
    with open(output_vcf, "w") as vcf_file:
        subprocess.run(
            cmd, stdout=vcf_file, check=True)
    print(f"Variants called successfully. "
        f"Output VCF: {output_ vcf}\n")
```

6. Finally, we define our `main` function, which will call the variant calling routines:

```python
def main():
    reference_fasta = "data/ecoli_genome/ecoli_reference.fasta"
    input_bam = "data/output/aligned_reads.sam"
    output_sorted_bam = "output/aligned_reads_sorted.bam"
    output_vcf = "output/variants.vcf"
    os.makedirs("output", exist_ok=True)

    try:
        index_reference(reference_fasta)
        sort_and_index_bam(input_bam, output_sorted_bam)
        call_variants_with_freebayes(
            reference_fasta,
            output_sorted_bam, output_vcf
        )
    except subprocess.CalledProcessError as e:
        print(f"Error occurred while running a command: {e}")
    except Exception as e:
        print(f"Unexpected error: {e}")

if __name__ == "__main__":
    main()
```

This function defines the location of the E. coli reference genome as well as the input SAM file coming from the alignment. It then outputs the data in the outputs directory in the `variants.vcf` file. It will index the reference file, sort and index the BAM, and then call the variants with FreeBayes using the default parameters.

Let's look at the `variants.vcf` file. In your terminal, type the following:

```
tail -100 output/variants.vcf
```

Here is a sample of the output:

```
NC_000913.3    4639632 .    A    G    2050.56 .    AB=0;ABP=0;AC=2;AF=1;AN=2;AO=65;CIGAR=1X;DP=65;DPB=65;DPRA=0;
0;PAO=0;PQA=0;PQR=0;PRO=0;QA=2334;QR=0;RO=0;RPL=28;RPP=5.71629;RPPR=0;RPR=37;RUN=1;SAF=21;SAP=20.6827;SAR=44;SRF=0;SRP=0;SRR=
NC_000913.3    4639719 .    C    T    1810.06 .    AB=0;ABP=0;AC=2;AF=1;AN=2;AO=55;CIGAR=1X;DP=55;DPB=55;DPRA=0;
0;PAO=0;PQA=0;PQR=0;PRO=0;QA=2050;QR=0;RO=0;RPL=31;RPP=4.94488;RPPR=0;RPR=24;RUN=1;SAF=23;SAP=6.20829;SAR=32;SRF=0;SRP=0;SRR=
NC_000913.3    4639764 .    G    A    1597.15 .    AB=0;ABP=0;AC=2;AF=1;AN=2;AO=49;CIGAR=1X;DP=49;DPB=49;DPRA=0;
0;PAO=0;PQA=0;PQR=0;PRO=0;QA=1808;QR=0;RO=0;RPL=20;RPP=6.59988;RPPR=0;RPR=29;RUN=1;SAF=23;SAP=3.40914;SAR=26;SRF=0;SRP=0;SRR=
NC_000913.3    4639956 .    T    C    1611.52 .    AB=0;ABP=0;AC=2;AF=1;AN=2;AO=50;CIGAR=1X;DP=50;DPB=50;DPRA=0;
0;PAO=0;PQA=0;PQR=0;PRO=0;QA=1825;QR=0;RO=0;RPL=25;RPP=3.8103;RPPR=0;RPR=25;RUN=1;SAF=30;SAP=7.35324;SAR=20;SRF=0;SRP=0;SRR=0
```

Figure 5.11 – Section of the VCF file

The tab-separated columns above contain the reference contig, the position of the variant, the reference base and changed (variant) base respectively, and lastly the CIGAR string. We'll cover the VCF format in more detail in the next chapter, but in the meantime, there is more information on the variant call format here: `https://gatk.broadinstitute.org/hc/en-us/articles/360035531692-VCF-Variant-Call-Format`.

There's more...

There are many other variant callers. One of the most popular is GATK from the Broad Institute (`https://gatk.broadinstitute.org/hc/en-us`). Variant calling is also being impacted by deep learning. For example, Google's DeepVariant (`https://github.com/google/deepvariant`) uses image tensors of the pileup variants and a **Convolutional Neural Network (CNN)** to make variant calls.

Variant calling is especially difficult in regions of the genome that are highly repetitive, or in organisms that have high **ploidy** (number of chromosomes). For a great review on this topic, read Fukasawa, *Genome complexity, not ploidy, dictates long-read variant calling accuracy*, bioRxiv, May 2025, `https://www.biorxiv.org/content/10.1101/2025.05.14.653922v1.abstract`.

It is also especially difficult to understand variation when it includes large changes to the genome, such as **structural variation**, **inversion**, or **repeat expansion**. Structural variations are large rearrangements in the genome, typically over 50 base pairs or even much larger. For example, an entire gene might be picked up and moved (inserted) into a wholly different chromosome. A gene could also be flipped around backward, which is referred to as an inversion (`https://en.wikipedia.org/wiki/Structural_variation`). These are very difficult to resolve accurately with short reads. As you have learned now by looking at alignments, you can readily see that short reads can only map within large regions and cannot tell you if those regions have been moved around in the genome – except by looking at the edges, and again, you would only have short reads to give you a limited amount of information there. Repeat expansions involve short sequences of a few nucleotides that are repeated over and over – they expand easily, causing genetic defects (Malik et al, *Molecular mechanisms underlying nucleotide repeat expansion disorders*, Nature Reviews Molecular Cell Biology, Jun 2021, `https://www.nature.com/articles/s41580-021-00382-6`). Again, such variations are difficult to map with short reads because they cannot span the length of the change – that is where long reads come in extremely handy, and we'll discuss this more in *Chapter 7, Genomes and Genome Assembly*.

In the next chapter, *Chapter 6, Annotation and Biological Interpretation*, we will learn about interpreting the variants in the VCF file. We'll discuss the VCF format in detail, and we'll learn what it means to interpret a variant and the potential biological impact it might have on an organism.

Let's clean up and close down our `conda` environment:

```
conda deactivate
```

See also

- For a great review of modern variant calling tools, read: Olson et al., *Variant calling and benchmarking in an era of complete human genome sequences*, Nature Reviews Genetics, April 2023, `https://pubmed.ncbi.nlm.nih.gov/37059810/`

- Koboldt presents a guide to variant calling for clinical purposes in *Best practices for variant calling in clinical sequencing*, Genome Medicine, October 2020, `https://genomemedicine.biomedcentral.com/articles/10.1186/s13073-020-00791-w`

Get This Book's PDF Version and Exclusive Extras

UNLOCK NOW

Scan the QR code (or go to `packtpub.com/unlock`). Search for this book by name, confirm the edition, and then follow the steps on the page.

Note: Keep your invoice handy. Purchases made directly from Packt don't require an invoice.

6

Annotation and Biological Interpretation

In this chapter, we will discuss how the results of bioinformatics analysis affect biology. We will first look at how to parse and use the variant calls we arrived at in *Chapter 5*. We will better understand the **Variant Call Format** (**VCF**) and how to parse it. Next, we will learn how to annotate the genes in a genome, using a simple prokaryote (a single-cell organism) as an example. Then, we'll learn about how gene structures impact the meaning of a variant. Finally, we'll explore how proteins are annotated and the importance of protein domains.

By the end of this chapter, you'll understand how to work with variant call files and interpret their contents. You will have learned how to annotate genomes and, most importantly, understand how gene structures are impacted by variants. This will give you a solid grounding in the interpretation of variants and how they can impact an organism. You will also learn about the proteins created from the genes and see how variants can impact important regions of a protein. This will give you a perspective on how we interpret genetic results and predict their impact.

In this chapter, we will cover the following topics:

- Parsing and filtering variant files
- Annotating a prokaryotic genome
- Interpreting variants on gene structures
- Annotating proteins

Technical requirements

We'll be using the following packages in this section:

- cyvcf2
- BioPython

- Matplotlib
- Prodigal
- HMMER

You'll be instructed on how to install these tools in the *Getting ready* section of each recipe.

You will find the code for this chapter at `https://github.com/PacktPublishing/Bioinformatics-with-Python-Cookbook-Fourth-Edition/tree/main/Ch06`.

Remember to activate your `conda` environment before beginning the recipes, like this:

```
conda activate bioinformatics_base
```

Or, if you would like to set up a `conda` environment specific to this chapter, before activating `bioinformatics_base`, run the following:

```
conda create -n ch06-annotation --clone bioinformatics_base
conda activate ch06-annotation
```

You will be able to install the packages for the chapter as you go, or you can use the YAML file provided in the repository:

```
conda env update --file ch06-annotation.yml
```

Parsing and filtering variant files

In this recipe, we will better understand the VCF file format and learn about some tools for parsing and filtering VCF files. We will discuss the meaning of variants and how to interpret their quality.

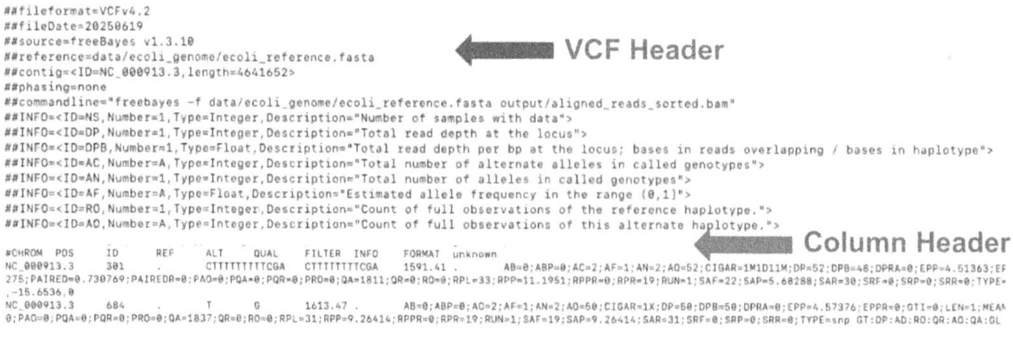

Figure 6.1 – Header section (top) and variant portions (bottom) of the variants VCF file

The first portion of the VCF file contains the VCF header information. We can see that the version of the VCF file format being used here is 4.2. We also see a file date and that the source of the VCF file was FreeBayes. We see the actual command used to generate the file. We then see definition lines, which show the meaning of various abbreviations used in the file. After the INFO lines, you may also find FILTER lines, which explain how the VCF file was filtered, and FORMAT fields with formatting info.

Next, we see the column header line, which explains the columns used below it. Here are the columns:

- CHROM: The chromosome or contig name; in this case, we see NC_000913.3, which is the E. coli chromosome

- POS: The position of the variant

- ID: A slot for a unique identifier, such as a dbSNP ID

- REF: The reference base (base in the original genome)

- ALT: The alternate base or bases (found in the sequencing alignment)

- QUAL: A Phred quality score indicating the confidence of the variant call

- FILTER: If this variant passes all filters, this will say PASS; otherwise, it will show why the variant failed to pass a filter

- INFO: A long string of additional information using various abbreviations

- FORMAT: Additional information, such as copy number

The VCF specification can be found here: https://samtools.github.io/hts-specs/. You can review specifications for the latest VCF format on this page. This book was written against version 4.2 of the

Another important concept we should cover is **ploidy**. Ploidy is a measure of the number of copies of a chromosome in an organism. In E. coli, which is a bacterium, there is only a single chromosome copy, so we say it is **haploid**. Humans have two copies of each chromosome, and we call this **diploid**. Some organisms can have even more copies of their chromosomes, such as plants, making variant interpretation even more difficult!

The sequence and structure of each individual chromosome in a region is called an **allele**. An allele typically represents a region of a gene on one copy of a chromosome, and so it is important to know which variants go with each allele. This is used to determine the potential biological impact of the variant.

Take a look at the following figure. It contains an illustration of a haploid organism with one chromosome and a diploid organism with two chromosomes. We'll evaluate the meaning of a variant analysis at a given position in each scenario:

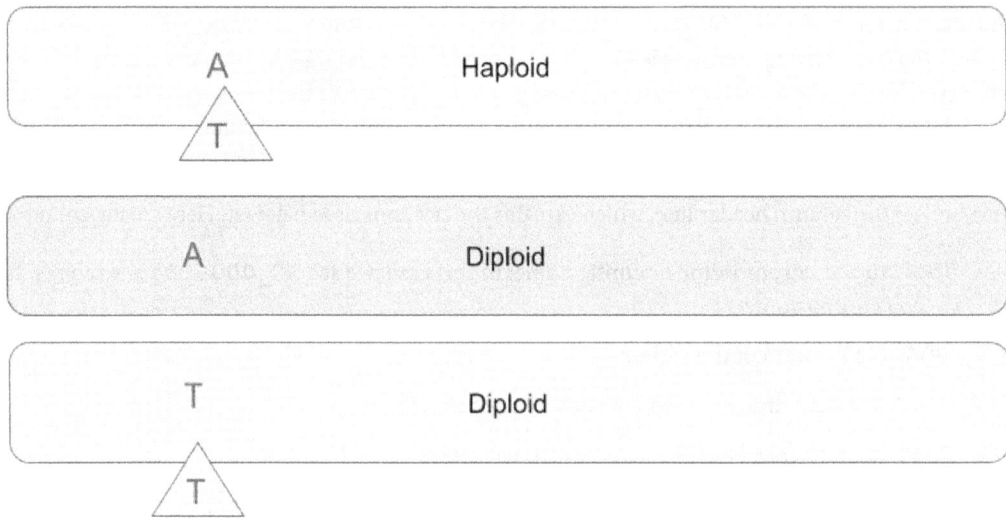

Figure 6.2 – Illustration of ploidy

In the figure, we see an example of a haploid genome on the top and a diploid genome with two chromosomes on the bottom. We see a variant in which a single base has changed. In the reference, the base is A. But in our sequencing data, we have seen a T (let's assume in this case that 100% of the reads at that position have a T). The term variant is often used interchangeably with the term **polymorphism**. This type of polymorphism involving a single base change is typically called a **single nucleotide polymorphism**, or **SNP** (pronounced "snip").

In the top example, if we have sufficient depth of reads with the T and we believe the alignment (meaning we're not worried the reads should have aligned somewhere else in the genome), then we can reliably say that this organism has a T in its genome instead of an A.

In the diploid case, we have the added problem of wondering which allele, or chromosome, these reads really came from. Most likely, the reads will align to either chromosome equally well, because the surrounding sequence is all identical. In this case, there was already a source of natural variation in the genome. One allele has A and one has T. So, in a normal sequencing run, we should expect to see 50% A and 50% T in the reads, and assuming we had mapped against the top version of the sequence, we would get a variant call of T with 50% **allele frequency**. The allele frequency is important to look at as it tells us about the frequency of the variant in the sequencing read population. In a diploid genome such as a human, there will be many 50% allele frequency SNPs, which represent the natural variation already present in the two chromosomes. We might want to filter those out by using a public reference source such as dbSNP (https://www.ncbi.nlm.nih.gov/snp/), which houses a huge database of known polymorphisms. On the other hand, a low allele frequency variant could represent some sort of sample contamination. We might want to use this as a QC check early on, but later in the process we will most likely want to filter out such low-frequency variants.

Again, if we see the T 100% of the time here, we would conclude that the top allele must have changed from A to T.

When two or more SNPs are near each other, we may want to know whether they are really lying together on the same allele (the same chromosome) or whether they might just be a mixture of variants coming from different chromosomes. If we have reads that contain these variants together in the same read, we can confirm that they are coming from the same chromosome. This is known as **phasing** (Browning and Browning, *Haplotype phasing: existing methods and new developments*, Nature Reviews Genetics, September 2011, https://www.nature.com/articles/nrg3054).

When multiple SNPs can be phased into a group, we refer to this as a **haplotype block**. A haplotype is a group of variants that are together on a chromosome.

As you can see, we will often want to filter our VCF file before further processing to remove noise and variants that might not be biologically relevant. Some of the most important sub-fields in the INFO field are DP (depth) and AF (allele frequency). The QUAL column is also typically used. OK, let's get started and see how to filter our VCF file before further processing!

Getting ready

Let's set up our directories:

```
! mkdir -p input
! mkdir -p output
! mkdir -p data
```

Copy over the variants.vcf file from Chapter 05 to the input folder:

```
! cp ../Ch05/output/variants.vcf input/
```

Next, we'll install cyvcf2. It is a program for fast parsing of VCF files (https://brentp.github.io/cyvcf2/):

```
! pip install cyvcf2
```

The code for this recipe can be found in Ch06/Ch06-1-variant-parsing.ipynb.

How to do it...

Let's look at how we can filter variants in a VCF file based on their quality or other criteria:

1. First, we will import the VCF and Writer classes from cyvcf2:

    ```
    from cyvcf2 import VCF, Writer
    ```

These classes assist with reading VCF and writing out VCF formats, respectively.

cyvfc2 is a **Cython** wrapper around htslib. Cython is a system for compiling Python code into C to make it faster, and htslib is a C library for parsing several common bioinformatics file formats. VCF also provides a compact binary format called **BCF**, and cyvfc2 can deal with this format as well:

```python
def filter_vcf(input_vcf, output_vcf,
               min_quality=30, chrom_filter=None):
    vcf = VCF(input_vcf)
    writer = Writer(output_vcf, vcf)
    for variant in vcf:
        if (variant.QUAL is not None and
            variant.QUAL < min_quality):
            continue
        if (chrom_filter and
            variant.CHROM not in chrom_filter):
            continue
        writer.write_record(variant)
    vcf.close()
    writer.close()
    print(f"Filtered VCF written to: {output_vcf}")
```

This code does the following:

- Defines a filter_vcf() function that takes an input VCF file, a designated output VCF file for the filtered results, a minimum quality score to filter on (defaults to Q30), and an optional chromosome (or contig) to filter on.

- Opens the vcf file using the VCF class. This class will parse out the header section and the variants and create an iterator for reading the variants. In Python, an iterator is an object that can be moved ahead to return the next element (https://wiki.python.org/moin/Iterator).

- Creates a writer for the output file using the Writer class.

- Loops over the variants using the iterator:

  - For each variant, we will retain it only if the quality meets the QUAL criteria

  - We will also check that the variant is on the right chromosome, if the CHROM criteria are defined

  - Variants passing the criteria will be written out

- We will clean up by closing our reader and writer. We do this by calling the .close() function of each object, which closes it and frees up the memory associated with the object. Closing the writer also ensures that all file writing operations have been completed.

2. Next, we'll set up our `main()` function to run the actual filtering:

```python
def main():
    input_vcf = "input/variants.vcf"
    output_vcf = "output/filtered_variants.vcf"
    min_quality = 30
    chrom_filter = ["NC_000913.3"]
    filter_vcf(input_vcf, output_vcf,
               min_quality, chrom_filter)
if __name__ == "__main__":
    main()
```

We set the input file to the variants file we generated for E. coli. We also set an output file called `filtered_variants.vcf`. We will use a quality cutoff of Q30, and we set the chromosome as the E. coli chromosome (if this were a multi-chromosome organism, we could also use this to filter variants down to a particular chromosome). We then call our function.

You can inspect `variants.vcf` and `filterered_variants.vcf` now and see that they are different, and some variants have been filtered out.

Now, let's look at more ways to use `cyvf2` for parsing VCF files. We can plot the allele frequency of variants across the genome:

3. First, we import our libraries:

```python
from cyvcf2 import VCF
import matplotlib.pyplot as plt
```

We will use `cyvcf2` for parsing and Matplotlib for graphing.

4. Now, we'll define our plotting function:

```python
def plot_allele_frequency(vcf_file, output_file=None):
    chrom_positions = []
    allele_frequencies = []
    vcf = VCF(vcf_file)
    for variant in vcf:
        af = variant.INFO.get("AF")
        if af is not None:
            if isinstance(af, (list, tuple)):
                for freq in af:
                    allele_frequencies.append(float(freq))
                    chrom_positions.append(
                        (variant.CHROM, variant.POS)
                    )
            else:
                allele_frequencies.append(float(af))
                chrom_positions.append(
                    (variant.CHROM, variant.POS)
```

```
            )
        vcf.close()
        chrom_names = sorted(
            set(chrom for chrom, _ in chrom_positions)
        )
        chrom_offsets = {
            chrom: i * 1e6
            for i, chrom in enumerate(chrom_names)
        }   # Chromosome offsets for spacing
        plot_positions = [
            chrom_offsets[chrom] + pos
            for chrom, pos in chrom_positions
        ]
        assert len(plot_positions) == len(allele_frequencies), (
            "Mismatch between positions and frequencies!"
        )
        plt.figure(figsize=(12, 6))
        plt.scatter(
            plot_positions, allele_frequencies,
            alpha=0.5, s=10, label="Allele Frequency"
        )
        plt.xlabel("Genomic Position (Chromosomes)")
        plt.ylabel("Allele Frequency")
        plt.title("Allele Frequency Across the Genome")
        plt.xticks(
            [chrom_offsets[chrom] for chrom in chrom_names],
            labels=chrom_names,
            rotation=45
        )
        plt.grid(True)
        plt.legend()
        if output_file:
            plt.savefig(
                output_file, dpi=300,
                bbox_inches="tight"
            )
            print(f"Plot saved to: {output_file}")
        else:
            plt.show()
```

This code does the following:

- Defines a function called `plot_allele_frequency()` that takes an input VCF file and an output file.

- Sets up lists to store the chromosome positions and allele frequencies.

- Uses the VCF class to open the VCF file.

- Loops over the variants and captures the allele frequencies (the AF field) and chromosome positions. The code isinstance(af, (list, tuple)) handles the case where multiple allele frequencies are embedded in the field.

- Closes the VCF file.

- Sorts the chromosome names.

- Creates chromosome offsets for the plot (we won't need this here since we have one chromosome).

- Creates the *x* coordinates for plotting.

- Sets up a plot with the *x* axis as the genomic coordinate and the *y* axis as the allele frequency.

- Sets labels, a title, and tick marks for the plot.

- Optionally saves the plot to an output PNG file or shows the plot.

5. Now, we'll define our main() function to run the code:

```
def main():
    vcf_file = "input/variants.vcf"
    output_file = "output/allele_frequency_plot.png"
    plot_allele_frequency(vcf_file, output_file)
if __name__ == "__main__":
    main()
```

We use the input variants.vcf file that we generated for E. coli. We set up an output PNG file in the output subdirectory and then call our function. This is what we get:

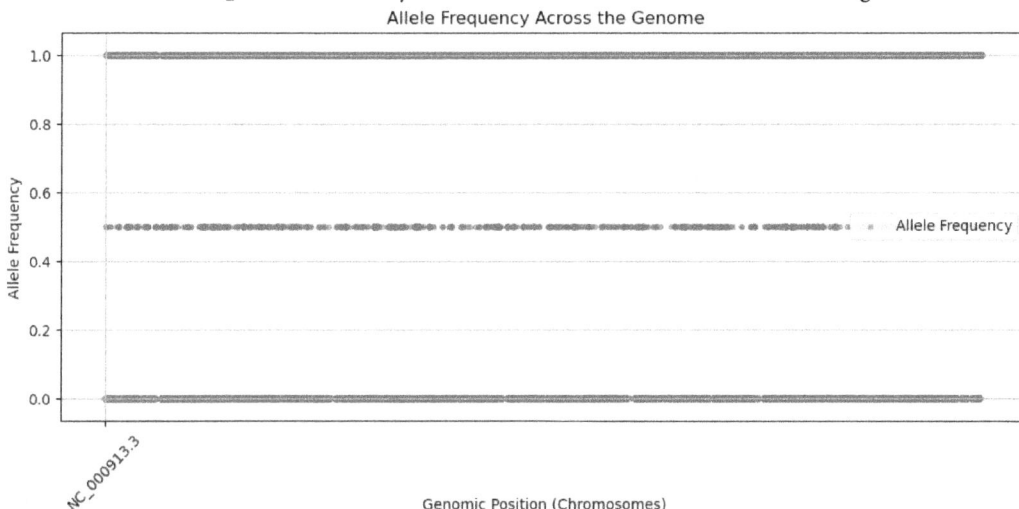

Figure 6.3 – Allele frequency across the E. coli genome

You can see the allele frequencies, ranging from 0 to 0.5 to 1, for each variant, mapped across the E. coli genome. Note that in a haploid organism such as E. coli, if you get a SNP, you might expect the allele frequency to be 1 or very close to 1 (this means 100% of the reads support an alternate base as compared to the reference allele). However, we do see some 0.5 alleles here. This can often be indicative of a mixed population. Indeed, if we look carefully at the entry for SRR31783077, which is the data we used for the alignment in *Chapter 5*, we will see that it came from a study that contained both E. coli and Shigella organisms, https://www.ncbi.nlm.nih.gov/sra/?term=SRR31783077.

So, perhaps this data contained a mixture of species, or even a mixture of different E. coli strains. FreeBayes by default assumes that it is analyzing a diploid organism. If it sees a mixture of reference and alternate bases, it will assume a heterozygous outcome, which is a 50% allele frequency. We could have used the -ploidy 1 option to force more allele calls to be either 1 or 0 in a haploid organism. The analysis presented here is for illustrative purposes, so it is not critical in this case. In fact, the data presented here is similar to what you might see in a human variant calling analysis, or any other diploid organism. You will notice that the number of 0.5 SNPs is much less than the 0 or 1 SNPs, though, because the mixed case is not the predominant one. An allele frequency of zero in this case just means that the reads support the reference base.

Next, let's look at the distribution of variant types in our VCF file:

1. First, we'll import our libraries:

    ```
    from cyvcf2 import VCF
    import matplotlib.pyplot as plt
    from collections import Counter
    ```

    This gives us cyvcf2 for VCF parsing and Matplotlib for graphing. We will use the Counter class from the collections module. This class helps us store a dictionary of our count results.

2. Next, we'll define a function to categorize the variant types:

    ```
    def categorize_variant(variant):
        ref_len = len(variant.REF)
        alt_len = max(len(alt) for alt in variant.ALT)
        if ref_len == 1 and alt_len == 1:
            return "SNP"
        elif ref_len < alt_len:
            return "Insertion"
        elif ref_len > alt_len:
            return "Deletion"
        else:
            return "Other"
    ```

    We define a categorize_variant() function, which pulls in the REF and ALT fields for the variant. If the REF and ALT lengths are the same, we have a SNP (a single base change). If REF is shorter than ALT, then we must have inserted an additional sequence in the genome. If REF is longer than ALT, that means that we saw a deletion or removal of sequence from the genome. Finally, if we see anything else, we categorize it as Other.

3. Now, we need to build our plotting function:

```python
def plot_variant_types(vcf_file, output_file=None):
    variant_counts = Counter()
    print(f"Processing VCF file: {vcf_file}")

    vcf = VCF(vcf_file)
    for variant in vcf:
        variant_type = categorize_variant(variant)
        variant_counts[variant_type] += 1
    vcf.close()

    labels = list(variant_counts.keys())
    sizes = list(variant_counts.values())
    plt.figure(figsize=(10, 8))
    wedges, _, autotexts = plt.pie(
        sizes, autopct="%1.1f%%", startangle=140,
        colors=plt.cm.tab10.colors,
        wedgeprops={
            "edgecolor": "black",
            "linewidth": 1.5
        }
    )
    plt.legend(
        wedges, labels,
        title="Variant Types",
        loc="center left",
        bbox_to_anchor=(1, 0.5),
        frameon=False
    )
    plt.title(
        "Variant Type Distribution",
        fontsize=14,
        fontweight="bold"
    )
    if output_file:
        plt.savefig(
            output_file, dpi=300,
            bbox_inches="tight"
        )
        print(f"Plot saved to: {output_file}")
    else:
        plt.show()
```

This code defines the `plot_variant_types()` function, which will do the following:

- Initialize a `Counter()` object to track the number of variants of each type
- Read in the variant file
- For each variant, loop through and categorize and count each variant
- Create the labels and sizes and initialize a pie chart plot
- Set the properties and legend for the pie chart and add a title
- Optionally save the output PNG file or show the plot

This is what we get:

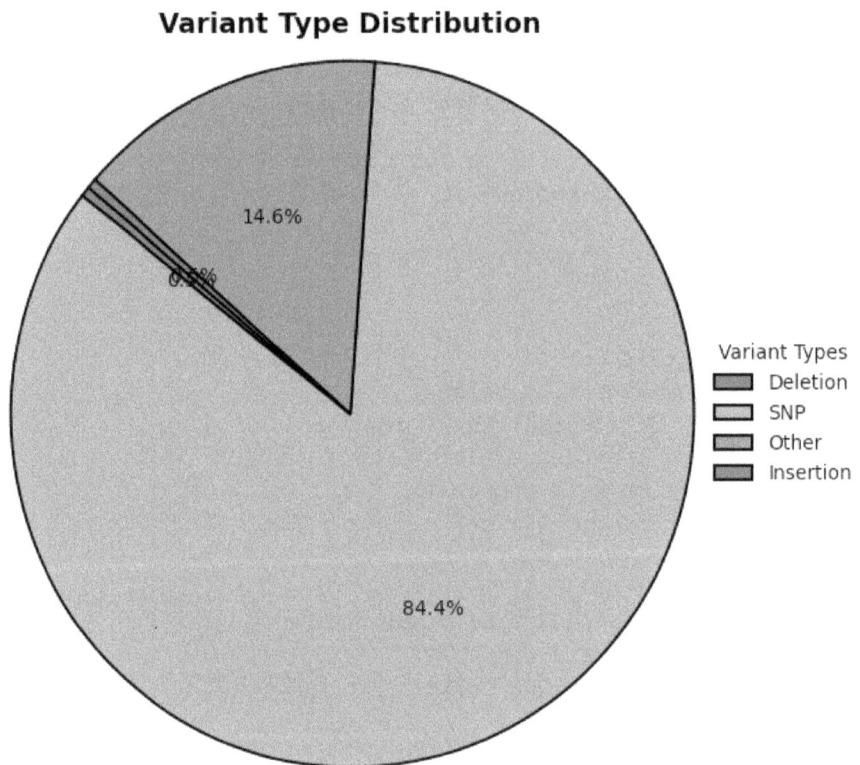

Figure 6.4 – Distribution of variant types found in our E. coli analysis

You can see that the majority of the variants are SNPs, and then we also have insertions, deletions, and others. Short insertions and deletions are often lumped together under the term **indels**. Short indels are typically classified as insertions or deletions under 50 bp in length. Beyond that, we typically call these **Structural Variations (SVs)**. We will talk more about SVs in *Chapter 7, Genomes and Genome Assembly*.

To recap, a deletion might have something in the REF and ALT fields such as "AT", "A", which would mean that a T has been removed from the genome. For an insertion, we might see something such as "G", "GC", meaning that a C has been inserted. If an indel happens in the coding portion of a gene, it will cause a **frameshift** mutation, which will almost certainly change the amino acids produced by the gene and may also introduce a premature **stop codon**. Stop codons cause ribosomes to stop translating amino acids. So, either of these effects would probably have a severe negative effect on the function of that protein.

There's more…

As you can see, we will typically want to filter our variants and make sure we have a set of high-quality variants before processing them further. Typically, the variants that matter the most will be the ones that impact genes and proteins.

You can take a quick look at *Optional Exercise #4* in the notebook to see how to simulate a simple change to a sequence called by a variant. When a SNP variant alters the coding sequence of a gene, it may change the codon used and thereby change the amino acid produced. But since there are multiple tRNA codons for the same amino acids (referred to as the degenerate genetic code), some mutations may change the DNA but not alter the amino acid produced. If the amino acid is altered, we call that a **missense mutation**. When the amino acid is not altered, we refer to this as a **silent mutation**.

Here is the output we get from *Optional Exercise #4* to illustrate:

```
Original Gene Sequence:
ATGGCATTTGACTGGTAA
Mutated Gene Sequence:
ATGTGTTTTGACTGGTAA
Original Protein Sequence:
MAFDW*
Mutated Protein Sequence:
MCFDW*
```

We can see that the alteration G->T in the fourth position in the DNA results in a change of amino acid from A (alanine) to C (cysteine) at the second position in the protein sequence. The stop codon is represented by a * and is unchanged at the end of the protein (a real protein would most likely be longer than this). This amino acid change would probably impact the structure of the protein, but it might not be radical. It would all depend on the structure of the protein. We'll discuss these matters more in *Chapter 9, Protein Structure and Proteomics*.

See also

- The `cyvcf2` paper can be found here: Pedersen and Quinlan, *Cyvcf2: fast, flexible variant analysis with Python*, BioInformatics, June 2017, `https://academic.oup.com/bioinformatics/article/33/12/1867/2971439`

- For more background on Cython, check out `https://cython.org/`

- Another variant parser is pyVCF: `https://pyvcf.readthedocs.io/en/latest/`

- Ploidy can be estimated using sequencing data – see Weib et al., `nQuire: a statistical framework for ploidy estimation using next generation sequencing`, BMC BioInformatics, April 2018

Annotating a prokaryotic genome

Once we have called and filtered variants, we want to understand the impact of these variants on the organism. The impact of the variant will typically depend on where it lands, for example, in a gene, promoter, or intergenic region. In this recipe, we will learn how to annotate a genome from scratch and begin to understand the core features within a genome. We will get more familiar with common annotation formats as well.

In the case of E. coli, the genome has already been annotated by a panel of experts. We can download the file in GenBank format:

```
wget ftp://ftp.ncbi.nlm.nih.gov/genomes/all/GCF/000/005/845/
GCF_000005845.2_ASM584v2/GCF_000005845.2_ASM584v2_genomic.gbff.gz
```

We could run any of our GenBank parsing code on this file as well. But what if we wanted to start with fresh sequencing data? Should we assemble the genome of an organism and then annotate it from scratch? In this recipe, we'll look at how to use gene prediction programs to do this (we'll cover assembly in *Chapter 7*, *Genomes and Genome Assembly*). But first, let's take a quick look at the GenBank format:

```
  LOCUS       NC_000913               4641652 bp    DNA      circular CON 09-MAR-2022
  DEFINITION  Escherichia coli str. K-12 substr. MG1655, complete genome.
  ACCESSION   NC_000913
  VERSION     NC_000913.3
  DBLINK      BioProject: PRJNA57779
              BioSample: SAMN02604091
  KEYWORDS    RefSeq.
  SOURCE      Escherichia coli str. K-12 substr. MG1655
    ORGANISM  Escherichia coli str. K-12 substr. MG1655
              Bacteria; Pseudomonadota; Gammaproteobacteria; Enterobacterales;
              Enterobacteriaceae; Escherichia.
  REFERENCE   1  (bases 1 to 4641652)
    AUTHORS   Riley,M., Abe,T., Arnaud,M.B., Berlyn,M.K., Blattner,F.R.,
              Chaudhuri,R.R., Glasner,J.D., Horiuchi,T., Keseler,I.M., Kosuge,T.,
              Mori,H., Perna,N.T., Plunkett,G. III, Rudd,K.E., Serres,M.H.,
              Thomas,G.H., Thomson,N.R., Wishart,D. and Wanner,B.L.
    TITLE     Escherichia coli K-12: a cooperatively developed annotation
              snapshot--2005
    JOURNAL   Nucleic Acids Res. 34 (1), 1-9 (2006)
    PUBMED    16397293
    REMARK    Publication Status: Online-Only
  REFERENCE   2  (bases 1 to 4641652)
    AUTHORS   Hayashi,K., Morooka,N., Yamamoto,Y., Fujita,K., Isono,K., Choi,S.,
              Ohtsubo,E., Baba,T., Wanner,B.L., Mori,H. and Horiuchi,T.
    TITLE     Highly accurate genome sequences of Escherichia coli K-12 strains
              MG1655 and W3110
    JOURNAL   Mol. Syst. Biol. 2, 2006 (2006)
    PUBMED    16738553
  REFERENCE   3  (bases 1 to 4641652)
    AUTHORS   Blattner,F.R., Plunkett,G. III, Bloch,C.A., Perna,N.T., Burland,V.,
              Riley,M., Collado-Vides,J., Glasner,J.D., Rode,C.K., Mayhew,G.F.,
              Gregor,J., Davis,N.W., Kirkpatrick,H.A., Goeden,M.A., Rose,D.J.,
```

Figure 6.5 – The GenBank header

The GenBank header consists of several important pieces of information. The Locus tag gives the organism or chromosome, length, type of molecule, and whether it is circular or linear. In this case, we have a linear E. coli chromosome called NC_000913 that is ~4.6 Mbp in length. We also have a definition line, accession number and version, and details of the organism. Then we have a series of publications related to the organisms. Shortly thereafter, we get into the core of the file, which includes the genetic features:

```
          Information (NCBI).
          COMPLETENESS: full length.
FEATURES           Location/Qualifiers
  source           1..4641652
                   /organism="Escherichia coli str. K-12 substr. MG1655"
                   /mol_type="genomic DNA"
                   /strain="K-12"
                   /sub_strain="MG1655"
                   /db_xref="taxon:511145"
  gene             190..255
                   /gene="thrL"
                   /locus_tag="b0001"
                   /gene_synonym="ECK0001"
                   /db_xref="ASAP:ABE-0000006"
                   /db_xref="ECOCYC:EG11277"
                   /db_xref="GeneID:944742"
  CDS              190..255
                   /gene="thrL"
                   /locus_tag="b0001"
                   /gene_synonym="ECK0001"
                   /codon_start=1
                   /transl_table=11
                   /product="thr operon leader peptide"
                   /protein_id="NP_414542.1"
                   /db_xref="UniProtKB/Swiss-Prot:P0AD86"
                   /db_xref="ASAP:ABE-0000006"
                   /db_xref="ECOCYC:EG11277"
                   /db_xref="GeneID:944742"
                   /translation="MKRISTTITTTITITTGNGAG"
  gene             337..2799
```

Figure 6.6 – Features section of the GenBank file

Here, we have repeating sections for different features in the genome. We have the source line, which contains the entire sequence. We then have a gene entry with a position (start..end) and gene name, along with other synonyms for the gene and cross-references to other public databases (the db_xref lines). We also have CDS entries for the coding portion of the gene, which include the translation or protein sequence.

If you run tail on the file, you see that the file ends with the complete genome sequence.

There are huge numbers of organisms already annotated at various levels of completeness in public databases. But there are also new organisms being discovered and sequenced all the time. Let's see how we can annotate a new prokaryotic genome!

Getting ready

For this recipe, we will use Prodigal, a popular genome annotation program:

```
! brew install prodigal
```

This will install Prodigal for you.

> **Tip**
>
> Some users may experience errors related to mixed architectures in `brew` installations, so we will provide alternatives, including installing via `conda`.

If you have trouble with the preceding `prodigal` command, you can try the following (from the terminal):

```
arch -arm64 brew install prodigal
```

Alternatively, try this:

```
conda install bioconda::prodigal
```

Let's also move over our E. coli reference genome from *Chapter 5*:

```
! cp ../Ch05/data/ecoli_genome/ecoli_reference.fasta input/
```

Alternatively, you could retrieve it again (from the terminal):

```
wget https://ftp.ncbi.nlm.nih.gov/genomes/all/GCF/000/005/845/
GCF_000005845.2_ASM584v2/GCF_000005845.2_ASM584v2_genomic.fna.gz
gunzip GCF_000005845.2_ASM584v2_genomic.fna.gz
```

The code for this recipe can be found in `Ch06/Ch06-2-genome-annotation.ipynb`.

How to do it...

Here are the steps to try this recipe:

1. We will write code to run Prodigal on a given input FASTA file. As output, it will provide an annotated GenBank file, a CDS file, and a Protein FASTA file. First, let's import our libraries:

    ```
    import subprocess
    import os
    ```

2. Next, we'll define a function to run Prodigal for us:

```python
def run_prodigal(input_fasta, output_gbk,
                 output_proteins, output_cds):
    if not os.path.exists(input_fasta):
        raise FileNotFoundError(
            f"Input FASTA file not found: {input_fasta}"
        )
    command = [
        "prodigal",
        "-i", input_fasta,
        "-o", output_gbk,
        "-a", output_proteins,
        "-d", output_cds,
        "-p", "single"
    ]
    try:
        print("Running Prodigal...")
        subprocess.run(command, check=True)
        print("Prodigal run completed.")
    except subprocess.CalledProcessError as e:
        print(f"Error running Prodigal: {e}")
    except FileNotFoundError:
        print(
            "Prodigal is not installed or not in your PATH."
        )
```

This function will do the following:

- Ensure the input FASTA file exists. If not, it will raise a `FileNotFound` error.

- Construct the `Prodigal` command with the input FASTA file and the desired output files. Note that Prodigal has a mode for metagenomes or single genomes – we'll use single here for E. coli.

- Run `subprocess` to run the `Prodigal` command and handle errors.

3. Finally, let's run Prodigal:

```python
if __name__ == "__main__":
    input_fasta = "input/ecoli_reference.fasta"
    output_gbk = "output/ecoli_genes.gbk"
    output_proteins = "output/ecoli_proteins.faa"
    output_cds = "output/ecoli_cds.fna"
    run_prodigal(input_fasta, output_gbk,
                 output_proteins, output_cds)
```

This code will run Prodigal on the E. coli reference genome and produce an output GenBank file and corresponding CDS and Protein files.

Let's look at the GenBank output:

```
DEFINITION  seqnum=1;seqlen=4641652;seqhdr="NC_000913.3 Escherichia coli str. K-12 substr. MG1655, complete genome";version=Prodigal.v2.6.3;run_type=Sin
gle;model="Ab initio";gc_cont=50.79;transl_table=11;uses_sd=1
FEATURES             Location/Qualifiers
     CDS             <3..98
                     /note="ID=1_1;partial=10;start_type=Edge;rbs_motif=None;rbs_spacer=None;gc_cont=0.427;conf=56.57;score=1.15;cscore=-1.57;sscore=2.
2;rscore=0.00;uscore=0.00;tscore=3.22;"
     CDS             337..2799
                     /note="ID=1_2;partial=00;start_type=ATG;rbs_motif=GGAG/GAGG;rbs_spacer=5-10bp;gc_cont=0.531;conf=99.99;score=336.87;cscore=320.95;
score=15.93;rscore=11.24;uscore=1.40;tscore=3.94;"
     CDS             2801..3733
                     /note="ID=1_3;partial=00;start_type=ATG;rbs_motif=AGGAG;rbs_spacer=5-10bp;gc_cont=0.563;conf=100.00;score=118.08;cscore=97.71;ssco
e=20.37;rscore=14.85;uscore=0.32;tscore=3.94;"
     CDS             3734..5020
                     /note="ID=1_4;partial=00;start_type=ATG;rbs_motif=GGA/GAG/AGG;rbs_spacer=5-10bp;gc_cont=0.528;conf=99.99;score=194.29;cscore=189.9
;sscore=2.99;rscore=-3.88;uscore=3.94;"
     CDS             5234..5530
                     /note="ID=1_5;partial=00;start_type=GTG;rbs_motif=AGGAG;rbs_spacer=5-10bp;gc_cont=0.539;conf=86.81;score=8.20;cscore=0.54;sscore=7
65;rscore=14.85;uscore=-0.06;tscore=-6.49;"
     CDS             complement(5683..6459)
                     /note="ID=1_6;partial=00;start_type=ATG;rbs_motif=AGGA;rbs_spacer=5-10bp;gc_cont=0.495;conf=100.00;score=117.71;cscore=100.93;ssco
```

Figure 6.7 – GenBank output from Prodigal

As you can see, this is a little different than our previous GenBank output; obviously, it is lacking the literature references and some other lines, but it is still in GenBank format. It includes the `Definition` line as well as CDS features called by Prodigal.

If you `tail` this file, you will see that it does not include the genome sequence. We'd like to include that as well.

4. Let's now add the FASTA sequence to the GenBank file:

```
from Bio import SeqIO
from Bio.Seq import Seq
from Bio.SeqRecord import SeqRecord
from Bio.SeqFeature import SeqFeature, FeatureLocation
```

We import sequence manipulation modules from BioPython.

5. Now, let's write a function to parse the Prodigal CDS file:

```
def parse_prodigal_header(header):
    parts = header.split(' # ')
    seqid = parts[0][1:]  # Remove '>'
    start = int(parts[1])
    end = int(parts[2])
    strand = 1 if parts[3] == '1' else -1
    return {
        'seqid': seqid,
        'start': start,
        'end': end,
        'strand': strand
    }
```

This code will split out the CDS lines and get the sequence ID, start, and stop position (and strand).

6. Now, we will write our function to create the final combined GenBank file:

```python
def create_genbank(genome_fasta, prodigal_fna, output_gb):
    genome_record = next(
        SeqIO.parse(genome_fasta, "fasta")
    )
    gb_record = SeqRecord(
        seq=genome_record.seq,
        id=genome_record.id,
        name=genome_record.id,
        description="Generated from Prodigal predictions"
    )
    gb_record.annotations["molecule_type"] = "DNA"
    gb_record.annotations["topology"] = "linear"
    gb_record.annotations["data_file_division"] = "BCT"
    gb_record.annotations["source"] = "Escherichia coli"
    gb_record.annotations["organism"] = "Escherichia coli"
    gb_record.annotations["taxonomy"] = [
        'Bacteria', 'Proteobacteria',
        'Gammaproteobacteria', 'Enterobacterales',
        'Enterobacteriaceae', 'Escherichia'
    ]
    feature_count = 0
    for record in SeqIO.parse(prodigal_fna, "fasta"):
        gene_info = parse_prodigal_header(
            record.description
        )
        feature = SeqFeature(
            location=FeatureLocation(
                gene_info['start'] - 1,
                gene_info['end'],
                strand=gene_info['strand']
            ),
            type="CDS",
            qualifiers={
                "locus_tag": f"CDS_{feature_count+1}",
                "translation": str(
                    record.seq.translate()
                ),
                "product": "hypothetical protein",
                "note": ["Predicted by Prodigal"]
            }
        )
```

```
        gb_record.features.append(feature)
        feature_count += 1
    SeqIO.write(gb_record, output_gb, "genbank")
    print(f"Created GenBank file with {feature_count} features")
```

This code will produce our desired GenBank file by doing the following:

I.   It uses the `SeqIO` module to parse the E. coli FASTA file.

II.   It uses BioPython's `SeqRecord` object to build up the GenBank file.

III.   It then adds some additional annotations to the GenBank record.

IV.   It uses our `parse_prodigal_header()` function to loop over the entries in the CDS file and add them as features to the GenBank record using the `SeqFeature()` function.

V.   Finally, it writes out the combined GenBank file with `SeqIO.write()`.

Let's use our functions!

```
def main():
    create_genbank(
        genome_fasta="input/ecoli_reference.fasta",
        prodigal_fna="output/ecoli_cds.fna",
        output_gb="output/ecoli_prodigal_combined.gb"
    )
if __name__ == "__main__":
    main()
```

This will give us our GenBank file with both sequence and gene features!

It is worth noting that Prodigal has several other useful options (you can see them by running `prodigal -h`). Using `-f gff`, you can get the output in **Genomic Feature Format** (**GFF**). You can find information on GFF format here: `https://useast.ensembl.org/info/website/upload/gff3.html`.

It is worth taking a moment to learn about GFF format. You should also take a look at the `gffutils` package: `https://gffutils.readthedocs.io/en/latest/`.

You can get a little more practice with GenBank files by trying out the *Optional Exercise #3* in the notebook, in which we parse back out the features in our new combined Prodigal file. If you want to try this, go into your notebook and review section 3, *(Optional Exercise) Parse a GenBank file to Extract Annotations*. You will define a new function called `annotate_ecoli()`, which takes in a GenBank file and parses it to provide the annotations in an output file, and then you'll run it to produce the parsed annotations.

We've now seen how to use annotation tools to find the gene structures in an organism from scratch. In the next section, we'll learn how to view our results!

There's more...

Once we have a GenBank file, we would like to review our genome in a more visual way than just inspecting the file. For this, we use a genome browser. One of the more popular genome browsers is the **Integrated Genomics Viewer** (**IGV**) (`https://igv.org/`):

Let's install IGV:

```
brew install --cask igv
```

If you have trouble with the above command, try:

```
brew install homebrew/cask/igv
```

OR

```
arch -arm64 brew install --cask igv
```

OR you can use conda:

```
conda install bioconda::igv
```

Note: Using conda may require a Java update! `brew link igv`

Now, open IGV using the Prodigal-annotated E. coli GenBank file:

```
! igv output/ecoli_prodigal_combined.gb
```

Here is an IGV view of our annotated E. coli genome:

Figure 6.8 – IGV view of the combined Prodigal E. coli genome annotation

This is a high-level view of the genome in IGV.

We can zoom in further to see gene-level features:

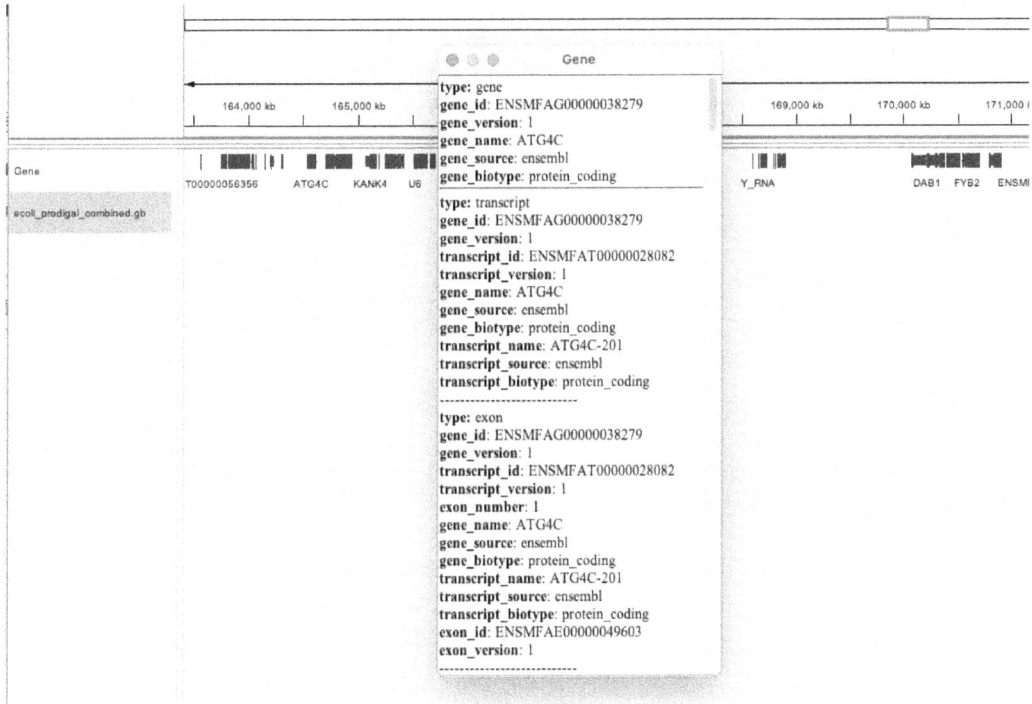

Figure 6.9 – Zoomed in view of IGV showing gene features derived from Prodigal

Here, we see an example of a zoomed-in view of the genome in IGV. We can see several gene annotations as well as a **Gene** informational popup for one of the genes.

There are other annotation programs out there you should be aware of. For eukaryotic annotation, BRAKER2 is popular (Bruna et al., *BRAKER2: automatic eukaryotic genome annotation with GeneMark-EP+ and AUGUSTUS supported by protein database*, NAR Genomics and BioInformatics, January 2021).

See also

- You can find a tutorial on the GenBank format here: `https://www.uvm.edu/~vgn/Archive/bioinf-outreach/2_entrez/2e_sequence-databases/nucleotides/genbank/flatfile/gb_flatfile.htm`

- Prodigal installation instructions: `https://github.com/hyattpd/Prodigal/wiki/installation`

- Prodigal paper: `https://bmcbioinformatics.biomedcentral.com/articles/10.1186/1471-2105-11-119`

- You can learn more about Python regular expressions here: `https://docs.python.org/3/howto/regex.html`

Interpreting variants on gene structures

Let's now dive deeper into the structure of genes and how variants affect them. Variants could occur throughout the genome, and one could argue that a variant almost anywhere could have a potential impact on an organism. But we should start by understanding the most obvious impacts, which are on genes and proteins.

Let's look at the core structures of a gene:

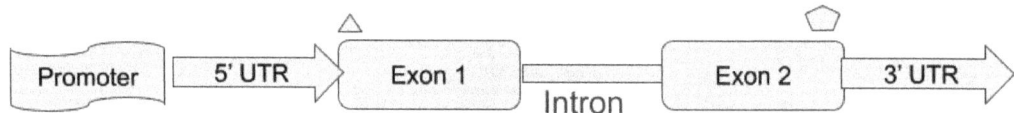

Figure 6.10 – Structure of a gene

The most important part of the gene is the exons. At the beginning of the first exon will be the **start codon**, typically ATG (the triangle in the preceding figure). The exons are separated by introns (prokaryotic genes are usually simpler and do not contain introns typically, but we'll focus on a more complex eukaryotic gene structure here). Introns are not translated and are spliced out. Once the exons are joined together after splicing, translation will begin at the start codon and continue until the end of the last exon, at the **stop codon**. The stop codon is shown as a pentagon in the preceding figure.

At the beginning and end of a gene, there is typically an **UnTranslated Region** (**UTR**). Since the gene is transcribed in the 5' to 3' direction (recall the directionality of DNA), we will refer to these as the 5' UTR and 3' UTR.

A variant would be expected to have the most impact if it occurs in the exons, as this will affect the coding part of the gene that will turn into a protein. The most severe effect would be introducing a premature stop codon or removing the start codon. Next would be changing the reading frame so that a large part of the protein is completely different. A lesser impact might be changing a single amino acid. Finally, we would have a silent mutation that would not change any amino acids but still might affect translation rate and thereby have a more subtle impact.

Variants in other locations near the gene can also have an impact, though. A mutation in the UTR can affect the regulation or translation of the gene (Li and Wang, *Predicting Functional UTR variants by integrating region-specific features*, Briefings in BioInformatics, Jul 2024, https://academic.oup.com/bib/article/25/4/bbae248/7680467).

Genes also have a **promoter region** upstream of them. Variants in the promoter can have impacts as well, such as up- or down-regulation of the gene.

The part of the gene that is read and translated into protein is called the **open reading frame** (**ORF**). Genes can also contain **upstream ORFs** (**uORFs**) located in the 5' UTR (Dasgupta and Prensner, *Upstream open reading frames: new players in the landscape of cancer gene regulation*, NAR Cancer, June 2024 https://academic.oup.com/narcancer/article/6/2/zcae023/7676824). These uORFS can be involved in many functions, from the regulation of RNA to the production of microproteins.

Genes can also contain a **signal peptide**, which is typically located in the first stretch of amino acids. Signal peptides control which parts of the cell a protein is directed to (for example, to the nucleus or mitochondria).

A variant in the intron could have an impact if it is near the **splice site** (the junction between the two exons). The base pairs right next to the splice site are both within the exonic region and also just within the border of the intronic region. Therefore, disruption of a base there could cause aberrant splicing. However, even variants more internal to the intron, which are called **deep intronic variants**, can cause disruptions in splicing.

We also need to consider that genes can have multiple **splice variants**. For example, a gene with exons A, B, C, D might produce **isoforms** ABCD, ABC, ABD, BCD, etc. In fact, there are on average about seven splice variants for any given gene (for more on this, read Leung et al., *Full-length transcript sequencing of human and mouse cerebral cortex identifies widespread isoform diversity and alternative splicing*, Cell Reports, November 2021, `https://www.cell.com/cell-reports/fulltext/S2211-1247(21)01504-7?uuid=uuid%3A815ec416-7e8c-401f-86bb-35d29fc13b75`. Genes may also have **alternative start sites**. This means multiple start codons are present within the gene. A common example would be to produce an isoform with a signal peptide and another isoform without the signal peptide by starting translation at a different point within the gene.

As you can see, understanding the impact of a variant within the genome can be quite complex. Knowing how to interpret gene structures and having access to the latest information are important for the proper analysis and interpretation of variants.

Let's take a look at how we can use the variants derived from our analysis with the gene structures we've annotated on the genome to better understand the potential impact of variants.

Getting ready

For this exercise, we'll be using Biopython, Matplotlib, and the Python `csv` module.

The code for this recipe can be found in `Ch06/Ch06-3-genes-variants.ipynb`.

How to do it...

Here are the steps to perform this recipe:

1. We will first import our libraries:

```
from Bio import SeqIO
from Bio.SeqFeature import SeqFeature, FeatureLocation
from Bio.Seq import Seq
import matplotlib.pyplot as plt
```

This will bring in the BioPython libraries we need, as well as Matplotlib.

2. Next, we will define our file paths:

```
genbank_file = "output/ecoli_prodigal_combined.gb"
vcf_file = "input/variants.vcf"
output_log_file = "output/variant_analysis.log"
```

We will use the combined Prodigal annotation file we produced in the *Annotating a prokaryotic genome* recipe and the variants file we made in the *Variant calling with GATK* recipe of *Chapter 5*. The output will go into the `variant_analysis.log` file.

3. Next, we will write a simple function to parse the VCF file:

```
def parse_vcf(vcf_file):
    variants = []
    with open(vcf_file, "r") as vcf:
        for line in vcf:
            if line.startswith("#"):
                continue
            fields = line.strip().split("\t")
            chrom = fields[0]
            pos = int(fields[1]) - 1
            ref = fields[3]
            alt = fields[4]
            variants.append((chrom, pos, ref, alt))
    return variants
```

This function loops over the variants in the input VCF file and splits the line. It parses out the chromosome and position fields as well as the reference and alternative bases.

4. Next is a simple function that checks whether the coding frame has been changed:

```
def changes_coding_frame(ref, alt):
    return (len(ref) - len(alt)) % 3 != 0
```

This is done by noting whether the REF and ALT fields have changed the length of the sequence by something that is not a multiple of 3.

5. Now, we'll define a function to check whether an amino acid will change because of this variant:

```
def introduces_amino_acid_change(cds_sequence,
                                 ref, alt,
                                 position_in_cds):
    try:
        if (cds_sequence[
                position_in_cds:position_in_cds
                + len(ref)] != ref):
            raise ValueError(
                "Reference allele does not match CDS"
```

```
                "at the specified position."
            )
        original_cds = cds_sequence
        mutated_cds = (
            cds_sequence[:position_in_cds]
            + alt +
            cds_sequence[position_in_cds + len(ref):]
        )
        original_protein = Seq(
            original_cds
        ).translate(to_stop=True)
        mutated_protein = Seq(
            mutated_cds
        ).translate(to_stop=True)
        return original_protein != mutated_protein

    except Exception as e:
        with open(output_log_file, "a") as log:
            log.write(
                f"Error processing variant at position"
                f"{position_in_cds + 1}: {e}\n"
            )
        return False
```

This function will create the mutated CDS sequence and then use the `Seq.translate()` function to check whether the two translated sequences are identical. If not, it will return `False`, meaning that the variant introduces an amino acid change in the protein.

6. Now, we build our plotting function:

```
def plot_variant_changes(variant_data):
    positions = [data[0] for data in variant_data]
    change_types = [data[1] for data in variant_data]
    plt.figure(figsize=(12, 6))
    plt.scatter(
        positions, change_types,
        alpha=0.7, edgecolors="k")
    plt.xlabel("Position in Genome", fontsize=12)
    plt.ylabel("Type of Change", fontsize=12)
    plt.title(
        "Variant Type vs. Position in Genome",
        fontsize=14, fontweight="bold"
    )
    plt.yticks(
        ticks=[0, 1, 2],
```

```
        labels=["No Change", "Frame Change", "AA Change"]
    )
    plt.grid(alpha=0.5)
    plt.show()
```

This uses Matplotlib to build our plot. We will set `positions` as the first column of the `variant_data` array. We then set `change_types` as the second column. We then create a scatter plot with the positions as the *x* axis and the change types on the *y* axis. Finally, we set the labels, title, tick marks, and grid, and then show the plot.

Now, we've defined our functions and will begin running our main code:

1. First, we will parse the VCF file:

    ```
    variants = parse_vcf(vcf_file)
    ```

2. Next, we will parse the GenBank file and look for variants that intersect with genes:

    ```
    variant_data = []
    with open(output_log_file, "w") as log:
      with open(genbank_file, "r") as gb_file:
        for record in SeqIO.parse(gb_file, "genbank"):
          for feature in record.features:
            if feature.type == "CDS":
              cds_start = int(
                feature.location.start)
                cds_end = int(
                  feature.location.end)
                  cds_sequence = str(
                    feature.extract(record.seq))
                  for chrom, pos, ref, alt in variants:
                    if (chrom == record.id and
                        cds_start <= pos < cds_end):
                      in_cds = True
                      frame_change = (
                        changes_coding_frame(
                          ref, alt)
                        )
                      position_in_cds = (
                        pos - cds_start)
                      amino_acid_change = (
                        introduces_amino_acid_change(
                          cds_sequence,
                          ref, alt,
                          position_in_cds)
    ```

```
                )
                change_type = (
                    2 if amino_acid_change else
                    (1 if frame_change else 0)
                )
                variant_data.append(
                    (pos + 1, change_type)
                )
                log.write(
                    f"Variant at position {pos + 1}"
                    f"(Ref: {ref}, Alt: {alt}) intersects"
                    f"CDS ({cds_start + 1}-"
                    f"{cds_end}).\n")
                if frame_change:
                    log.write(
                        "\tThis variant changes"
                        "the coding frame.\n")
                else:
                    log.write(
                        "\tThis variant does not"
                        "change the coding frame.\n"
                    )
                if amino_acid_change:
                    log.write(
                        "\tThis variant introduces an"
                        "amino acid change.\n")
                else:
                    log.write(
                        "\tThis variant does not"
                        "introduce an amino acid change.\n")
```

This code does the following:

- Creates an empty list to hold the variant data.

- Opens our output log file for writing and our GenBank file for parsing.

- Loops over the records in the GenBank file. For each record, it checks whether it is a CDS type and, if so, it is processed:

    i. Get the start, end, and sequence for the CDS

    ii. Now we do an inner loop over the variants:

    - If the position of the variant is inside our CDS, we perform a series of checks to determine whether the frame or amino acid has been changed by the variant

- Assign a `change_type` value: 0 means no significant change, 1 is a frame shift, and 2 is an amino acid change
- Append the change information to the `variant_data` list
- Log the information to our output file, including position, REF, ALT, region of the intersection, and the type of change

3. Finally, let's plot our variant changes. This is what we get:

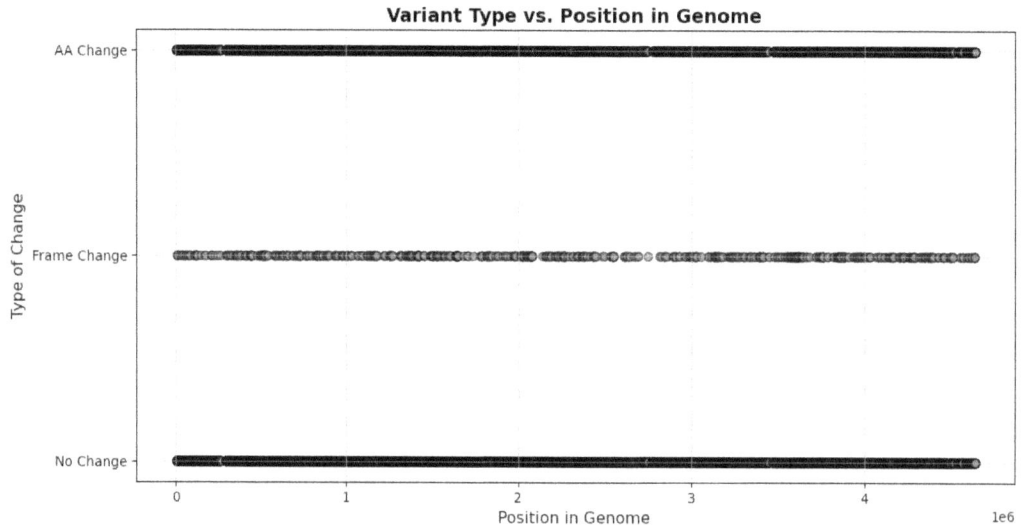

Figure 6.11 – Variant-induced changes mapped along the genome

We can see variants with no change, a frame shift, or an amino acid change mapped along the E. coli genome.

This was just a simple example of looking at how variants can induce potential biological changes by focusing on the genes and their coding into proteins. More sophisticated modern programs may also look at whether variants fall in promoters, splice sites, UTRs of genes, and so forth. It is through understanding the effects of these variants (and often, combinations of variants) that we can begin to predict the outcome of sequence changes on the phenotype of the organism.

There's more...

We've touched on variants that introduce obvious impacts, such as frame shifts and amino acids in proteins. Let's discuss more aspects of variant interpretation. **SnpEff** is a popular program for determining the impacts of variants (https://pcingola.github.io/SnpEff/). It supports numerous genomes and interprets several different kinds of impacts. It also comes with SnpSift, which is a tool for filtering variants.

If you want, you can run the optional exercise at the end of the notebook for this *Interpreting variants on gene structures* recipe. This will install SnpEff and run it on the human genome using a test file that comes with SnpEff:

```
! wget https://snpeff.blob.core.windows.net/versions/snpEff_latest_
core.zip
! unzip snpEff_latest_core.zip


! java -Xmx4g -jar snpEff/snpEff.jar download GRCh38.99
! java -Xmx4g -jar snpEff/snpEff.jar GRCh38.99 snpEff/examples/test.
vcf > output/human_annotated_variants.vcf
```

Let's take a look at the output:

```
##SnpEffVersion="5.2e (build 2024-10-04 18:09), by Pablo Cingolani"
##SnpEffCmd="SnpEff  GRCh38.99 snpEff/examples/test.vcf "
##INFO=<ID=ANN,Number=.,Type=String,Description="Functional annotations: 'Allele | Annotation | Annotation_Impact | Gene_Name
ure_ID | Transcript_BioType | Rank | HGVS.c | HGVS.p | cDNA.pos / cDNA.length | CDS.pos / CDS.length | AA.pos / AA.length | Di
FO' ">
##INFO=<ID=LOF,Number=.,Type=String,Description="Predicted loss of function effects for this variant. Format: 'Gene_Name | Ger
_gene | Percent_of_transcripts_affected'">
##INFO=<ID=NMD,Number=.,Type=String,Description="Predicted nonsense mediated decay effects for this variant. Format: 'Gene_Nam
ipts_in_gene | Percent_of_transcripts_affected'">
1       10469   .       C       G       365.78  PASS    AC=30;AF=0.0732;ANN=G|upstream_gene_variant|MODIFIER|DDX11L1|ENSG0000€
28.2|processed_transcript||n.-1400C>G|||||1400|,G|upstream_gene_variant|MODIFIER|DDX11L1|ENSG00000223972|transcript|ENST000004
_pseudogene||n.-1541C>G|||||1541|,G|downstream_gene_variant|MODIFIER|WASH7P|ENSG00000227232|transcript|ENST00000488147.1|unprc
|||3935|,G|intergenic_region|MODIFIER|CHR_START-DDX11L1|CHR_START-ENSG00000223972|intergenic_region|CHR_START-ENSG00000223972|
```

Figure 6.12 – SnpEff human output

SnpEff has added a long line with additional variant information. It shows that there is a variant that is a potential modifier of the DOX11L1 gene, as well as some other potential impacts. SnpEff will tend to call variants upstream of genes as potential modifiers because they might impact promoters, and will also report downstream, intergenic, and other types of variants. It will classify numerous types of variants into the following general categories of impact:

- **HIGH**: Disruptive effects on proteins, such as frameshifts or introduction of stop codons

- **MODERATE**: Effects that might disrupt a protein, such as an amino acid change

- **LOW**: Changes that are unlikely to have much impact, such as a silent mutation in a protein (does not change an amino acid)

- **MODIFIER**: Variants that may be upstream or downstream of genes and could potentially impact promoters or other aspects of gene regulation

SnpEff is one of several programs that can help you interpret the impact of a variant on a gene structure. The study of gene structures, their variants and isoforms, and the way they impact cellular metabolism and organismal phenotype is called **functional annotation**. Functional annotation can be performed by searching for similar genes using programs such as BLAST and by studying **orthologs** (genes from another species that are similar to the gene of interest, having descended from a common ancestor). See Tegenfeldt et al., *OrthoDB and BUSCO update: annotation of orthologs with wider sampling of genomes*, Nucleic Acids Research, January 2025, https://academic.oup.com/nar/

article/53/D1/D516/7899526. A popular tool for this is eggNOG-mapper. See Cantalipiedra et al., *eggNOG-mapper v2: Functional Annotation, Orthology Assignments, and Domain Prediction at the Metagenomic Scale*" Molecular Biology and Evolution, Dec 2021. https://academic.oup.com/mbe/article/38/12/5825/6379734. It uses the popular eggNOG database, which is a huge collection of orthologs across all domains of life. See Hernandez-Plaza et al., *eggNOG 6.0: enabling comparative genomics across 12,535 organisms*, Nucleic Acids Research, November 2022, https://academic.oup.com/nar/article/51/D1/D389/6833261.

Another popular method involves looking at sets of genes to see whether they are enriched for certain functions. The DAVID system is a great example of this; see Sherman et al., *DAVID: a web server for functional enrichment analysis and functional annotation of genes lists (2021 update)*, Nucleic Acids Research, July 2022, https://academic.oup.com/nar/article/50/W1/W216/6553115.

The **Gene Ontology** (**GO**) system is also a great method for annotating the functions of genes; see *The Gene Ontology resource: enriching a GOld mine*, The Gene Ontology Consortium, Nucleic Acids Research, Jan 2021. https://academic.oup.com/nar/article/49/D1/D325/6027811.

Tools also exist for performing functional annotation by mapping genes to reaction networks such as KEGG (https://www.genome.jp/kegg/) and Reactome (https://reactome.org/). Read more in Palu et al., *KEMET – A Python tool for KEGG module evaluation and microbial genome annotation expansion*, Computational and Structural Biotechnology Journal, Vol 20, Mar 2022.

Many genes are completely novel and have no known function. For these genes, it is often dangerous to simply assign a function based on a BLAST hit to a nearby organism with a related gene as the hit may be weak, and the other gene may have itself been simply assigned a function computationally. This is known as the problem of **transitive annotation**. For this, there are newer algorithms that apply more sophisticated approaches, such as PANNZER; see Koskinen et al., *PANNZER: high-throughput functional annotation of uncharacterized proteins in an error-prone environment*, BioInformatics, May 2015, https://academic.oup.com/bioinformatics/article/31/10/1544/176441.

TALE uses a transformer (an AI-based approach that we'll cover in *Chapter 17*) to annotate unknown genes; see Cao and Shen, *TALE: Transformer-based protein function annotation with joint sequence-label embedding*", BioInformatics, September 2021, https://academic.oup.com/bioinformatics/article/37/18/2825/6182677.

Finally, unknown genes and their function can be directly interrogated via a variety of high-throughput experimental methods. Rocha et al. created an **Unknome** database cataloging large numbers of unknown genes, and then identified a subset of genes from humans that are conserved in Drosophila. They then applied a high-throughput CRISPRi screen in Drosophila to identify phenotypic effects of knockdowns on those genes. This then helps us infer putative functions for the human genes; see Rocha et al., *Functional unknomics: Systematic screening of conserved genes of unknown function*, Plos Biology, August 2023, https://journals.plos.org/plosbiology/article?id=10.1371/journal.pbio.3002222&trk=public_post_comment-text.

We can also try to understand the function of unknown genes by seeing whether they are co-expressed with other known genes and then looking at their expression patterns, pathway enrichment, and treatment response information; see Horan et al., *Annotating Genes of Known and Unknown Function by Large-Scale Coexpression Analysis*, Plant Physiology, May 2008, `https://academic.oup.com/plphys/article/147/1/41/6107500`.

Advances in high-throughput technologies such as microfluidics and imaging are rapidly transforming our ability to identify and annotate unknown genes. Screening large numbers of perturbed genes in an organism, identifying the desired phenotype, and mapping back to the gene(s) of interest is known as **forward genetics**. As an example, Yu et al. performed this type of screening using microfluidics to find high-performing strains of *C. glutamicum* that could secrete their products at high titer. A CRISPRi genomic library perturbs large numbers of genes, and this is combined with a titer readout on a microfluidic device that can screen incredibly large numbers of samples. When genes of interest are found, they can be readily identified by reading out the barcode in the guide RNA cassette using NGS to identify the locus that was perturbed; see Yu et al., *CRISPRi-microfluidics screening enables genome-scale target identification for high-titer protein production and secretion*, Metabolic Engineering, January 2023, `https://pubmed.ncbi.nlm.nih.gov/36572334/`.

High-throughput imaging is another emerging technology that can be leveraged to understand genes of unknown function. Perrin et al. fluorescently tagged many proteins in the cyanobacteria *Synechococcus elongatus* to determine their subcellular location and association in complexes with known proteins to enhance their functional annotation; see Perrin et al., *CyanoTag: Discovery of protein function facilitated by high-throughput endogenous tagging in a photosynthetic prokaryote*, Science Advances, February 2025, `https://www.science.org/doi/full/10.1126/sciadv.adp6599`.

AI tip

Background: Let's write an example of performing a gene set enrichment using some popular tools: GSEApy (`https://gseapy.readthedocs.io/en/latest/`) and GOATOOLS (`https://github.com/tanghaibao/goatools`).

Prompt: Write an example using Python libraries for GSEApy and GOATOOLS that performs a functional enrichment of a sample gene set; provide sample output reports and visualizations for the results.

What you should see: Code to provide a sample gene list with enrichment analysis. GO analysis and reports on top of enriched genes. Visualizations will be included as well as installation instructions.

As you can see, there are many applications for studying gene structures and their annotations, and for deriving functional annotations for unknown genes, ranging from microbial natural product development to plant traits and physiology, through human disease and clinical study. These areas are rapidly advancing as we see the combination of high-throughput screening technologies, large knowledge bases, and AI.

Before we continue, let's talk a little about a non-traditional gene structure, known as non-coding RNA. This is RNA that is transcribed but not translated into proteins. As we learn more about the genome, we find out that more of it is functional than was previously thought, meaning variants once dismissed as harmless must be looked at more closely. For example, not long ago, non-coding RNA was typically thought to be junk related to spurious transcription. We now know that this **non-coding RNA (ncRNA)** is often functional (Cao et al., *Very long intergenic non-coding (vlinc) RNAs directly regulate multiple genes in cis and trans*, BMC Biology, May 2021, `https://bmcbiol.biomedcentral.com/articles/10.1186/s12915-021-01044-x`).

Because of the complex nature of interpreting variants, AI is beginning to have a big impact in this field. SpliceAI (`https://github.com/Illumina/SpliceAI`) uses deep learning to predict the impact of variants on gene splicing. PrimateAI-3D (`https://www.illumina.com/science/genomics-research/articles/primateai-3d.html`) leverages information from closely related primate species to predict the pathogenicity of variants in humans. New advanced knowledge platforms are being developed to help researchers and clinicians prioritize variants, such as Illumina Connected Insights (`https://www.illumina.com/products/by-type/informatics-products/connected-insights.html`). OpenCravat (`https://www.opencravat.org/`) is another useful platform for variant annotation.

A variety of databases and resources exist to help with variant interpretation in humans. The **American College of Medical Geneticists** (**ACMG**) sets guidelines on variant interpretation in genetic testing. For example, ClinGen (`https://clinicalgenome.org/`) and ClinVar (`https://www.ncbi.nlm.nih.gov/clinvar/`) provide curation of clinically relevant genes and variants. CANVAR is a Python program for annotation of variants in the ClinVar database (Vestergaard et al., *CANVAR: A Tool for Clinical Annotation of Variants using ClinVar Databases*, Molecular Genetics & Genomic Medicine, October 2024). Genopyc is a Python library for investigating the effects of variants on complex diseases (Gualdi et al., *Genopyc: a Python library for investigating the functional effects of genomic variants associated to complex diseases*, BioInformatics, June 2024).

Now, we've talked about the effects variants can have on gene structures. In the *Annotating proteins* recipe, we'll take a close look at proteins and understand how we might determine which regions of a protein a variant falls into.

See also

- Variant Effect Predictor (VEP) is another popular program for variant impact prediction: `https://www.ensembl.org/info/docs/tools/vep/index.html`

- VarCards2 is an integrated database to help with human clinical variant interpretation: `https://academic.oup.com/nar/article/52/D1/D1478/7416810`

- Learn more about KEGG in Kanehisa et al., *KEGG: a biological systems database as a model of the real world*, Nucleic Acids Research, January 2025, `https://academic.oup.com/nar/article/53/D1/D672/7824602`

Annotating proteins

Let's next discuss proteins in more detail. Each protein has regions or **domains** that can be annotated and give us further insight into the biological function of that portion of the protein. With this information, we can further reason about the potential impact of variants in that domain. For now, we'll focus on 2D protein structure, meaning we look at the primary amino acid sequence without trying to fold or understand the 3D structure of the protein. We'll talk more about structural prediction in *Chapter 9, Protein Structure and Proteomics*.

Much like the structure of a gene, proteins have various regions that perform different functions. Let's take a look at the 2D structure of a typical protein:

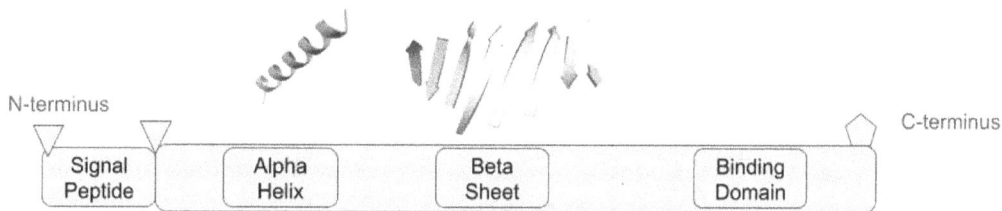

Figure 6.13 – Protein structure overview

Proteins in 2D are read from left to right, and the left side is called the **N-terminus**. Just like DNA, amino acids have a directionality, with an NH_2 group at the beginning and a COOH or carboxyl group at the end, which is called the **C-terminus**.

The most basic structures proteins fold into are called **secondary structures** and include things such as the **alpha helix** and **beta sheet** depicted in the preceding figure.

Proteins may often have a short peptide at the beginning, sometimes called a leader peptide. This may be a **signal peptide**. Signal peptides typically help direct a protein to either be secreted outside the cell or inserted into the cell membrane. They are typically at the N-terminus but sometimes can be found in the C-terminus or even internally to the protein.

Proteins can also have a **transit peptide**, which may direct the protein to a particular organelle such as the peroxisome or mitochondria. Because of this, proteins may also have **alternative start codons** (triangles above), which are often used as a way to translate two versions of a protein, one with a transit peptide and one without. For instance, in plants, this is a common way to make a version of a protein that is destined for the chloroplast and another version that is destined for the cytoplasm.

Proteins may also have a **nuclear localization signal** (**NLS**). This type of short peptide directs proteins to go back from the cytoplasm (where they are translated on ribosomes) to the nucleus, where they may perform various roles in the structure and function of chromosomes.

Proteins can also have alternative lengths for their C-terminus. The C-terminus contains many important motifs that help determine protein stability, trafficking, and regulation.

Proteins also have a variety of other functional domains that we can identify, such as binding domains. These can be important clues to the function of a protein, especially when trying to annotate and understand the functions of novel and newly discovered proteins. Finally, proteins can often have a variety of disordered or repetitive domains, much like repetitive DNA.

Let's look at how we can predict the domains in a protein sequence. We will use HMMER (`http://hmmer.org/`), which is a program that uses hidden Markov models. It can use databases of protein domain profiles such as PFAM (`http://pfam.xfam.org/`) that contain large databases of protein families that have been aligned to find common domains.

Getting ready

First, we will install HMMER:

```
! brew install hmmer
```

Alternatively, you can run the following:

```
! conda install bioconda::hmmer
```

Next, we need to download the PFAM database and index it:

```
! mkdir -p pfam
! wget ftp://ftp.ebi.ac.uk/pub/databases/Pfam/current_release/Pfam-A.hmm.gz
! gunzip Pfam-A.hmm.gz
! mv Pfam-A.hmm pfam/
! hmmpress pfam/Pfam-A.hmm
```

This will place our indexed PFAM database in the `pfam/` subdirectory.

The code for this recipe can be found in `Ch06/Ch06-4-protein-domains.ipynb`.

How to do it...

Before we get started, note that HMM searches on proteins are quite compute intensive. So, let's first make a short script to just take the top 10 proteins from our E. coli FASTA (`.faa`) file:

```
from Bio import SeqIO
input_fasta = "output/ecoli_proteins.faa"
output_fasta = "output/ecoli_proteins.top10.faa"
def select_first_n_entries(input_file,
                           output_file, n=10):
    with open(input_file, "r") as infile, \
        open(output_file, "w") as outfile:
        records = SeqIO.parse(infile, "fasta")
        limited_records = (
```

```
        record for i, record in enumerate(records)
        if i < n
    )
    SeqIO.write(limited_records, outfile, "fasta")
    print(
        f"First {n} entries written to {output_file}"
    )
select_first_n_entries(
    input_fasta, output_fasta, n=10
)
```

This code will take the first 10 entries from the E. coli protein FASTA file that we created using Prodigal and deposit them in the ecoli_proteins.top10.faa file in our output subdirectory.

Now, we can begin writing our code to run HMMER:

1. First, we will import our libraries:

    ```
    import os
    import subprocess
    from Bio import SeqIO
    ```

    We will use the os and subprocess libraries to call HMMER, as well as Biopython for sequence processing.

2. Next, we define our file paths:

    ```
    fasta_file = "output/ecoli_proteins.top10.faa"
    pfam_hmm_db = "pfam/Pfam-A.hmm"
    output_domtblout = "output/ecoli.pfam_domains.top10.out"
    ```

    This will set up our input file as the top 10 proteins in E. coli and point to our PFAM database. The output will go into the ecoli.pfam_domains.top10.out file.

3. Now, we will define a function to run HMMER:

    ```
    def run_hmmsearch(input_fasta, hmm_db, output_file):
        command = [
            "hmmsearch",
            "--domtblout", output_file,
            hmm_db,
            input_fasta
        ]
        print(f"Running HMMER with command: {
            ' '.join(command)}")
        subprocess.run(command, check=True)
        print(f"HMMER search completed."
            f"Results saved to {output_file}")
    ```

This is a pretty straightforward function that will construct a command using `hmmsearch` with the input FASTA file and the output file, and we will use our PFAM database for the profile search.

4. Here is the function to parse our `hmmsearch` output:

```
def parse_hmmsearch_output(domtbl_file):
    annotations = []
    with open(domtbl_file, "r") as file:
        for line in file:
            if (line.startswith("#") or not line.strip()):
                continue
            fields = line.split()
            query_name = fields[0]
            domain_name = fields[3]
            e_value = float(fields[6])
            annotations.append((
                query_name, domain_name, e_value
            ))
    return annotations
```

This function will initialize a list of annotations and then open the `domtbl` (domain table) file. It will parse and split the lines in the file, grabbing the query name, domain name, and e-value. It will append these to the annotations list and return it.

5. We will also write a function to enrich our protein FASTA file with the PFAM annotations:

```
def annotate_fasta_with_pfam(fasta_file, annotations):
    annotated_file = "annotated_protein.fasta"
    seq_annotations = {
        query: [] for query, _, _ in annotations
    }
    for query, domain, e_value in annotations:
        seq_annotations[query].append(
            f"{domain} (E-value: {e_value:.2e})"
        )
    with open(annotated_file, "w") as output:
        for record in SeqIO.parse(fasta_file, "fasta"):
            domains = seq_annotations.get(
                record.id, []
            )
            record.description += (
                " | Domains: " + ", ".join(domains)
            ) if domains else ""
            SeqIO.write(record, output, "fasta")
    print(f"Annotated FASTA file saved to {annotated_file}")
```

This function will take a FASTA file and a set of annotations and build an annotated file. Note the line for `seq_annotations`; this uses a syntax to build a dictionary in which the keys are queries and the values are empty lists for storing the domain annotations.

6. For each annotation, we will next format the e-value. We will then open the output file and loop over each record in the FASTA file. For each record, we will grab the corresponding domain annotation and append them to the description for the record. We will then write out the record using the SeqIO class.

7. Let's now execute our functions:

```python
if __name__ == "__main__":
    if not os.path.exists(f"{pfam_hmm_db}.h3f"):
        print("Indexing Pfam HMM database...")
        subprocess.run(
            ["hmmpress", pfam_hmm_db],
            check=True
        )
    run_hmmsearch(
        fasta_file, pfam_hmm_db,
        output_domtblout
    )
    pfam_annotations = parse_hmmsearch_output(
        output_domtblout)
    annotate_fasta_with_pfam(
        fasta_file, pfam_annotations)
```

This will run indexing if we haven't already done it, and then run the `hmmsearch` function using our input FASTA file and PFAM database. Next, it will parse the `hmmsearch` results and annotate the FASTA file.

One last thing; let's move our annotated FASTA file back into our `output` directory to keep things clean:

```
! mv annotated_protein.fasta output
```

That's it! We now have a PFAM domain output for our top 10 proteins and an annotated FASTA file. Let's see what they look like.

Here is the PFAM domain file:

```
# target name         accession  tlen query name           accession  qlen  E-value  score
   to  from   to  from   to  acc description of target
#-------------------- ---------- ----- -------------------- ---------- ----- --------- -------
---- ----- ----- ----- ----- ---- ----------------------
NC_000913.3_2           -        821 AA_kinase            PF00696.33   234   3.4e-49  157.0
  234     2   284     1   284 0.88 # 337 # 2799 # 1 # ID=1_2;partial=00;start_type=ATG;rbs_mot
NC_000913.3_2           -        821 AA_kinase            PF00696.33   234   3.4e-49  157.0
  233   317   349   305   350 0.78 # 337 # 2799 # 1 # ID=1_2;partial=00;start_type=ATG;rbs_mot
NC_000913.3_2           -        821 ACT                  PF01842.31    66   3.6e-17   51.4
   59   325   375   323   380 0.87 # 337 # 2799 # 1 # ID=1_2;partial=00;start_type=ATG;rbs_mot
NC_000913.3_2           -        821 ACT                  PF01842.31    66   3.6e-17   51.4
   49   404   446   400   454 0.78 # 337 # 2799 # 1 # ID=1_2;partial=00;start_type=ATG;rbs_mot
```

Figure 6.14 – PFAM domain output

This file contains the domain information for each accession, the e-value, start-stop coordinates, and so on. For example, for gene NC_000913.3_2, we have an AA_kinase domain as the first line. If you perform an Internet search for the AA Kinase domain, you will quickly find an entry like this in Interpro: https://www.ebi.ac.uk/interpro/entry/pfam/PF00696/. This is PFAM domain number PF00696. You can learn more about this domain and see that it phosphorylates amino acids. You will also see the query length (qlen) field, the e-value (which is the probability of the hit being correct), the hit score, and a few other fields. You can read more about HMMER results here: http://hmmer.org/documentation.html.

If you look at the third line down in the PFAM results, we see that this same gene NC_000913.3_2 is hitting a protein domain called ACT or PF01842 (https://www.ebi.ac.uk/interpro/entry/pfam/PF01842/). This domain plays a regulatory role in metabolism and is named after three key proteins that contain this domain: **Aspartate Kinase**, **Chorismate Mutase**, and **TyrA**.

There is an extensive HMMER user guide here: http://eddylab.org/software/hmmer/Userguide.pdf.

You can also check out the annotated_proteins.fasta file, which is a little less interesting. It is basically just our original FASTA file with the domains mentioned earlier appended to each entry.

As you can see, this approach allows us to more deeply understand and annotate our proteins. This type of investigation can lead to further insights about the functions of individual proteins, or even into aspects of the organism as a whole.

There's more…

The function of many protein domains remains poorly understood. These unannotated protein domains are referred to as **domains of unknown function (DUFs)**. Much as we discussed in the previous recipe, they can be addressed by a variety of bioinformatics approaches, including AI and high-throughput functional screening.

ProtNote uses AI to perform protein function prediction on unknown proteins; see Char et al., *ProtNote: a multimodal method for protein-function annotation*, Bioinformatics, May 2025, https://academic.oup.com/bioinformatics/article/41/5/btaf170/8113843.

Microfluidics was used to screen for and annotate proteins, including unknown proteins that would improve amylase secretion in *Aspergillus oryzae*; see Li et al., *High-throughput droplet microfluidics screening and genome sequencing analysis for improved amylase-producing Aspergillus oryzae*, Biotechnology for Biofuels and Bioproducts, November 2023, https://link.springer.com/article/10.1186/s13068-023-02437-6.

PIFia performs high-throughput imaging analysis to infer functional annotations of proteins; see Razdaibiedina et al., *PIFia: self-supervised approach for protein functional annotation from single-cell imaging data*, Molecular Systems Biology, Mar 2024, https://www.embopress.org/doi/full/10.1038/s44320-024-00029-6.

Another important aspect of protein function is where in the cell it is targeted, known as subcellular localization. Many tools exist for predicting domains and targeting peptides in proteins. One key one to know about is **SignalP**, which predicts signalling peptides (Teufel et al., `SignalP 6.0 predicts all` *five types of signal peptides using protein language models*, Nature Biotechnology, January 2022). Another is **TargetP**, which predicts targeting peptides (`https://services.healthtech.dtu.dk/services/TargetP-2.0/`).

AI is rapidly impacting the field of protein domain prediction. Much like the language models used in programs such as ChatGPT, researchers are building language models based on protein or DNA sequences. You may want to check out PSALM (`https://github.com/Protein-Sequence-Annotation/PSALM`), which uses protein language models for domain annotation.

As you can see, we can take what we learned in the `Interpreting variants in gene structure` recipe and combine that knowledge with an even deeper understanding of proteins by looking at their domains and secondary structure. For example, if we can overlay an amino acid mutation into a binding domain, we can expect that it may have a big impact on the protein's ability to bind its substrate. On the other hand, introducing an early stop codon very near the C-terminus may have little impact on protein function.

Let's clean up and close down our `conda` environment:

```
conda deactivate
```

See also

- Interpro contains a wealth of protein domain information. You may want to try manually pasting some of your protein sequences into the web interface and see how they match up with your PFAM annotations: `https://www.ebi.ac.uk/interpro/`.

- We mentioned GO annotations in the last recipe – PFresGO uses deep learning and GO annotations to provide accurate functional annotation of proteins; see Pan et al., *PFresGO: an attention mechanism-based deep-learning approach for protein annotation by integrating gene ontology inter-relationships*, BioInformatics, March 2023, `https://academic.oup.com/bioinformatics/article/39/3/btad094/7043095`.

- This article gives a great review of DUFs: Lv et al., *Unraveling the Diverse Roles of Neglected Genes Containing Domains of Unknown Function (DUFs): Progress and Perspective*, International Journal of Molecular Sciences, February 2023, `https://www.mdpi.com/1422-0067/24/4/4187`.

- OpenProt provides a comprehensive resource on alternative proteins coming from unusual reading frames and includes AI-based annotation and structural prediction; see LeBlanc et al., *OpenProt 2.0 builds a path to the functional characterization of alternative proteins*, Nucleic Acids Research, January 2024, `https://academic.oup.com/nar/article/52/D1/D522/7416803`.

Get This Book's PDF Version and Exclusive Extras

UNLOCK NOW

Scan the QR code (or go to `packtpub.com/unlock`). Search for this book by name, confirm the edition, and then follow the steps on the page.

Note: Keep your invoice handy. Purchases made directly from Packt don't require an invoice.

Genomes and Genome Assembly

Understanding the quality of your genome reference sequence and knowing how to assemble high-quality genomes are core bioinformatics skills. So far, we have learned a great deal about variant calling and alignment, as well as gene annotation. The quality of these steps is greatly impacted by the underlying reference genome. As you might imagine, variants cannot be mapped effectively to poorly assembled regions, and genes cannot be called well when you have incorrect bases or other issues in your assembly.

In the past few years, there have been tremendous advances in genome sequencing and assembly that have transformed the field. In this chapter, we'll learn about how those advances have impacted bioinformatics and the resulting benefits.

In this chapter, we will cover:

- Accessing genome assemblies
- Working with graph genomes
- Long-read assembly with Raven
- Assessing genome quality with QUAST

Technical requirements

For this chapter, we will be using the following tools and packages:

- Pyfastx
- NetworkX
- Matplotlib
- Raven
- QUAST

You will be instructed on how to install each package as we go. You can find the code for this chapter at `https://github.com/PacktPublishing/Bioinformatics-with-Python-Cookbook-Fourth-Edition/tree/main/Ch07`.

Remember to activate your conda environment before beginning the recipes, like this:

```
conda activate bioinformatics_base
```

Or, if you would like to set up a conda environment specific to this chapter, before activating bioinformatics_base, run the following:

```
conda create -n ch07-genomes --clone bioinformatics_base
conda activate ch07-genomes
```

You will be able to install the packages for the chapter as you go, or you can use the YAML file provided in the repository:

```
conda env update --file ch07-genomes.yml
```

Accessing genome assemblies

In the early days of genome sequencing, short reads were primarily used to assemble genomes, leading to short, fractured assemblies. Scientists would go to conferences and talk about how they had gotten the assembly for their favorite organism down to 10,000 **contigs** instead of 50,000! This meant it was often unclear which chromosomes these contigs belonged to, or what order they went in. Scientists would try to **scaffold** the contigs by looking for markers to place them in general areas on chromosomes, and they would engage in painstaking efforts to close the gaps. But if you have a human genome of 3 Gb and 24 chromosomes, you would imagine that you might like to get 24 contigs if you could! (Or, better yet, 48 contigs in total by resolving the haplotypes of the diploid genome).

With the advent of long-read sequencing, this has changed dramatically. We are now able to routinely assemble entire microbial chromosomes with ease (often called "closed genomes" in the case of organisms with circular chromosomes). We can assemble yeast and many other organisms quite well, also. And in the case of the human genome, we are now producing what are called "Telomere-to-Telomere" or T2T genomes. The **Telomere** is a long, highly repetitive structure at the end of a chromosome. In the center of the chromosome is another highly repetitive structure called the **Centromere**. Getting through both of these regions was always difficult, and one of the main reasons genome assemblies remained fragmented. That's why getting all the way across from end to end is quite a feat! Have a look at the following figure:

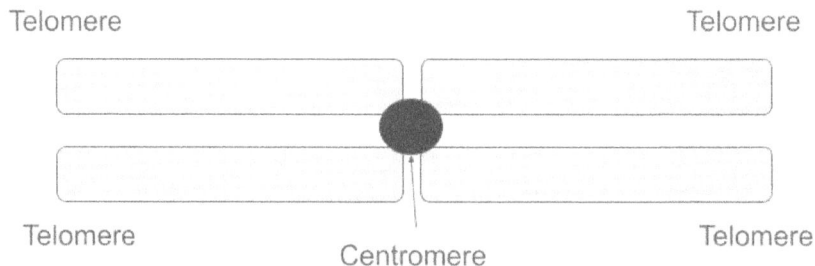

Figure 7.1 – Major features of a chromosome

In 2022, scientists completed the first T2T genome called CHM13 (Nurk et al, *The complete sequence of a Human genome*, Science, Mar 2022: `https://www.science.org/doi/10.1126/science.abj6987`). This assembly has no gaps in any chromosome except the Y chromosome, which is especially repetitive and difficult to assemble. They used a combination of PacBio HiFi reads (~20 kb circular reads with a low error rate), Oxford Nanopore ultra-long reads (reaching up to 100 kb!), and 3D genome technologies to achieve this result. The assembly is represented as a graph genome, which we'll talk about in the *Working with Graph Genomes* recipe. You can find more information in the T2T consortium (`https://sites.google.com/ucsc.edu/t2tworkinggroup`).

Since then, researchers have not slowed down and are embarking on ambitious plans to sequence hundreds more T2T genomes. The Human Pangenome Research Consortium (`https://humanpangenome.org/`) aims to produce hundreds of T2T genomes. With continued advances (and cost reductions) coming in long-read technology, we can expect T2T genomes to become more and more standard.

Getting ready

First, we will download a T2T reference genome. Let's grab the T2T CHM13 genome:

```
! mkdir -p data
! wget https://ftp.ncbi.nlm.nih.gov/genomes/all/GCA/009/914/755/
GCA_009914755.4_T2T-CHM13v2.0/GCA_009914755.4_T2T-CHM13v2.0_genomic.
fna.gz
! gunzip GCA_009914755.4_T2T-CHM13v2.0_genomic.fna.gz
! mv GCA_009914755.4_T2T-CHM13v2.0_genomic.fna data/T2T_genome.fasta
```

This will download the CHM13 version 2.0 genome FASTA file, unzip it, and then move it into our data subdirectory. We will rename it as `T2T_genome.fasta` for convenience.

Let's also install a tool to investigate the genome, `pyfastx`:

```
! pip install pyfastx
```

How to do it...

Let's take a quick look at this genome:

1. We can use `pyfastx` to investigate it further (`https://github.com/lmdu/pyfastx`). This library provides fast access to FASTA and FASTQ files. Let's import it:

    ```
    import pyfastx
    ```

2. Now we will set the input FASTA file we will use as the CHM13 genome that we downloaded:

    ```
    genome_fasta = "data/T2T_genome.fasta"
    ```

3. Next, we define a function to compute the genome size:

```
def compute_genome_size(fasta_file):
    genome_size = 0
    genome = pyfastx.Fasta(
        fasta_file, build_index=False)
    for _, seq in genome:
        genome_size += len(seq)
    return genome_size
```

This function starts by using `pyfastx` to open the FASTA file. Note the code fragment `for _, seq in genome`, which loops over the sequences in the file and returns them as a tuple – the underscore causes us to ignore the first element in the tuple, which is the sequence ID. The second element is the sequence that we will use. We then keep incrementing the genome size variable by the length of the sequence until we have calculated the total length of the genome, which we return.

4. Next, we will define a function to compute the GC content of the genome:

```
def compute_gc_content(fasta_file):
    total_bases = 0
    gc_count = 0
    genome = pyfastx.Fasta(
        fasta_file, build_index=False)
    for _, seq in genome:
        total_bases += len(seq)
        gc_count += (
            seq.upper().count('G') +
            seq.upper().count('C')
        )
    return (
        (gc_count / total_bases) * 100
        if total_bases > 0 else 0
    )
```

GC content is an important genome feature. G and C base pairs are held together by three hydrogen bonds, whereas A and T are held together by 2. This makes GC bonds harder to break. GC-rich regions of the genome may have important properties, including higher stability, gene richness, and recombination frequency. GC content is also a key signature of different organisms, varying widely between microbes, plants, and humans.

This function again uses `pyfastx` to open the genome and process each sequence as before. This time, we count up the total bases as well as any G or C bases and return the percentage. Note that we use `upper()` to make sure that both uppercase and lowercase G and C are counted.

5. Next up is a function to compute the N50 value:

```
def compute_n50(fasta_file):
    lengths = []
    genome = pyfastx.Fasta(
        fasta_file, build_index=False)
    lengths = sorted(
        [len(seq) for _, seq in genome],
        reverse=True
    )
    cumulative_length = 0
    total_length = sum(lengths)
    for length in lengths:
        cumulative_length += length
        if cumulative_length >= total_length / 2:
            return length
    return 0
```

N50 is an important value in genomics. It is the length of the shortest contig in a list of contigs, starting from the longest, that make up 50% of the genome. In other words, if you take the longest contigs and start adding them up in order until you get to 50% genome coverage, the last contig length you get to is the N50 value. In this way, it represents the completeness of the assembly. A highly fragmented assembly with a lot of contigs will have a small N50 value, and a good assembly with few gaps will have a high N50 value. Some closely related metrics that are important to know are the **N90**, which is the length of the shortest contig needed to get up to 90% coverage of the genome, and the **L50**, which is the number of contigs needed to reach the N50 value.

This function uses `pyfastx` to build a list of contig lengths in sorted order. It then iterates through the list, building up `cumulative_length` until it exceeds 50% of `total_length`. It returns this as the N50 value.

6. Okay, let's evaluate our genome!

```
def assess_quality(fasta_file):
    genome_size = compute_genome_size(fasta_file)
    gc_content = compute_gc_content(fasta_file)
    n50 = compute_n50(fasta_file)

    print(f"Genome Quality Metrics for {fasta_file}:")
    print(f"Total Genome Size: {genome_size:,} bp")
    print(f"GC Content: {gc_content:.2f}%")
    print(f"N50: {n50:,} bp")


if __name__ == "__main__":
```

```
assess_quality(genome_fasta)
```

This code defines a master function, `assess_quality()`, which runs our three calculation functions and prints out their results. Here are the results we see:

```
Genome Quality Metrics for data/T2T_genome.fasta:
Total Genome Size: 3,117,292,070 bp
GC Content: 40.75%
N50: 150,617,247 bp
```

Figure 7.2 – Genome statistics for T2T genome

We've now seen that we can get access to a wide variety of high-quality genome assemblies and manipulate them. In the coming recipes, we'll deepen our knowledge of genome assembly and analysis.

There's more...

Recall that we used the `upper()` function in the preceding GC calculation to make sure we count both uppercase and lowercase Gs and Cs. Genomes are often **repeat-masked**, which means that repetitive regions of the genome are written as lowercase.

A common tool for repeat masking is RepeatMasker (`https://www.repeatmasker.org/`). This tool scans the genome and masks repetitive or low-complexity regions.

In **soft masking**, the bases in the genome are converted to lowercase so that they can still be read but can be treated differently by programs. In **hard masking**, we replace the bases with Ns so that the sequence is not readable by a program (X is used in the case of protein sequences).

When two or more **SNVs** (short for **Single Nucleotide Variants**) can be confirmed to be within the same read of a genome, we can **phase** them. This means we know those variants go on the same allele or chromosome half. When we can phase two or more variants, we know we have a whole block of variants that go together. This is called a **haplotype block** as shown here:

Haplotype 1	ACGT**GGG**ACTGCCCAC**A**

Haplotype 2	ACGTACTGCCCAC**T**

Figure 7.3 – Illustration of haplotype blocks

In the preceding example, we see two haplotype blocks where variants have been phased together during assembly. We can see that GGG is inserted in the preceding sequence. At the end of the sequence, an SNV is shown with a T->A change. By knowing the phasing structure of the haplotype block, we know that the A change goes with the GGG insertion in the same sequence.

In typical approaches to genomics, the genome assembly is represented as a collapsed haplotype sequence that represents a mixture of the two alleles. This can be convenient for mapping because we only have to think about aligning our reads to one combined allele sequence, and then we just calculate SNVs against that. An SNV between the two alleles will be seen as 50%/50% variant frequency.

But to better represent a Diploid genome, we would want to provide a **diploid assembly**. In this representation, there would be two separate alleles. Some key diploid genome assemblers include Falcon (`https://pb-falcon.readthedocs.io/en/latest/`), hifiasm (`https://github.com/chhylp123/hifiasm`), and Verkko (`https://github.com/marbl/verkko`).

See also

- More information on the CHM13 genome can be found here: `https://github.com/marbl/CHM13`

- The pyfastx paper is here: Du et al, *Pyfastx: a robust Python package for fast random access to sequences from plain and gzipped FASTA/Q files*, Briefings in Bioinformatics, Jul 2021: `https://academic.oup.com/bib/article/22/4/bbaa368/6042388`

- The first T2T genomes for Chinese populations have been produced – Xiao & Yu, *T2T-YAO, T2T-SHUN and More*, Genomics, Proteomics & BioInformatics, Dec 2023: `https://pmc.ncbi.nlm.nih.gov/articles/PMC11082254/`

- Read about Verkko here: Rautiainen et al, *Telomere-to-Telomere assembly of diploid chromosomes with Verkko*, Nature Biotechnology, Feb 2023: `https://www.nature.com/articles/s41587-023-01662-6`

Working with graph genomes

As we discussed in the *Accessing genome assemblies* recipe of this chapter, traditionally, genomes have been represented as linear, haploid structures. However, this comes with significant limitations. Although you can represent an SNV this way, you cannot represent significant structural variations or major haplotype changes. Organisms have a wide array of variation of this nature, and so there is no single, true "reference genome." Graph genomes were introduced to handle this problem.

Graph genomes represent genomes as a series of nodes and edges, which allows us to traverse from common areas into alternative areas of a genome and helps us better represent diversity in the population. This improves alignment and variant calling because we can align reads to the most likely representation of their reference. Have a look at the following figure:

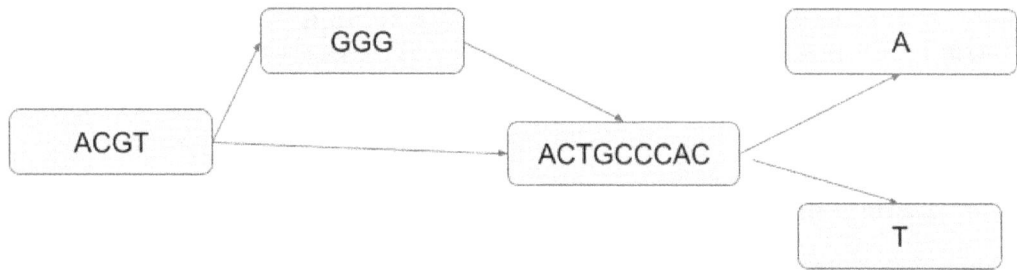

Figure 7.4 – Illustration of a graph genome

In the preceding example, we start out with ACGT but then have a choice of going through GGG first (the insertion) or straight into the next sequence. We then see ACTGCCCAC, which is common for all sequences, and then finally we have a choice of an A or T (SNV).

It should be noted that graph genomes are closely related (but not identical) to **de Bruijn graphs**, which are used to efficiently represent the k-mer paths in a genome assembly. Whereas graph genomes represent potential variation in a reference genome, de Bruijn graphs represent possible paths in a putative genome assembly and are used to help make decisions about the correct assembly. This is an important concept in bioinformatics, and you are encouraged to learn more about de Bruijn graphs here: Compeau et al, *How to apply de Bruijn graphs to genome assembly*, Nature Biotechnology, Nov 2011: `https://www.nature.com/articles/nbt.2023`.

Let's look at a simple example of a graph genome representation. We will use a package called **NetworkX** (`https://networkx.org/`), which is designed to create and manipulate complex networks.

Getting ready

Let's install NetworkX:

```
! pip install networkx
```

This will set up the library for you. We'll also use matplotlib in this recipe, which you should already have installed.

How to do it...

Here are the steps of this recipe:

1. Let's start by importing the libraries we need:

    ```
    import networkx as nx
    import matplotlib.pyplot as plt
    ```

    We'll use NetworkX to represent our graph genome, and matplotlib to plot it.

2. Now we'll write a function to set up our variation graph:

```python
def create_variation_graph():
    G = nx.DiGraph()

    G.add_node("1", seq="ATGCG")
    G.add_node("2", seq="A")
    G.add_node("2_alt", seq="T")
    G.add_node("3", seq="C")
    G.add_node("4", seq="G")
    G.add_node("5", seq="GTT")
    G.add_node("6", seq="TAA")

    G.add_edge("1", "2")
    G.add_edge("2", "3")
    G.add_edge("3", "4")
    G.add_edge("4", "6")
    G.add_edge("1", "2_alt")
    G.add_edge("2_alt", "3")
    G.add_edge("4", "5")
    G.add_edge("5", "6")

    return G
```

This code starts out by creating a **directed graph** using the NetworkX DiGraph() function. In a directed graph, the edges between the nodes have a direction, so in this case, going from left to right through the genome.

3. Next, we start building up our example graph. We use the add_node() function to set up the primary nodes containing the sequences in our example genome, including some alternate sequences at certain positions.

Then we add the connections between the nodes using the add_edge() function. We first build the reference path, which you can think of as the default genome sequence. Finally, we add edges for the alternate sequences, which show where variants may occur, creating an alternate path through the genome. We return the graph object.

4. Next, we will build a function to find the haplotypes using the graph we just made:

```python
def find_haplotypes(G, start="1", end="6"):
    paths = list(nx.all_simple_paths(
        G, source=start, target=end))
    haplotypes = []
    for path in paths:
        seq = "".join(G.nodes[node]["seq"] for node in path)
        haplotypes.append(seq)
    return haplotypes
```

This function does the following:

- Takes in a graph object and a start and end parameter.

- Uses the NetworkX `all_simple_paths()` function to return a **generator**. In Python, generators are used to return iterators. This type of function returns multiple values over time by using the `yield` statement instead of `return`. We then wrap this command in a `list()` function to turn the results into a list.

- Initializes a haplotype list.

- Loops over the paths. For each path, join the `seq` attributes of the nodes into a sequence, and then append it to the haplotypes list.

- Finally, we return the haplotypes list.

5. Now let's write a function to visualize our graph genome:

```
def visualize_graph(G):
    pos = nx.spring_layout(G)
    labels = {node: G.nodes[node]["seq"] for node in G.nodes}
    plt.figure(figsize=(8, 6))
    nx.draw(
        G, pos, with_labels=True,
        labels=labels, node_size=2000,
        node_color="skyblue", edge_color="black"
    )
    plt.title("Genome Variation Graph")
    plt.show()
```

This function first uses the NetworkX `spring_layout()` function to create a layout for the graph. It uses a **force-directed** algorithm in which nodes that have more connections are closer together. It will return the results in the `pos` variable, which will contain a **dictionary** of the nodes and their positions as (x,y) values.

6. We then create a dictionary of labels by mapping each node to its corresponding `seq` attribute.

7. We then initialize a plot using matplotlib. We use the NetworkX `draw()` function to draw the graph, giving it the following:

- Our input graph object

- The positions of the nodes

- A flag to display labels

- The labels (sequences) of the nodes

- Node size and color parameters

8. Finally, we add a title to the plot and show it.

Okay, we are ready to put it all together!

```
if __name__ == "__main__":
    graph = create_variation_graph()
    haplotypes = find_haplotypes(graph)
    print("All possible haplotypes:")
    for haplotype in haplotypes:
        print(haplotype)
    visualize_graph(graph)
```

We will call our function to create the variation graph and then use the find_haplotypes() function to return all our haplotypes, loop over them, and print them out. Lastly, we will run our plotting function. Here is what we get:

```
All possible haplotypes:
ATGCGACGTAA
ATGCGACGGTTTAA
ATGCGTCGTAA
ATGCGTCGGTTTAA
```

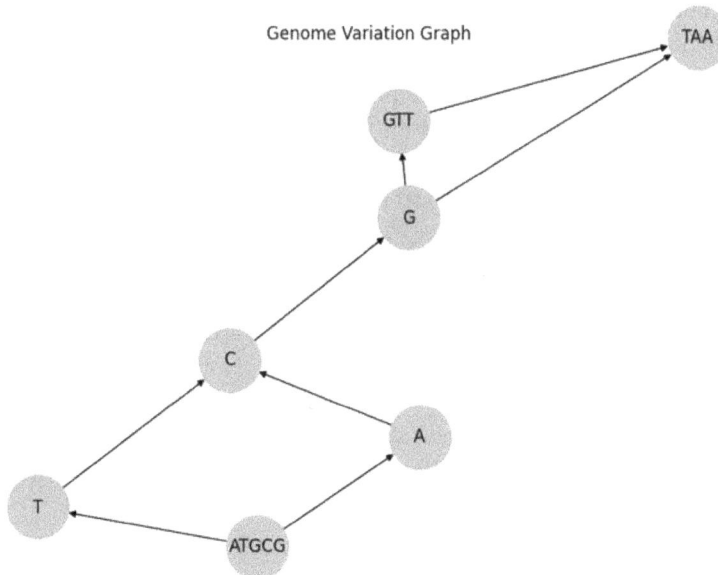

Figure 7.5 – Haplotypes and network output for the genome graph

Now you can see what the different potential haplotypes look like for each scenario. If we traverse this graph in all possible ways from left to right, we get the preceding sequences as potential outcomes. These are the haplotypes. We can also see the graphical representation of this and look at the possible paths.

There's more...

Graph genomes are stored in file formats specialized for their representation. The most widely used is **Graphical Fragment Assembly** (**GFA**). Others include FASTG and VG, as well as XG/GWBT for compressed graph formats.

A **pangenome** represents a graph genome that goes beyond a single individual to record variation within a species. In this way, we can represent variation that is present in some individuals in a species but not others. This gives us a compact representation of many genomes.

See also

- GenGraph is a Python library for dealing with Graph genomes – Ambler et al, *GenGraph: a Python module for the simple generation and manipulation of genome graphs*, BMC BioInformatics, Oct 2019: https://bmcbioinformatics.biomedcentral.com/articles/10.1186/s12859-019-3115-8

- To learn more about generators, read here: https://realpython.com/introduction-to-python-generators/

- Go deeper into the science by reading: Liao et al, *A draft human pangenome reference*, Nature, May 2023: https://www.nature.com/articles/s41586-023-05896-x

- Verkko2 integrates de Bruijn graph approaches to further improve assembly contiguity – Antipov et al, *Verkko2 integrates proximity-ligation data with long-read de Bruijn graphs for efficient telomere-to-telomere genome assembly, phasing, and scaffolding*, Genome Research, May 2025: https://genome.cshlp.org/content/35/7/1583.short

Long-read assembly with Raven

Next, we'll look at assembly in more detail. In assembly, we are tasked with finding reads that go together by overlapping them and continuously extending them until we can build a single contiguous region, or **contig**. When differences are found, we have to decide whether those differences should cause us to break up the contig into different alleles, or whether we are seeing a completely distinct (but highly similar) region of the genome. **Paralogs**, which are nearly identical copies of a gene that have been copied into different regions of the genome, can confuse this process.

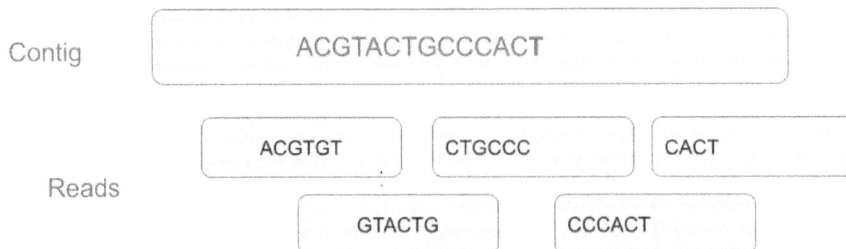

Figure 7.6 – Illustration of genome assembly

In the preceding example, you can see how we can walk across a series of reads and find overlapping sequences. We keep going until we can build up the contig sequence.

In this recipe, we'll use Raven (`https://github.com/lbcb-sci/raven`) to assemble a genome. Raven excels at using the type of error-prone long reads produced by third-generation sequencing technologies (Vaser & Sikic, *Time- and memory-efficient genome assembly with Raven*, Nature Computational Science, May 2021: `https://www.nature.com/articles/s43588-021-00073-4`).

Getting ready

First, let's download some long-read data that we can work with:

```
! wget https://nanopore.s3.climb.ac.uk/MAP006-1_2D_pass.fasta
! mv MAP006-1_2D_pass.fasta data/
```

Next, let's install Raven:

```
! git clone https://github.com/lbcb-sci/raven.git && cd raven
! cmake -S ./ -B./build -DRAVEN_BUILD_EXE=1 -DCMAKE_BUILD_TYPE=Release
! cmake --build build
```

You can also put Raven in your path (note that you may need to update the command below based on your particular path to Raven):

```
! echo 'export PATH=$PATH:~/work/CookBook/Ch07/raven/build/bin' >>
~/.zshrc
! source ~/.zshrc
```

As an alternative method for installing Raven, you can try this:

```
conda install bioconda::raven-assembler
```

When using the `conda` method, Raven should automatically be in your path.

You can now check that Raven is installed by running the following:

```
! raven -help
```

You should see the `help` output for Raven.

How to do it...

Here are the steps to try this recipe:

1. We will set up a function to run Raven on our input long-read data:

```python
def run_raven(input_fasta, output_fasta):
    try:
        print(f"Running Raven on {input_fasta}...")
        with open(output_fasta, "w") as output_file:
            command = ["raven", input_fasta]
            subprocess.run(
                command, stdout=output_file,
                check=True
            )
        print(
            f"Assembly completed."
            f"Output saved to {output_fasta}"
        )
    except FileNotFoundError:
        print(
            "Error: Raven is not installed or "
            "not found in the system PATH."
        )
    except subprocess.CalledProcessError as e:
        print(f"Error running Raven: {e}")
    except Exception as e:
        print(f"Unexpected error: {e}")
```

This function will open our input FASTA file and construct a command for Raven. The command is then run using the subprocess() function. If the command is successful, we will display a message. We also include some error handling in case the file is not found, Raven is not installed, or we encounter an error.

2. Now let's run our assembly function:

```python
if __name__ == "__main__":
    input_fasta = "data/MAP006-1_2D_pass.fasta"
    output_fasta = "assembly.fasta"
    run_raven(input_fasta, output_fasta)
```

3. Move the result to our output subdirectory:

```
! mkdir -p output
! mv assembly.fasta output/ecoli-assembly.fasta
```

That's it! We now have our assembly. In the next recipe, we'll examine the assembly results further.

There's more...

There are many other genome assemblers. Spades (`https://github.com/ablab/spades`) is a popular one aimed primarily at short-read data, which can also take in long-read data for **hybrid assembly**. Canu (`https://canu.readthedocs.io/en/latest/quick-start.html`) specializes in using PacBio or Oxford Nanopore long-reads – Koren et al, "Canu: scalable and accurate long-read assembly via adaptive k-mer weighting and repeat separation," Genome Research, Mar 2017: `https://genome.cshlp.org/content/27/5/722.short`. Canu is descended from the **Celera Assembler**, which was used heavily in the original human genome effort. Recently, HiCanu was introduced, which utilizes the latest PacBio HiFi reads, which are higher-accuracy consensus-based long reads – see Nurk et al, *HiCanu: accurate assembly of segmental duplications, satellites, and allelic variants from high-fidelity long reads*, Genome Research, Aug 2020: `https://genome.cshlp.org/content/30/9/1291.short`.

More and more, we see assemblers being able to take in multiple types of data, including Hi-C data, which provides 3D information on genome structure by obtaining information on chromosomal contacts in the cell – Simkova et al, *Hi-C techniques: from genome assemblies to transcription regulation*, Journal of Experimental Botany, September 2024: `https://academic.oup.com/jxb/article/75/17/5357/7617848`. Hifiasm and Verkko are some of the best examples of these types of assemblers. Scientists are even beginning to build three-dimensional genome assemblies – Wang & Cheng, *Reconstructing 3D chromosome structures from single-cell Hi-C data with SO(3)-equivariant graph neural networks*, NAR Genomics and Bioinformatics, March 2025: `https://academic.oup.com/nargab/article/7/1/lqaf027/8090331`.

Given the advances in sequencing technology, it is now quite routine to obtain fully closed bacterial assemblies (meaning the circular bacterial contig is completely assembled with no gaps) – Moss et al, *Complete, closed bacterial genomes from microbiomes using nanopore sequencing*, Nature Biotechnology, February 2020: `https://www.nature.com/articles/s41587-020-0422-6`. Eukaryotic genomes are more complex, but the use of long-read technology is leading to the creation of more and more T2T genomes even in complex species such as plants – Garg et al, *Unlocking plant genetics with telomere-to-telomere genome assemblies*, Nature Genetics, July 2024: `https://www.nature.com/articles/s41588-024-01830-7`.

In the next section, we'll see how to assess the quality of our genome assemblies.

See also

- B-assembler is specialized for circular bacterial genome assembly: Huang et al, *B-assembler: a circular bacterial genome assembler, BMC Genomics, May 2022*: `https://bmcgenomics.biomedcentral.com/articles/10.1186/s12864-022-08577-7`

- Read a paper about using Hi-C data in genome assembly: Kronenberg et al, *Extended haplotype-phasing of long-read de novo genome assemblies using Hi-C*, Nature Communications, Apr 2021: `https://www.nature.com/articles/s41467-020-20536-y`

Assessing genome quality with QUAST

Once a genome assembly has been produced, we need to understand its quality. This is typically assessed through a variety of metrics such as the number of contigs, N50, or the number of genes predicted within the genome.

One of the key tools used for genome assessment is **QUAST** (**QUality ASessement Tool**). QUAST (https://quast.sourceforge.net/) can evaluate the quality of an assembly using a variety of metrics and can support microbial genomes, metagenomes, and larger genomes. In many cases, we may also have a **reference genome** available. This would either be a gold standard public genome or perhaps just the best genome you have been able to assemble so far. If given a reference genome, QUAST can also help you compare your current assembly to it, to see if you have structural differences, have made progress in joining contigs, or have mis-assembled regions.

Okay, let's see what your Raven E. coli assembly looks like with QUAST!

Getting ready

Let's install QUAST first:

```
! pip install quast
```

Check that it is working:

```
! quast.py --version
```

Now we are ready to work with QUAST!

How to do It...

Here are the steps to try this recipe:

1. Now that we have QUAST installed, let's write a function to use it. First, we will import our library:

    ```
    import subprocess
    ```

2. We only need the subprocess library for this work. Let's write a function to run QUAST:

```python
def run_quast(
    assembly_file,
    reference_file=None,
    output_dir="quast_output"
):
    try:
        command = [
            "quast.py", assembly_file,
            "-o", output_dir
        ]
        if reference_file:
            command.extend(["-r", reference_file])

        print(f"Running QUAST...\nCommand: {' '.join(command)}")
        subprocess.run(command, check=True)
        print(f"QUAST analysis complete. "
                f"Results saved in: {output_dir}")
    except FileNotFoundError:
        print("QUAST is not installed or "
                "not found in the system PATH.")
    except subprocess.CalledProcessError as e:
        print(f"Error running QUAST: {e}")
    except Exception as e:
        print(f"Unexpected error: {e}")
```

This function takes in an assembled genome file, a reference file for comparison (which we won't use in this case), and a directory to save the outputs in. We then build a QUAST command and run it with some basic error handling in place to check if QUAST is installed and for any errors when running the command.

Okay, let's run our function!

```python
if __name__ == "__main__":
    assembly = "output/ecoli-assembly.fasta"
    reference = None
    output = "quast_results"
    run_quast(assembly, reference, output)
```

We supply our E. coli assembly from the previous recipe. We won't be using a reference sequence for comparison in this example. We'll save our outputs into the quast_results subdirectory.

Let's review our output!

```
! open quast_results/report.html
```

This will open the QUAST report HTML file. Let's look:

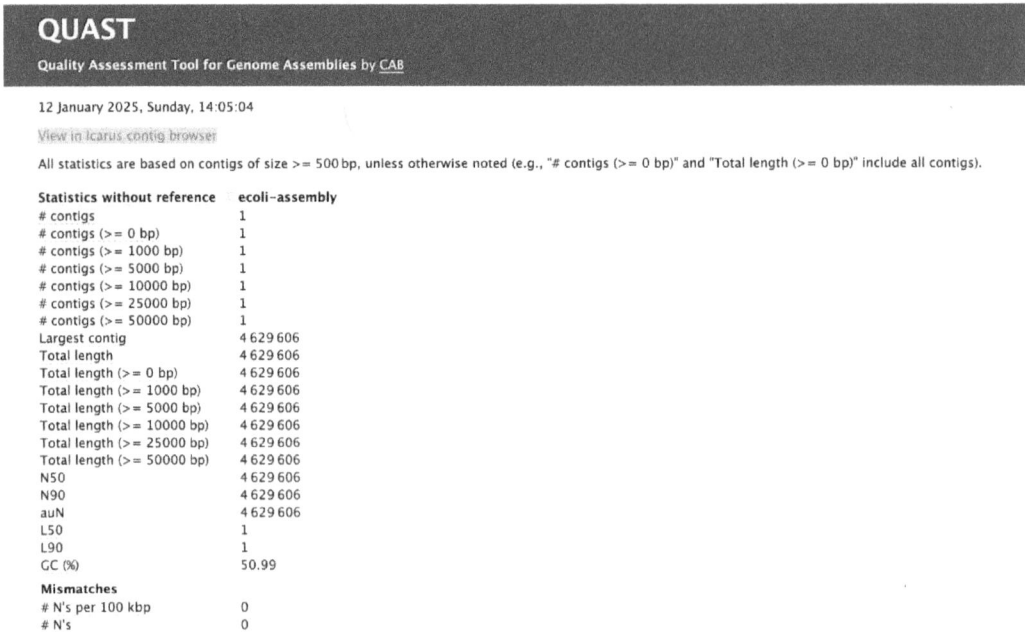

Figure 7.7 – QUAST results for E. coli

Here, we can see the QUAST output. We report a number of contigs – one in this case, as we were able to completely assemble a microbial genome! Keep in mind that we were able to do this because of the power of long reads. If this had been done using short reads, we may have ended up with a much more fragmented assembly!

You will see QUAST reports several other metrics, such as the N50, L50, GC percentage, and so on. Several graphs are also provided.

If you look at the upper left, you can click on a link to take you to the **Icarus Contig Browser** view:

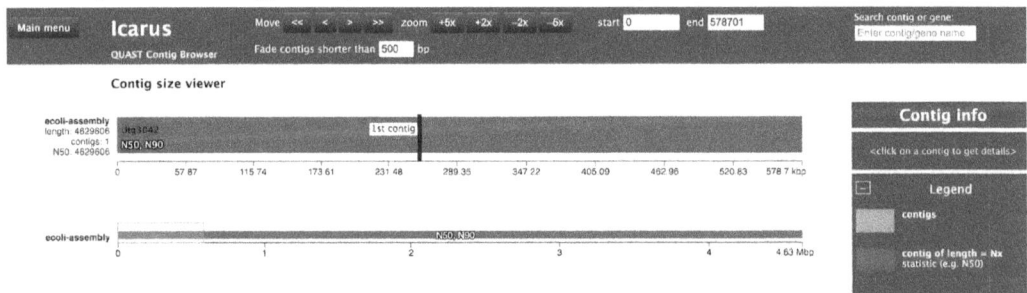

Figure 7.8 – Icarus Contig Browser for E. coli assembly

Icarus Contig Browser (`https://quast.sourceforge.net/icarus.html`) allows you to review your assembly. If a reference genome is provided, you can inspect how your contigs align to the reference.

Using QUAST, you can play around with assembly parameters, add more data, or take other approaches to build and refine your genome assembly.

There's more...

The example presented here was simple, of course, to make it approachable in this lesson. In a real-life example, you might be dealing with a complex plant genome, which would be larger and harder to get into full-size contigs. You might try to visually examine the placement of contigs against a reference and order them. You might also be actively engaged with lab staff to generate more data, or different types of data, to combine them into the assembly and improve it further. These are good examples of the tools at your disposal for these types of tasks.

There are many other tools for assessing genome quality. **Benchmarking Universal Single-Copy Orthologs** (**BUSCO**) looks for genes that are expected to be found in a given species (`https://busco.ezlab.org/`). In this way, it goes beyond just looking at assembly continuity to assess whether the assembly would support gene calling.

There are also tools for evaluating graph genome assemblies. Gretl is a tool for evaluating and visualizing the structures of graph genomes – Vorbrugg et al, *Gretl - variation GRaph Evaluation TooLkit*, BioInformatics, January 2025: `https://academic.oup.com/bioinformatics/article/41/1/btae755/7932228`

As we mentioned, human genome assemblies are continuing to improve dramatically, largely due to long-read technology and the use of graph genomes to better represent variation in the population. As the assemblies improve, the variant calling approaches we discussed in *Chapter 5* will continue to improve and allow for better resolution of not only SNVs, but also repeat expansions and structural variations. For this type of assessment, the **Genome In A Bottle** (**GIAB**) consortium provides standard reference materials with known variants (`https://www.nist.gov/programs-projects/genome-bottle`).

See also

- A good article on genome quality assessment: Jauhal & Newcomb, *Assessing genome assembly quality prior to downstream analysis: N50 versus BUSCO*, Molecular Ecology Resources, February 2021: `https://onlinelibrary.wiley.com/doi/abs/10.1111/1755-0998.13364`

- This article discusses new human genome assemblies and GIAB – Dwarshuis et al, *The GIAB genomic stratifications resource for human reference genomes*, Nature Communications, October 2024: `http://nature.com/articles/s41467-024-53260-y`

Get This Book's PDF Version and Exclusive Extras

Scan the QR code (or go to `packtpub.com/unlock`). Search for this book by name, confirm the edition, and then follow the steps on the page.

Note: Keep your invoice handy. Purchases made directly from Packt don't require an invoice.

8

Accessing Public Databases

In this chapter, we will see how to access the tremendous volume of information from previous sequencing and genome annotation projects. We'll cover how to get access to genomic, **RiboNucleic Acid (RNA)**, and protein data.

There is a wealth of public data sources available to bioinformaticians these days. The **National Center for Biotechnology Information (NCBI)** houses GenBank, RefSeq, and other key sequence data sources. It holds protein structural data, taxonomies, variant information, and scientific references as well - `https://www.ncbi.nlm.nih.gov/`. It also houses Entrez - `https://www.ncbi.nlm.nih.gov/search/`, which provides a unified search across numerous NCBI databases.

The UCSC Genome Database - `https://genome.ucsc.edu/` houses a popular genome browser for major organisms, comparative genomics data, and tracks for regulatory elements, clinical variations, and more.

Ensembl - `https://www.ensembl.org/index.html?redirect=no` - provides genomic data, annotations, and comparative genomics viewers for numerous organisms. It includes data on regulatory elements and gene expression patterns as well.

The **Protein Data Bank (PDB)** and **UniProt (Universal Protein Resource)** hold a wealth of protein sequences. Find them at: `https://www.rcsb.org/` and `https://www.uniprot.org/` respectively.

In this chapter, we will learn about the sources of public data available for bioinformatics and look at tools to programmatically access this data.

We will cover the following recipes in this chapter:

- Accessing GenBank and navigating NCBI databases
- Using the Sequence Read Archive
- Using PDB and UniProt

Technical requirements

In this chapter, we'll use the following tools and packages:

- Biopython
- SRA tools
- NCBI BLAST

You'll be instructed on how to install the relevant tools in each Getting ready section. The code for this chapter can be found at `https://github.com/PacktPublishing/Bioinformatics-with-Python-Cookbook-Fourth-Edition/tree/main/Ch08`.

Remember to activate your conda environment before beginning the recipes, like this:

```
conda activate bioinformatics_base
```

Or, if you would like to set up a conda environment specific to this chapter, before activating bioinformatics_base, run the following:

```
conda create -n ch08-databases --clone bioinformatics_base
conda activate ch08-databases
```

You will be able to install the packages for the chapter as you go, or you can use the YAML file provided in the repository:

```
conda env update --file ch08-databases.yml
```

Accessing GenBank and navigating NCBI databases

Although you may have your own data to analyze, you will probably need existing genomic datasets. Here, we will look at how to access such databases from NCBI. We will not only discuss GenBank but also other databases from NCBI. Many people refer (wrongly) to the whole set of NCBI databases as GenBank, but NCBI includes the nucleotide database and many others, for example, PubMed.

In this recipe, we will see how to use Entrez and EFetch to search for and retrieve genetic sequences.

Getting ready

We will use Biopython, which you installed in *Chapter 1*, *Computer Specifications and Python Setup*. Biopython provides an interface to **Entrez**, the data retrieval system made available by NCBI.

This recipe is made available in the Ch08/Ch08-1-genbank-ncbi.ipynb file.

How to do it...

Here are the steps to try this recipe:

1. We will start by importing the relevant module and configuring the email address:

    ```
    from Bio import Entrez, SeqIO
    Entrez.email = 'put@your.email.here'
    ```

    Remember to set your email address in the section above where you see put@your.email.here.

2. Next, let's make our output directory:

    ```
    ! mkdir -p output
    ```

3. We will now try to find the **chloroquine resistance transporter** (**CRT**) gene in *Plasmodium falciparum* (the parasite that causes the deadliest form of malaria) from the nucleotide database:

    ```
    handle = Entrez.esearch(
        db='nucleotide',
        term='CRT[Gene Name] AND "Plasmodium falciparum"[Organism]'
    )
    rec_list = Entrez.read(handle)
    if int(rec_list['RetMax']) < int(rec_list['Count']):
        handle = Entrez.esearch(
            db='nucleotide',
            term=('CRT[Gene Name] AND'
                '"Plasmodium falciparum"[Organism]'),
            retmax=rec_list['Count']
        )
        rec_list = Entrez.read(handle)
    ```

    We will search the nucleotide database for our gene and organism (for the syntax of the search string, check the NCBI website). Then, we will read the result that is returned. Note that the standard search will limit the number of record references to 20, so if you have more, you may want to repeat the query with an increased maximum limit. In our case, we will actually override the default limit with retmax. The Entrez system provides quite a few sophisticated ways to retrieve a large number of results (for more information, check the Biopython or NCBI Entrez documentation). Although you now have the **identifiers** (**IDs**) of all of the records, you still need to retrieve the records properly.

4. Now, let's try to retrieve all of these records. The following query will download all matching nucleotide sequences from GenBank, which is 1,374 at the time of writing this book:

    ```
    id_list = rec_list['IdList']
    hdl = Entrez.efetch(
        db='nucleotide', id=id_list, rettype='gb'
    )
    ```

Be careful with this technique, because you will retrieve a large number of complete records, and some of them will have fairly large sequences inside. You risk downloading a lot of data (which would be a strain both on your side and on the NCBI servers).

There are several ways around this. One way is to make a more restrictive query and/or download just a few at a time, and stop when you have found the one that you need. The precise strategy will depend on what you are trying to achieve. In any case, we will retrieve a list of records in the GenBank format (which includes sequences, plus a lot of interesting metadata).

5. Let's read and parse the result:

    ```
    recs = list(SeqIO.parse(hdl, 'gb'))
    ```

    Note that we have converted an iterator (the result of SeqIO.parse) to a list. The advantage of doing this is that we can use the result as many times as we want (for example, iterate many times over), without repeating the query on the server.

    This saves time, bandwidth, and server usage if you plan to iterate many times over. The disadvantage is that it will allocate memory for all records.

    If you are doing interactive computing, you will probably prefer to have a list (so that you can analyze and experiment with it multiple times), but if you are developing a library, an iterator will probably be the best approach.

6. We will now just concentrate on a single record. This will only work if you used the exact same preceding query:

    ```
    for rec in recs:
        if rec.name == 'KM288867':
            break
    print(rec.name)
    print(rec.description)
    ```

    The rec variable now has our record of interest. The rec.description file will contain its human-readable description.

7. Now, let's extract some sequence features that contain information such as gene products and exon positions on the sequence:

    ```
    for feature in rec.features:
        if feature.type == 'gene':
            print(feature.qualifiers['gene'])
        elif feature.type == 'exon':
            loc = feature.location
            print(loc.start, loc.end, loc.strand)
        else:
            print('not processed:\n%s' % feature)
    ```

If the `feature.type` value is `gene`, we will print its name, which will be in the qualifiers dictionary. We will also print all the locations of exons. Exons, as with all features, have locations in this sequence: a start, an end, and the strand from which they are read. While all the start and end positions for our exons are `ExactPosition`, note that Biopython supports many other types of positions. One type of position is `BeforePosition`, which specifies that a location point is before a certain sequence position. Another type of position is `BetweenPosition`, which gives the interval for a certain location start/end. There are quite a few more position types; these are just some examples.

Coordinates will be specified in such a way that you will be able to easily retrieve the sequence from a Python array with ranges, so generally, the start will be one before the value on the record, and the end will be equal. The issue of coordinate systems will be revisited in future recipes.

For other feature types, we simply print them. Note that Biopython will provide a human-readable version of the feature when you print it.

We will now look at the annotations on the record, which are mostly metadata that is not related to the sequence position:

```
for name, value in rec.annotations.items():
    print('%s=%s' % (name, value))
```

The syntax of the `print` statement above `%s=%s` means to create two string placeholders, and then fill them in with values, in this case, the `name` and `value` variables. So we'll get a printout of `name=value...` from this.

Note that some values are not strings; they can be numbers or even lists (for example, the taxonomy annotation is a list).

Last but not least, you can access a fundamental piece of information—the sequence:

```
print(len(rec.seq))
```

This will print out the length of your sequence record.

There's more...

Here are a few more interesting genes you might want to try downloading:

- The human TP53 tumor gene: `https://www.ncbi.nlm.nih.gov/datasets/gene/7157/`

- The BRCA1 breast cancer gene: `https://www.ncbi.nlm.nih.gov/datasets/gene/672/`

- The ApoE gene, which is involved in Alzheimer's disease: `https://www.ncbi.nlm.nih.gov/datasets/gene/348/`

There are many more databases at NCBI. You will probably want to check the **Sequence Read Archive** database (previously known as **Short Read Archive**) if you are working with **NGS** (short for **Next Generation Sequencing**) data. (We'll cover this more in the next recipe.) The **SNP** (short for **Single Nucleotide Polymorphism**) database contains information on SNPs, whereas the protein database has protein sequences, and so on. A full list of databases in Entrez is linked in the *See also* section of this recipe.

Another database that you probably already know about with regard to NCBI is PubMed, which includes a list of scientific and medical citations, abstracts, and even full texts. You can also access it via Biopython. Furthermore, GenBank records often contain links to PubMed. For example, we can perform this on our previous record, as shown here:

```
from Bio import Medline
refs = rec.annotations['references']
for ref in refs:
    if ref.pubmed_id != '':
        print(ref.pubmed_id)
        handle = Entrez.efetch(
            db='pubmed',
            id=[ref.pubmed_id],
            rettype='medline',
            retmode='text'
        )
        records = Medline.parse(handle)
        for med_rec in records:
            for k, v in med_rec.items():
                print('%s: %s' % (k, v))
```

This will take all reference annotations, check whether they have a PubMed ID, and then access the PubMed database to retrieve the records, parse them, and then print them.

The output per record is a Python dictionary. Note that there are many references to external databases on a typical GenBank record.

Another important tool we used here is EFetch, which is part of the Entrez E-utilities (https://www.ncbi.nlm.nih.gov/books/NBK25501/). These tools can be used to programmatically access sequences, annotations, and publications.

Another great genomics resource is the **Genomes Online Database** (GOLD). GOLD (https://gold.jgi.doe.gov/) tracks the status of numerous sequencing projects and houses data from metagenomics projects and environmental samples.

There are also numerous organism-specific databases and resources out there for you. For example, MaizeGDB (https://www.maizegdb.org/) organizes data for Corn. FlyBase (https://flybase.org/) has information on *Drosophila* genomes. These databases can be a tremendous resource for you when studying particular organisms.

Each year, **Nucleic Acids Research** (**NAR**) publishes an excellent review of bioinformatics databases. They maintain a comprehensive list here: `https://www.oxfordjournals.org/nar/database/c/`.

See also

- You can find more examples on the Biopython tutorial at `https://biopython.org/docs/latest/Tutorial/index.html`.

- A list of accessible NCBI databases can be found at `https://www.ncbi.nlm.nih.gov/search/`.

- A great **question and answer** (**Q&A**) site where you can find help for your problems with databases and sequence analysis is Biostars (`https://www.biostars.org/`). You can use it for all of the content in this book, not just for this recipe.

- GenomeKit is a Python library for accessing genomic data: `https://github.com/deepgenomics/GenomeKit`.

- BioTite also provides tools for accessing genomic databases: `https://www.biotite-python.org/latest/index.html`.

Using the Sequence Read Archive

We often want to retrieve raw data, such as FASTQ data, either for testing purposes or to obtain data for an organism, or from publicly available experiments. The **Sequence Read Archive** (**SRA**) from the NCBI provides a huge collection of sequencing data from numerous studies, and includes DNA, RNA, and metagenomic data from multiple types of platforms.

Getting ready

You will want to make sure that the SRA tools and `fasterq-dump` are installed and in your `PATH`. We briefly covered this in *Chapter 5*, *Alignment and Variant Calling*. If you have not already performed this installation, please refer back to that recipe and install the SRA Toolkit now.

If `fasterq-dump` is not in your `PATH`, you may have trouble with the code in this recipe. To make sure the SRA Toolkit and `fasterq-dump` are in your `PATH`, you can do the following. Remember, you can do this from the terminal. If you prefer to run it in your notebook, just add an exclamation point in front of the command.

```
echo 'export PATH=$PATH:~/Software/sratoolkit.3.1.1-mac-x86_64/bin' >>
~/.zshrc
source ~/.zshrc
```

Note that the preceding method for setting the `PATH` is macOS-specific. It may be different depending on your operating system.

Let's check that `fasterq-dump` is working:

```
fasterq-dump -h
```

You should see the help output from `fasterq-dump` describing its options.

Next, we will install the `pysradb` library (`https://github.com/saketkc/pysradb`). This provides a Python wrapper to obtain raw data and metadata from the SRA (jumping back into our notebook now):

```
! pip install pysradb
```

You can find the code for this recipe in `Ch08/Ch08-2-using-sra.ipynb`.

How to do it...

We will write some wrappers around `pysradb` to search for and download data from SRA. Let's see the steps to do this:

1. First, let's import our libraries:

    ```
    import os
    import subprocess
    from pysradb.sraweb import SRAweb
    ```

    This brings in the `SRAweb` module for us.

2. Now, let's define a function to fetch the metadata, given an SRA accession number, using this module:

    ```
    def fetch_sra_metadata(sra_accession):
        db = SRAweb()
        metadata = db.sra_metadata(
            sra_accession, detailed=True)
        return metadata
    ```

    This function will first initialize a connection to the `SRAweb` database. This database, called `SRAdb` is part of a project to provide a SQLite database that tracks the metadata of objects stored in the sequence archive. We can then use this to make a metadata retrieval call, giving the desired accession number. Finally, we return the metadata as a pandas DataFrame.

3. Next, we will define a function to perform a download of SRA data:

    ```
    def download_sra_run(run_accession, output_dir="sra_data"):
        if not os.path.exists(output_dir):
            os.makedirs(output_dir)
        try:
            print(f"Downloading SRA run {run_accession}...")
    ```

```
        subprocess.run([
            "fasterq-dump", run_accession,
            "--outdir", output_dir, "--split-files"
        ], check=True)
        print(f"Download complete. Files saved in {output_dir}")
    except subprocess.CalledProcessError as e:
        print(f"Error downloading {run_accession}: {e}")
```

This function will first create the output directory if it does not already exist. It then runs `fasterq-dump` to download the file for the accession number provided and puts it into the output directory.

4. Now, let's try out our functions:

```
def main():
    sra_accession = "SRR536546"
    metadata = fetch_sra_metadata(sra_accession)
    print("Metadata for the accession:")
    print(metadata)
        if not metadata.empty:
        first_run = metadata["run_accession"].iloc[0]
        download_sra_run(first_run)
    else:
        print("No runs found for this accession.")
if __name__ == "__main__":
    main()
```

We will supply an accession to try out. This very small accession should download in a reasonable time of a few minutes or less.

Accession numbers

Accession numbers are unique IDs or record numbers that are assigned to help retrieve specific information. They are used in places such as libraries or museums, but here they have been extended to include records from sequence databases. These include GenBank, NCBI, PDB, and so on. They can be used to refer to nucleotide sequences, transcripts, protein structures, and many other entities. Throughout this book, we will use the general term *accession number*.

We first fetch the metadata for the accession. Assuming that we get a valid result, we will grab the first accession number for the related data and put it into the `first_run` variable. We then call our download function for the accession. The output for this is shown here:

```
Metadata for the accession:
  run_accession study_accession  \
0     SRR536546       SRP014780


                                  study_title experiment_accession  \
0  Single Neuron Sequencing Quantifies L1 Retrotr...          SRX175596

                                  experiment_title  \
0  Genome-wide L1 insertion profiling of single n...

                                  experiment_desc organism_taxid  \
0  Genome-wide L1 insertion profiling of single n...          9606

  organism_name                library_name library_strategy  ...  \
0  Homo sapiens  4638-cortex_1-neuron_MDA_28         AMPLICON  ...

              experiment_alias    label body_site sample-type  \
0  4638-cortex_1-neuron_MDA_28  UMB4638    cortex  1-cell_MDA

                                  ena_fastq_http ena_fastq_http_1  \
0  http://ftp.sra.ebi.ac.uk/vol1/fastq/SRR536/SRR...              <NA>

  ena_fastq_http_2                                      ena_fastq_ftp  \
0              <NA>  era-fasp@fasp.sra.ebi.ac.uk:vol1/fastq/SRR536/...

  ena_fastq_ftp_1 ena_fastq_ftp_2
0            <NA>            <NA>

[1 rows x 53 columns]
Downloading SRA run SRR536546...
Download complete. Files saved in sra_data

spots read      : 3,512,181
reads read      : 3,512,181
reads written   : 3,512,181
```

Figure 8.1 – Results of the SRA query

You should see the output metadata showing the results of our query. If you look in your terminal, you should also see in your directory an `sra_data` subdirectory containing the associated FASTQ data. Let's put it in our output directory:

```
! mv sra_data output/
```

> **Note**
>
> The data for this example is ~1 GB in size. In most cases, it should take on the order of a few minutes to download.

There's more...

It is also worth taking a moment to learn about the **Basic Local Alignment Search Tool** (**BLAST**). This is a tool for aligning sequences (`https://blast.ncbi.nlm.nih.gov/Blast.cgi`). You can use it to search for your query of interest against major organism genomes. The major flavors of BLAST are as follows:

- `Blastn`: Compares nucleotides to nucleotides
- `Blastx`: Compares translations of a nucleotide query to proteins
- `Tblastn`: Compares a back-translated protein query to nucleotides
- `Blastp`: Compares proteins to proteins

Here is an example of searching using BLAST via the NCBI API:

```
from Bio.Blast import NCBIWWW
from Bio.Blast import NCBIXML

query_sequence = ">test_query\nATGGCCATTGTAATCATGTTCTAATAGTGTTCA"
result_handle = NCBIWWW.qblast("blastn", "nt", query_sequence)

with open("blast_result.xml", "w") as out_file:
    out_file.write(result_handle.read())

print("BLAST search completed! Results saved in 'blast_result.xml'")
```

This code imports the NCBI web library for using BLAST and the XML library. We run BLAST with a query sequence against the nt (nucleotide) database and save the results as an XML file.

> **Note**
> Historically these terms have been used for NCBI sequence collections - "nt" is short for the nucleotide collection, and "nr" is short for the "Non-Redundant" protein collection.

We can also parse the results of BLAST:

```
with open("blast_result.xml") as result_file:
    blast_records = NCBIXML.read(result_file)
for alignment in blast_records.alignments[:5]:
    print(f"Hit: {alignment.title}")
    for hsp in alignment.hsps:
        print(f"  Score: {hsp.score}, E-value: {hsp.expect}")
```

This is what we get:

```
Hit: gi|1338838386|ref|XM_023806334.1| PREDICTED: Paramormyrops kingsleyae T-box transcription factor TBX5-like (LO
C111840956), transcript variant X2, mRNA
; double click to hide , E-value: 1.12623
Hit: gi|1338838384|ref|XM_023806333.1| PREDICTED: Paramormyrops kingsleyae T-box transcription factor TBX5-like (LO
C111840956), transcript variant X1, mRNA
   Score: 45.0, E-value: 1.12623
Hit: gi|2647104289|gb|CP141595.1| Rossellomorea aquimaris strain Rossellomorea aquimaris S-2 chromosome, complete g
enome
   Score: 44.0, E-value: 1.12623
Hit: gi|1190964948|ref|XR_002333164.1| PREDICTED: Arabidopsis lyrata subsp. lyrata uncharacterized LOC110229766 (LO
C110229766), ncRNA
   Score: 43.0, E-value: 3.93094
Hit: gi|891573148|ref|XM_013165470.1| Schizosaccharomyces cryophilus OY26 RNA polymerase II associated Paf1 complex
(SPOG_02808), mRNA
   Score: 42.0, E-value: 3.93094
```

Figure 8.2 – BLAST results from the XML file

Let's move the file to our output directory:

```
! mv blast_result.xml output/
```

Note that the NCBI limits the number of API calls you can make to BLAST to reduce the chance of network traffic overwhelming their servers (https://docs.blastapi.io/blastdocumentation/things-you-need-to-know/limits). This should be fine for the limited example presented here. But if you are interested in running BLAST jobs with large numbers of API calls, you should check out Elastic BLAST: https://blast.ncbi.nlm.nih.gov/doc/elastic-blast/.

As you can see, there are numerous powerful tools and databases at your disposal! The information in genomics is growing exponentially, and knowing how to access and search these resources is critical to your role as a bioinformatician.

See also

- Choudhary, *pysradb: A Python package to query next-generation sequencing metadata and data from the NCBI Sequence Read Archive*, F1000 Research, April 2019: https://pmc.ncbi.nlm.nih.gov/articles/PMC6505635/

- You can learn more about BLAST here: https://conmeehan.github.io/blast+tutorial.html

Using PDB and UniProt

Proteomics is the study of proteins, including their function and structure. One of the main objectives of this field is to characterize the three-dimensional structure of proteins. One of the most widely known computational resources in the proteomics field is the **PDB**, a repository with the structural data of large biomolecules.

Let's start with something that you should be more familiar with by now: accessing databases, especially for a protein's primary structure (as in, sequences of amino acids). We'll use UniProt, a large repository of protein sequences.

You can find the code for this recipe in `Ch08/Ch08-3-pdb-uniprot.ipynb`.

How to do it...

Here are the steps to try this recipe:

1. Let's import the libraries we need:

    ```
    import requests
    import sys
    import json
    ```

    We will use the `requests` library, which provides support for HTTP requests. We will also be using **JavaScript Object Notation (JSON)**. JSON (`https://www.json.org/json-en.html`) is a very popular and lightweight format for storing information that is both human-readable and easy to parse.

2. Let's start by retrieving an example protein from UniProt. We'll get the protein for accession **P21802**. Go to `uniprot.org` and type `P21802` into the search bar. You will see this web page:

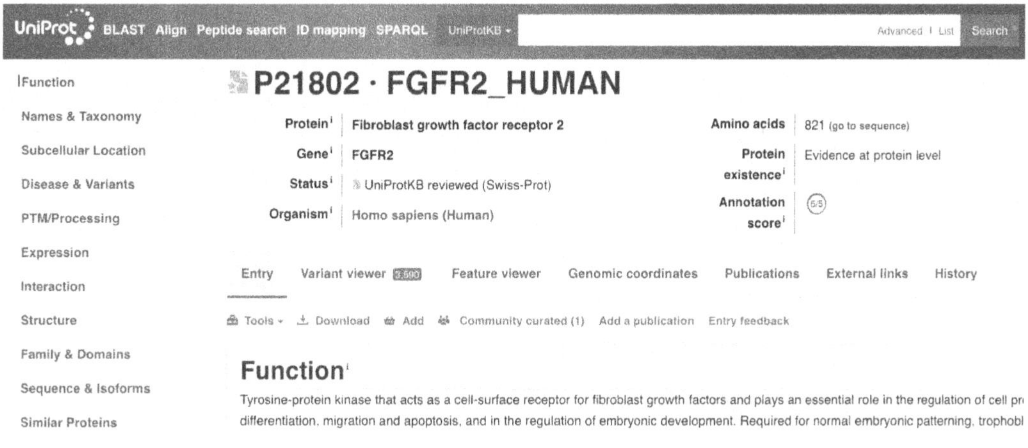

Figure 8.3 – UniProt entry for P21802

The preceding web page result shows our protein through the online interface of UniProt. It is a human fibroblast growth factor. These proteins are involved in signaling for tissue repair and wound healing, but the exact nature of the protein does not matter that much here; we are using it as an example. We can see a lot of information on the protein, such as its length in amino acids and the known evidence for it. There are also links to viewers for the protein, variants known in the protein, and more. The information you can find on UniProt can be tremendously useful to understand the protein's function, active sites, literature about the protein, and much more.

3. Next, we will define a function for retrieving protein information using an accession:

```python
def fetch_protein_data_json(accession):
    request_url = (
        f"https://www.ebi.ac.uk/proteins/api/proteins?"
        f"offset=0&size=100&accession={accession}"
    )
    headers = {"Accept": "application/json"}

    try:
        print(f"Fetching data for accession: {accession}")
        response = requests.get(
            request_url, headers=headers, timeout=30)
        response.raise_for_status()
        return response.json()

    except requests.exceptions.RequestException as e:
        print(f"Error fetching protein data: {e}")
        sys.exit(1)
```

This function first sets `request_url` to point at the UniProt database hosted at the **European Bioinformatics Institute (EBI)**. By using `offset=0`, we start with the first result, and by using `size=100`, we will return a maximum of 100 results.

4. Next, we specify our `request` header so that we will use JSON.

Then, we use `requests.get()` to retrieve the data with a timeout of 30 seconds. The result will go into the `response` variable.

We use `response.raise_for_status()` to check for any HTTP error status codes. We then return the response in JSON format. All of this is wrapped in a `try..except` block to catch any errors.

5. Next, let's create a function to save our JSON response to a file:

```python
def save_json_to_file(data, filename):
    try:
        with open(filename, "w") as json_file:
            json.dump(data, json_file, indent=4)
        print(f"Protein data saved to {filename}")
    except IOError as e:
        print(f"Error saving data to file: {e}")
        sys.exit(1)
```

This function opens the JSON output file for writing. It then uses the `json.dump()` function to take the input data and write it out to the JSON file.

6. Now, we will run our functions:

```python
def main():
    accession = "P21802"
    output_file = "protein_data.json"
    protein_data = fetch_protein_data_json(accession)
    print("Protein Data (JSON):")
    print(protein_data)
    save_json_to_file(protein_data, output_file)
if __name__ == "__main__":
    main()
```

This code will first set our accession to be P21802, which is our FGFR2 protein. We then set our output file. Next, we use our function to fetch the protein data, and we print it out. Finally, we save the protein data into our JSON file. You can use less to examine the JSON file:

```
[
    {
        "accession": "P21802",
        "id": "FGFR2_HUMAN",
        "proteinExistence": "Evidence at protein level",
        "info": {
            "type": "Swiss-Prot",
            "created": "1991-05-01",
            "modified": "2024-11-27",
            "version": 277
        },
        "organism": {
            "taxonomy": 9606,
            "names": [
                {
                    "type": "scientific",
                    "value": "Homo sapiens"
                },
                {
                    "type": "common",
                    "value": "Human"
                }
            ],
            "lineage": [
                "Eukaryota",
                "Metazoa",
                "Chordata",
                "Craniata",
                "Vertebrata",
```

Figure 8.4 – Snippet of the JSON file for P21802

You can see information on the protein from the UniProt database.

7. Finally, let's move our file into the output directory:

```
! mv protein_data.json output/
```

Now, we've seen how we can query data from UniProt and store it in a convenient and parsable format (JSON), let's next look at another very popular protein database, PDB.

We will look for a protein called 1A8M, which is a tumor necrosis factor. You can look it up on the PDB website (`https://www.rcsb.org/structure/1A8M`):

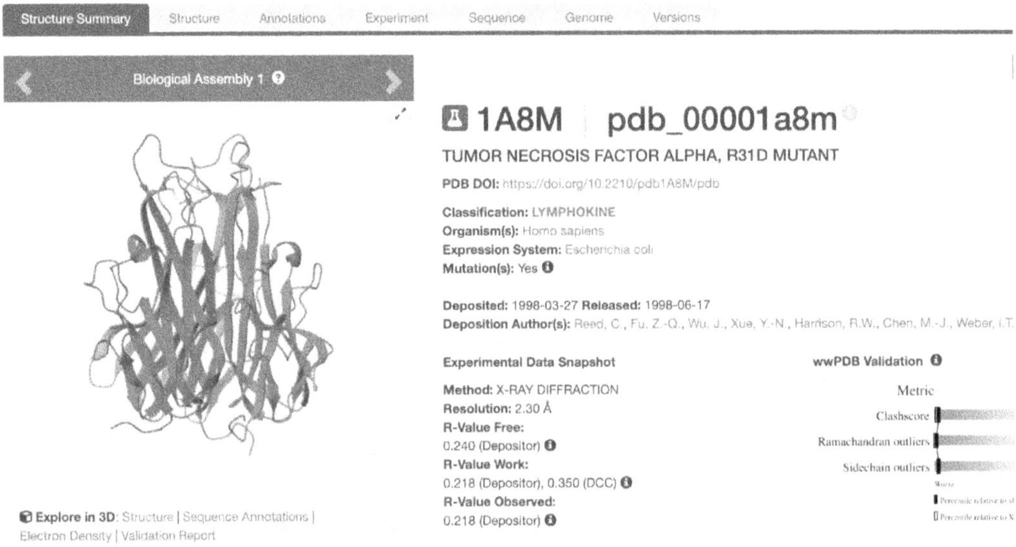

Figure 8.5 – PDB entry for tumor necrosis factor 1A8M

You can see a wealth of information about the protein here, including the 3D structure, methods used to characterize the protein, and related literature.

8. Let's see how to download data from the PDB. We will import our libraries:

```
import os
import requests
from Bio import PDB
```

We use the Biopython PDB module. This provides functions for parsing PDB files and related useful methods.

9. Next, we will define our download function:

```
def download_pdb(pdb_id, output_dir="output"):
    pdb_id = pdb_id.lower()
    base_url = "https://files.rcsb.org/download"
    metadata_url = (
        f"https://data.rcsb.org/rest/v1/core/entry/{pdb_id}")
    pdb_url = f"{base_url}/{pdb_id}.pdb"
    os.makedirs(output_dir, exist_ok=True)
    pdb_file_path = os.path.join(output_dir, f"{pdb_id}.pdb")
    response = requests.get(pdb_url)

    if response.status_code == 200:
        with open(pdb_file_path, "w") as file:
            file.write(response.text)
        print(f"PDB file saved at: {pdb_file_path}")
    else:
        print(f"Failed to download PDB file for {pdb_id}.")
    metadata_file_path = os.path.join(
        output_dir, f"{pdb_id}_metadata.json")
    response = requests.get(metadata_url)

    if response.status_code == 200:
        with open(metadata_file_path, "w") as file:
            file.write(response.text)
        print(f"Metadata saved at: {metadata_file_path}")
    else:
        print(f"Failed to download metadata for {pdb_id}.")
```

This function will take as input the four-letter PDB code and will put our files in the output directory. It ensures that the PDB ID is lowercase. It then sets the URLs for the download, including the PDB file itself and the annotation metadata. We create the output directory if needed, and then use the `requests` library as before to download the file. Finally, we download the metadata.

10. Okay, let's run the function on our protein:

```
pdb_id = "1A8M"
download_pdb(pdb_id)
```

Here is what we see when we examine the PDB file:

```
HEADER    LYMPHOKINE                              27-MAR-98   1A8M
TITLE     TUMOR NECROSIS FACTOR ALPHA, R31D MUTANT
COMPND    MOL_ID: 1;
COMPND    2 MOLECULE: TUMOR NECROSIS FACTOR ALPHA;
COMPND    3 CHAIN: A, B, C;
COMPND    4 SYNONYM: TNF-ALPHA;
COMPND    5 ENGINEERED: YES;
COMPND    6 MUTATION: YES
SOURCE    MOL_ID: 1;
SOURCE    2 ORGANISM_SCIENTIFIC: HOMO SAPIENS;
SOURCE    3 ORGANISM_COMMON: HUMAN;
SOURCE    4 ORGANISM_TAXID: 9606;
SOURCE    5 EXPRESSION_SYSTEM: ESCHERICHIA COLI;
SOURCE    6 EXPRESSION_SYSTEM_TAXID: 562
KEYWDS    LYMPHOKINE, CYTOKINE, CYTOTOXIN
EXPDTA    X-RAY DIFFRACTION
AUTHOR    C.REED,Z.-Q.FU,J.WU,Y.-N.XUE,R.W.HARRISON,M.-J.CHEN,I.T.WEBER
REVDAT   5   23-OCT-24 1A8M    1       REMARK
REVDAT   4   02-AUG-23 1A8M    1       REMARK
REVDAT   3   03-NOV-21 1A8M    1       SEQADV
REVDAT   2   24-FEB-09 1A8M    1       VERSN
REVDAT   1   17-JUN-98 1A8M    0
JRNL        AUTH   C.REED,Z.Q.FU,J.WU,Y.N.XUE,R.W.HARRISON,M.J.CHEN,I.T.WEBER
JRNL        TITL   CRYSTAL STRUCTURE OF TNF-ALPHA MUTANT R31D WITH GREATER
JRNL        TITL 2 AFFINITY FOR RECEPTOR R1 COMPARED WITH R2.
JRNL        REF    PROTEIN ENG.                  V.  10  1101 1997
JRNL        REFN                   ISSN 0269-2139
JRNL        PMID   9488135
JRNL        DOI    10.1093/PROTEIN/10.10.1101
```

Figure 8.6 – PDB file for 1A8M

You can also take a look at the metadata file. As you can see, we have now retrieved the PDB file and information for this protein. In the next chapter, we will take a deeper look at the PDB format and how to interrogate protein structures.

There's more...

Here are some other interesting proteins to download:

- The aldehyde decarbonylase from Oryza sativa (rice) - Q9AV39 - https://www.uniprot.org/uniprotkb/Q9AV39/entry

- P68871 – human hemoglobin subunit: https://www.uniprot.org/uniprotkb/P68871/entry

- P42212 – green fluorescent protein (GFP): https://www.uniprot.org/uniprotkb/P42212/entry

There are many other useful protein databases. The **Integrated Microbial Genomes** (**IMG**) database houses a large number of microbial genomes and metagenomes: `https://img.jgi.doe.gov/`. It includes an extensive set of tools for comparing microbial genomes, discovering **biosynthetic gene clusters** (**BGCs**), and performing pathway analysis.

This is the front page of the IMG website. From here, you can explore and learn about multiple functions of the IMG system:

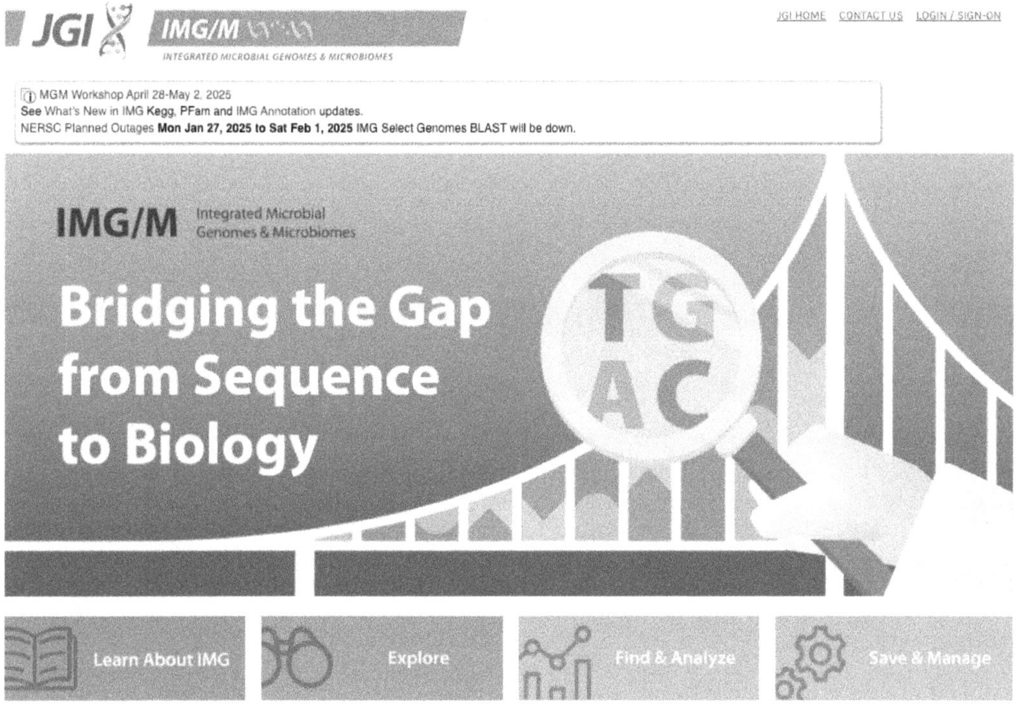

Figure 8.7 – The IMG home page

See also

- Burley et al, *RCSB Protein Data Bank: Celebrating 50 years of the PDB with new tools for understanding and visualizing biological macromolecules in 3D*, Protein Science, October 2021: `https://pubmed.ncbi.nlm.nih.gov/34676613/`

- The UniProt Consortium, *UniProt: The Universal Protein Knowledgebase in 2025*, Nucleic Acids Research, January 2025: `https://academic.oup.com/nar/article/53/D1/D609/7902999`

Get This Book's PDF Version and Exclusive Extras

Scan the QR code (or go to packtpub.com/unlock). Search for this book by name, confirm the edition, and then follow the steps on the page.

Note: Keep your invoice handy. Purchases made directly from Packt don't require an invoice.

9

Protein Structure and Proteomics

In this chapter, we will learn more about the exciting world of protein databases, structural analysis, and proteomics!

The fields of **Structural BioInformatics** and **Enzyme Engineering** have made tremendous advances in the past few years, and in this chapter, we'll get a good overview of what is happening.

Protein structures are typically determined by X-ray crystallography. In this method, proteins are first crystallized, and then an X-ray beam is shone through at different angles to build up a picture of the bonds between the atoms. There are several other methods, such as **Cryo-Electron Microscopy (Cryo-EM)**, which is good for large macromolecular complexes, and **Nuclear Magnetic Resonance (NMR)**, which is good for small molecules in solution.

Because of the intense experimental work needed when obtaining protein structures, this area has lagged behind genomics. As such, many structures were not known – enter the world of protein folding. Programs were developed that could attempt to look at a 2-D amino acid sequence and then predict the nature of the folded 3-D protein. At first, these predictions were weak, but in the last few years, AI programs have stunned the world by successfully making accurate predictions. AlphaFold (`https://deepmind.google/technologies/alphafold/`) was able to determine the structures of millions of proteins computationally (Jumper et al, "Highly accurate protein structure prediction with AlphaFold," Nature, Jul 2021): `https://www.nature.com/articles/s41586-021-03819-2`.

Because of advances in both high-throughput structure determination and the advent of AI-driven protein folding, the number of protein structures available in public databases has grown exponentially – Ahmad et al, "RCSB Protein Data Bank: revolutionising drug discovery and design for over five decades," Medical Data Mining, Vol 8, 2025: `https://www.tmrjournals.com/public/articlePDF/20250220/23eeff3cc1890ab75e547508f407cd54.pdf`.

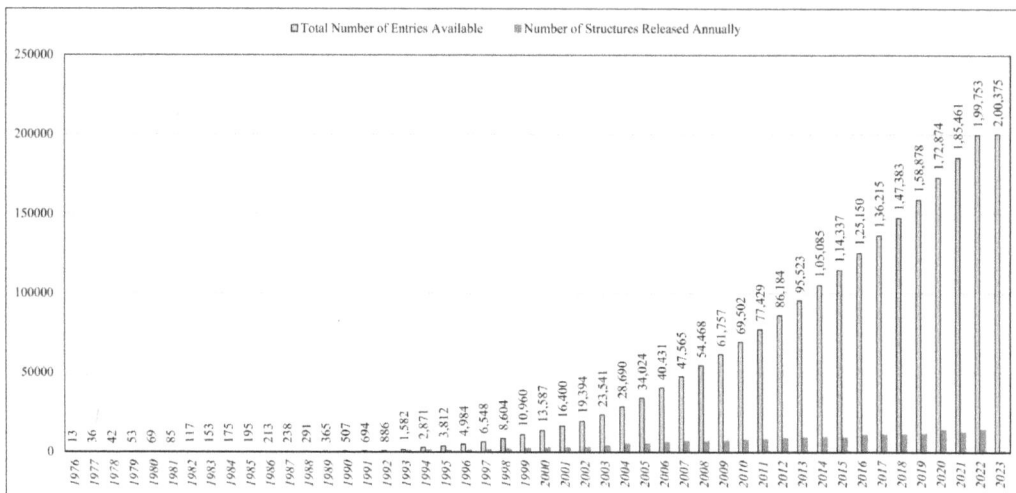

Figure 9.1 – Growth of available protein structures (Source: Ahmad et al, 2025)

With the wealth of structures and tools available, scientists can now routinely predict new protein structures and design enzymes computationally to enable capabilities never-before known.

The **PDB** or **Protein Data Bank** is a large public repository of protein structures obtained from experimental techniques. In this chapter, we will mostly focus on processing data from the PDB. We will look at how to parse PDB files, perform some geometric computations, and visualize molecules. We will use the old PDB file format because, conceptually, it allows you to perform most necessary operations within a stable environment. We will also touch briefly on the newer mmCIF format. We will use Biopython and introduce tools for protein visualization. We will also cover the exciting field of proteomics, in which protein expression levels can be measured to obtain biological insights. We will not discuss molecular docking here because that is probably more suited to a book about chemoinformatics.

In this chapter, we will cover the following recipes:

- Extracting information from PDB files
- Computing molecular distances in proteins
- Performing geometric operations
- Animating proteins
- Performing proteomics analysis

Technical requirements

In this chapter, we will use the following:

- BioPython
- NGLview
- Pyteomics

You will be instructed on how to install the necessary packages in each recipe.

You can find the code for this chapter at `https://github.com/PacktPublishing/Bioinformatics-with-Python-Cookbook-Fourth-Edition/tree/main/Ch09`.

You will want to create a `Ch09` working folder.

Remember to activate your `conda` environment before beginning the recipes, like this:

```
conda activate bioinformatics_base
```

Or, if you would like to set up a `conda` environment specific to this chapter, before activating `bioinformatics_base`, run the following:

```
conda create -n ch09-proteins --clone bioinformatics_base
conda activate ch09-proteins
```

You will be able to install the packages for the chapter as you go, or you can use the YAML file provided in the repository:

```
conda env update --file ch09-proteins.yml
```

Extracting information from PDB files

In the previous chapter, we got a brief introduction to PDB files. Let's now go deeper and look at how to extract more information from these files.

In this recipe, we will use Biopython's PDB module to examine protein structure files in detail. We'll focus on one particular protein involved in cancer. We'll see how to retrieve and parse these files and examine their contents.

We will learn how to traverse the amino acid residues in a protein model, find out their characteristics, and retrieve the atoms that make them up. We will learn about **protein chains**, which are large subsections of a protein structure. Finally, we will see how to plot the contents of the PDB file.

Getting ready

First, let's set up:

```
! mkdir -p output
```

To access the data, we will use the PDB package from BioPython (https://biopython.org/docs/1.75/api/Bio.PDB.html). This module includes functions for parsing PDB and mmCIF format files.

For this recipe, we'll work with the p53 tumor suppressor protein, which has the PDB code **1TUP**.

You should already be aware of the basic PDB data model of model, chain, residue, and atom objects. A good explanation of the Bio.PDB module can be found in *Biopython's Structural Bioinformatics FAQ* can be found at http://biopython.org/wiki/The_Biopython_Structural_Bioinformatics_FAQ.

This recipe is made available in the Ch09/Ch09-1-extracting-from-pdb.ipynb file.

How to do it...

Take a look at the following steps:

Note that Bio.PDB will take care of downloading files for you. Moreover, these downloads will only occur if no local copy is already present:

1. First, let's import our libraries:

    ```
    from Bio import PDB
    import os
    import gzip
    import shutil
    from collections import defaultdict
    ```

2. Next, we'll download our protein:

    ```
    repository = PDB.PDBList()
    parser = PDB.PDBParser(QUIET=True)
    pdb_id = "1TUP"
    repository.retrieve_pdb_file(
        pdb_id, pdir=".", file_format="pdb")
    ```

    This code will first create a PDBList object, which allows us to download PDB structures. Next, we create a PDBParser object, which allows us to read and parse PDB files. We then use the retrieve_pdb_file() function to download the 1TUP protein, in PDB format, into the current folder.

    The file should now exist in your working folder as pdb1tup.ent.

3.   Next, we'll retrieve a variety of atom-related statistics from our PDB file:

```
pdb_file = "pdb1tup.ent"
if os.path.exists(pdb_file):
    p53_1tup = parser.get_structure("P53", pdb_file)
else:
    print("Error: PDB file not found!")
    exit()
atom_cnt = defaultdict(int)
atom_chain = defaultdict(int)
atom_res_types = defaultdict(int)

if not list(p53_1tup.get_atoms()):
    print("Error: No atoms found in the structure!")
    exit()

for atom in p53_1tup.get_atoms():
    my_residue = atom.parent
    my_chain = my_residue.parent
    atom_chain[my_chain.id] += 1

    if my_residue.resname != "HOH":
        atom_cnt[atom.element] += 1

    atom_res_types[my_residue.resname] += 1

print("Residue Types:", dict(atom_res_types))
print("Chain Atom Counts:", dict(atom_chain))
print("Element Counts:", dict(atom_cnt))
```

This will print information on the atom's residue type, the number of atoms per chain, and the quantity per element, as follows:

```
{' DT': 257, ' DC': 152, ' DA': 270, ' DG': 176, 'HOH': 384,
 'SER': 323, 'VAL': 315, 'PRO': 294, 'GLN': 189, 'LYS': 135,
 'THR': 294, 'TYR': 288, 'GLY': 156, 'PHE': 165, 'ARG': 561,
 'LEU': 336, 'HIS': 210, 'ALA': 105, 'CYS': 180, 'ASN': 216,
 'MET': 144, 'TRP': 42, 'ASP': 192, 'ILE': 144, 'GLU': 297,
 'ZN': 3}
{'E': 442, 'F': 449, 'A': 1734, 'B': 1593, 'C': 1610}
{'O': 1114, 'C': 3238, 'N': 1001, 'P': 40, 'S': 48, 'ZN': 3}
```

Note that the preceding number of residues is not the proper number of residues, but the number of times that a certain residue type is referred to (it adds up to the number of atoms, not residues).

Notice the water (W), nucleotide (DA, DC, DG, and DT), and zinc (ZN) residues, which add to the amino acid ones. These elements represent additional parts of the protein structure in addition to the amino acids that make up the main protein. Water is often part of a protein structure – it can stabilize a protein via hydrogen bonds. Nucleotides are seen because this structure (1TUP) includes the p53 tumor suppressor complexed to DNA that it binds to. Zinc is especially important as this ion coordinates amino acids in the **DNA Binding Domain (DBD)** of p53 and stabilizes its activity – Ha et al, *p53 and Zinc: A Malleable Relationship*, Frontiers in Molecular Biosciences, Apr 2022: https://pmc.ncbi.nlm.nih.gov/articles/PMC9043292/.

4.   Now, let's count the instance per residue and the number of residues per chain:

```
res_types = defaultdict(int)
res_per_chain = defaultdict(int)
for residue in p53_1tup.get_residues():
    res_types[residue.resname] += 1
    res_per_chain[residue.parent.id] +=1
print(dict(res_types))
print(dict(res_per_chain))
```

The following is the output:

```
{' DT': 13, ' DC': 8, ' DA': 13, ' DG': 8, 'HOH': 384, 'SER':
 54, 'VAL': 45, 'PRO': 42, 'GLN': 21, 'LYS': 15, 'THR': 42,
 'TYR': 24, 'GLY': 39, 'PHE': 15, 'ARG': 51, 'LEU': 42, 'HIS':
 21, 'ALA': 21, 'CYS': 30, 'ASN': 27, 'MET': 18, 'TRP': 3,
 'ASP': 24, 'ILE': 18, 'GLU': 33, ' ZN': 3}
{'E': 43, 'F': 35, 'A': 395, 'B': 265, 'C': 276}
```

We can also get the bounds of a set of atoms:

```
import sys
def get_bounds(my_atoms):
    my_min = [sys.maxsize] * 3
    my_max = [-sys.maxsize] * 3
    for atom in my_atoms:
        for i, coord in enumerate(atom.coord):
            if coord < my_min[i]:
                my_min[i] = coord
            if coord > my_max[i]:
                my_max[i] = coord
    return my_min, my_max

chain_bounds = {}
for chain in p53_1tup.get_chains():
    print(chain.id, get_bounds(chain.get_atoms()))
    chain_bounds[chain.id] = get_bounds(chain.get_atoms())

print(get_bounds(p53_1tup.get_atoms()))
```

A set of atoms can be a whole model, a chain, a residue, or any subset that you are interested in. In this case, we will print boundaries for all the chains and the whole model. Numbers don't convey it so intuitively, so we will get a little bit more graphical.

To get a notion of the size of each chain, a plot is probably more informative than the numbers in the following code:

```python
import matplotlib.pyplot as plt
from mpl_toolkits.mplot3d import Axes3D
from Bio import PDB

fig = plt.figure(figsize=(16, 9))
ax3d = fig.add_subplot(111, projection='3d')
ax_xy = fig.add_subplot(331)
ax_xy.set_title('X/Y')
ax_xz = fig.add_subplot(334)
ax_xz.set_title('X/Z')
ax_zy = fig.add_subplot(337) ax_zy.set_title('Z/Y')
color = {'A': 'r', 'B': 'g', 'C': 'b',
         'E': '0.5', 'F': '0.75'}
zx, zy, zz = [], [], []

for chain in p53_1tup.get_chains():
    xs, ys, zs = [], [], []
        for i, residue in enumerate(chain.get_residues()):
            if i % 10 != 0:
                continue
            if "CA" in residue:
                ref_atom = residue["CA"]
            elif len(residue) > 0:
                ref_atom = list(residue.get_atoms())[0]
            else:
                continue
            x, y, z = ref_atom.coord
            if ref_atom.element == 'ZN':
                zx.append(x)
                zy.append(y)
                zz.append(z)
                continue

        xs.append(x)
        ys.append(y)
        zs.append(z)
```

```
    if chain.id in color:
        ax3d.scatter(xs, ys, zs, color=color[chain.id], s=10)
        ax_xy.scatter(xs, ys, marker='.',
                      color=color[chain.id], s=10)
        ax_xz.scatter(xs, zs, marker='.',
                      color=color[chain.id], s=10)
        ax_zy.scatter(zs, ys, marker='.',
                      color=color[chain.id], s=10)
ax3d.set_xlabel('X')
ax3d.set_ylabel('Y')
ax3d.set_zlabel('Z')

if zx:
    ax3d.scatter(zx, zy, zz, color='k', marker='v', s=80)
    ax_xy.scatter(zx, zy, color='k', marker='v', s=40)
    ax_xz.scatter(zx, zz, color='k', marker='v', s=40)
    ax_zy.scatter(zz, zy, color='k', marker='v', s=40)

for ax in [ax_xy, ax_xz, ax_zy]:
    ax.get_yaxis().set_visible(False)
    ax.get_xaxis().set_visible(False)

plt.show(block=False)
```

There are plenty of molecular visualization tools. Indeed, we will discuss NGLview and PyMOL later. However, `matplotlib` is enough for simple visualization. The most important point about `matplotlib` is that it's stable and very easy to integrate into reliable production code.

In the following chart, we performed a three-dimensional plot of chains, with the DNA in gray and the protein chains in different colors. We also plot planar projections (X/Y, X/Z, and Z/Y) on the left-hand side of the chart:

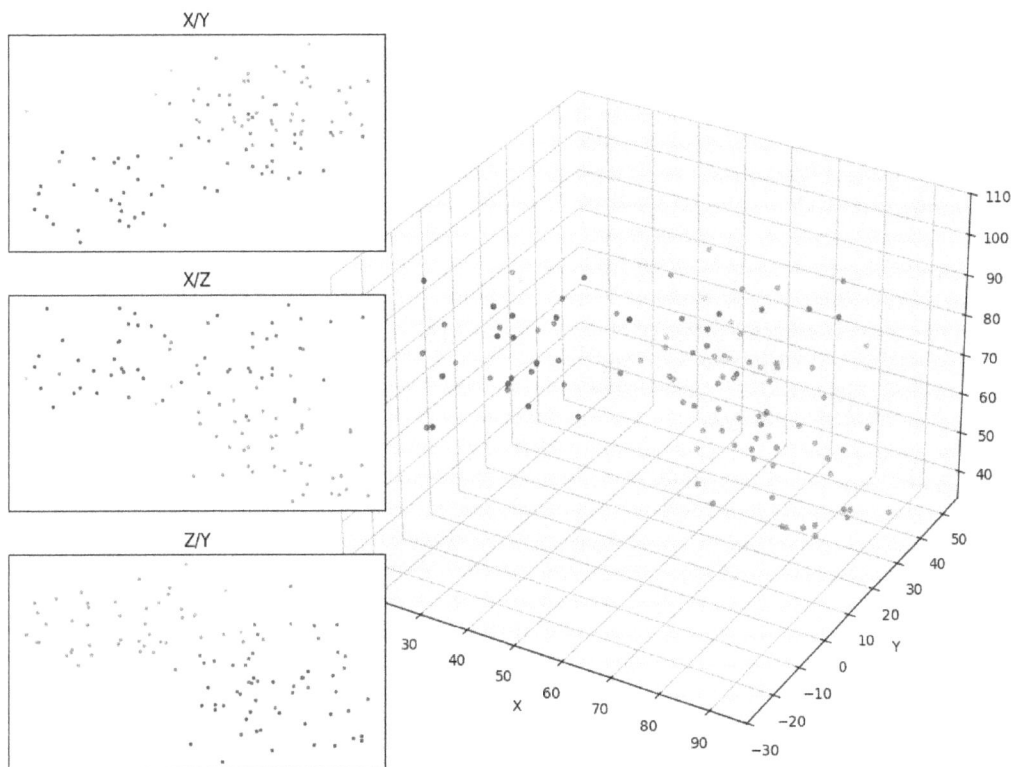

Figure 9.2 – The spatial distribution of the protein chains – the main figure is a
3D plot, and the left subplots are planar views (X/Y, X/Z, and Z/Y)

There's more...

The PDB parser is incomplete. It's not very likely that a complete parser will be seen soon, as the community is migrating to the mmCIF format.

Although the future is the mmCIF format (`https://mmcif.wwpdb.org/`), PDB files are still around. Conceptually, many operations are similar after you have parsed the file.

Computing molecular distances on a PDB file

Here, we will find atoms closer to three zincs in the 1TUP model. We will consider several distances to these zincs. We will take this opportunity to discuss the performance of algorithms.

In this recipe, we will use the Biopython PDB module to parse and interact with the structure of a protein. We'll learn how to define a simple distance function to see how close various atoms in the model are from key points in the structure. Finally, we'll dive into some compute optimization and see how we can make our calculations faster!

Knowing how to calculate the molecular distances between various atoms in the protein structure model is important and can be useful for several reasons. First off, it can be used to check various constraints to validate the accuracy of the model. It can be especially important in **Protein Engineering** when we need to try and find what residues might be close to the **Active Site**. It can also be used to help understand protein-protein interactions and perform comparative structural analysis across different organisms, focusing on 3-D structure instead of sequence-level differences.

Getting ready

This recipe is made available in the Ch09/Ch09-2-molecular-distances.ipynb file.

How to do it...

1. Let's load our model, as follows:

```
from Bio import PDB
repository = PDB.PDBList()
parser = PDB.PDBParser()
repository.retrieve_pdb_file('1TUP', pdir='.',
                              file_format='pdb')
p53_1tup = parser.get_structure('P 53', 'pdb1tup.ent')
```

2. We will now get our zincs, against which we will perform comparisons later:

```
zns = []
for atom in p53_1tup.get_atoms():
    if atom.element == 'ZN':
        zns.append(atom)
for zn in zns:
    print(zn, zn.coord)
```

You should see three zinc atoms.

3. Now, let's define a function to get the distance between one atom and a set of other atoms, as follows:

```
import math
def get_closest_atoms(pdb_struct, ref_atom, distance):
    atoms = {}
    rx, ry, rz = ref_atom.coord
    for atom in pdb_struct.get_atoms():
        if atom == ref_atom:
            continue
        x, y, z = atom.coord
        my_dist = math.sqrt((x - rx)**2 +
```

```
                        (y - ry)**2 +
                        (z - rz)**2)
        if my_dist < distance:
            atoms[atom] = my_dist
    return atoms
```

We get coordinates for our reference atom and then iterate over our desired comparison list. If an atom is close enough, it's added to the `return` list.

4. We now compute the atoms near our zincs, the distance of which can be up to 4 angstroms for our model:

```
for zn in zns:
    print()
    print(zn.coord)
    atoms = get_closest_atoms(p53_1tup, zn, 4)
    for atom, distance in atoms.items():
        print(atom.element, distance, atom.coord)
```

Here, we show the result for the first zinc, including the element, distance, and coordinates:

```
[58.108 23.242 57.424]
C 3.4080117696286854 [57.77  21.214 60.142]
S 2.3262243799594877 [57.065 21.452 58.482]
C 3.4566537492335123 [58.886 20.867 55.036]
C 3.064120559761192 [58.047 22.038 54.607]
N 1.9918273537290707 [57.755 23.073 55.471]
C 2.9243719601324525 [56.993 23.943 54.813]
C 3.857729198122736 [61.148 25.061 55.897]
C 3.62725094648044 [61.61  24.087 57.001]
S 2.2789209624943494 [60.317 23.318 57.979]
C 3.087214470667822 [57.205 25.099 59.719]
S 2.2253158446520818 [56.914 25.054 57.917]
```

Let's briefly review the preceding partial output. The first line shows the XYZ coordinates of the first zinc atom. We then call `get_closest_atoms()` with a parameter of 4 to get all the atoms that are within 4 angstroms of the zinc. The next section lists the elements nearby, the distance from the zinc, and the XYZ coordinates of that element. For example, the first element found nearby is a carbon (C) atom, which is 3.4 angstroms away from the zinc, at coordinates X=57.77, Y=21.214, Z=60.142. We only show the output section for one zinc atom here, but in your notebook, you will see repeating blocks for all the zinc atoms.

We only have three zinc atoms in this example, so the number of computations is quite significantly reduced. However, imagine that we had more, or that we were doing a pairwise comparison among all the atoms in the set (remember that the number of comparisons grows quadratically with the number of atoms in a pairwise case). Although our case is small, it's not difficult to forecast use cases, while more comparisons take a lot of time. We will get back to this soon.

5. Let's see how many atoms we get as we increase the distance:

```
for distance in [1, 2, 4, 8, 16, 32, 64, 128]:
    my_atoms = []
    for zn in zns:
        atoms = get_closest_atoms(
            p53_1tup, zn, distance)
        my_atoms.append(len(atoms))
    print(distance, my_atoms)
```

The result is as follows:

```
1 [0, 0, 0]
2 [1, 0, 0]
4 [11, 11, 12]
8 [109, 113, 106]
16 [523, 721, 487]
32 [2381, 3493, 2053]
64 [5800, 5827, 5501]
128 [5827, 5827, 5827]
```

As we have seen previously, this specific case is not very expensive, but let's time it anyway:

```
import timeit
nexecs = 10
print(
    timeit.timeit(
        'get_closest_atoms(p53_1tup, zns[0], 4.0)',
        'from __main__ import get_closest_atoms, p53_1tup, zns',
        number=nexecs
    ) / nexecs * 1000
)
```

Here, we will use the `timeit` module to execute this function 10 times and then print the result in milliseconds. We pass the function as a string and pass yet another string with the necessary imports to make this function work. On a notebook, you are probably aware of the `%timeit` magic and how it makes your life much easier in this case. This takes roughly 40 milliseconds on the machine where the code was tested. Obviously, on your computer, you will get somewhat different results.

There's more...

Can we do better? Let's consider a different `distance` function, as shown in the following code:

```
def get_closest_alternative(pdb_struct, ref_atom, distance):
    atoms = {}
    rx, ry, rz = ref_atom.coord
```

```
for atom in pdb_struct.get_atoms():
    if atom == ref_atom:
        continue
    x, y, z = atom.coord
    if abs(x - rx) > distance or
        abs(y - ry) > distance or
        abs(z - rz) > distance:
        continue
    my_dist = math.sqrt((x - rx)**2 +
                        (y - ry)**2 +
                        (z - rz)**2)
    if my_dist < distance:
        atoms[atom] = my_dist
return atoms
```

So, we take the original function and add a very simplistic `if` statement with the distances. The rationale for this is that the computational cost of the square root, and the exponentiation operation, are very expensive, so we will try to avoid them. However, for all atoms that are closer than the target distance in any dimension, this function will be more expensive. To recap, the key innovation in this function is to only check for atoms that are clearly within a cube bounded by the distance. This reduces unnecessary computation by not calling `sqrt()` as many times.

Now, let's time it:

```
print(
    timeit.timeit(
        'get_closest_alternative(p53_1tup, zns[0], 4.0)',
        'from __main__ import get_closest_alternative,
        p53_1tup, zns', number=nexecs
    ) / nexecs * 1000
)
```

On the same machine that we used in the preceding example, it takes 16 milliseconds, which means that it is roughly three times faster.

However, is this always better? Let's compare the cost with different distances, as follows:

```
print('Standard')
for distance in [1, 4, 16, 64, 128]:
    print(
        timeit.timeit(
            'get_closest_atoms(p53_1tup, zns[0], distance)',
            'from __main__ import get_closest_atoms,
            p53_1tup, zns, distance', number=nexecs
        ) / nexecs * 1000
    )
```

```
print('Optimized')
for distance in [1, 4, 16, 64, 128]:
    print(
        timeit.timeit(
            'get_closest_alternative(p53_1tup, zns[0], distance)',
            'from __main__ import get_closest_alternative,
            p53_1tup, zns, distance', number=nexecs
        ) / nexecs * 1000
    )
```

The result is shown in the following output:

```
Standard
85.08649739999328
86.50681579999855
86.79630599999655
96.95437099999253
96.21982420001132
Optimized
30.253444099980698
32.69531210000878
52.965772600009586
142.53310030001103
141.26269519999823
```

Note that the cost of the Standard version is mostly constant, whereas the Optimized version varies depending on the distance of the closest atoms; the larger the distance, the more cases that will be computed using the extra if statement, plus the square root, making the function more expensive.

The larger point here is that you can probably code functions that are more efficient using smart computation shortcuts, but the complexity cost may change qualitatively. In the preceding case, I suggest that the second function is more efficient for all realistic and interesting cases when you're trying to find the closest atoms. However, you have to be careful while designing your own versions of optimized algorithms.

Performing geometric operations

We will now perform computations with geometry information, including computing the center of mass of chains and whole models.

In this recipe, we will learn how to use Biopython's PDB module to calculate the masses of structures in our model. We'll then use operations to find the geometric center of mass.

Getting ready

This recipe is made available in the `Ch09/Ch09-3-geometric-operations.ipynb` file.

How to do it...

Here are the steps to try this recipe:

1. First, let's retrieve the data:

```
from Bio import PDB
repository = PDB.PDBList()
parser = PDB.PDBParser()
repository.retrieve_pdb_file(
    '1TUP', pdir='.', file_format='pdb')
p53_1tup = parser.get_structure('P 53', 'pdb1tup.ent')
```

2. Then, let's recall the type of residues that we have with the following code:

```
my_residues = set()
for residue in p53_1tup.get_residues():
    my_residues.add(residue.id[0])
print(my_residues)
```

So, we have H_ ZN (zinc) and W (water), which are HETATM types; the vast majority are standard PDB atoms.

3. Let's compute the masses for all chains, zinc atoms, and water atom instances using the following code:

```
import numpy as np
def get_mass(atoms, accept_fun=lambda x: True):
    return sum([atom.mass for atom in atoms
                if accept_fun(atom)])
chain_names = [chain.id for chain in p53_1tup.get_chains()]
my_mass = np.ndarray((len(chain_names), 3))

for i, chain in enumerate(p53_1tup.get_chains()):
    my_mass[i, 0] = get_mass(chain.get_atoms())
print("Mass array:", my_mass)
```

The `get_mass` function returns the mass of all atoms in the list that pass an acceptance criterion function. Here, the default acceptance criterion involves not being a water residue.

We then compute the mass for all chains. We have three versions: just amino acids, zincs, and water. Zinc does nothing more than detect a single atom per chain in this model. The output is as follows:

```
Mass Distribution by Chain (Daltons)
=============================================
Chain  No water     Zincs        Water
---------------------------------------------
E      6068.04      0.00         351.99
F      6258.20      0.00         223.99
A      20548.26     65.39        3167.88
B      20368.19     65.39        1119.96
C      20466.23     65.39        1279.95
---------------------------------------------
Total  73708.93     196.17       6143.77
```

Figure 9.3 – The mass for all protein chains

4. Let's compute the geometric center and the center of mass of the model, as follows:

```
def get_center(
    atoms,
    weight_fun=lambda atom: 1 if atom.parent.id[0] != 'W'
    else 0
):
    xsum = ysum = zsum = 0.0
    acum = 0.0
    for atom in atoms:
        x, y, z = atom.coord
        weight = weight_fun(atom)
        acum += weight
        xsum += weight * x
        ysum += weight * y
        zsum += weight * z
    return xsum / acum, ysum / acum, zsum / acum

print(get_center(p53_1tup.get_atoms()))
print(get_center(
    p53_1tup.get_atoms(),
    weight_fun=lambda atom: atom.mass
    if atom.parent.id[0] != 'W' else 0
))
```

First, we define a weighted function to get the coordinates of the center. The default function will treat all atoms as equal, as long as they are not a water residue.

We then compute the geometric center and the center of mass by redefining the weight function with a value of each atom equal to its mass. The geometric center is computed, irrespective of its molecular weight.

For example, you may want to compute the center of mass of the protein without DNA chains.

5. Let's compute the center of mass and the geometric center of each chain, as follows:

```
import pandas as pd
my_center = np.ndarray((len(chain_names), 6))
for i, chain in enumerate(p53_1tup.get_chains()):
    x, y, z = get_center(chain.get_atoms())
    my_center[i, 0] = x
    my_center[i, 1] = y
    my_center[i, 2] = z
    x, y, z = get_center(
        chain.get_atoms(),
        weight_fun=lambda atom: atom.mass
        if atom.parent.id[0] != 'W' else 0
    )
    my_center[i, 3] = x
    my_center[i, 4] = y
    my_center[i, 5] = z

weights = pd.DataFrame(
    my_center, index=chain_names,
    columns=['X', 'Y', 'Z',
             'X (Mass)', 'Y (Mass)', 'Z (Mass)'])
print(weights)
```

The result is shown here:

	X	Y	Z	X (Mass)	Y (Mass)	Z (Mass)
E	49.727215	32.744877	81.253433	49.708504	32.759739	81.207359
F	51.982376	33.843376	81.578781	52.002220	33.820042	81.624405
A	72.990845	28.825418	56.714001	72.822617	28.810341	56.716141
B	67.810066	12.624434	88.656647	67.729164	12.724131	88.545692
C	38.221588	−5.010491	88.293114	38.169384	−4.915401	88.166679

Figure 9.4 – The center of mass and the geometric center of each protein chain

There's more...

Although this is not a book based on the protein structure determination technique, it's important to remember that X-ray crystallography methods cannot detect hydrogens, so computing the mass of residues might be based on very inaccurate models; refer to `https://www.umass.edu/microbio/chime/pe_beta/pe/protexpl/help_hyd.htm` for more information.

Animating proteins

One of the most fun and exciting things to do with proteins is to view and animate their structures. Proteins are the workhorses of the cell. They are essentially nanobots that can create structure in a cell or perform enzymatic reactions to convert one chemical into another. Indeed, for a long time, DNA and RNA were viewed somewhat as just instructions on the way toward making a protein, consistent with the **Central Dogma**. We now know that DNA and RNA can also perform important structural roles and will often form a complex with proteins (i.e., a DNA-protein structure) to complement or enhance the function of the protein. Ribozymes are RNA molecules that can perform enzymatic reactions and hence are viewed as a likely starting point for life.

In this recipe, we'll learn how to animate a protein structure in our notebook using a library called NGLview (`https://github.com/nglviewer/nglview`). We'll also check out another library, py3Dmol: `https://pypi.org/project/py3Dmol/`.

Getting ready

For this recipe, you will need Biopython, which we'll assume is already installed, and `nglview`. To install it, run the following:

```
! pip install nglview
```

This recipe is made available in the `Ch09/Ch09-4-nglview.ipynb` file.

How to do it...

Here are the steps to perform this recipe:

1. First, we will import our libraries:

    ```
    import nglview as nv
    from Bio import PDB
    ```

2. Next, let's download our protein:

    ```
    pdb_id = "1TUP"
    pdb_list = PDB.PDBList()
    pdb_list.retrieve_pdb_file(
        pdb_id, pdir=".", file_format="pdb")
    ```

3. We set `pdb_id` to the 1TUP protein and set up a `PDBList` object to download the protein. We then load the PDB file into NGLview:

```
view = nv.show_file(f"pdb{pdb_id.lower()}.ent")
```

4. Now we will set some visualization parameters:

```
view.clear_representations()
view.add_cartoon(color="spectrum")
view.add_spacefill("ZN")
```

This clears any settings and then sets up a cartoon style, which means that we highlight secondary structures such as alpha helices and beta sheets. We also highlight zinc atoms:

```
view.camera = "perspective"
view.center()
view.animate = True
```

This sets our camera mode to `perspective`, which gives a depth effect. We center the view on the protein and allow animation, which means we can move around and manipulate our view of the protein using the mouse.

5. Finally, we trigger our view:

```
view
```

Here is what you will see:

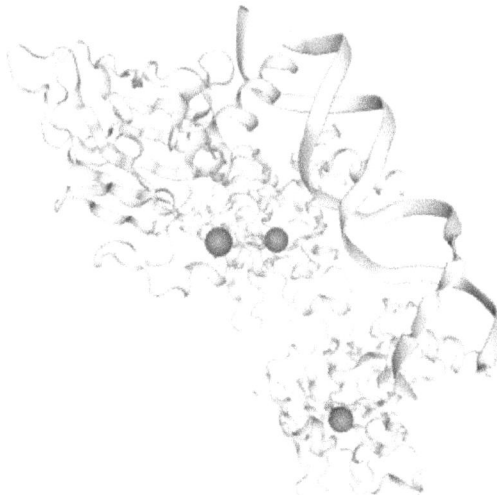

Figure 9.5 – NGLview picture of the 1TUP protein

We can see the 1TUP protein with secondary structures and zincs highlighted. We can rotate it with our mouse and hover over individual atoms and see their respective amino acid codes.

> **Tip**
>
> Some of you may experience problems with NGLview not displaying the protein in the notebook. This may be related to your version of `ipywidgets` or Jupyter Notebook. If you are having issues, there are a couple of things you can try. Leave your notebook and, from the terminal, run the following:
>
> ```
> pip install --upgrade notebook
> ```
> ```
> pip install --upgrade ipywidgets
> ```
>
> Then, go back into your notebook and try the code again. Remember, you can run `pip show` to see the version of any package. There is some additional test code in the notebook to help with debugging.
>
> If you are still having trouble with NGLview, you can move on to learn about py3Dmol in the next section, which is an alternative protein viewer.

We can explore more capabilities of NGLview. For instance, we can highlight certain residues:

> **AI tip**
>
> **Try this prompt:** Write code using NGLview to visualize the 1TUP protein and highlight the Cysteine residues.
>
> **You should see:** Code to download and visualize the 1TUP protein with CYS residues highlighted in a particular color.

We can also use NGLview to visualize the differences between a native version of a protein and a mutated copy:

> **Try this prompt:** Write code using NGLview to compare the 6GFG protein to the 1TUP protein.
>
> **You should see:** Code to compare 6GFG, which is a mutated version of the p54 protein, to its wild-type version, 1TUP. You will see two side-by-side, interactive 3-D models.

Okay, let's try another handy library for protein structure viewing, py3Dmol. Take a look at the `Ch09-4-py3dmol.ipynb` notebook. First, we will install `py3dmol`:

```
! pip install py3Dmol
```

Now we import our library:

```
import py3Dmol
```

Now we run our viewer:

```
view = py3Dmol.view(query='pdb:1crn')
view.setStyle({'cartoon': {'color': 'spectrum'}})
view.zoomTo()
view.show()
```

This code sets up our viewer with the 1CRN protein (Crambin). It then sets style and zoom parameters and finally calls `view.show()` to display the protein.

Here is what we see:

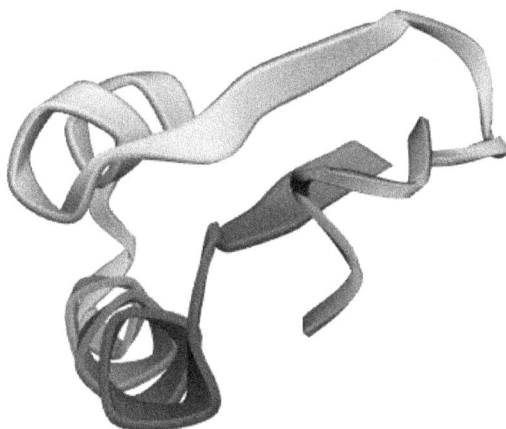

Figure 9.6 – The 1CRN Crambin protein displayed by py3Dmol

There's more...

PyMOL is another useful and important structure viewer to know about. It has recently been integrated with several structural bioinformatics tools using PyMod – Janson & Paiardini, "PyMod 3: a complete suite for structural bioinformatics in PyMOL," Bioinformatics, May 2021: `https://academic.oup.com/bioinformatics/article/37/10/1471/5917627`.

Using structure viewers, scientists can gain insights into the function of the protein. In particular, the **Active Site** of the enzyme can be modified to increase or decrease the rate of activity of the protein or alter the substrate selectivity.

Enzyme Engineering is the art of changing enzymes by mutating various amino acid residues, or in some cases, introducing entirely new loops or pieces of structure. This was traditionally done primarily through random mutagenesis or focused libraries that were screened in the lab. But with the advances in AI, mutant enzymes can also be designed and tested *in silico*.

See also

- PyMOL is here: `https://www.pymol.org/`

- Read more about NGLview in Nguyen et al, *NGLview - interactive molecular graphics for Jupyter notebooks*, Bioinformatics, Apr 2018: `https://academic.oup.com/bioinformatics/article/34/7/1241/4721781`

- This article gives a good tutorial on py3Dmol: `https://william-dawson.github.io/using-py3dmol.html`

- For a good review on the current state of the art in structural bioinformatics, read: Rosignoli et al, "An outlook on structural biology after AlphaFold: tools, limits, and perspectives," FEBS open bio, Sep 2024: `https://febs.onlinelibrary.wiley.com/doi/10.1002/2211-5463.13902`

- Machine learning can be used for enzyme engineering: Landwehr et al, "Accelerated enzyme engineering by machine-learning guided cell-free expression," Nature Communications, Jan 2025: `https://www.nature.com/articles/s41467-024-55399-0`

- SeaMoon uses language models to simulate the motions of proteins: Lombard et al, *SeaMoon: From protein language models to continuous structural heterogeneity*, Structure, Jul 2025: `https://www.sciencedirect.com/science/article/abs/pii/S0969212625002448?dgcid=author`

- La-Proteina can generate protein structures with atom-level accuracy: `https://research.nvidia.com/labs/genair/la-proteina/`

Performing proteomics analysis

Proteomics is the study of protein expression. Much as we learned that we can analyze the expression of genes by RNA-Seq, it is also possible to look at the expression of proteins.

Typically, this is done by **Mass Spectrometry**. In this technique, proteins are first fragmented by digestion, and then small protein fragments are run through a spectrometer, so we can determine their mass and charge. We then work backward with a set of fragments to understand what protein they may have come from. To do this, we first need to perform an *in silico* digestion of the proteins that are potentially present in the sample. We do this by taking the putative proteome of the organism in the sample and calculating where the enzyme used for digestion will cut it. For example, trypsin cleaves amino acids between lysine and arginine, unless they are followed by a proline.

Once we have a database of our potential fragments, we can feed this into our proteomics software to help us determine the identity and level of various proteins in the sample.

In this recipe, we'll learn how to use a package called `pyteomics`, which specializes in proteomics analysis. We'll learn how to calculate key properties of proteins and perform a simulated digestion of a protein to understand how a typical proteomics analysis would work. Finally, we'll learn how to visualize the peptide mass distributions of a digested protein, as it would be used in typical proteomics workflows.

Getting ready

In this recipe, we will use the `pyteomics` package (https://pyteomics.readthedocs.io/en/latest/). You will also need `matplotlib`, Biopython, pandas, and seaborn. The latter are probably already installed for you, but this line will make sure everything is installed:

```
! pip install biopython matplotlib pandas seaborn pyteomics
```

This recipe is made available in the `Ch09/Ch09-5-proteomics.ipynb` file.

How to do it...

Here are the steps to try this recipe:

1. First, we will import our libraries:

    ```python
    import pandas as pd
    import matplotlib.pyplot as plt
    import seaborn as sns
    from Bio.SeqUtils.ProtParam import ProteinAnalysis
    from pyteomics import parser, mass
    ```

    From the `pyteomics` package, we will use the `parser` class for digesting protein sequences and the `mass` class for computing molecular mass.

2. Now let's define our protein sequence:

    ```python
    protein_sequence = (
        "MEEPQSDPSVEPPLSQETFSDLWKLLPENNVLSPLPSQAMDDLMLSPDD"
        "IEQWFTEDPGPDEAPRMPEAAPPVAPAPAAPTPAAPAPAPSWPLSSSVPSQ"
        "KTYQGSYGFRLGFLHSGTGFVKVGQSTSRHKKLMFKTEGPDSD"
    )
    ```

    We will use the 1TUP p53 tumor suppressor protein.

3. Next, let's analyze some key properties of this protein:

    ```python
    protein = ProteinAnalysis(protein_sequence)
    molecular_weight = protein.molecular_weight()
    hydrophobicity = protein.gravy()
    isoelectric_point = protein.isoelectric_point()
    amino_acid_composition = protein.count_amino_acids()

    print(f"Protein Molecular Weight: {molecular_weight:.2f} Da")
    print(f"Protein Hydrophobicity (GRAVY): {hydrophobicity:.2f}")
    print(f"Protein Isoelectric Point (pI): {isoelectric_point:.2f}")
    ```

We create a `ProteinAnalysis` object to enable calculations on the protein sequence. We then calculate the molecular weight of the protein in Daltons using the standard molecular weights of the amino acids.

We also calculate the hydrophobicity or **gravy** score. This is the **Grand Average Hydropathy** score, which calculates the sum of the hydropathy values of the amino acids in the protein. The higher the value, the greater the hydrophobicity.

We also calculate the **Isoelectric Point** for the protein, which is the pH at which the protein has a net charge of 0. The amino acid composition is also calculated, which reflects the number of each type of amino acid in the protein.

4. Next, we'll perform an *in silico* digestion of the protein using trypsin as the enzyme:

```
peptides = sorted(list(parser.cleave(
    protein_sequence, parser.expasy_rules['trypsin'])))
peptide_masses = [
    mass.calculate_mass(sequence=p)
    for p in peptides
]
df = pd.DataFrame({
    'Peptide': peptides,
    'Mass (Da)': peptide_masses
})
df = df[df['Mass (Da)'] > 500]
print("\nTop 10 Peptides:")
print(df.head(10))
```

This will first perform a simulated trypsin digest on our protein sequence using the `parser.cleave()` function. We return this as a sorted list of peptide fragments. We then calculate the mass of each peptide using the `mass.calculate_mass()` function. We create a pandas dataframe to store the peptides and their masses. Finally, we filter out small peptides. Here is an example of what we get:

```
Top 10 Peptides:
                                       Peptide     Mass (Da)
2                                  LGFLHSGTGFVK   1261.681907
3   LLPENNVLSPLPSQAMDDLMLSPDDIEQWFTEDPGPDEAPR   4591.141076
4                                          LMFK    537.298491
5                      MEEPQSDPSVEPPLSQETFSDLWK   2775.258542
6         MPEAAPPVAPAPAAPTPAAPAPAPSWPLSSSVPSQK   3442.759512
7                                        TEGPDSD    719.260978
8                                      TYQGSYGFR   1077.487959
9                                        VGQSTSR    733.371866
```

Figure 9.7 – Top 10 peptides from digestion and their masses

Okay, now let's visualize our results!

```
plt.figure(figsize=(10, 5))
sns.histplot(df['Mass (Da)'], bins=30, kde=True, color="blue")
plt.xlabel("Peptide Mass (Da)")
plt.ylabel("Frequency")
plt.title("Peptide Mass Distribution (Trypsin Digest)")
plt.show()
```

We set up a plot using the seaborn histogram function. We plot the mass of the peptides and add a density curve. We then set the labels and title for the plot. Here is what we get:

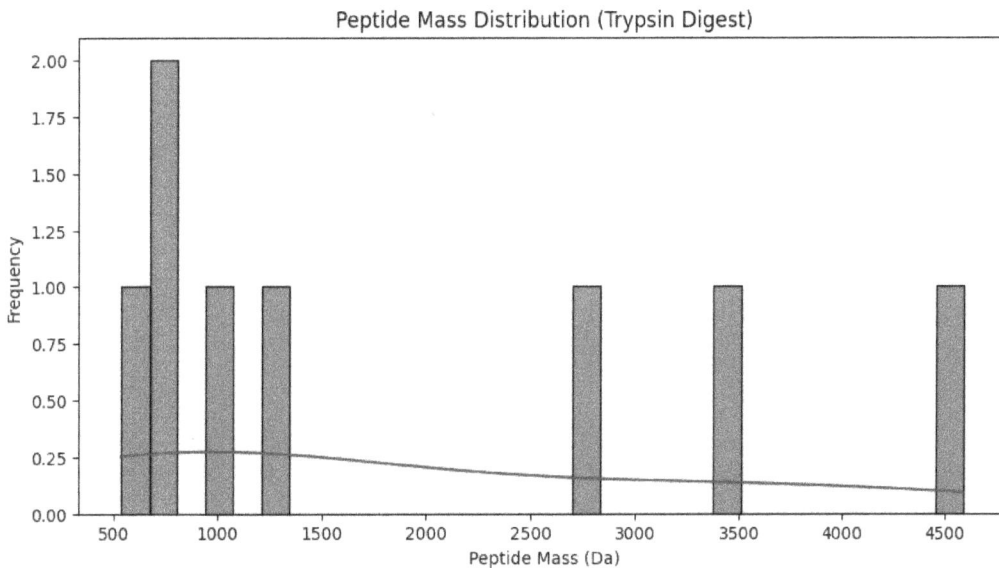

Figure 9.8 – Peptide mass distribution for digested protein

This gives you a basic sense of how we can digest proteins for analysis by proteomics software! This approach allows us to look at a proteomics sample and make calculations about the identity and abundance of various proteins in the sample.

There's more...

Once proteins have been digested, they can be built into a database to analyze a proteomics dataset. Special software is used to look at the spectra generated by peptide fragments in a proteomics experiment to determine the identity of the original protein. We can then calculate the expression levels of proteins as well. Much like RNA-Seq, we can start to make inferences about what is going on biologically. Several pieces of software exist to process proteomics data. For example, OMSSA (`https://bioinformaticshome.com/tools/proteomics/descriptions/OMSSA.html`) is open-source software for identifying proteins in mass spectrometry data. A variety of proteomics tools are available via OHSU here: `https://www.ohsu.edu/proteomics-shared-resource/proteomics-tools`. NIST maintains spectral libraries for proteins: `https://www.nist.gov/programs-projects/peptide-mass-spectral-libraries`.

This was just a brief introduction to the basics of proteomics, which is a large field. You may wish to follow up by diving deeper into some of the materials provided here. This video is a good place to start: `https://www.youtube.com/watch?v=wx4F6kGy1Fs`. Next, get a broad overview of proteomics in Al-Amrani et al, *Proteomics: concepts and applications in human medicine*, World Journal of Biological Chemistry, Sep 2021: `https://pmc.ncbi.nlm.nih.gov/articles/PMC8473418/`. Another great review is available in Jiang et al, *Comprehensive Overview of Bottom-Up Proteomics using Mass Spectrometry*, ACS Measurement Science Au, Jun 2024: `https://pubs.acs.org/doi/full/10.1021/acsmeasuresciau.3c00068`. Read about proteomics techniques in Cui et al, *High-throughput proteomics: a methodological mini-review*, Laboratory Investigation, Aug 2022: `https://www.nature.com/articles/s41374-022-00830-7`. You can brush up on the very latest advances in proteomics in Guo et al, *Mass-spectrometry-based proteomics: from single cells to clinical applications*, Nature, Feb 2025: `https://www.nature.com/articles/s41586-025-08584-0`.

Protein abundance can also be assayed by protein sequencing. Protein post-translational modifications can be assayed by nanopore sequencing (Lu et al, *Toward single-molecule protein sequencing using nanopores*, Nature Biotechnology, Mar 2025 - `https://www.nature.com/articles/s41587-025-02587-y`). Companies such as Nautilus Biotechnology (`https://www.nautilus.bio/`) use protein arrays to assay proteomes.

Okay, let's clean up and close down our `conda` environment:

```
conda deactivate
```

See also

- Learn about one of the largest proteomics studies ever here: *Largest proteome study enlists 14 biopharmas*, Nature Biotechnology, Feb 2025: `https://www.nature.com/articles/s41587-025-02577-0`

- Learn more about predicting protein properties with ProtParam here: `https://web.expasy.org/protparam/`

- This tool provides a web interface to analyze proteomics data: Didusch et al, *amica: an interactive and user-friendly web-platform for the analysis of proteomics data*, BMC Genomics, Dec 2022: `https://bmcgenomics.biomedcentral.com/articles/10.1186/s12864-022-09058-7`

- Microfluidics can be used in proteomics: Steinbach et al, *Digital Microfluidics for Sample Preparation in Low-Input Proteomics*, Small Methods, Aug 2024: `https://www.nature.com/articles/s43586-024-00318-2`

- Learn about a proteomics method where proteins are not digested first – Roberts et al, *Top-down proteomics*, Nature Reviews Methods Primers, Jun 2024: `https://www.nature.com/articles/s43586-024-00318-2`

- Learn how nanopores can be used for protein sequencing – Dorey & Howorka, *Nanopore DNA sequencing technologies and their applications toward single-molecule proteomics*, Nature Chemistry, Mar 2024: `https://www.nature.com/articles/s41557-023-01322-x`

Get This Book's PDF Version and Exclusive Extras

UNLOCK NOW

Scan the QR code (or go to `packtpub.com/unlock`). Search for this book by name, confirm the edition, and then follow the steps on the page.

Note: Keep your invoice handy. Purchases made directly from Packt don't require an invoice.

10
Phylogenetics

Phylogenetics is the application of molecular sequencing that is used to study the evolutionary relationship among organisms. The typical way to illustrate this process is through the use of phylogenetic trees. The computation of these trees from genomic data is an active field of research with many real-world applications.

In this book, we will take the practical approach that is mentioned to a new level: most of the recipes here are inspired by a study on the Ebola virus, researching the Ebola outbreak in Africa. This study is called *Genomic surveillance elucidates Ebola virus origin and transmission during the 2014 outbreak*, by *Gire et al.*, published in *Science*. It is available at `https://pubmed.ncbi.nlm.nih.gov/25214632/`. Here, we will try to follow a similar methodology to arrive at similar results to the paper.

In this chapter, we will use DendroPy (a phylogenetics library) and Biopython.

We will cover the following recipes in this chapter:

- Preparing a dataset for phylogenetic analysis
- Aligning genetic and genomic data
- Comparing sequences
- Reconstructing phylogenetic trees
- Playing recursively with trees
- Visualizing phylogenetic data

Phylogenetics is used extensively in bioinformatics to understand the relationships of genomes, genes, and proteins throughout evolution. It can be useful to track the evolution of a virus or to establish the identity of a strain for intellectual property purposes – see Allaby and Woodwark, *Phylogenetics in the bioinformatics culture of understanding*, Comparative and Functional Genomics, March 2004 – `https://onlinelibrary.wiley.com/doi/full/10.1002/cfg.381`.

By the end of this chapter, you'll have a good understanding of how to create, traverse, and visualize phylogenetic trees. We'll take a look at an example involving viral outbreaks to tie this back to real-world uses.

Technical requirements

The code for this chapter can be found in `https://github.com/PacktPublishing/Bioinformatics-with-Python-Cookbook-Fourth-Edition/tree/main/Ch10`.

You will want to create a `Ch10` folder and set up your notebooks there.

Remember to activate your `conda` environment before beginning the recipes, like this:

```
conda activate bioinformatics_base
```

If you would like to set up a `conda` environment specific to this chapter, before activating `bioinformatics_base`, run the following:

```
conda create -n ch10-phylogenetics --clone bioinformatics_base
conda activate ch10-phylogenetics
```

You will be able to install the packages for the chapter as you go, or you can use the YAML file provided in the repository:

```
conda env update --file ch10-phylogenetics.yml
```

Preparing a dataset for phylogenetic analysis

In this recipe, we will download and prepare the dataset to be used for our analysis. The dataset contains complete genomes of the Ebola virus. We will use DendroPy (`https://jeetsukumaran.github.io/DendroPy/`) to download and prepare the data. DendroPy offers Python functions for phylogenetic computing. It includes a variety of methods for reading and writing phylogenetic trees in popular formats such as NEWICK, NEXUS, and Phylip. It can also generate and compare phylogenetic trees.

Here, we will first use DendroPy to download and format several Ebola genomes. We will then create FASTA files, which will be used throughout the recipes in this chapter to examine the phylogenetic relationships between the different species of Ebola. We will learn about the `DnaCharacterMatrix` class in DendroPy, which is a useful container class for storing and manipulating your sequences.

Next, we'll see how to extract a subset of genes from the alignment and calculate basic statistics on them. We'll also see how to determine the number of distinct taxons present in the alignment and calculate general statistics on the alignment.

By the end of this recipe, you will be comfortable with using the DendroPy alignment class and extracting information from it. You will then be prepared to dive further into phylogenetic analysis with the data.

Getting ready

We will download complete genomes from GenBank; these genomes were collected from various Ebola outbreaks, including several from the 2014 outbreak. Note that there are several virus species that cause the Ebola virus disease; the species involved in the 2014 outbreak (the EBOV virus, which was formally known as the Zaire Ebola virus) is the most common, but this disease is caused by more species of the genus Ebolavirus. Four others are also available in a sequenced form. You can read more at https://en.wikipedia.org/wiki/Ebolavirus.

If you have already gone through the previous chapters, you might panic looking at the potential data sizes involved here; this is not a problem at all because these are genomes of viruses that are each around 19 kbp in size. So, our approximately 100 genomes are actually quite light.

We will be using DendroPy in this recipe. Let's install it:

```
! pip install dendropy
```

As usual, this information is available in the corresponding Jupyter Notebook file, which is available at Ch10/Ch10-1-preparing-dataset.ipynb.

How to do it...

Take a look at the following steps:

1. First, let's start by specifying our data sources using DendroPy, as follows:

    ```
    import dendropy
    from dendropy.interop import genbank

    def get_ebov_2014_sources():
        # EBOV_2014
        # yield 'EBOV_2014', genbank.GenBankDna(
        #     id_range=(233036, 233118), prefix='KM')
        yield 'EBOV_2014', genbank.GenBankDna(
            id_range=(34549, 34563), prefix='KM0')

    def get_other_ebov_sources():
        # EBOV other
        yield 'EBOV_1976', genbank.GenBankDna(
            ids=['AF272001', 'KC242801'])
        yield 'EBOV_1995', genbank.GenBankDna(
            ids=['KC242796', 'KC242799'])
        yield 'EBOV_2007', genbank.GenBankDna(
            id_range=(84, 90), prefix='KC2427')
    ```

```
def get_other_ebolavirus_sources():
    # BDBV
    yield 'BDBV', genbank.GenBankDna(
        id_range=(3, 6), prefix='KC54539')
    yield 'BDBV', genbank.GenBankDna(
        ids=['FJ217161'])
    # RESTV
    yield 'RESTV', genbank.GenBankDna(
        ids=['AB050936', 'JX477165', 'JX477166',
            'FJ621583', 'FJ621584', 'FJ621585'])
    # SUDV
    yield 'SUDV', genbank.GenBankDna(
        ids=['KC242783', 'AY729654', 'EU338380',
            'JN638998', 'FJ968794', 'KC589025', 'JN638998'])
    # yield 'SUDV', genbank.GenBankDna(
    #     id_range=(89, 92), prefix='KC5453')
    # TAFV
    yield 'TAFV', genbank.GenBankDna(ids=['FJ217162'])
```

Here, we have three functions: one to retrieve data from the most recent EBOV outbreak, another to retrieve data from the previous EBOV outbreaks, and one to retrieve data from the outbreaks of other species.

Note that the DendroPy GenBank interface provides several different ways to specify lists or ranges of records to retrieve. Some lines are commented out. These include the code to download more genomes. For our purpose, the subset that we will download is enough.

2. Now, we will create a set of FASTA files; we will use these files here and in future recipes:

```
other = open('other.fasta', 'w')
sampled = open('sample.fasta', 'w')

for species, recs in get_other_ebolavirus_sources():
    tn = dendropy.TaxonNamespace()
    char_mat = recs.generate_char_matrix(
        taxon_namespace=tn,
        gb_to_taxon_fn=lambda gb: tn.require_taxon(
            label='%s_%s' % (species, gb.accession)))
    char_mat.write_to_stream(other, 'fasta')
    char_mat.write_to_stream(sampled, 'fasta')
other.close()

ebov_2014 = open('ebov_2014.fasta', 'w')
ebov = open('ebov.fasta', 'w')
```

```
for species, recs in get_ebov_2014_sources():
    tn = dendropy.TaxonNamespace()
    char_mat = recs.generate_char_matrix(
        taxon_namespace=tn,
        gb_to_taxon_fn=lambda gb: tn.require_taxon(
            label='EBOV_2014_%s' % gb.accession))
    char_mat.write_to_stream(ebov_2014, 'fasta')
    char_mat.write_to_stream(sampled, 'fasta')
    char_mat.write_to_stream(ebov, 'fasta')
ebov_2014.close()
ebov_2007 = open('ebov_2007.fasta', 'w')

for species, recs in get_other_ebov_sources():
    tn = dendropy.TaxonNamespace()
    char_mat = recs.generate_char_matrix(
        taxon_namespace=tn,
        gb_to_taxon_fn=lambda gb: tn.require_taxon(
            label='%s_%s' % (species, gb.accession)))
    char_mat.write_to_stream(ebov, 'fasta')
    char_mat.write_to_stream(sampled, 'fasta')
    if species == 'EBOV_2007':
        char_mat.write_to_stream(ebov_2007, 'fasta')

ebov.close()
ebov_2007.close()
sampled.close()
```

We will generate several different FASTA files, which include either all genomes, just EBOV, or just EBOV samples from the 2014 outbreak. In this chapter, we will mostly use the sample. fasta file with all genomes.

Note the use of the dendropy functions to create FASTA files that are retrieved from GenBank records through conversion. The ID of each sequence in the FASTA file is produced by a lambda function that uses the species and the year, alongside the GenBank accession number.

3. Let's extract four (of the total seven) genes in the virus, as follows:

```
my_genes = ['NP', 'L', 'VP35', 'VP40']
def dump_genes(species, recs, g_dls, p_hdls):
    for rec in recs:
        for feature in rec.feature_table:
            if feature.key == 'CDS':
                gene_name = None
                for qual in feature.qualifiers:
                    if qual.name == 'gene':
                        if qual.value in my_genes:
```

```
                            gene_name = qual.value
                    elif qual.name == 'translation':
                        protein_translation = qual.value
                if gene_name is not None:
                    locs = feature.location.split('.')
                    start, end = int(
                        locs[0]), int(locs[-1])
                    g_hdls[gene_name].write(
                        '>%s_%s\n' % (
                            species, rec.accession))
                    p_hdls[gene_name].write(
                        '>%s_%s\n' % (
                            species, rec.accession))
                    g_hdls[gene_name].write(
                        '%s\n' % rec.sequence_text[
                            start - 1 : end])
                    p_hdls[gene_name].write(
                        '%s\n' % protein_translation)
g_hdls = {}
p_hdls = {}

for gene in my_genes:
    g_hdls[gene] = open('%s.fasta' % gene, 'w')
    p_hdls[gene] = open('%s_P.fasta' % gene, 'w')
for species, recs in get_other_ebolavirus_sources():
    if species in ['RESTV', 'SUDV']:
        dump_genes(species, recs, g_hdls, p_hdls)
for gene in my_genes:
    g_hdls[gene].close()
    p_hdls[gene].close()
```

We start by searching the first GenBank record for all gene features (please refer to the **National Center for Biotechnology Information** (**NCBI**) documentation for further details; although we will use DendroPy and not Biopython here, the concepts are similar) and write to the FASTA files in order to extract the genes. We put each gene into a different file and only take two virus species. We also get translated proteins, which are available in the records for each gene.

4. Let's create a function to get the basic statistical information from the alignment, as follows:

```
def describe_seqs(seqs):
    print('Number of sequences: %d' % len(seqs.taxon_namespace))
    print(
        'First 10 taxon sets: %s' %
        ' '.join([taxon.label for taxon in
            seqs.taxon_namespace[:10]])
    )
```

```
lens = []
for tax, seq in seqs.items():
    lens.append(len(
        [x for x in seq.symbols_as_list() if x != '-']
    ))
print(
    'Genome length: min %d, mean %.1f, max %d' %
    (min(lens), sum(lens) / len(lens), max(lens))
)
```

The `describe_seqs()` function takes a `DnaCharacterMatrix` DendroPy class and counts the number of taxons. Then, we extract all the amino acids per sequence (we exclude gaps identified by -) to compute the length and report the minimum, mean, and maximum sizes. Take a look at the DendroPy documentation for additional details regarding the API (`https://jeetsukumaran.github.io/DendroPy/primer/index.html`).

5. Let's inspect the sequence of the EBOV genome and compute the basic statistics, as shown earlier:

```
ebov_seqs = dendropy.DnaCharacterMatrix.get_from_path(
    'ebov.fasta', schema='fasta', data_type='dna')
print('EBOV')
describe_seqs(ebov_seqs)
del ebov_seqs
```

We then call a function and get 26 sequences with a minimum size of 18,700, a mean size of 18,925.2, and a maximum size of 18,959. This is a small genome when compared to eukaryotes.

Note that at the very end, the memory structure has been deleted. This is because the memory footprint is still quite big (DendroPy is a pure Python library and has some costs in terms of speed and memory). Be careful with your memory usage when you load full genomes.

6. Now, let's inspect the other Ebola virus genome file and count the number of different species:

```
print('ebolavirus sequences')
ebolav_seqs = dendropy.DnaCharacterMatrix.get_from_path(
    'other.fasta', schema='fasta', data_type='dna')
describe_seqs(ebolav_seqs)
from collections import defaultdict
species = defaultdict(int)
for taxon in ebolav_seqs.taxon_namespace:
    toks = taxon.label.split('_')
    my_species = toks[0]
    if my_species == 'EBOV':
        ident = '%s (%s)' % (my_species, toks[1])
    else:
        ident = my_species
    species[ident] += 1
```

```
for my_species, cnt in species.items():
    print("%20s: %d" % (my_species, cnt))
del ebolav_seqs
```

The name prefix of each taxon is indicative of the species, and we leverage that to fill a dictionary of counts.

The output for the species and the EBOV breakdown is detailed next (with the legend as Bundibugyo virus=BDBV, Tai Forest virus=TAFV, Sudan virus=SUDV, and Reston virus=RESTV; we have 1 TAFV, 6 SUDV, 6 RESTV, and 5 BDBV).

7. Let's extract the basic statistics of a gene in the virus:

```
gene_length = {}
my_genes = ['NP', 'L', 'VP35', 'VP40']
for name in my_genes:
    gene_name = name.split('.')[0]
    seqs = dendropy.DnaCharacterMatrix.get_from_path(
        '%s.fasta' % name,
        schema='fasta',
        data_type='dna'
    )
    gene_length[gene_name] = []
    for tax, seq in seqs.items():
        gene_length[gene_name].append(
            len([x for x in seq.symbols_as_list() if x != '-'])
        )
for gene, lens in gene_length.items():
    print ('%6s: %d' % (gene, sum(lens) / len(lens)))
```

This code gives you an overview of the basic gene information (that is, the name and the mean size), as follows:

```
NP: 2218
L: 6636
VP35: 990
VP40: 988
```

There's more...

Most of the work here can probably be performed with Biopython, but DendroPy has additional functionalities that will be explored in later recipes. Furthermore, as you will discover, it's more robust with certain tasks (such as file parsing). More importantly, there is another Python library to perform phylogenetics that you should consider. It's called ETE and is available at http://etetoolkit.org/.

See also

- The US **Center for Disease Control** (**CDC**) has a good introductory page on the Ebola virus disease at `https://www.cdc.gov/ebola/about/index.html?CDC_AAref_Val`

- The reference application in phylogenetics is Joe Felsenstein's *Phylip*, which can be found at `https://phylipweb.github.io/phylip/`

- We will use the Nexus and Newick formats in future recipes (`https://plewis.github.io/nexus/`), but do also check out the PhyloXML format (`https://en.wikipedia.org/wiki/Newick_format`)

- Read more about DendroPy in this paper: Moreno et al, "DendroPy 5: a mature Python library for phylogenetic computing," The Journal of Open Source Software, May 2024 – `https://joss.theoj.org/papers/10.21105/joss.06943`

Aligning genetic and genomic data

Before we can perform any phylogenetic analysis, we need to align our genetic and genomic data. Here, we will use MAFFT (`http://mafft.cbrc.jp/alignment/software/`) to perform the genome analysis. The gene analysis will be performed using MUSCLE (`http://www.drive5.com/muscle/`).

Alignment is a key step in any phylogenetic analysis. When we align whole genomes, we are aligning the entire nucleotide sequences of the genomes against each other. When we align genes or the genetic content of an organism, we are aligning one or more genes against each other from a gene family. This could potentially be done either at the nucleotide or amino acid (protein) level. There are even **structure-guided** alignments, which take place in 3D protein space (see Ghaly et al., "EcoFoldDB: Protein-structure guided functional profiling of ecologically relevant microbial traits at the metagenome scale," bioRxiv, April 2025 – `https://www.biorxiv.org/content/10.1101/2025.04.02.646905v1.abstract`).

Alignments find the most common regions of a gene or genome and line them up. Where gaps need to be introduced to make the best fit, we put a - character.

When blocks of genes are present in the same region and in the same order in two different organisms, we call this **synteny**. Synteny is a powerful tool for elucidating gene function, because blocks of genes that stay together during evolution tend to have important, conserved functions. The genes that stay together are likely to operate in the same biochemical pathway, be regulated as a unit, or may even form a protein complex with one another.

Here is a diagram to help you understand these key concepts of synteny and alignment:

Genome vs Gene Phylogenetic Alignments

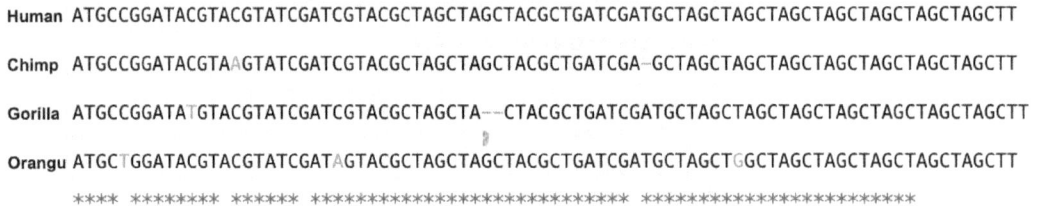

Genome Alignment

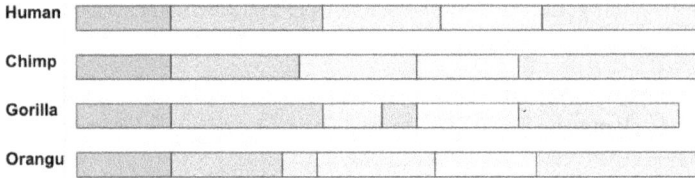

Human	
Chimp	
Gorilla	
Orangu	

Colored blocks represent syntenic regions with homology across species
Different block lengths indicate insertions/deletions; unique colors show different genomic regions

Gene Alignment

Human ATGCCGGATACGTACGTATCGATCGTACGCTAGCTAGCTACGCTGATCGATGCTAGCTAGCTAGCTAGCTAGCTAGCTAGCTT

Chimp ATGCCGGATACGTAAGTATCGATCGTACGCTAGCTAGCTACGCTGATCGA-GCTAGCTAGCTAGCTAGCTAGCTAGCTAGCTT

Gorilla ATGCCGGATATGTACGTATCGATCGTACGCTAGCTA--CTACGCTGATCGATGCTAGCTAGCTAGCTAGCTAGCTAGCTAGCTT

Orangu ATGCTGGATACGTACGTATCGATAGTACGCTAGCTAGCTACGCTGATCGATGCTAGCTGGCTAGCTAGCTAGCTAGCTAGCTT

     **** ******* ****** ***************************** ************************

Red letters show mutations (substitutions); red dashes indicate gaps (insertions/deletions)
Asterisks below show conserved positions across all sequences

Figure 10.1 – Illustration of synteny and gene alignment concepts

As you look at the preceding diagram, you can see first of all that blocks of genes in a genome are conserved to varying degrees in different primate or human genomes. For example, the light-orange block (second block from the left) is quite conserved, whereas gorillas have a small, light-blue group of genes (fourth block over for Gorillas) that must be unique to them. In the second half of the diagram, we see an example of a sequence-level gene alignment. Genes are lined up across the species, showing many positions where they are in common (asterisks, bottom row). When a base is different from the common base, we call that a **substitution** and highlight it in red (for example, the second T in the Organutan row). Substitution rates are used to calculate the lengths of branches in phylogenetic trees. The more substitutions occur in a gene, the greater its evolutionary distance, and hence the longer the branch length in the tree. Rates of substitution can also be used to model how quickly different organisms or areas of a genome are evolving.

Getting ready

To perform the genomic alignment, you will need to install MAFFT. Additionally, to perform the genic alignment, MUSCLE will be used. Also, we will use trimAl (https://vicfero.github.io/trimal/) to remove spurious sequences and poorly aligned regions in an automated manner. Let's install our packages:

```
! brew install trimal
! brew install mafft
! brew install muscle
```

Alternatively, you can install using (From the terminal type:)

```
conda install trimal muscle mafft
```

As usual, this information is available in the corresponding Jupyter Notebook file at `Ch10/Ch10-2-aligning-genetic-data.ipynb`. You will need to run the previous notebook beforehand, as it will generate the files that are required here. In this chapter, we will use Biopython.

How to do it...

1. Let's get started by importing our libraries:

   ```python
   import subprocess
   from Bio.Align.Applications import MafftCommandline
   ```

2. Next, we need to set up our MAFFT command line and run it (this may take 30–60 minutes):

   ```python
   mafft_cline = MafftCommandline(
       input="sample.fasta", ep=0.123,
       reorder=True, maxiterate=1000,
       localpair=True
   )
   print("Running MAFFT with command:", mafft_cline)
   process = subprocess.run(
       str(mafft_cline),
       shell=True,
       capture_output=True,
       text=True
   )
   if process.returncode != 0:
       print("Error running MAFFT:", process.stderr)
   else:
       with open("align.fasta", "w") as w:
           w.write(process.stdout)
   print("Alignment completed and saved to align.fasta")
   ```

   The preceding parameters are standard for MAFFT. We set the gap extension penalty to a reasonable value, keep the aligned sequences in the order they are input, and iterate up to 1,000 times. We will use the Biopython interface to call MAFFT.

3. Let's use trimAl to trim sequences, as follows:

```
import os
os.system(
    'trimal -automated1 -in align.fasta -out trim.fasta -fasta'
)
```

Here, we just call the application using `os.system`. The `-automated1` parameter will automatically select trimming parameters based on the characteristics of your alignment.

4. Additionally, we can run MUSCLE to align the proteins:

```
import subprocess
import os
my_genes = ['NP', 'L', 'VP35', 'VP40']
for gene in my_genes:
    input_file = f"{gene}_P.fasta"
    output_file = f"{gene}_P_align.fasta"

    if not os.path.exists(input_file):
        print(f"Error: Input file '{input_file}' not found.")
    else:
        muscle_cmd = (
            f"muscle -align {input_file} -output {output_file}"
        )
        print(f"Running MUSCLE with command: {muscle_cmd}")
        process = subprocess.run(
            muscle_cmd, shell=True,
            capture_output=True, text=True
        )
        if process.returncode != 0:
            print("Error running MUSCLE:", process.stderr)
        else:
            print(f"Alignment completed and saved to"
                    f"{output_ file}")
```

We use Biopython to call an external application. Here, we will align a set of proteins.

Note that to make some analysis of molecular evolution, we have to compare aligned genes, not proteins (for example, comparing synonymous and nonsynonymous mutations). However, we have just aligned the proteins. Therefore, we have to convert the alignment into the gene sequence form.

5. Let's align the genes by finding three nucleotides that correspond to each amino acid:

```
from Bio import SeqIO
from Bio.Seq import Seq
from Bio.SeqRecord import SeqRecord
for gene in my_genes:
    gene_seqs = {}
    unal_gene = SeqIO.parse('%s.fasta' % gene, 'fasta')
    for rec in unal_gene:
        gene_seqs[rec.id] = rec.seq
    al_prot = SeqIO.parse('%s_P_align.fasta' % gene, 'fasta')
    al_genes = []

    for protein in al_prot:
        my_id = protein.id
        seq = ''
        pos = 0
        for c in protein.seq:
            if c == '-':
                seq += '---'
            else:
                seq += str(
                    gene_seqs[my_id][pos:pos + 3])
                pos += 3
        al_genes.append(SeqRecord(Seq(seq), id=my_id))
    SeqIO.write(al_genes, '%s_align.fasta' % gene, 'fasta')
```

The code gets the protein and the gene coding. If a gap is found in a protein, three gaps are written; if an amino acid is found, the corresponding nucleotides of the gene are written.

Comparing sequences

Here, we will compare the sequences we aligned in the previous recipe. We will perform gene-wide and genome-wide comparisons.

In this recipe, we'll use DendroPy to organize and process the sequence data across species of the Ebola virus. We will first set up a data structure to help us organize data by gene and species. We will use DendroPy's popgenstats module to calculate population genetics statistics on our sequences. This module can calculate the number of **segregating sites** – positions where the sequences differ in the alignment, a measure of evolutionary distance. We will learn about **Tajima's D** value, a measure of genetic diversity (https://en.wikipedia.org/wiki/Tajima%27s_D). This measure of substitution rates helps us see whether a genetic region is just evolving randomly, or may be evolving in a particular direction under selection. We will also learn about **Watterson's theta**, which measures rates of population mutation (https://en.wikipedia.org/wiki/Watterson_estimator).

By the end of this recipe, you will be able to determine how genetically distinct different Ebola outbreaks are, and you'll have an understanding of how to determine whether selection pressures occurred during Ebola outbreaks using phylogenetic analysis. This will give you a good understanding of key statistics and approaches that are applicable to any phylogenetic analysis.

Getting ready

We will use DendroPy and will require the results from the previous two recipes. As usual, this information is available in the corresponding notebook at Ch10/Ch10-3-comparing-sequences.ipynb.

How to do it...

Take a look at the following steps:

1. Let's start analyzing the gene data. For simplicity, we will only use data from two other species of the genus Ebola virus that are available in the extended dataset, that is, the Reston virus (RESTV) and the Sudan virus (SUDV):

```
import os
from collections import OrderedDict
import dendropy
from dendropy.calculate import popgenstat

genes_species = OrderedDict()
my_species = ['RESTV', 'SUDV']
my_genes = ['NP', 'L', 'VP35', 'VP40']

for name in my_genes:
    gene_name = name.split('.')[0]
    char_mat = dendropy.DnaCharacterMatrix.get_from_path(
        '%s_align.fasta' % name, 'fasta')
    genes_species[gene_name] = {}

    for species in my_species:
        genes_species[gene_name][species] = (
            dendropy.DnaCharacterMatrix())

    for taxon, char_map in char_mat.items():
        species = taxon.label.split('_')[0]
        if species in my_species:
            genes_species[gene_name][species].\
                taxon_namespace.add_taxon(taxon)
            genes_species[gene_name][species][taxon] = char_map
```

We get four genes that we stored in the first recipe and aligned in the second.

We load all the files (which are FASTA-formatted) and create a dictionary with all of the genes. Each entry will be a dictionary itself with the RESTV or SUDV species, including all reads. This is not a lot of data, just a handful of genes.

2. Let's print some basic information for all four genes, such as the number of segregating sites (seg_sites), nucleotide diversity (nuc_div), Tajima's D (taj_d), and Watterson's theta (wat_theta):

```python
import numpy as np
import pandas as pd
summary = np.ndarray(
    shape=(len(genes_species), 4 * len(my_species))
)
stats = ['seg_sites', 'nuc_div', 'taj_d', 'wat_theta']

for row, (gene, species_data) in enumerate(
    genes_species.items()
):
    for col_base, species in enumerate(my_species):
        summary[row, col_base * 4] = (
            popgenstat.num_segregating_sites(
                species_data[species])
        )
        summary[row, col_base * 4 + 1] = (
            popgenstat.nucleotide_diversity(
                species_data[species])
        )
        summary[row, col_base * 4 + 2] = (
            popgenstat.tajimas_d(
                species_data[species])
        )
        summary[row, col_base * 4 + 3] = (
            popgenstat.wattersons_theta(
                species_data[species])
        )
columns = []
for species in my_species:
    columns.extend([
        '%s (%s)' % (stat, species) for stat in stats
    ])
df = pd.DataFrame(
    summary,
    index=genes_species.keys(),
    columns=columns
)
df
```

3. First, let's look at the output, and then we'll explain how to build it:

	seg_sites (RESTV)	nuc_div (RESTV)	taj_d (RESTV)	wat_theta (RESTV)	seg_sites (SUDV)	nuc_div (SUDV)	taj_d (SUDV)	wat_theta (SUDV)
NP	113.0	0.020659	-0.482275	49.489051	118.0	0.029630	1.203522	56.64
L	288.0	0.018143	-0.295386	126.131387	282.0	0.024193	1.412350	135.36
VP35	43.0	0.017427	-0.553739	18.832117	50.0	0.027761	1.069061	24.00
VP40	61.0	0.026155	-0.188135	26.715328	41.0	0.023517	1.269160	19.68

Figure 10.2 – A DataFrame for the virus dataset

We used a `pandas` DataFrame to print the results because it's really tailored to deal with an operation like this. We will initialize our DataFrame with a NumPy multidimensional array with four rows (genes) and four statistics times the two species.

The statistics, such as the number of segregating sites, nucleotide diversity, Tajima's D, and Watterson's theta, are computed by DendroPy. Note the placement of individual data points in the array (the coordinate computation).

Look at the very last line: if you are in Jupyter, just putting `df` at the end will render the DataFrame and the cell output, too. If you are not in a notebook, use `print(df)` (you can also perform this in a notebook, but it will not look as pretty).

4. Now, let's extract similar information, but genome-wide instead of only gene-wide. In this case, we will use a subsample of two EBOV outbreaks (from 2007 and 2014). We will perform a function to display basic statistics, as follows:

```
def do_basic_popgen(seqs):
    num_seg_sites = popgenstat.num_segregating_sites(seqs)
    avg_pair = popgenstat.\
        average_number_of_pairwise_differences(seqs)
    nuc_div = popgenstat.nucleotide_diversity(seqs)

    print(
        'Segregating sites: %d, Avg pairwise diffs: %.2f,'
        'Nucleotide diversity %.6f' % (
            num_seg_sites, avg_pair, nuc_div))
    print(
        "Watterson's theta: %s"
        % popgenstat.wattersons_theta(seqs))
    print(
        "Tajima's D: %s" % popgenstat.tajimas_d(seqs))
```

By now, this function should be easy to understand, given the preceding examples.

5. Now, let's extract a subsample of the data properly and output the statistical information:

```python
import dendropy
taxon_namespace = dendropy.TaxonNamespace()
ebov_seqs = dendropy.DnaCharacterMatrix.get(
    path="trim.fasta", schema="fasta",
    taxon_namespace=taxon_namespace
)
sl_2014 = []
drc_2007 = []
ebov2007_set = dendropy.DnaCharacterMatrix(
    taxon_namespace=taxon_namespace)
ebov2014_set = dendropy.DnaCharacterMatrix(
    taxon_namespace=taxon_namespace)

for taxon, char_map in ebov_seqs.items():
    print(taxon.label)
    if taxon.label.startswith("EBOV_2014") and len(sl_2014) < 8:
        sl_2014.append(char_map)
        new_taxon = taxon_namespace.require_taxon(
            label=taxon.label)
        ebov2014_set[new_taxon] = char_map

    elif taxon.label.startswith("EBOV_2007"):
        drc_2007.append(char_map)
        new_taxon = taxon_namespace.require_taxon(
            label=taxon.label)
        ebov2007_set[new_taxon] = char_map

del ebov_seqs
print("2007 outbreak:")
print(
    f"Number of individuals: "
    f"{len(ebov2007_set.taxon_namespace)}"
)
do_basic_popgen(ebov2007_set)
print("\n2014 outbreak:")
print(
    f"Number of individuals:"
    f"{len(ebov2014_set.taxon_namespace)}"
)
do_basic_popgen(ebov2014_set)
```

Here, we will construct two versions of two datasets: the 2014 outbreak and the 2007 outbreak. We will generate one version as DnaCharacterMatrix and another as a list. We will use this list version at the end of this recipe.

As the dataset for the EBOV outbreak of 2014 is large, we subsample it with just eight individuals, which is a comparable sample size to the dataset of the 2007 outbreak.

Again, we delete the ebov_seqs data structure to conserve memory (these are genomes, not only genes).

If you perform this analysis on the complete dataset for the 2014 outbreak available on GenBank (99 samples), be prepared to wait for quite some time.

The output is shown here:

```
2007 outbreak:
Number of individuals: 7
Segregating sites: 25, Avg pairwise diffs: 7.71, Nucleotide
diversity 0.000412
Watterson's theta: 10.204081632653063
Tajima's D: -1.383114157484101
2014 outbreak:
Number of individuals: 8
Segregating sites: 6, Avg pairwise diffs: 2.79, Nucleotide
diversity 0.000149
Watterson's theta: 2.31404958677686
Tajima's D: 0.9501208027581887
```

6. Finally, we perform some statistical analysis on the two subsets of 2007 and 2014, as follows:

```
pair_stats = popgenstat.PopulationPairSummaryStatistics(
    sl_2014, drc_2007)

print('Average number of pairwise differences irrespective of'
      'population: %.2f'
      '% pair_stats.average_number_of_pairwise_differences')
print('Average number of pairwise differences between '
      'populations: %.2f' %
      'pair_stats.average_number_of_pairwise_differences_
between')
print('Average number of pairwise differences within '
      'populations: %.2f' %
      'pair_stats.average_number_of_pairwise_differences_
within')
print('Average number of net pairwise differences : %.2f' %
      'pair_stats.average_number_of_pairwise_differences_net')
```

```
print('Number of segregating sites: %d' %
      'pair_stats.num_segregating_sites')
print("Watterson's theta: %.2f" %
      'pair_stats.wattersons_theta')
print("Wakeley's Psi: %.3f" % pair_stats.wakeleys_psi)
print("Tajima's D: %.2f" % pair_stats.tajimas_d)
```

Note that we will perform something slightly different here; we will ask DendroPy (`popgenstat.PopulationPairSummaryStatistics`) to directly compare two populations so that we get the following results:

```
Average number of pairwise differences irrespective of
population: 284.46
Average number of pairwise differences between populations:
535.82
Average number of pairwise differences within populations: 10.50
Average number of net pairwise differences : 525.32
Number of segregating sites: 549
Watterson's theta: 168.84
Wakeley's Psi: 0.308
Tajima's D: 3.05
```

Now the number of segregating sites is much bigger because we are dealing with data from two different populations that are reasonably diverged. The average number of pairwise differences among populations is quite large. As expected, this is much larger than the average number for the population, irrespective of the population information.

Reconstructing phylogenetic trees

Here, we will construct phylogenetic trees for the aligned dataset for all Ebola species. We will follow a procedure that's quite similar to the one Gire et al. used in their paper (referenced at the beginning of this chapter).

We will use a program called RAxML-ng, which is an updated version of the original RAxML program. It uses **maximum likelihood** phylogenetic analysis. In this method, the best tree is the one that has the highest likelihood of explaining the underlying observed evolutionary structure, given the data we have. We'll then use Matplotlib to make a visualization of the relationships between the sequences.

By the end of this recipe, you'll have produced your maximum likelihood phylogenetic tree and will have your first visualization of a tree!

Getting ready

This recipe requires RAxML-ng (`https://github.com/amkozlov/raxml-ng`), the updated version of RAxML, a program for maximum likelihood-based inference of large phylogenetic trees. You may also want to check out the collection of phylogenetic programs at the Exelixis website: `https://cme.h-its.org/exelixis/software.html`. Bioconda also includes it, but it is named `raxml`. Note that the binary is called `raxmlHPC`. You can perform the following command to install it:

```
! brew install raxml-ng
```

Alternatively, you can install using conda (from the terminal, type):

```
conda install raxml-ng
```

You can verify that raxml-ng is installed by typing (from the terminal):

```
raxml-ng --version
```

The following code is simple, but it will take time to execute because it will call RAxML (which is computationally intensive). Although there is a recipe for visualization later in this chapter, we will, nonetheless, plot one of our generated trees here.

As usual, this information is available in the corresponding notebook at `Ch10/Ch10-4-reconstructing-trees.ipynb`. You will need the output of the previous recipe to complete this one.

How to do it...

Take a look at the following steps:

1. First, we will reconstruct the genus dataset, as follows:

   ```
   import os
   import subprocess
   ```

   We import the libraries we will use to call RAxML-ng.

2. Next, we define our input file and output prefix:

   ```
   data_path = "trim.fasta"
   output_prefix = "ebola_tree"
   ```

3. Next, let's check that the input file exists:

   ```
   if not os.path.exists(data_path):
       raise FileNotFoundError(
           f"Error: The file {data_path} does not exist!")
   ```

4. Let's construct our `raxml-ng` command:

```
cmd = [
    "raxml-ng",
    "--msa", data_path,
    "--model", "GTR+G",
    "--prefix", output_prefix,
    "--search"
]
```

This will use the input sequence alignment file with the **GTR+G** substitution model. This is the **General Time Reversible plus Gamma** model, which looks at how substitutions of amino acids are expected to occur over evolutionary time. You can read about it here: `https://academic.oup.com/bioinformatics/article/36/Supplement_2/i884/6055914`.

5. Now we will run RAxML-NG:

```
try:
    subprocess.run(cmd, check=True)
    print(f"RAxML-NG completed successfully. Output "
            f"files are saved with prefix '{output_ prefix}'")
except subprocess.CalledProcessError as e:
    print(f"Error running RAxML-NG: {e}")
```

This executes our command using `os subprocess` and prints out a success or error message.

6. We will also clean up temporary files:

```
for ext in [
    ".raxml.log", ".raxml.bestTree",
    ".raxml.rba", ".raxml.rfdist"
]:
    file_path = f"{output_prefix}{ext}"
    if os.path.exists(file_path):
        os.remove(file_path)
print("Temporary files cleaned up.")
```

Great! Now we've run RAxML-ng on our sequences and are ready to visualize our tree!

You should at this point see `ebola_tree.raxml*` files in your working directory.

7. Now we will visualize our tree!

```
import matplotlib.pyplot as plt
from Bio import Phylo
tree_file = "ebola_tree.raxml.bestTreeCollapsed"
# Based on the raxml-ng output from the previous step
```

```
my_ebola_tree = Phylo.read(tree_file, "newick")
my_ebola_tree.name = "Our Ebolavirus Tree"

fig = plt.figure(figsize=(16, 18))
ax = fig.add_subplot(1, 1, 1)
Phylo.draw(my_ebola_tree, axes=ax)
plt.show()
```

This code uses Matplotlib and the Biopython Phylo module. The Phylo module is great for handling phylogenetic trees (`https://biopython.org/wiki/Phylo`).

We read our tree in from the previous step, which will be in the Newick format. We'll use the `Phylo.read()` function to do this. We assign a name to the tree, and then set up our plot. We then use the `Phylo.draw()` function to draw our tree, and we show the plot.

Here is the phylogenetic tree we get:

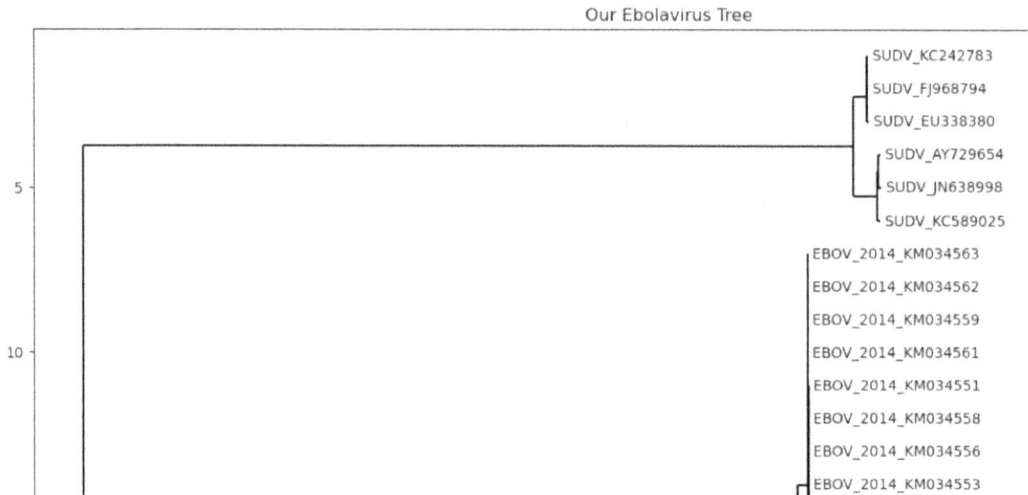

Figure 10.3 – Phylogenetic tree visualization

That's it! We have now seen how we can reconstruct and visualize phylogenetic trees. Next, we'll look at how to iterate recursively over trees.

There's more…

Although the purpose of this book is not to teach phylogenetic analysis, it's important to know why we do not inspect consensus and support information in the tree topology. You should research this in your dataset. For more information, refer to `https://www.geol.umd.edu/~tholtz/G331/lectures/cladistics5.pdf`.

See also

- Learn about the latest updates to RAxML-NG here: Togkouisidis et al., "Adaptive RAxML-NG: Accelerating Phylogenetic Inference under Maximum Likelihood using Dataset Difficulty," Molecular Biology and Evolution, October 2023 – `https://academic.oup.com/mbe/article/40/10/msad227/7296053`

- Read more about tree formats here: `https://evomics.org/resources/tree-formats/`

Playing recursively with trees

This is not a book about programming in Python, as the topic is vast. Having said that, it's not common for introductory Python books to discuss recursive programming at length. Usually, recursive programming techniques are well tailored to deal with trees. It is also a required programming strategy with functional programming dialects, which can be quite useful when you perform concurrent processing. This is common when processing very large datasets.

The phylogenetic notion of a tree is slightly different from that in computer science. Phylogenetic trees can be rooted (if so, then they are normal tree data structures) or unrooted, making them undirected acyclic graphs. Additionally, phylogenetic trees can have weights on their edges. Therefore, be mindful of this when you read the documentation; if the text is written by a phylogeneticist, you can expect the tree (rooted and unrooted), while most other documents will use undirected acyclic graphs for unrooted trees. In this recipe, we will assume that all of the trees are rooted.

Finally, note that while this recipe is mostly devised to help you understand recursive algorithms and tree-like structures, the final part is actually quite practical and fundamental for the next recipe to work.

Getting ready

You will need to have the files from the previous recipe. As usual, you can find this content in the `Ch10/Ch10-5-recursive-trees.ipynb` notebook file. Here, we will use DendroPy's tree representations. Note that most of this code is easily generalizable compared to other tree representations and libraries (phylogenetic or not).

How to do it...

Take a look at the following steps:

1. First, let's load the RAxML-generated tree for all Ebola viruses, as follows:

```
import dendropy
tree_file = "ebola_tree.raxml.bestTreeCollapsed"
ebola_raxml = dendropy.Tree.get_from_path(
    tree_file, schema="newick")
print(ebola_raxml.as_string(schema="newick"))
```

2. Then, we need to compute the level of each node (the distance to the root node):

```
def compute_level(node, level=0):
    for child in node.child_nodes():
        compute_level(child, level + 1)
    if node.taxon is not None:
        print("%s: %d %d" % (node.taxon, node.level(), level))
compute_level(ebola_raxml.seed_node)
```

DendroPy's node representation has a level method (which is used for comparison), but the point here is to introduce a recursive algorithm, so we will implement it anyway.

Note how the function works; it's called seed_node (which is the root node, since the code works under the assumption that we are dealing with rooted trees). The default level for the root node is 0. The function will then call itself for all its child nodes, increasing the level by 1. Then, for each node that is not a leaf (that is, it is internal to the tree), the calling will be repeated, and this will recurse until we get to the leaf nodes.

For the leaf nodes, we then print the level (we could have done the same for the internal nodes) and show the same information computed by DendroPy's internal function.

3. Now, let's compute the height of each node. The height of the node is the number of edges of the maximum downward path (going to the leaves), starting on that node, as follows:

```
def compute_height(node):
    children = node.child_nodes()
    if len(children) == 0:
        height = 0
    else:
        height = 1 + max(map(
                lambda x: compute_height(x), children))
    desc = node.taxon or 'Internal'
    print("%s: %d %d" % (desc, height, node.level()))
    return height
compute_height(ebola_raxml.seed_node)
```

Here, we will use the same recursive strategy, but each node will return its height to its parent. If the node is a leaf, then the height is 0; if not, then it's 1 plus the maximum height of its entire offspring.

Note that we use a map over a lambda function to get the heights of all the children of the current node. Then, we choose the maximum (the max function performs a reduce operation here because it summarizes all of the values that are reported). If you are relating this to MapReduce frameworks, you are correct; they are inspired by functional programming dialects like these.

4. Now, let's compute the number of offspring for each node. By now, this should be quite easy to understand:

```
def compute_nofs(node):
    children = node.child_nodes()
    nofs = len(children)
    map(lambda x: compute_nofs(x), children)
    desc = node.taxon or 'Internal'
    print("%s: %d %d" % (desc, nofs, node.level()))
compute_nofs(ebola_raxml.seed_node)
```

5. Now we will print all of the leaves (this is, apparently, trivial):

```
def print_nodes(node):
    for child in node.child_nodes():
        print_nodes(child)
    if node.taxon is not None:
        print('%s (%d)' % (node.taxon, node.level()))
print_nodes(ebola_raxml.seed_node)
```

Note that all the functions that we have developed so far impose a very clear traversal pattern on the tree. It calls its first offspring, then that offspring will call their offspring, and so on; only after this will the function be able to call its next offspring in a depth-first pattern. However, we can do things differently.

6. Now, let's print the leaf nodes in a breadth-first manner, that is, we will print the leaves with the lowest level (closer to the root) first, as follows:

```
from collections import deque
def print_breadth(tree):
    queue = deque()
    queue.append(tree.seed_node)
    while len(queue) > 0:
        process_node = queue.popleft()
        if process_node.taxon is not None:
            print('%s (%d)' % (process_node.taxon,
                               process_node.level()))
        else:
            for child in process_node.child_nodes():
                queue.append(child)
print_breadth(ebola_raxml)
```

Before we explain this algorithm, let's look at how different the result from this run will be compared to the previous one. For starters, take a look at the following diagram. If you print the nodes by depth-first order, To match the updated diagram for this figure, we need to change this to the following:

You will get one result, but with breadth-first traversal, you will get a different result. Tree traversal will have an impact on how the nodes are visited; more often than not, this is important.

Regarding the preceding code, here, we will use a completely different approach, as we will perform an iterative algorithm. We will use a **first-in, first-out** (**FIFO**) queue to help order our nodes. Note that Python's deque can be used as efficiently as FIFO, as well as in **last-in, first-out** (**LIFO**). That's because it implements an efficient data structure when you operate at both extremes.

The algorithm starts by putting the root node onto the queue. While the queue is not empty, we will take the node out front. If it's an internal node, we will put all of its children into the queue.

We will iterate the preceding step until the queue is empty. I encourage you to take a pen and paper and see how this works by performing the example shown in the following diagram. The code is small, but not trivial:

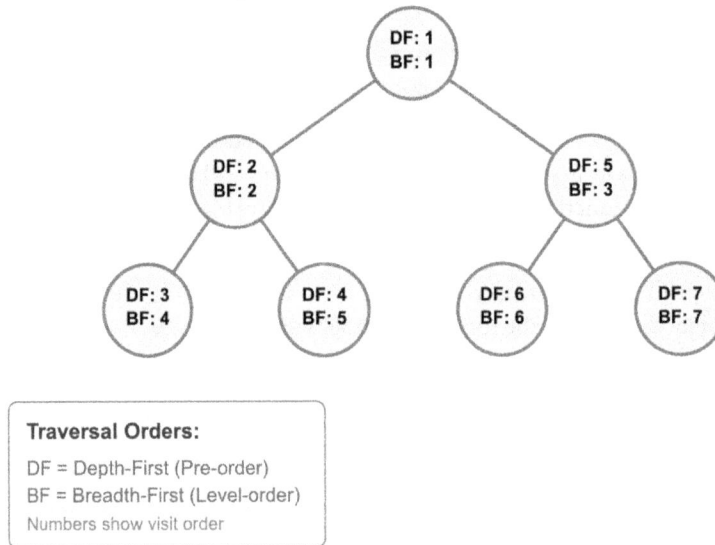

Figure 10.4 – Visiting a tree; the first number indicates the order in which that node is visited by traversing depth-first, while the second assumes breadth-first

7. Let's get back to the real dataset. As we have a bit too much data to visualize, we will generate a trimmed-down version, where we remove the subtrees that have a single species (in the case of EBOV, they have the same outbreak). We will also ladderize the tree, that is, sort the child nodes in order of the number of children:

```
from copy import deepcopy
simple_ebola = deepcopy(ebola_raxml)
def simplify_tree(node):
    prefs = set()
    for leaf in node.leaf_nodes():
```

```
        my_toks = leaf.taxon.label.split(' ')
        if my_toks[0] == 'EBOV':
            prefs.add('EBOV' + my_toks[1])
        else:
            prefs.add(my_toks[0])
    if len(prefs) == 1:
        print(prefs, len(node.leaf_nodes()))
        node.taxon = dendropy.Taxon(label=list(prefs)[0])
        node.set_child_nodes([])
    else:
        for child in node.child_nodes():
            simplify_tree(child)

simplify_tree(simple_ebola.seed_node)
simple_ebola.ladderize()
simple_ebola.write_to_path('ebola_simple.nex', 'nexus')
```

We will perform a deep copy of the tree structure. As our function and the ladderization are destructive (they will change the tree), we will want to maintain the original tree.

DendroPy is able to enumerate all the leaf nodes (at this stage, a good exercise would be to write a function to perform this). With this functionality, we will get all the leaves for a certain node. If they share the same species and outbreak year as in the case of EBOV, we remove all of the child nodes, leaves, and internal subtree nodes.

If they do not share the same species, we recurse down until that happens. The worst case is that when you are already at a leaf node, the algorithm trivially resolves to the species of the current node.

There's more...

There is a massive amount of computer science literature on the topic of trees and data structures; if you want to read more, Wikipedia provides a great introduction at https://en.wikipedia.org/wiki/Tree_(abstract_data_type).

Note the use of the map and lambda functions in this recipe. The map function applies a function to each item in an iterator. Lambdas are small, concise functions that can be written in one line. They are often used together to apply simple functions and can be especially useful in data transformation operations. You can read more about them here: https://www.analyticsvidhya.com/blog/2021/10/an-explanation-to-pythons-lambda-map-filter-and-reduce/. Although we only touch on functional programming briefly in this book, it is an important area for further study. To learn more read "Functional Python Programming: Use a functional approach to write succinct, expressive, and efficient Python code", 3rd Edition, by Steven F. Lott, Packt Publishing, Dec 2022 - https://a.co/d/3PM1yjh

Visualizing phylogenetic data

In this recipe, we will discuss how to visualize phylogenetic trees. DendroPy only has simple visualization mechanisms based on drawing textual ASCII trees, but Biopython has quite a rich infrastructure, which we will leverage here. We'll use the Biopython Phylo module (`https://biopython.org/wiki/Phylo`), which has nice functions for visualizing trees.

Visualizing phylogenetic trees is an important part of the field and can be extremely useful for understanding the relationships between DNA sequences, genes, proteins, or entire organisms. Several popular software packages exist for visualizing phylogenetic trees. **Geneious** (`https://www.geneious.com/features/phylogenetic-tree-building`) contains some great visualization tools for phylogenetics (as well as tree building and alignment algorithms). **Interactive Tree of Life (iTOL)** also provides some excellent tools for phylogenetics (`https://itol.embl.de/`).

In this recipe, we will load up our phylogenetic trees and compare ASCII versus graphical representations. We'll also play around with coloring and ladderization, in which more diverse taxons are placed to the right of less diverse taxons, making the tree more readable.

By the end of this recipe, you'll be a master of phylogenetic tree visualization and interpretation!

Getting ready

This will require you to have completed all of the previous recipes. Remember that we have the files for the whole genus of the Ebola virus, including the RAxML tree. Furthermore, a simplified genus version will have been produced in the previous recipe. As usual, you can find this content in the `Ch10/Ch10-6-visualizing-phylogenetics.ipynb` notebook file.

How to do it...

Take a look at the following steps:

1. Let's load all of the phylogenetic data:

    ```
    from copy import deepcopy
    from Bio import Phylo

    best_tree_file = "ebola_tree.raxml.bestTreeCollapsed"
    ebola_tree = Phylo.read(best_tree_file, "newick")
    ebola_tree.name = "Ebolavirus Tree"
    ```

    For all of the trees that we read, we will change the name of the tree, as the name will be printed later.

2. Now, we can draw ASCII representations of the trees:

```
Phylo.draw_ascii(ebola_tree)
```

The ASCII representation of the simplified genus tree is shown in the following diagram:

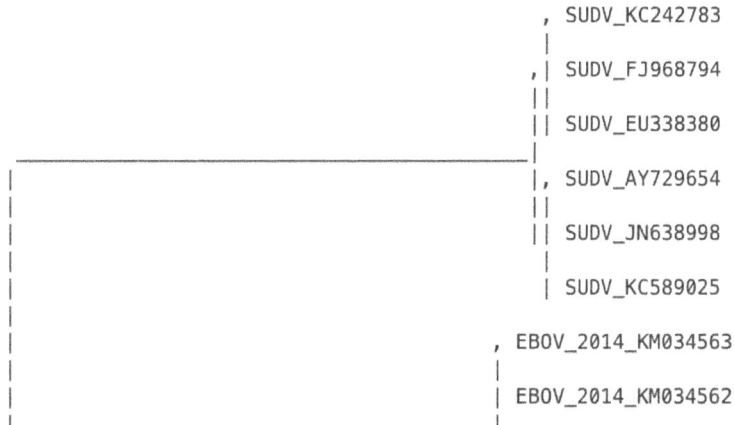

```
                                                        , SUDV_KC242783
                                                        |
                                                      ,| SUDV_FJ968794
                                                      ||
                                                      || SUDV_EU338380
          _____|
         |                                            |, SUDV_AY729654
         |                                            ||
         |                                            || SUDV_JN638998
         |                                            |
         |                                            | SUDV_KC589025
         |
         |                                          , EBOV_2014_KM034563
         |                                          |
         |                                          | EBOV_2014_KM034562
         |
```

Figure 10.5 – The ASCII representation of a simplified Ebola virus dataset

Here, we will not print the complete version because it will take several pages. But if you run the preceding code, you will be able to see that it's actually quite readable.

Bio.Phylo allows for the graphical representation of trees by using Matplotlib as a backend:

```python
import matplotlib.pyplot as plt
from Bio import Phylo
simplified_tree_file = "ebola_tree.raxml.bestTreeCollapsed"
ebola_simple_tree = Phylo.read(simplified_tree_file, "newick")
fig = plt.figure(figsize=(16, 22))
ax = fig.add_subplot(111)

def label_branches(clade):
    if clade.branch_length and clade.branch_length > 0.02:
        return f"{clade.branch_length:.3f}"
    return None
Phylo.draw(
    ebola_simple_tree,
    branch_labels=label_branches,
    axes=ax
)
plt.show()
```

In this case, we will print the branch lengths at the edges, but we will remove all of the lengths that are less than 0.02 to avoid clutter. Below you will see a closeup of a portion of the resulting phylogenetic tree (you can review the entire tree in your notebook):

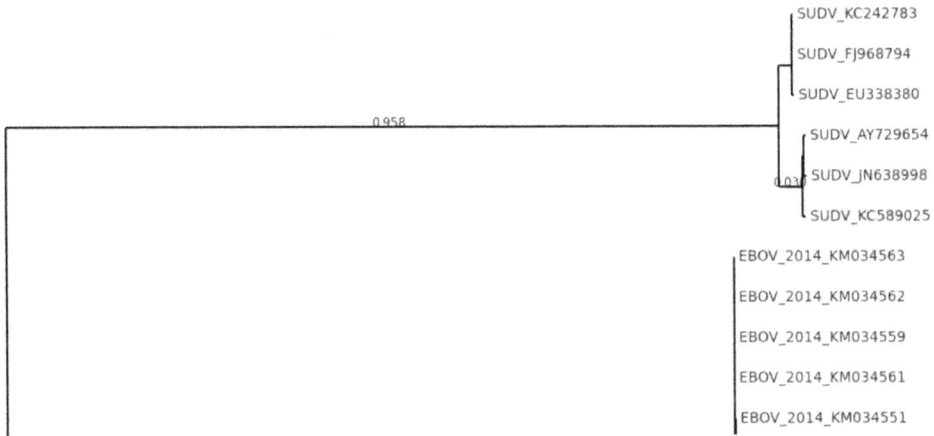

Figure 10.6 – A Matplotlib-based version of the simplified dataset with branch lengths added

3. Now we will plot the complete dataset, but we will color each bit of the tree differently. If a subtree only has a single virus species, it will get its own color. EBOV will have two colors, that is, one for the 2014 outbreak and one for the others, as follows:

```
fig = plt.figure(figsize=(16, 22))
ax = fig.add_subplot(111)
from collections import OrderedDict
my_colors = OrderedDict({
    'EBOV_2014': 'red', 'EBOV': 'magenta',
    'BDBV': 'cyan', 'SUDV': 'blue',
    'RESTV' : 'green', 'TAFV' : 'yellow'
})
def get_color(name):
    for pref, color in my_colors.items():
        if name.find(pref) > -1:
            return color
    return 'grey'
```

```
def color_tree(node, fun_color=get_color):
    if node.is_terminal():
        node.color = fun_color(node.name)
    else:
        my_children = set()
        for child in node.clades:
            color_tree(child, fun_color)
            my_children.add(child.color.to_hex())
        if len(my_children) == 1:
            node.color = child.color
        else:
            node.color = 'grey'
ebola_color_tree = deepcopy(ebola_tree)
color_tree(ebola_color_tree.root)
Phylo.draw(
    ebola_color_tree, axes=ax,
    label_func=lambda x: x.name.split(' ')[0][1:]
        if x.name is not None else None
)
```

This is a tree-traversing algorithm, not unlike the ones presented in the previous recipe. As a recursive algorithm, it works in the following way. If the node is a leaf, it will get a color based on its species (or the EBOV outbreak year). If it's an internal node and all the descendant nodes below it are of the same species, it will get the color of that species; if there are several species after that, it will be colored in gray. Actually, the color function can be changed and will be changed later. Only the edge colors will be used (the labels will be printed in black).

Note that ladderization (performed in the previous recipe with DendroPy) helps quite a lot in terms of a clear visual appearance.

We also deep copy the genus tree to color a copy; remember from the previous recipe that some tree traversal functions can change the state, and in this case, we want to preserve a version without any coloring.

Note the usage of the lambda function to clean up the name that was changed by trimAl, as shown in the following diagram.

A portion of the phylogenetic tree we get is shown in the following figure:

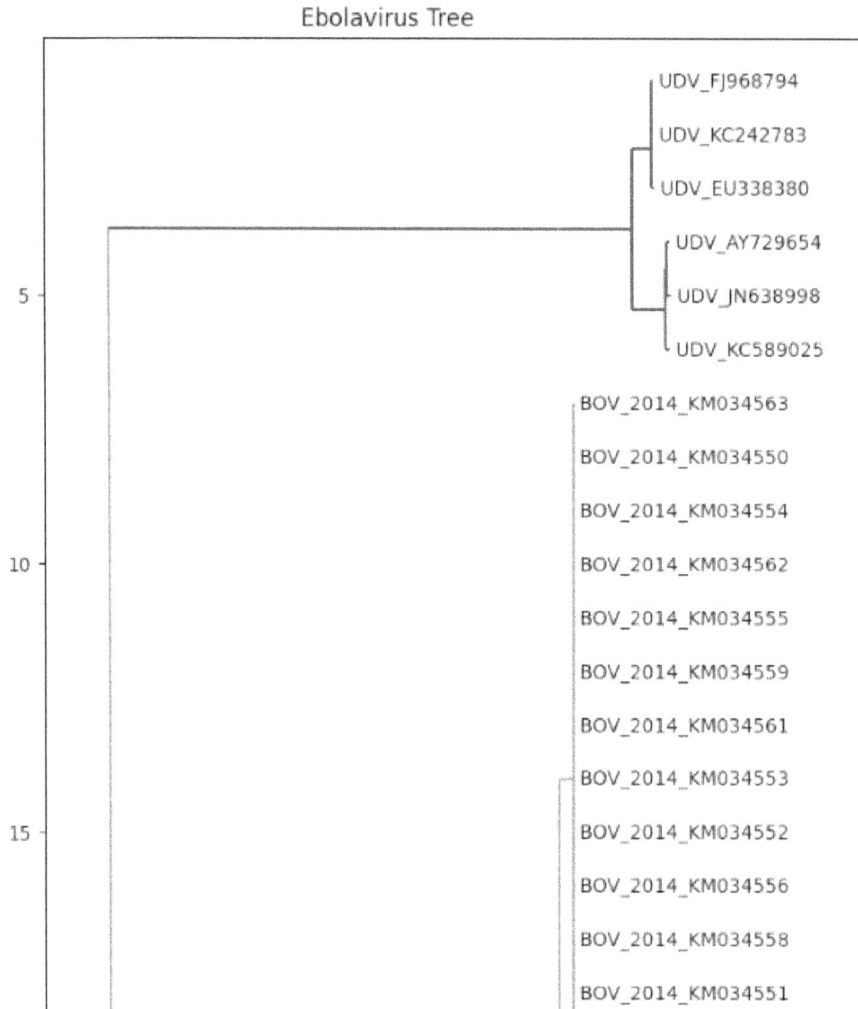

Figure 10.7 – A ladderized and colored phylogenetic tree with the complete Ebola virus dataset

There's more...

Tree and graph visualization is a complex topic; arguably, here, the tree's visualization is rigorous but far from pretty. One alternative to DendroPy, which has more visualization features, is ETE (https://etetoolkit.org/). General alternatives for drawing trees and graphs include Cytoscape (https://cytoscape.org/) and Gephi (https://gephi.org/). If you want to know more about the algorithms for rendering trees and graphs, check out the Wikipedia page at https://en.wikipedia.org/wiki/Graph_drawing for an introduction to this fascinating topic.

Be careful not to trade style for substance, though. For example, the second edition of this book had a pretty rendering of a phylogenetic tree using a graph-rendering library. While it was clearly the most beautiful image in that chapter, it was misleading in terms of branch lengths.

OK, let's clean up and close down our `conda` environment:

```
conda deactivate
```

Get This Book's PDF Version and Exclusive Extras

UNLOCK NOW

Scan the QR code (or go to `packtpub.com/unlock`). Search for this book by name, confirm the edition, and then follow the steps on the page.

Note: Keep your invoice handy. Purchases made directly from Packt don't require an invoice.

11
Population Genetics

Population genetics is the study of the changes in the frequency of alleles in a population on the basis of selection, drift, mutation, and migration.

In this chapter, we will focus on studying population genetics using **Single-Nucleotide Polymorphism (SNPs)**. We will learn about **genetic association** analysis, which focuses on whether certain traits in a larger population can be tied back to particular genetic variants in those individuals. First, a word on file formats in this field. There is no such thing as a default file format for population genetics data. There is a plenitude of formats in this field, most of them developed with a specific application in mind; therefore, none are generically applicable. Some of the efforts to create a more general format (or even just a file converter to support many formats) have had limited success. Furthermore, as our knowledge of genomics increases, we will require new formats anyway (for example, to support some kind of previously unknown genomic structural variation). Here, we will work with PLINK (`https://www.cog-genomics.org/plink/2.0/`), which was originally developed to perform **Genome-Wide Association Studies (GWASs)** with human data but has many more applications. If you have **Next-Generation Sequencing (NGS)** data, you may ask the question, why not use the **Variant Call Format (VCF)**? Well, a VCF file is normally annotated to help with sequencing analysis, which you do not need at this stage (you should now have a filtered dataset). If you convert your SNP calls from VCF to PLINK, you will get roughly a 95 percent reduction in terms of size (this is in comparison to a compressed VCF). More importantly, the computational cost of processing a VCF file is much higher (think of processing all this highly structured text) than the cost of the other two formats.

In this chapter, we will cover the following recipes:

- Managing datasets with PLINK
- Using `sgkit` for population genetics analysis with `xarray`
- Exploring a dataset with `sgkit`
- Analyzing population structure

First, let's start with a discussion on file format issues and then continue to discuss interesting data analysis.

Technical requirements

In this chapter, we'll use the following tools and packages:

- `PLINK`
- `sgkit`
- `Dask`

You'll be instructed on how to install the relevant tools in each *Getting ready* section. The code for this chapter can be found at `https://github.com/PacktPublishing/Bioinformatics-with-Python-Cookbook-Fourth-Edition/tree/main/Ch11`.

You will want to create a `Ch11` folder and set up your notebooks there.

Remember to activate your `conda` environment before beginning the recipes, like this:

```
conda activate bioinformatics_base
```

Alternatively, if you would like to set up a `conda` environment specific to this chapter, before activating `bioinformatics_base`, run the following:

```
conda create -n ch11-population-genomics --clone bioinformatics_base
conda activate ch11-population-genomics
```

You will be able to install the packages for the chapter as you go, or you can use the YAML file provided in the repository:

```
conda env update --file ch11-population-genomics.yml
```

Managing datasets with PLINK

Here, we will manage our dataset using PLINK, a toolset for whole-genome association analysis. We will create subsets of our main dataset (from the HapMap Project) that are suitable for analysis in the following recipes.

In this recipe, we'll go over installing PLINK (`https://www.cog-genomics.org/plink/2.0/`). We'll then prepare our dataset from the HapMap Project. We will learn how to manipulate and subsample our data. We will also learn how to convert our data into different formats or extract individual chromosomes from the data. This will give us a solid grounding for subsequent chapters.

> **Warning**
>
> Note that neither PLINK nor any similar programs were developed for their file formats. There was probably no objective to create a default file standard for population genetics data. In this field, you will need to be ready to convert from format to format (for this, Python is quite appropriate) because every application that you will use will probably have its own quirky requirements. The most important point to learn from this recipe is not the formats being used, although these are relevant, but a "file conversion mentality." Beyond this, some of the steps in this recipe also convey genuine analytical techniques that you may want to consider using, for example, subsampling or **Linkage Disequilibrium** (**LD**)-pruning.

Getting ready

Throughout this chapter, we will use data from the International HapMap Project. The HapMap Project is in many ways the precursor to the 1000 Genomes Project; instead of whole-genome sequencing, genotyping was used. You can refer to the HapMap site (`https://www.genome.gov/10001688/international-hapmap-project`) for more information on the dataset. Remember that we have genotyping data for many individuals split across populations around the globe. We will refer to these populations by their acronyms. Here is the list taken from `https://www.broadinstitute.org/medical-and-population-genetics/hapmap-3`:

Acronym	Population
ASW	African ancestry in Southwest US
CEU	Utah residents with Northern and Western European ancestry from the CEPH collection
CHB	Han Chinese in Beijing, China
CHD	Chinese in Metropolitan Denver, Colorado
GIH	Gujarati Indians in Houston, Texas
JPT	Japanese in Tokyo, Japan
LWK	Luhya in Webuye, Kenya
MXL	Mexican ancestry in Los Angeles, California
MKK	Maasai in Kinyawa, Kenya
TSI	Toscani in Italy
YRI	Yoruba in Ibadan, Nigeria

Table 11.1 – The populations in the 1000 Genomes Project

> **Note**
>
> We will be using data from the HapMap Project that has, in practice, been replaced by the 1000 Genomes Project. For the purpose of teaching population genetics programming techniques in Python, the HapMap Project dataset is more manageable than the 1000 Genomes Project, as the data is considerably smaller. The HapMap samples are a subset of the 1000 Genomes Project samples. If you do research in human population genetics, you are strongly advised to use the 1000 Genomes Project as a base dataset.

This will require a fairly big download (approximately 1 GB), which will have to be uncompressed. Make sure that you have approximately 20 GB of disk space for this chapter. The files can be found at `https://ftp.ncbi.nlm.nih.gov/hapmap/genotypes/hapmap3_r3/plink_format/` Place the files in a "data" subdirectory under your Ch11 working directory.

Decompress the PLINK files using the following commands:

```
! gunzip data/hapmap3_r3_b36_fwd.consensus.qc.poly.map.gz
! gunzip data/hapmap3_r3_b36_fwd.consensus.qc.poly.ped.gz
```

Now, we have PLINK files; the `.map` file has information on the marker position across the genome, whereas the PED file has actual markers for each individual, along with some pedigree information. We also downloaded a metadata file that contains information about each individual. Take a look at all these files and familiarize yourself with them.

We can also download the relationships file:

```
! wget https://ftp.ncbi.nlm.nih.gov/hapmap/genotypes/hapmap3_r3/
relationships_w_pops_041510.txt
! mv relationships_w_pops_041510.txt data/
```

Again, this recipe will make heavy usage of PLINK. Python will mostly be used as the glue language to call PLINK. So, you will want to make sure you have PLINK installed. First, download the appropriate binary file from here: `https://www.cog-genomics.org/plink/2.0/`

Then, move the file from your `Downloads` directory to your Ch11 working directory.

Test that PLINK is working by running this command from the terminal:

```
./plink2
```

You should see the following help message:

```
PLINK v2.0.0-a.6.9 64-bit (29 Jan 2025)          cog-genomics.org/plink/2.0/
(C) 2005-2025 Shaun Purcell, Christopher Chang   GNU General Public License v3

  plink2 <input flag(s)...> [command flag(s)...] [other flag(s)...]
  plink2 --help [flag name(s)...]

Commands include --rm-dup list, --make-bpgen, --export, --freq, --geno-counts,
--sample-counts, --missing, --hardy, --het, --fst, --indep-pairwise,
--r2-phased, --sample-diff, --make-king, --king-cutoff, --pmerge, --pgen-diff,
--check-sex, --write-samples, --write-snplist, --make-grm-list, --pca, --glm,
--adjust-file, --gwas-ssf, --pheno-svd, --clump, --score-list, --variant-score,
--genotyping-rate, --pgen-info, --validate, and --zst-decompress.

"plink2 --help | more" describes all functions.
```

Figure 11.1 – Help output for PLINK2

You may get a message saying that your Mac cannot trust this file.

If so, do the following:

1. Go into your Mac settings.

2. Open **System Preferences | Security & Privacy**.

3. Click on the **General** tab.

4. Look for a message saying **plink2 was blocked because it is from an unidentified developer**.

5. Click **Allow Anyway**.

Now you should be able to run `./plink2`.

You may need to click **Allow All** again and provide the administrator password for your Mac.

As usual, this is also available in the `Ch11/Ch11-1-plink.ipynb` notebook file, where everything has been taken care of.

OK, now we are ready to start our first recipe!

How to do it...

Take a look at the following steps:

1. Let's get the metadata for our samples. We will load the population of each sample and note all the individuals that are offspring of others in the dataset:

```
import os
from collections import defaultdict
f = open('data/relationships_w_pops_041510.txt')
pop_ind = defaultdict(list)
f.readline()
offspring = []

for l in f:
    toks = l.rstrip().split('\t')
    fam_id = toks[0]
    ind_id = toks[1]
    mom = toks[2]
    dad = toks[3]
    if mom != '0' or dad != '0':
        offspring.append((fam_id, ind_id))
    pop = toks[-1]
    pop_ind[pop].append((fam_id, ind_id))
f.close()
```

This will load a dictionary where the population is the key (CEU, YRI, and so on), and its value is the list of individuals in that population. This dictionary will also store information on whether the individual is the offspring of another. Each individual is identified by the family and individual ID (information that can be found in the PLINK file). The file provided by the HapMap Project is a simple tab-delimited file, which is not difficult to process. While we are reading the files using standard Python text processing, this is a typical example where pandas would help.

There is an important point to make here: the reason this information is provided in a separate, ad hoc file is that the PLINK format makes no provision for the population structure (this format makes provision only for the case and control information for which PLINK was designed). This is not a flaw of the format, as it was never designed to support standard population genetic studies (it's a GWAS tool). However, this is a general feature of data formats in population genetics: whichever you end up working with, there will be something important missing.

We will use this metadata in other recipes in this chapter. We will also perform some consistency analysis between the metadata and the PLINK file, but we will defer this to the next recipe.

2. Now, let's subsample the dataset at 10 percent and 1 percent of the number of markers, as follows:

```
! ./plink2 --pedmap data/hapmap3_r3_b36_fwd.consensus.qc.poly
--out hapmap10 --thin 0.1 --geno 0.1 --export ped
! ./plink2 --pedmap data/hapmap3_r3_b36_fwd.consensus.qc.poly
--out hapmap1 --thin 0.01 --geno 0.1 --export ped
```

> **Note**
>
> If you have trouble running PLINK with the . / relative path notation in the preceding code, try putting the full path to it in front of the command.

Note the subtlety that you will not really get 1 or 10 percent of the data; each marker will have a 1 or 10 percent chance of being selected, so you will get approximately 1 or 10 percent of the markers.

Obviously, as the process is random, different runs will produce different marker subsets. This will have important implications further down the road. If you want to replicate the exact same result, you can nonetheless use the - - seed option.

We will also remove all SNPs that have a genotyping rate lower than 90 percent (with the --geno 0.1 parameter).

> **Note**
>
> There is nothing special about Python code used here, but there are two reasons why you may want to subsample your data. First, if you are performing an exploratory analysis of your own dataset, you may want to start with a smaller version because it will be easy to process. Also, you will have a broader view of your data. Second, some analytical methods may not require all your data (indeed, some methods might not even be able to use all of your data). Be very careful with the last point, though; that is, for every method that you use to analyze your data, be sure that you understand the data requirements for the scientific questions you want to answer. Feeding too much data may be okay normally (even if you pay a time and memory penalty), but feeding too little will lead to unreliable results.

3. Now, let's generate subsets with just the **autosomes** (that is, let's remove the sex chromosomes and mitochondria), as follows:

```
def get_non_auto_SNPs(map_file, exclude_file):
    f = open(map_file)
    w = open(exclude_file, 'w')
    for l in f:
        toks = l.rstrip().split('\t')
        try:
            chrom = int(toks[0])
        except ValueError:
            rs = toks[1]
```

```
                    w.write('%s\n' % rs)
        w.close()

    get_non_auto_SNPs('hapmap1.map', 'exclude1.txt')
    get_non_auto_SNPs('hapmap10.map', 'exclude10.txt')

    os.system(
        './plink2 --pedmap hapmap1 --out hapmap1_auto'
        '--exclude exclude1.txt --export ped')
    os.system(
        './plink2 --pedmap hapmap10 --out hapmap10_auto'
        '--exclude exclude10.txt --export ped')
```

This will create a function that generates a list with all the SNPs not belonging to autosomes. With human data, that means all non-numeric chromosomes. If you use another species, be careful with your chromosome coding because PLINK is geared toward human data. If your species are diploid, have fewer than 23 autosomes, and have a sex determination system, that is, X/Y, this will be straightforward; if not, refer to https://www.cog-genomics.org/plink2/input#allow_extra_chr for some alternatives (such as the --allow-extra-chr flag).

We then create autosome-only PLINK files for subsample datasets of 10 and 1 percent (prefixed as hapmap10_auto and hapmap1_auto, respectively).

4. Let's create some datasets without offspring. These will be needed for most population genetic analysis, which requires unrelated individuals to a certain degree:

    ```
    os.system('./plink2 --pedmap hapmap10_auto --filter-founders
    --out hapmap10_auto_noofs --export ped')
    ```

    Let's take a quick look at the file we created. Type the following (from the terminal):

    ```
    head hapmap10_auto_noofs.ped
    ```

    Here is what we see:

```
2431    NA19916 0    0    1    -9    A    A    A    A    G    G    G    G
2424    NA19835 0    0    2    -9    A    A    A    A    G    G    G    G
2469    NA20282 0    0    2    -9    A    A    A    G    A    G    G    G
2368    NA19703 0    0    1    -9    A    A    A    G    G    G    A    G
2425    NA19901 0    0    2    -9    A    A    A    A    G    G    G    G
2427    NA19908 0    0    1    -9    A    A    A    G    G    G    G    G
2430    NA19914 0    0    2    -9    A    A    G    G    A    G    G    G
2470    NA20287 0    0    2    -9    A    A    A    A    G    G    G    G
2436    NA19713 0    0    2    -9    A    A    A    A    G    G    A    G
2426    NA19904 0    0    1    -9    A    A    A    A    G    G    G    G
```

Figure 11.2 – The first few lines of the .ped file

This shows the family ID and individual ID first, followed by two columns of zeros for the father and mother, which we don't have in this example. Next is the sex (1 = male, 2 = female). Then comes a column to represent the phenotype. After that, we have multiple columns of genotypes, with two columns per allele. For example, A G means that this allele has an A as the reference and a G as the alternate allele.

The .ped file is paired with a .map file, which contains the chromosome code, variant ID, genetic distance, and position in the genome – https://www.cog-genomics.org/plink/1.9/formats.

Note

This step is representative of the fact that most population genetic analyses require samples to be unrelated to a certain degree. Obviously, as we know that some offspring are in HapMap, we remove them.

However, note that with your dataset, you are expected to be much more refined than this. For instance, run plink --genome or use another program to detect related individuals. The fundamental point here is that you have to dedicate some effort to detecting related individuals in your samples; this is not a trivial task.

5. We will also generate an LD-pruned dataset, as required by many PCA and admixture algorithms, as follows:

```
os.system(
    './plink2 --pedmap hapmap10_auto_noofs '
    '--indeppairwise 50 10 0.1 --out keep --export ped')
os.system(
    './plink2 --pedmap hapmap10_auto_noofs --extract '
    'keep.prune.in --out hapmap10_auto_noofs_ld --export ped')
```

The first step generates a list of markers to be kept if the dataset is LD-pruned. This uses a sliding window of 50 SNPs, advancing by 10 SNPs at a time with a cut value of 0.1. The second step extracts SNPs from the list that was generated earlier.

6. Let's recode a couple of cases in different formats:

```
os.system(
    './plink2 --pedmap hapmap10_auto_noofs_ld '
    '--export ped --out hapmap10_auto_noofs_ld_12')
os.system(
    './plink2 --pedmap hapmap10_auto_noofs_ld '
    '--make-bed --out hapmap10_auto_noofs_ld')
```

The first operation will convert a PLINK format that uses nucleotide letters from the ACTG to another, which recodes alleles with 1 and 2. The second operation recodes a file in a binary format. If you work inside PLINK (using the many useful operations that PLINK has), the binary format is probably the most appropriate (offering, for example, a smaller file size). We will also extract a single chromosome (2) for analysis. We will start with the autosome dataset, which has been subsampled at 10 percent:

```
os.system(
    './plink2 --pedmap hapmap10_auto_noofs --chr 2 '
    '--out hapmap10_auto_noofs_2 --export ped')
```

Great work! We have now seen how to use PLINK to manipulate our files and perform basic operations on the data.

There's more...

There are many reasons why you might want to create different datasets for analysis. You may want to perform some fast initial exploration of data – for example, if the analysis algorithm that you plan to use has some data format requirements or a constraint on the input, such as the number of markers or relationships between individuals. Chances are that you will have lots of subsets to analyze (unless your dataset is very small to start with, for instance, a microsatellite dataset).

This may seem to be a minor point, but it's not: be very careful with file naming (note that I have followed some simple conventions while generating filenames). Make sure that the name of the file gives some information about the subset options. When you perform the downstream analysis, you will want to be sure that you choose the correct dataset; you will want your dataset management to be agile and reliable, above all. The worst thing that can happen is that you create an analysis with an erroneous dataset that does not obey the constraints required by the software.

The LD-pruning that we used is somewhat standard for human analysis, but be sure to check the parameters, especially if you are using non-human data.

The HapMap file that we downloaded is based on an old version of the reference genome (build 36). Be sure to use annotations from build 36 if you plan to use this file for more analysis of your own.

This recipe sets the stage for the following recipes, and its results will be used extensively.

See also

- For a good primer on genetic association studies, read Uffelmann et al., *Genome-wide association studies*, Nature Reviews Methods Primers, Aug 2021 – https://www.nature.com/articles/s43586-021-00056-9

- The Wikipedia page https://en.wikipedia.org/wiki/Linkage_disequilibrium on LD is a good place to start to learn more about the concept

- The website of PLINK, https://www.cog-genomics.org/plink/2.0/, is very well documented, something lacking in most genetics software

- Check out pandasGWAS, a Python package for retrieving GWAS data: Cao et al., *pandasGWAS: a Python package for easy retrieval of GWAS catalog data*, BMC Genomics, May 2023 – https://link.springer.com/article/10.1186/s12864-023-09340-2

Using sgkit for population genetics analysis with xarray

sgkit (https://sgkit-dev.github.io/sgkit/latest/) is the most advanced Python library for doing population genetics analysis. It's a modern implementation, leveraging almost all of the fundamental data science libraries in Python. When I say almost all, I am not exaggerating; it uses NumPy, pandas, xarray, Zarr, and Dask.

Here, we will introduce xarray as the main data container for sgkit. Because I feel that I cannot ask you to get to know data engineering libraries to an extreme level, I will gloss over the Dask part (mostly by treating Dask structures as equivalent NumPy structures). Dask is covered in more depth here: https://www.dask.org/. You have already been learning about NumPy and pandas. It will not be necessary to cover Zarr for this recipe. It is a library for handling large N-dimensional arrays, also known as tensors. If you want to learn more about Zarr you can find more here: https://zarr.dev/

Getting ready

You will need to run the previous recipe because its output is required for this one; we will be using one of the PLINK datasets. You will need to install sgkit:

```
! pip install sgkit
! pip install 'sgkit[plink]'
```

As usual, this is available in the Ch11/Ch11-2-using-sgkit.ipynb notebook file, but it will still require you to run the previous notebook file in order to generate the required files.

How to do it...

Take a look at the following steps:

1. Let's load the hapmap10_auto_noofs_ld dataset generated in the previous recipe:

   ```
   import numpy as np
   from sgkit.io import plink
   data = plink.read_plink(
       path='hapmap10_auto_noofs_ld',
       fam_sep='\t'
   )
   ```

   Remember that we are loading a set of PLINK files. It turns out that sgkit creates a very rich and structured representation for that data. That representation is based on an xarray dataset (https://xarray.dev/).

2. Let's check the structure of our data – if you are in a notebook, just enter the following:

   ```
   data
   ```

Note that `sgkit` – if you are printing the "data" structure in a notebook as above – will generate the following representation:

xarray.Dataset

▸ Dimensions: (contigs: 22, variants: 54719, alleles: 2, samples: 1198, ploidy: 2)

▸ Coordinates: (0)

▼ Data variables:

contig_id	(contigs)	<U2	'1' '2' '3' '4' ... '20' '21' '22'	
variant_contig	(variants)	int16	0 0 0 0 0 0 0 ... 21 21 21 21 21 21	
variant_position	(variants)	int32	dask.array<chunksize=(54719,), meta=n...	
variant_allele	(variants, alleles)	|S1	dask.array<chunksize=(54719, 1), meta=...	
sample_id	(samples)	<U7	dask.array<chunksize=(1198,), meta=np....	
call_genotype	(variants, samples, ploidy)	int8	dask.array<chunksize=(54719, 1198, 2), ...	
call_genotype_...	(variants, samples, ploidy)	bool	dask.array<chunksize=(54719, 1198, 2), ...	
variant_id	(variants)	<U10	dask.array<chunksize=(54719,), meta=n...	
sample_family_id	(samples)	<U7	dask.array<chunksize=(1198,), meta=np....	
sample_membe...	(samples)	<U7	dask.array<chunksize=(1198,), meta=np....	
sample_paterna...	(samples)	<U1	dask.array<chunksize=(1198,), meta=np....	
sample_matern...	(samples)	<U1	dask.array<chunksize=(1198,), meta=np....	
sample_sex	(samples)	int8	dask.array<chunksize=(1198,), meta=np....	
sample_phenot...	(samples)	int8	dask.array<chunksize=(1198,), meta=np....	

▸ Indexes: (0)

▼ Attributes:

contigs :   [np.str_('1'), np.str_('2'), np.str_('3'), np.str_('4'), np.str_('5'), np.str_('6'), np.str_('7'), np.str_('8'), np.str_('9'), np.str_('10'), np.str_('11'), np.str_('12'), np.str_('13'), np.str_('14'), np.str_('15'), np.str_('16'), np.str_('17'), np.str_('18'), np.str_('19'), np.str_('20'), np.str_('21'), np.str_('22')]

source :   sgkit-0.10.0

Figure 11.3 – An overview of the xarray data loaded by sgkit for our PLINK file

`data` is an `xarray` dataset. An `xarray` dataset is essentially a dictionary in which each value is a Dask array. For our purposes, you can assume it is a NumPy array. In this case, we can see that we have 54,719 variants for 1198 samples. We have 2 alleles per variant for a ploidy of 2.

Note

The exact numbers may vary depending on the version of libraries you are using and whether any files have been updated.

This is a three-dimensional array, with variants, samples, and ploidy dimensions. In the notebook, we can expand some entries. In our case, we expanded `Data variables` and `Attributes`. You can see the data types of elements in the array such as `int8`, `int16`, etc. Another way to get summary information, which is especially useful if you are not using notebooks, is by inspecting the `dims` field:

```
print(data.dims)
```

The output shows the size of each dimension in the array:

```
FrozenMappingWarningOnValuesAccess({'contigs': 22, 'variants': 54719, 'alleles': 2, 'samples': 1198, 'ploidy': 2})
```

Figure 11.4 – Data dimensions output

3. Don't worry about the "warning" in this message. This output simply shows that we have a "frozen" (read-only) dictionary of the dimensions of our dataset. Let's extract some information about the samples:

    ```
    print(len(data.sample_id.values))
    print(data.sample_id.values)
    print(data.sample_family_id.values)
    print(data.sample_sex.values)
    ```

    The output is as follows:

    ```
    1198
    ['NA19916' 'NA19835' 'NA20282' ... 'NA18915' 'NA19250'
     'NA19124']
    ['2431' '2424' '2469' ... 'Y029' 'Y113' 'Y076']
    [1 2 2 ... 1 2 1]
    ```

We have `1198` samples. The first one has a sample ID of `NA19916`, a family ID of `2431`, and a sex of `1` (male). Remember that, given PLINK as the data source, a sample ID is not enough to be a primary key (you can have different samples with the same sample ID). The primary key is a composite of the sample ID and sample family ID.

Tip

You might have noticed that we add `.values` to all the data fields; this is actually rendering a lazy Dask array into a materialized NumPy one. The `.values` property is akin to the `compute` method in Dask.

The `.values` call is not just about formatting – it strips away the xarray metadata and returns a raw NumPy array – the reason our code works is that our dataset is small enough to fit into memory, which is great for our teaching example. But if you have a very large dataset, the preceding code is too naive. In a case like that, you may need to preserve the xarray metadata and operate on it natively. For now, the simplicity used here is for pedagogical purposes.

Before we look at the variant data, we have to be aware of how sgkit stores contigs (keep in mind that contigs are contiguous regions of a genome; in this case we are working with human chromosomes):

```
print(data.contigs)
```

The output is as follows:

```
['1', '2', '3', '4', '5', '6', '7', '8', '9', '10', '11', '12',
 '13', '14', '15', '16', '17', '18', '19', '20', '21', '22']
```

The contigs here are the human autosomes, which are chromosomes 1 through 22, excluding the sex chromosomes X and Y (you will not be so lucky if your data is based on most other species – you will probably have some ugly identifier here).

4. Now, let's look at the variants:

```
print(len(data.variant_contig.values))
print(data.variant_contig.values)
print(data.variant_position.values)
print(data.variant_allele.values)
print(data.variant_id.values)
```

Here is the output:

```
54719
[ 0  0  0 ... 21 21 21]
[  557616    871896    894491 ... 49360864 49413787 49503799]
[[b'G' b'A']
 [b'A' b'G']
 [b'A' b'G']

 ...
 [b'T' b'C']
 [b'A' b'G']
 [b'G' b'A']]
['rs11510103' 'rs2272756' 'rs28562326' ... 'rs1557502' 'rs131715'
 'rs6010063']
```

Figure 11.5 – Output from printing variant data structure information

We have 54,719 variants (your output may vary slightly). The contig index is 0, which, if you look at *step 3* of this recipe, is chromosome 1. The variant is in position 557,616 (against build 36 of the human genome) and has possible alleles G and A. It has a SNP ID of rs11510103.

Finally, let's look at the genotype data:

```
call_genotype = data.call_genotype.values
print(call_genotype.shape)
```

```
first_individual = call_genotype[:,0,:]
first_variant = call_genotype[0,:,:]
first_variant_of_first_individual = call_genotype[0,0,:]

print(first_variant_of_first_individual)
print(data.sample_family_id.values[0],
      data.sample_id.values[0])
print(data.variant_allele.values[0])
```

call_genotype has a shape of 54,719 x 1198 x 2, which represents the number of dimensioned variants, samples, and ploidy respectively (again, your results could vary slightly, for example if the underlying files were to be updated).

To get all variants for the first individual, you fix the second dimension. To get all the samples for the first variant, you fix the first dimension.

If you print the first individual's details (sample and family ID), you get 2431 and NA19916 – as expected, exactly as in the first case in the previous sample exploration.

There's more...

This recipe is mostly an introduction to xarray, disguised as an sgkit tutorial. There is much more to be said about xarray – be sure to check out https://docs.xarray.dev/. It is worth reiterating that xarray depends on a plethora of Python data science libraries and that we are glossing over Dask for now.

Exploring a dataset with sgkit

In this recipe, we will perform an initial exploratory analysis of one of our generated datasets. Now that we have some basic knowledge of xarray, we can actually try to do some data analysis. In this recipe, we will ignore population structure, an issue we will return to in the following recipe.

Getting ready

You will need to run the first recipe and should have the hapmap10_auto_noofs_ld files available. There is a notebook file with this recipe called Ch11/Ch11-3-exploring-with-sgkit. ipynb. You will need the software that you installed for the previous recipe.

How to do it...

Take a look at the following steps:

1. We start by loading the PLINK data with sgkit, exactly as in the previous recipe:

```
import numpy as np
import xarray as xr
import sgkit as sg
from sgkit.io import plink
data = plink.read_plink(
    path='hapmap10_auto_noofs_ld',
    fam_sep='\t'
)
```

Let's ask `sgkit` for `variant_stats`:

```
variant_stats = sg.variant_stats(data)
variant_stats
```

The output is the following (remember that you can expand `Data variables` if you want):

xarray.Dataset

Dimensions: (variants: 54719, alleles: 2, contigs: 22, samples: 1198, ploidy: 2)

Coordinates: (0)

▶ Data variables:
(23)

Indexes: (0)

▼ Attributes:

contigs :   [np.str_('1'), np.str_('2'), np.str_('3'), np.str_('4'), np.str_('5'), np.str_('6'), np.str_('7'),
                np.str_('8'), np.str_('9'), np.str_('10'), np.str_('11'), np.str_('12'), np.str_('13'), np.str_('1
                4'), np.str_('15'), np.str_('16'), np.str_('17'), np.str_('18'), np.str_('19'), np.str_('20'), np.
                str_('21'), np.str_('22')]
source :   sgkit-0.10.0

Figure 11.6 – The variant statistics provided by sgkit's variant_stats

Let's now look at the statistic, `variant_call_rate`:

```
variant_stats.variant_call_rate.to_series().describe()
```

There is more to unpack here than it may seem. The fundamental part is the `to_series()` call. `sgkit` is returning a pandas `series` to you – remember that `sgkit` is highly integrated with Python data science libraries; you can learn more about panda series functionality here: `https://pandas.pydata.org/docs/reference/api/pandas.Series.html` After you get the `series` object, you can call the pandas `describe` function and get the following:

```
count    54719.000000
mean         0.997200
std          0.003947
min          0.966611
25%          0.996661
50%          0.998331
75%          1.000000
max          1.000000
Name: variant_call_rate, dtype: float64
```

Figure 11.7 – Variant call rate summary statistics

Our variant call rate is quite good, which is not shocking because we are looking at array data – you would have worse numbers if you had a dataset based on NGS.

2. Let's now look at sample statistics:

```
sample_stats = sg.sample_stats(data)
sample_stats
```

Again, sgkit provides a lot of sample statistics out of the box:

xarray.Dataset

▸ Dimensions: (samples: 1198, contigs: 22, variants: 54719, alleles: 2, ploidy: 2)

▸ Coordinates: (0)

▸ Data variables:
(20)

▸ Indexes: (0)

▾ Attributes:

contigs :   [np.str_('1'), np.str_('2'), np.str_('3'), np.str_('4'), np.str_('5'), np.str_('6'), np.str_('7'),
                 np.str_('8'), np.str_('9'), np.str_('10'), np.str_('11'), np.str_('12'), np.str_('13'), np.str_('1
                 4'), np.str_('15'), np.str_('16'), np.str_('17'), np.str_('18'), np.str_('19'), np.str_('20'), np.
                 str_('21'), np.str_('22')]
source :   sgkit-0.10.0

Figure 11.8 – The sample statistics obtained by calling sample_stats

3.   We will now have a look at sample call rates:

```
sample_stats.sample_call_rate.to_series().hist()
```

This time, we plot a histogram of sample call rates. Again, sgkit allows us to do this by leveraging pandas:

```
<Axes: >
```

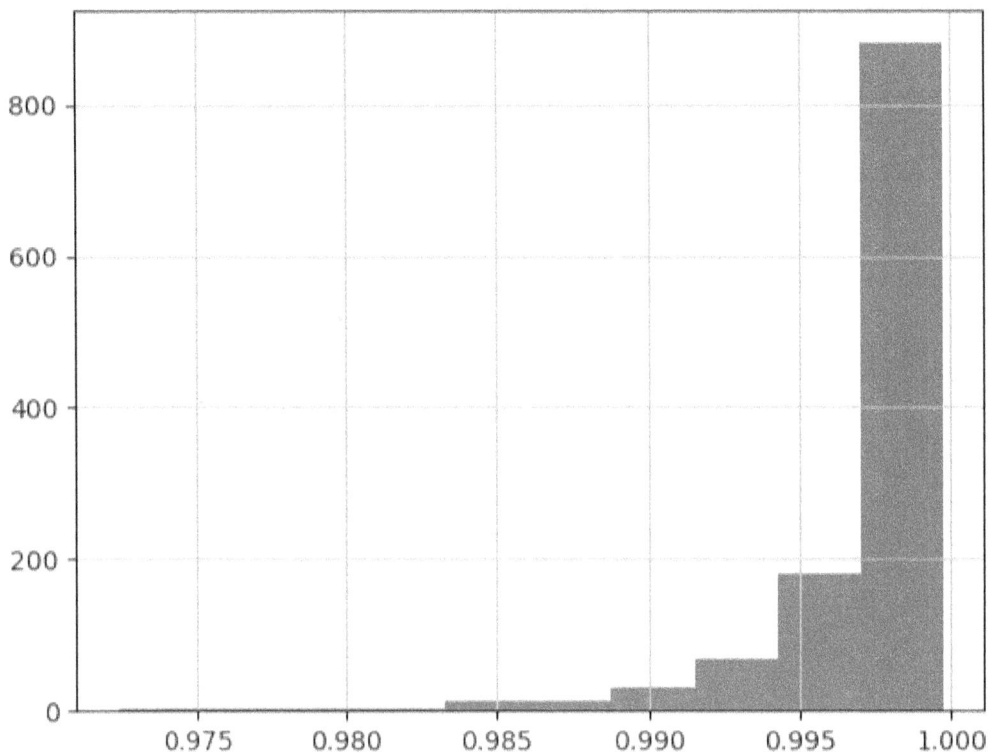

Figure 11.9 – The histogram of sample call rates

There's more...

The truth is that for population genetic analysis, nothing beats R; you are definitely encouraged to take a look at the existing R libraries for population genetics.

Most of the analysis presented here will be computationally costly if done on bigger datasets. Indeed, sgkit is prepared to deal with that because it leverages Dask.

See also

- A list of R packages for statistical genetics is available at `https://cran.r-project.org/web/packages/genetics/index.html`

- If you need to know more about population genetics, I recommend the book *Principles of Population Genetics*, by *Daniel L. Hartl and Andrew G. Clark*, *Sinauer Associates*

Analyzing population structure

Previously, we introduced data analysis with `sgkit`, ignoring the population structure. Most datasets, including the one we are using, actually do have a population structure. `sgkit` provides functionality to analyze genomic datasets with population structure, and that is what we are going to investigate here.

Getting ready

You will need to run the first recipe and should have the `hapmap10_auto_noofs_ld` data we produced, as well as the original population metadata `relationships_w_pops_041510.txt` file downloaded. There is a notebook file with the `Ch11/Ch11-4-population-structure.ipynb` recipe in it.

How to do it...

Take a look at the following steps:

1. First, let's load the PLINK data with `sgkit`:

    ```
    from collections import defaultdict
    from pprint import pprint
    import numpy as np
    import matplotlib.pyplot as plt
    import seaborn as sns
    import pandas as pd
    import xarray as xr
    import sgkit as sg
    from sgkit.io import plink
    data = plink.read_plink(
        path='hapmap10_auto_noofs_ld',
        fam_sep='\t'
    )
    ```

2. Now, let's load the data assigning individuals to populations:

```
f = open('data/relationships_w_pops_041510.txt')
pop_ind = defaultdict(list)
f.readline()
for line in f:
    toks = line.rstrip().split('\t')
    fam_id = toks[0]
    ind_id = toks[1]
    pop = toks[-1]
    pop_ind[pop].append((fam_id, ind_id))
pops = list(pop_ind.keys())
```

We end up with a dictionary, `pop_ind`, where the key is the population code, and the value is a list of samples. Remember that a sample primary key is the family ID and the sample ID.

We also have a list of populations in the `pops` variable.

3. We now need to inform `sgkit` about which population or cohort each sample belongs to:

```
def assign_cohort(
    pops, pop_ind, sample_family_id, sample_id
):
    cohort = []
    for fid, sid in zip(
        sample_family_id,
        sample_id
    ):
        processed = False
        for i, pop in enumerate(pops):
            if (fid, sid) in pop_ind[pop]:
                processed = True
                cohort.append(i)
                break
        if not processed:
            raise Exception(
                f'Not processed {fid}, {sid}')
    return cohort
cohort = assign_cohort(
    pops, pop_ind,
    data.sample_family_id.values,
    data.sample_id.values
)
data['sample_cohort'] = xr.DataArray(cohort, dims='samples')
```

Remember that each sample in `sgkit` has a position in an array. So, we must create an array where each element refers to a specific population or cohort within a sample. The `assign_cohort` function does exactly that: it takes the metadata that we loaded from the `relationships` file and the list of samples from the `sgkit` file and gets the population index for each sample.

4. Now that we have loaded population information structure into the `sgkit` dataset, we can start computing statistics at the population or cohort level. Let's start by getting the number of monomorphic loci per population:

```python
cohort_allele_frequency = sg.cohort_allele_frequencies(
    data
)['cohort_allele_frequency'].values
monom = {}
for i, pop in enumerate(pops):
    monom[pop] = len(list(filter(
        lambda x: x,
        np.isin(
            cohort_allele_frequency[:, i, 0],
            [0, 1]
        )
    )))
pprint(monom)
```

We start by asking `sgkit` to calculate the allele frequencies per cohort or population. After that, we filter all loci per population where the allele frequency of the first allele is either 0 or 1 (that is, there is the fixation of one of the alleles). Finally, we print it. Incidentally, we use the `pprint.pprint` function to make it look a bit better (the function is quite useful for more complex structures if you want to render the output in a readable way):

```
{'ASW': 3263,
 'CEU': 8981,
 'CHB': 11108,
 'CHD': 12151,
 'GIH': 9026,
 'JPT': 12855,
 'LWK': 3845,
 'MEX': 6421,
 'MKK': 3346,
 'TSI': 8680,
 'YRI': 5081}
```

Figure 11.10 – Monomorphic loci per population

Let's get the minimum allele frequency for all loci per population. This is still stored in `cohort_allele_frequency` – so no need to call `sgkit` again:

```
mafs = {}
for i, pop in enumerate(pops):
    min_freqs = map(
        lambda x: x if x < 0.5 else 1 - x,
        filter(
            lambda x: x not in [0, 1],
            cohort_allele_frequency[:, i, 0]
        )
    )
    mafs[pop] = pd.Series(min_freqs)
```

We create pandas `series` objects for each population, as this permits lots of helpful functions, such as plotting.

We will now print the MAF histograms for the `YRI` and `JPT` populations. We will leverage pandas and Matplotlib for this:

```
maf_plot, maf_ax = plt.subplots(nrows=2, sharey=True)
mafs['YRI'].hist(ax=maf_ax[0], bins=50)
maf_ax[0].set_title('YRI')
mafs['JPT'].hist(ax=maf_ax[1], bins=50)
maf_ax[1].set_title('JPT')
maf_ax[1].set_xlabel('MAF')
```

We get pandas to generate the histograms and put the results in a Matplotlib plot. The result is the following:

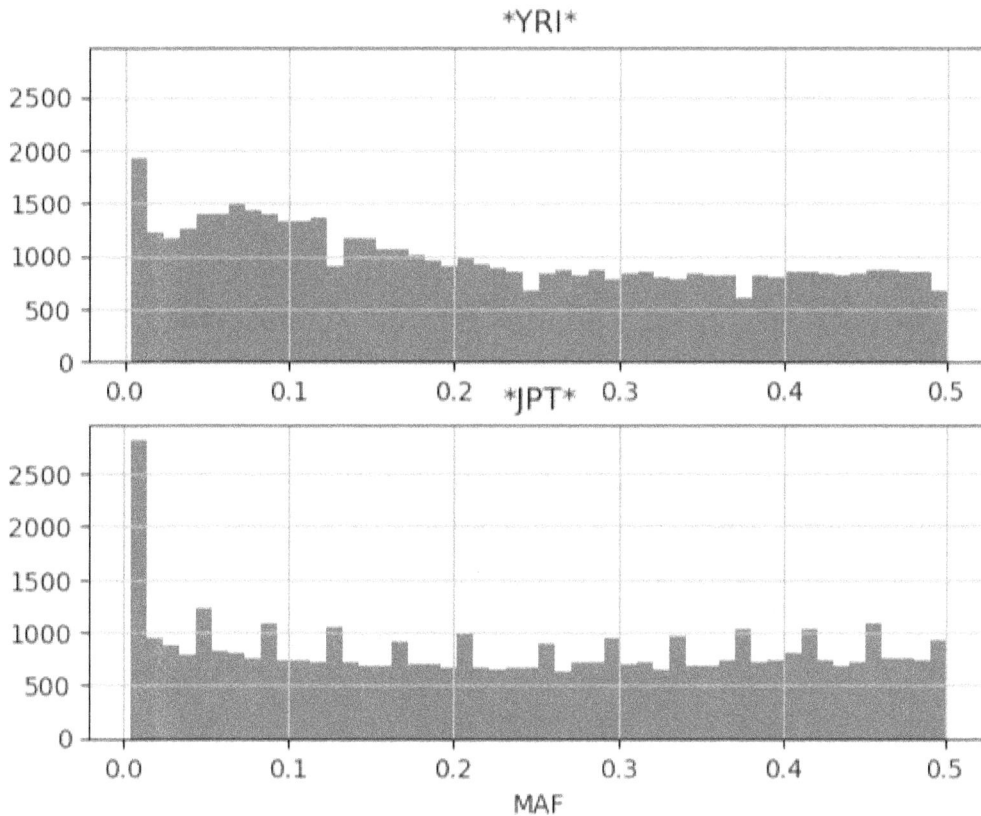

Figure 11.11 – A MAF histogram for the YRI and JPT populations

5. We are now going to concentrate on computing the F_{ST}. The F_{ST} (pronounced "F sub ST") is a special type of F-statistic known as the **fixation index**. It is a widely used statistic that tries to represent the genetic variation created by population structure. You can read more about it here: https://en.wikipedia.org/wiki/Fixation_index. Let's compute it with sgkit:

```
fst = sg.Fst(data)
fst = fst.assign_coords({
    "cohorts_0": pops,
    "cohorts_1": pops
})
```

The first line computes fst, which, in this case, will be pairwise fst across cohorts or populations. The second line assigns names to each cohort by using the xarray coordinates feature. This makes it easier and more declarative.

Let's compare fst between the CEU and CHB populations with CHB and CHD:

```
remove_nan = lambda data: filter(
    lambda x: not np.isnan(x), data
)
ceu_chb = pd.Series(remove_nan(fst.stat_Fst.sel(
    cohorts_0='CEU', cohorts_1='CHB'
).values))
chb_chd = pd.Series(remove_nan(fst.stat_Fst.sel(
    cohorts_0='CHB', cohorts_1='CHD'
).values))
ceu_chb.describe()
chb_chd.describe()
```

We take the pairwise results returned by the sel function from stat_FST to both compare and create a pandas series with it. Note that we can refer to populations by name, as we have prepared the coordinates in the previous step.

6. Let's plot the distance matrix across populations based on the multi-locus pairwise. Before we do it, we will prepare the computation:

```
mean_fst = {}
for i, pop_i in enumerate(pops):
    for j, pop_j in enumerate(pops):
        if j <= i:
            continue
        pair_fst = pd.Series(remove_nan(
            fst.stat_Fst.sel(
                cohorts_0=pop_i, cohorts_1=pop_j
            ).values
        ))
        mean = pair_fst.mean()
        mean_fst[(pop_i, pop_j)] = mean
min_pair = min(mean_fst.values())
max_pair = max(mean_fst.values())
```

We compute all the F_{ST} values for the population pairs. The execution of this code will be demanding in terms of time and memory, as we are actually requiring Dask to perform a lot of computations to render our NumPy arrays.

7. We can now do a pairwise plot of all means across populations:

```
sns.set_style("white")
num_pops = len(pops)
arr = np.ones(
    (num_pops - 1, num_pops - 1, 3),
    dtype=float
)
fig = plt.figure(figsize=(16, 9))
ax = fig.add_subplot(111)

for row in range(num_pops - 1):
    pop_i = pops[row]
    for col in range(row + 1, num_pops):
        pop_j = pops[col]
        val = mean_fst[(pop_i, pop_j)]
        norm_val = (
            val - min_pair) / (max_pair - min_pair
        )
        ax.text(col - 1, row, '%.3f' % val, ha='center')
        if norm_val == 0.0:
            arr[row, col - 1, 0] = 1
            arr[row, col - 1, 1] = 1
            arr[row, col - 1, 2] = 0
        elif norm_val == 1.0:
            arr[row, col - 1, 0] = 1
            arr[row, col - 1, 1] = 0
            arr[row, col - 1, 2] = 1
        else:
            arr[row, col - 1, 0] = 1 - norm_val
            arr[row, col - 1, 1] = 1
            arr[row, col - 1, 2] = 1

ax.imshow(arr, interpolation='none')
ax.set_title('Multilocus Pairwise FST')
ax.set_xticks(range(num_pops - 1))
ax.set_xticklabels(pops[1:])
ax.set_yticks(range(num_pops - 1))
ax.set_yticklabels(pops[:-1])
```

In the following diagram, we will draw an upper triangular matrix, where the background color of a cell represents the measure of differentiation; white (lighter) means less different (a lower level of differentiation) and blue (darker) means more different (a higher level of differentiation). The lowest value between CHB and CHD is represented in yellow (the 0.000 value in the image), and the biggest value between JPT and YRI is represented in magenta (the 0.101 value in the image). The value on each cell is the average pairwise between these two populations:

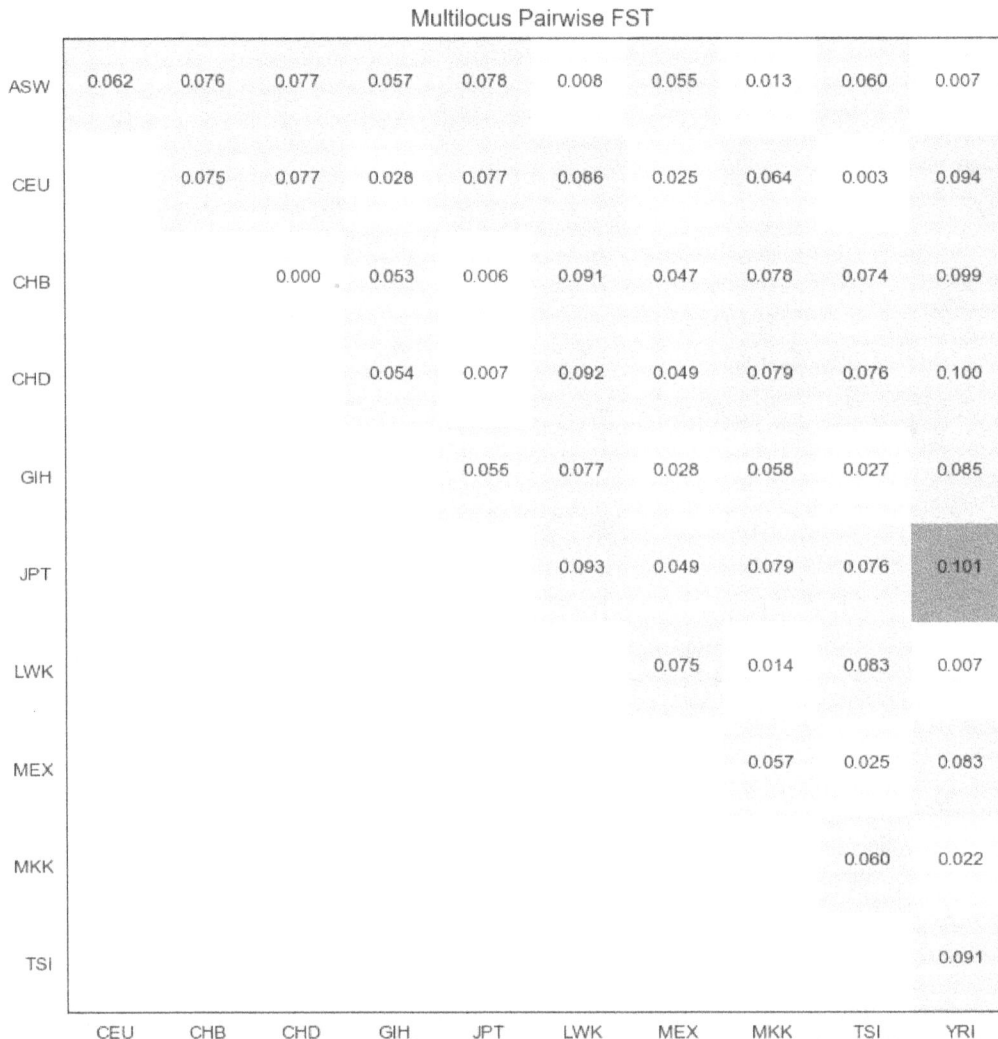

Multilocus Pairwise FST

	CEU	CHB	CHD	GIH	JPT	LWK	MEX	MKK	TSI	YRI
ASW	0.062	0.076	0.077	0.057	0.078	0.008	0.055	0.013	0.060	0.007
CEU		0.075	0.077	0.028	0.077	0.086	0.025	0.064	0.003	0.094
CHB			0.000	0.053	0.006	0.091	0.047	0.078	0.074	0.099
CHD				0.054	0.007	0.092	0.049	0.079	0.076	0.100
GIH					0.055	0.077	0.028	0.058	0.027	0.085
JPT						0.093	0.049	0.079	0.076	0.101
LWK							0.075	0.014	0.083	0.007
MEX								0.057	0.025	0.083
MKK									0.060	0.022
TSI										0.091

Figure 11.12 – The average pairwise FST across the 11 populations in the HapMap Project for all autosomes

OK, let's clean up and close down our conda environment:

```
conda deactivate
```

See also

- F-statistics is an immensely complex topic, so I will direct you firstly to the Wikipedia page at `https://en.wikipedia.org/wiki/F-statistics`

- A great explanation of FST can be found in Holsinger and Weir's paper: *Genetics in geographically structured populations: defining, estimating, and interpreting FST*, Nature Reviews Genetics, September 2009 - `https://www.nature.com/nrg/journal/v10/n9/abs/nrg2611.html`

Get This Book's PDF Version and Exclusive Extras

UNLOCK NOW

Scan the QR code (or go to `packtpub.com/unlock`). Search for this book by name, confirm the edition, and then follow the steps on the page.

Note: Keep your invoice handy. Purchases made directly from Packt don't require an invoice.

12

Metabolic Modeling and Other Applications

We have now obtained a solid grounding in the core areas of bioinformatics. In this chapter, we will see how this field has expanded over the years to impact numerous areas of science.

Today, bioinformatics professionals work in many diverse areas, including neuroscience, imaging, agriculture, food and nutrition, biofuels, and more. Artificial intelligence and data science are finding increasing overlap with bioinformatics.

In this chapter, we will work through the following recipes:

- Metabolic modeling with COBRApy
- Designing siRNAs using BioPython and ViennaRNA
- Predicting food properties using bioinformatics
- Discovering genes to make novel molecules

Technical requirements

In this chapter, we'll use the following tools and packages:

- BioPython
- COBRApy
- ViennaRNA
- Entrez
- Prodigal
- HMMER

You'll be instructed on how to install the relevant tools in each *Getting ready* section. The code for this chapter can be found at `https://github.com/PacktPublishing/Bioinformatics-with-Python-Cookbook-Fourth-Edition/tree/main/Ch12`.

You will want to create a `Ch12` folder and set up your notebooks there.

Remember to activate your `conda` environment before beginning the recipes, like this:

```
conda activate bioinformatics_base
```

If you would like to set up a `conda` environment specific to this chapter, before activating `bioinformatics_base`, run the following:

```
conda create -n ch12-applications --clone bioinformatics_base
conda activate ch12-applications
```

You will be able to install the packages for the chapter as you go, or you can use the YAML file provided in the repository to set up the environment all at once by running:

```
conda env update --file ch12-applications.yml
```

Metabolic modeling with COBRApy

The computational simulation of biochemical pathways in cells is called **metabolic modeling**. In this approach, a matrix of reactions in a cell is defined, and a mass balance is set up so that we achieve a steady state. We can then try to optimize one of the outputs so that fluxes are diverted toward it – this is called **Flux Balance Analysis** (**FBA**). A **constraint-based** model provides reasonable ranges for fluxes and so requires less information, and models the cell at a steady state. A model with reaction rates, called a **kinetic model**, provides more detail but is more complex to build.

In FBA, we typically optimize an **objective function**. This is the output the model is trying to maximize. Typically, by default, we use the growth of the organism for this function. But often, we use something else, such as the output of a metabolite. For instance, if we were making biofuels, we might try to maximize the output of that molecule, even at the expense of growth.

Typically, a metabolic model includes reactions for a smaller set of pathways that focus on carbon and energy metabolism. But researchers are increasingly using **Genome Scale Models** (**GSMs**) in which nearly all of the reactions that can be predicted from the genome of the organism are present.

Once we have a model, we can simulate changes to the organism. The model starts out with all reactions, which we call a **wild-type** model. When we simulate a gene **knockout**, we are looking at what would happen if we eliminate the gene. This is typically done through introducing a foreign sequence into the location of the gene to disrupt it. This will obviously have an effect on the organism – it might either increase or decrease the flux through your pathway of interest, or it could just be neutral. We could also simulate gene **knockdown**, which is just a reduction of the gene's expression, often down through using an siRNA (a short RNA that interferes with the gene) or altering its codons so that it translates less effectively. There is also gene **upregulation**, in which we increase the expression of a particular gene. This is typically done by inserting a stronger promoter in front of it. Finally, there is gene **knock-in**, which involves adding a foreign gene sequence into the organism. You won't be able to simulate this unless you add those reactions to your model.

COBRApy (`https://opencobra.github.io/cobrapy/`) is a Python package for simulating metabolic models. It has functions to load and represent models, perform FBA, simulate changes to the model, determine which genes are essential, and more. It can read several different file formats, but one of the most popular is the **Systems Biology Markup Language** (**SBML**). You can read more about SBML here: `https://sbml.org/`.

Since metabolic modeling is a large topic, you may want to review some additional background before continuing:

- This video series provides background on metabolic modeling: `https://www.youtube.com/watch?v=fuVVCvDKcKg&list=PLbUnrAlFxLTFpmy-GiWT7Atlm85WFS786`
- This tutorial goes over COBRApy: `https://cnls.lanl.gov/external/qbio2018/Slides/FBA%202/qBio-FBA-lab-slides.pdf`
- More tutorials can be found here: `https://docs.kbase.us/workflows/metabolic-models`
- A great in-depth review can be found here: Passi et al *Genome-Scale Metabolic Modeling Enables In-Depth Understanding of Big Data*, Metabolites, Dec 2021: `https://www.mdpi.com/2218-1989/12/1/14`

In this recipe, we'll learn how to set up a metabolic model and simulate various changes to it. We will use E. coli as the example organism. You'll get a sense of how strain engineers utilize these models to make decisions and come up with strategies to improve their strains.

For this exercise, we'll use the SBML model of E. coli. We can download it from a public repository such as the **Biochemically, Genetically, and Genomically** (**BiGG**) models repository (`http://bigg.ucsd.edu/`).

In this recipe, we will use the E. coli model and learn how to perform FBA. We'll learn how to calculate growth rates and simulate the effects of gene knockouts.

Getting ready

First, let's install COBRApy:

```
! pip install cobra
```

Now we will download our E. coli model from BiGG:

```
! wget http://bigg.ucsd.edu/static/models/e_coli_core.xml
```

The code for this recipe can be found in the GitHub repository for this book, under `Ch12/Ch12-1-cobrapy.ipynb`.

That's it! We are ready to get started with COBRApy!

How to do it...

Here are the steps to perform this recipe:

1. First, we will import the COBRApy library and read in the model:

```
import cobra
model = cobra.io.read_sbml_model("e_coli_core.xml")
```

This code uses the `read_sbml_model()` method to read the SMBL file that we downloaded from BiGG. This model contains the metabolites, reactions, and interactions needed to simulate E. coli metabolism.

2. Next, we will run the FBA on our model:

```
solution_wt = model.optimize()
print("Wild-type growth rate (objective value):",
        solution_wt.objective_value)
print("Flux distribution for key reactions:")

for rxn in model.reactions[:10]:
    print(f"{rxn.id}: {solution_wt.fluxes[rxn.id]}")
```

This code uses the `optimize()` function to find the flux distribution that will maximize the objective function, which in this case will be the growth rate. It does this by using **linear programming**, which is a mathematical technique for finding the optimal outcome given a series of constrained linear relationships. We'll print the growth rate based on the wild-type model.

We'll also print out our top 10 reactions.

Here is what we get:

```
Wild-type growth rate (objective value): 0.8739215069684303
Flux distribution for key reactions:
PFK: 7.477381962160286
PFL: 0.0
PGI: 4.860861146496822
PGK: -16.02352614316761
PGL: 4.959984944574652
ACALD: 0.0
AKGt2r: 0.0
PGM: -14.716139568742836
PIt2r: 3.2148950476848035
ALCD2x: 0.0
```

We can see the current growth rate (without any changes) is 0.87. We also see the flux rates for the top 10 reactions in our model.

How do we know that the model is optimizing growth here? Let's take a quick detour and find out:

```
from cobra.util.solver import linear_reaction_coefficients
print("Objective direction:", model.objective.direction)
for rxn, coef in linear_reaction_coefficients(model).items():
    print(f"{rxn.id}: {coef}")
```

When we optimize the model, it is optimizing the flux of reactions toward whatever output is in the model.objective variable. If the variable is BIOMASS, then we are optimizing toward growth, meaning the production of carbon to support the mass of the cell, as opposed to some other product. We also see the direction is max, so we are trying to maximize growth. You can set the objective function to any reaction by setting model.objective. For example, we might want to maximize the secretion of acetate, ATP production, or nutrient uptake. You can even set the objective to a linear combination of multiple reactions to try and maximize combinatorial outcomes.

3. Next, let's simulate knocking out a gene in our model:

```
gene_to_knockout = "b0351"
with model:
    model.genes.get_by_id(gene_to_knockout).knock_out()
    solution_ko = model.optimize()
    print(
        f"\nGrowth rate after knocking out gene "
        f"{gene_to_knockout}:",
        solution_ko.objective_value
    )
```

This code will simulate knocking out the gene b0351, which is a thymidate synthase gene involved in DNA synthesis. We use the with statement on our model to make changes temporarily in a context, without making the changes permanent. The purpose of using the with statement here is to be able to check what happens when we knock out the gene, but without actually changing our model except within the context of this with block. When we go back to the next step, outside of the with block, the gene will no longer be knocked out.

We next get the gene based on its ID and use the knock_out() function to disable all the reactions related to that gene in the model.

4. Next, we use optimize() to run our flux balance again. Here is what we get:

```
Growth rate after knocking out gene b0351: 0.8739215069684303
```

Knocking out this gene didn't really change our growth rate!

5. Let's try some other genes and see whether they have an effect. We can print out a list of possible genes to knock out:

```
print("Available gene IDs in the model:")
for gene in model.genes:
    print(gene.id)
```

This code will simply loop over the genes in the `model.genes` property and print them out. Let's try another gene:

```
gene_to_knockout = "s0001"
with model:
    model.genes.get_by_id(gene_to_knockout).knock_out()
    solution_ko = model.optimize()
    print(
        f"\nGrowth rate after knocking out gene "
        f"{gene_to_knockout}:",
        solution_ko.objective_value
    )
```

Again, we set a gene to knockout – this gene, `s0001`, is a special one meant to represent spontaneous reactions in the cell, so it has a big impact:

Growth rate after knocking out gene s0001: 0.21114065257211664

Here, we see a much lower growth rate.

6. We can now also review the impact on the flux rates of our top 10 reactions:

```
print("\nChange in fluxes for selected reactions after
knockout:")
for rxn in model.reactions[:10]:
    flux_change = (
        solution_wt.fluxes[rxn.id] - solution_ko.fluxes[rxn.id]
    )
    print(f"{rxn.id}: Δ flux = {flux_change:.2f}")
```

Here is what we see:

```
Change in fluxes for selected reactions after knockout:
PFK: Δ flux = -2.31
PFL: Δ flux = -17.43
PGI: Δ flux = -5.10
PGK: Δ flux = 3.42
PGL: Δ flux = 4.96
ACALD: Δ flux = 8.47
AKGt2r: Δ flux = 0.00
PGM: Δ flux = 4.41
PIt2r: Δ flux = 2.44
ALCD2x: Δ flux = 8.47
```

The fluxes for many reactions have changed, some in a positive direction and some negative.

You can play around with other genes in the list and see how they impact the model!

Now we've seen how to set up and implement a metabolic model and knock out genes to see their effect on the model.

There's more...

Strain engineering is the art of modifying organisms, usually microbes, to make valuable products. Typically, these might be biofuels or medicines. Strain engineers use metabolic modeling to simulate flux through pathways and maximize their products. They use it to simulate gene knockouts or overexpression, or to optimize media and fermentation parameters. Metabolic modeling can be used to optimize cofactors needed by the cell, such as NADH and NADPH – see King et al, *Optimizing Cofactor Specificity of Oxidoreductase Enzymes for the Generation of Microbial Production Strains – OptSwap*, Industrial Biotechnology, Aug 2013: `https://www.liebertpub.com/doi/abs/10.1089/ind.2013.0005`.

Several popular packages exist to optimize genetic engineering efforts using metabolic modeling. For example, OptKnock is a tool for suggesting gene deletion strategies for the overproduction of molecules. Other packages specialize in optimizing cofactor usage or provide regulatory control strategies.

Here is a table of some key packages and their uses. More details of the references are provided in the *See also* section.

Package	Usage	Reference
OptKnock	Gene knockout	Burgard 2003
PyCoMo	Community modeling	Predl 2024
COBREXA	Modular modeling	Kratochvil 2025
k-OptForce	Kinetics with FBA	Chowdhury 2014
FastKnock	Identify all possible knockout strategies	Hassani 2024
MOMA	Prediction based on minimal reorganization of fluxes	Segre 2002
OptORF	Gene knockout and upregulation	Kim 2010
OptDesign	Combines regulation and knockout	Jiang 2022
StrainDesign	Python package that integrates multiple approaches	Schneider 2022

Table 12.1 – Major metabolic modeling packages

When building metabolic models, it can be incredibly useful to integrate other data sources for the organism into the model, such as RNA-Seq (transcriptomics), proteomics, or small metabolite data (metabolomics). These are collectively referred to as "omics" data (because in most cases you end the word with "omics"). Metabolic models can be greatly enhanced with **omics data**. For example, RNA-Seq data could be used to help estimate the level of genes under different conditions or to identify regulatory sequences in an organism. Proteomics can be used in a similar manner to help determine the actual levels of proteins in fermentation processes. Metabolomics data can be used to measure actual concentrations of metabolites.

> **Tip**
>
> For a short review of integrating multi-omics data into systems models, read *Approaches to Computational Strain Design in the Multiomics Era*, John & Bomble, Frontiers in Microbiology, Apr 2019: `https://www.frontiersin.org/journals/microbiology/articles/10.3389/fmicb.2019.00597/full`.

Many tools have been developed to integrate omics data into metabolic models. For example, METAFlux can infer fluxes from bulk or single-cell RNA-Seq data – refer to Huang et al, *Characterizing cancer metabolism from bulk and single-cell RNA-seq data using METAflux*, Nature Communications, Aug 2023: `https://www.nature.com/articles/s41467-023-40457-w`.

Metabolomics is the study of small molecules in cells. Compared to transcriptomics and proteomics, its capabilities have lagged behind somewhat. One common approach has been **13C metabolic flux analysis**, in which isotopically labeled carbon sources, such as glucose, are fed to the organism to analyze where carbon goes in cellular pathways. **Mass spectrometry** and **nuclear magnetic resonance** are also commonly used. Metabolomics can be **untargeted**, looking at all metabolites that can be detected by an instrument, or **targeted**, in which a subset of metabolites is measured with higher accuracy. NEXT-FBA uses metabolomic data to improve flux estimation in models – refer to Morrissey et al, *NEXT-FBA: A hybrid stoichiometric/data-driven approach to improve intracellular flux predictions*, Metabolic Engineering, Mar 2025: `https://www.sciencedirect.com/science/article/pii/S1096717625000461`.

MiNEApy can integrate multiple types of omics data, including transcriptomic, proteomic, and metabolomic – refer to Pandey, *MiNEApy: enhancing enrichment network analysis in metabolic networks*, BioInformatics, Mar 2025: `https://academic.oup.com/bioinformatics/article/41/3/btaf077/8030213`.

Machine learning is being used increasingly within the field of metabolic modeling. It can be used to upgrade and fill in gaps in models, estimate thermodynamic constraints, optimize media conditions, and much more. For a good review of machine learning methods in metabolic modeling, read Kundu et al, *Machine learning for the advancement of genome-scale metabolic modeling*, Biotechnology Advances, Sep 2024: `https://www.sciencedirect.com/science/article/pii/S0734975024000946?casa_token=d6K79DgkHFIAAAAA:ZLIcmC6Dn4u4vE8eDjw3m1ufUiSSoyKA7NNGaWE2LsEzNGAQo5JR4ck0e-XLu9L04bojFiKa`.

Taken together, strain engineers can incorporate advice from metabolic models and more general-purpose approaches, such as randomly generated library screening, to make multiple changes to a strain as part of an iterative **Design-Build-Test-Learn** (**DBTL**) strain optimization program. The **Automated Recommendation Tool** (**ART**) is a good example of a system that uses machine learning to guide strain engineers as part of the DBTL cycle – Radivojevic et al, *A machine learning Automate Recommendation Tool for synthetic biology*, Nature Communications, Sep 2020: `https://www.nature.com/articles/s41467-020-18008-4`.

In addition to flux balance models, **kinetic** models add dynamic information on metabolite fluxes and protein levels – Hu et al, *KETCHUP: Parameterizing of large-scale kinetic models using multiple datasets with different reference states*. Metabolic Engineering, Mar 2024: `https://www.sciencedirect.com/science/article/pii/S1096717624000181`.

More advanced models also add thermodynamic constraints, resource constraints on protein production, and extracellular factors to model strain performance in reactor tanks – Moimenta et al, *Temperature-Dependent Kinetic Modeling of Nitrogen-Limited Batch Fermentation by Yeast Species*, Mathematics, Apr 2025: `https://www.mdpi.com/2227-7390/13/9/1373`.

Genome-scale metabolic models now exist for a huge number of organisms – Gu et al, *Current status and applications of genome-scale metabolic models*, Genome Biology, Jun 2019: `https://genomebiology.biomedcentral.com/articles/10.1186/s13059-019-1730-3`.

Scientists are actively using these models to build better strains for applications in food, fuel, agriculture, and industrial products.

As you can see, metabolic modeling is an exciting field that will only continue to advance. In the future, we should see increasingly sophisticated models that apply to whole cells, incorporate numerous omics data sources, and can make increasingly accurate predictions. Let's try one last example before we move on to the next topic!

AI tip

Prompt: Write a Python script using MiNEApy that integrates RNA-Seq transcriptomic data with an E. coli genome-scale metabolic model to create context-specific models under two different conditions. The script should do the following:

Load the iML1515 E. coli model using MiNEApy's model-loading functionality

Import real or simulated RNA-Seq data from a CSV file with gene IDs matching the model

Preprocess the expression data (normalization, thresholding)

Implement the **Gene Inactivity Moderated by Metabolism and Expression** (GIMME) algorithm to create condition-specific models

Run flux balance analysis on both condition-specific models

Identify and analyze differential flux patterns between conditions

Visualize key pathway differences using MiNEApy's visualization tools

Generate a comprehensive output report including growth rates, flux distributions, and reaction essentiality changes

Include detailed comments explaining the biological significance of each step in the workflow

Please include error handling and ensure the code follows best practices for reproducibility in computational biology research

You should see: Code to import the MiNEApy package and simulate RNA-Seq data for incorporation into a metabolic model. It can compare the outputs of the model under different conditions and visualize the results using Matplotlib.

See also

- The COBRApy paper is here: Ebrahim et al, *COBRApy: Constraints-Based Reconstruction and Analysis for Python*, BMC Systems Biology, Aug 2013: `https://link.springer.com/article/10.1186/1752-0509-7-74`

- A good primer on FBA can be found in Orth et al, *What is flux balance analysis?*, Nature Biotechnol, Mar 2010: `https://www.nature.com/articles/nbt.1614.pdf`

- COBREXA focuses on modular modeling – Kratochvil et al, *COBREXA 2: tidy and scalable construction of complex metabolic models*, BioInformatics, Feb 2025: `https://academic.oup.com/bioinformatics/article/41/2/btaf056/8005852`

- PyCoMo is used to model communities of microbes – Predl et al, *PyCoMo: a Python package for community metabolic model creation and analysis*, BioInformatics, Apr 2024: `https://academic.oup.com/bioinformatics/article/40/4/btae153/7635576`

- A review of linear programming can be found here: `https://sites.math.washington.edu/~burke/crs/409/LP-rev/lp_rev_notes.pdf`

- OptKnock is discussed in Burgard et al, *OptKnock: a bilevel programming framework for identifying gene knockout strategies for microbial strain optimization*, Biotechnol Bioeng, Dec 2003: `https://pubmed.ncbi.nlm.nih.gov/14595777/`

- OptORF generates both gene deletion and up-regulation strategies: Kim & Reed, *OptORF: Optimal metabolic and regulatory perturbations for metabolic engineering of microbial strains*, BMC Systems Biology, Apr 2010: `https://link.springer.com/article/10.1186/1752-0509-4-53`

- k-OptForce integrates kinetics with FBA: Chowdhury et al, *k-OptForce: Integrating Kinetics with Flux Balance Analysis for Strain Design*, PLOS Computational Biology, Feb 2014: `https://journals.plos.org/ploscompbiol/article?id=10.1371/journal.pcbi.1003487`

- FastKnock identifies all possible combinations of knockouts for strain design: Hassani et al, *FastKnock: an efficient next-generation approach to identify all knockout strategies for strain optimization*, Microbial Cell Factories, Jan 2024: `https://link.springer.com/article/10.1186/s12934-023-02277-x`

- OptDesign uses a two-stage strain design strategy, combining regulatory and knockout approaches: Jiang et al, *OptDesign: Identifying Optimum Design Strategies in Strain Engineering for Biochemical Production*, ACS Synthetic Biology, Apr 2022: `https://pubs.acs.org/doi/full/10.1021/acssynbio.1c00610`

- StrainDesign is a Python package that integrates multiple packages – Schneider et al, *StrainDesign: a comprehensive Python package for computational design of metabolic networks*, BioInformatics, Nov 2022: `https://academic.oup.com/bioinformatics/article/38/21/4981/6701962`

- **MOMA** (short for **Minimization of Metabolic Adjustment**) is described in Segre et al, *Analysis of optimality in natural and perturbed metabolic networks*, Biological Sciences, Nov 2002: `https://www.pnas.org/doi/10.1073/pnas.232349399`

- There is more information on the E. coli core model here: `https://systemsbiology.ucsd.edu/Downloads/E_coli_Core`

Designing siRNAs with BioPython and ViennaRNA

In this recipe, we'll learn how to design **siRNAs** (short for **short interfering RNAs**). These RNAs are typically short (~21nt long) and interfere with genes by binding to a target gene and causing it to be cut up by a protein complex, thereby **silencing** the gene.

siRNAs can be used in a wide variety of applications. For example, a cell expressing a siRNA could be used to kill pathogens of an aquatic food product, increasing its yield (Huang et al, *RNA-Based Biopesticides: Pioneering Precision Solutions for Sustainable Aquaculture in China*, AROH, Feb 2025: `https://onlinelibrary.wiley.com/doi/full/10.1002/aro2.70000`). siRNA is also being explored for use in gene silencing to treat human diseases – Friedrich & Agner, *Therapeutic siRNA: State-of-the-Art and Future Perspectives*, BioDrugs, Aug 2022: `https://link.springer.com/article/10.1007/s40259-022-00549-3`.

When scoring siRNA candidates, there are a variety of criteria that can be used. Algorithms are used to look at the thermal properties of the RNA, how it folds, and how stable it is. We can examine the free energy of the siRNA-target duplex. Sequence rules have been developed to understand key determinants of how a siRNA sequence will bind with its target and induce silencing by protein complexes. We also need to examine the folding of the target RNA to see how accessible it will be for binding with the siRNA.

We look for **off-target** effects by examining the target genomes to see whether the siRNA candidate might also target other genes. This helps determine whether unexpected side effects might occur when introducing the siRNA.

ViennaRNA (`https://www.tbi.univie.ac.at/RNA/`) is a powerful package for predicting RNA structure. In this recipe, we'll use it to design a class for siRNA prediction.

Getting ready

First, we will install the ViennaRNA package. This command will install it along with any other packages you need for this recipe:

```
! pip install biopython pandas matplotlib seaborn ViennaRNA requests
```

Here is an alternative method for installing ViennaRNA in case you run into any issues (from the terminal):

```
conda install bioconda::viennarna
```

You will also want to make sure BLAST is installed:

```
! brew install blast
```

The code for this recipe can be found in the GitHub repository for this book, under `Ch12/Ch12-2-sirna.ipynb`.

How to do it...

Here are the steps to perform this recipe:

1. First, let's import our libraries:

```
import os
import pandas as pd
import matplotlib.pyplot as plt
import seaborn as sns
from Bio import SeqIO
from Bio.Seq import Seq
import numpy as np
import requests
import tempfile
import re
from io import StringIO
import RNA
import subprocess
import sys
import warnings
warnings.filterwarnings('ignore')
```

This will set up the **RNA** package, which contains ViennaRNA. We also bring in our data manipulation and visualization libraries. We will ignore any warnings that come up.

Then, we will define a siRNADesigner() class. In Python, a class is used for **Object-Oriented Programming (OOP)**. In this approach, we bundle data and the methods needed to act on that data together in objects. This keeps things more organized compared to just using functions.

Our class for designing siRNAs will have several functions. It will extract sequences, calculate properties of RNAs, generate and score candidates, and check for off-target effects, as shown here:

```
class SiRNADesigner:
    def __init__(self):
        self.parameters = {
            'length': 21,
            'min_gc': 30,
            'max_gc': 60,
            'seed_max_gc': 60,
            'check_off_targets': True
        }
```

```
        self.weights = {
            'gc_content': 0.20,
            'seed_gc': 0.15,
            'thermo_asymmetry': 0.25,
            'secondary_structure': 0.20,
            'motif_penalty': 0.20
        }

        self.avoid_patterns = [
            'AAAA', 'CCCC', 'GGGG', 'TTTT',
            'GUCCUUCAA', 'UGUGU',
            'TAAAA', 'AAAAA'
        ]

        self.blast_available = self._check_blast_installed()
```

This code first defines an initialization routine for the class. The first function defined in a class is the special __init__(self) method. This is also often called a **constructor** function. It gets automatically called when you create an instance of the class and is used to set up basic parameters.

We have default parameters for the length of siRNA we want to make, minimum GC content allowed (the percentage of G and C bases versus A and T), and so forth. We also create default weight penalties for scoring. Next, we define key sequence patterns to avoid:

- Homopolymers – runs of the same base, which can cause problems with folding

- Immune stimulatory motifs – sequences that might activate the host immune system

- Termination signals – patterns that could lead to premature termination of transcription of the siRNA

2. Finally, we'll run a routine to check whether BLAST is installed, which looks like this:

```
def _check_blast_installed(self):
    try:
        result = subprocess.run(
            ['blastn', '-version'],
            stdout=subprocess.PIPE,
            stderr=subprocess.PIPE
        )
        return result.returncode == 0
    except FileNotFoundError:
        return False
```

This code just attempts to run BLAST via the OS. If it works and the return code is 0, we return True; if there is any other return code, we get an error and we return it, or if BLAST is not found, we catch that error and return False.

3. Next, we'll define a handy function to fetch a sequence from NCBI using EFetch and the `requests` library. We won't use it in this example, but it will give you an option to use the class that way in the future:

```
def fetch_sequence_from_ncbi(self, accession):
    url = (
        f"https://eutils.ncbi.nlm.nih.gov/entrez/eutils/"
        f"efetch.fcgi?db=nucleotide&id={accession}"
        "&rettype=fasta&retmode=text"
    )
    response = requests.get(url)
    if response.status_code == 200:
        fasta_io = StringIO(response.text)
        for record in SeqIO.parse(fasta_io, "fasta"):
            return str(record.seq)
    else:
        raise Exception(
            f"Failed to fetch sequence from NCBI: "
            f"{response.status_code}"
        )
```

This code defines a URL for the EFetch request and sends it ahead. If it works, it then parses the FASTA and returns it.

4. Next, we'll define three functions to calculate important parameters:

```
def calculate_gc_content(self, sequence):
    gc_count = sequence.count('G') + sequence.count('C')
    return (gc_count / len(sequence)) * 100
```

This function calculates the percentage of G and C bases in the sequence.

```
def calculate_seed_region_gc(self, antisense):
    seed_region = antisense[1:8]
    return self.calculate_gc_content(seed_region)
```

This function simply calculates the GC content of the antisense seed region. The **seed region** is typically the first 8 nucleotides of the siRNA.

```
def check_forbidden_patterns(self, sequence):
    rna_seq = sequence.replace('T', 'U')
    for pattern in self.avoid_patterns:
        pattern = pattern.replace('T', 'U')
        if pattern in rna_seq:
            return False
    return True
```

This function checks for forbidden patterns by checking the `avoid_patterns` variable (defined in the preceding during initialization of the class).

5. Now we will define a function to calculate the thermal asymmetry:

```python
def calculate_thermodynamic_asymmetry(self, sense, antisense):
    sense_rna = sense.replace('T', 'U')
    antisense_rna = antisense.replace('T', 'U')
    five_prime_sense = sense_rna[:4]
    five_prime_antisense = antisense_rna[-4:]
    delta_g_5prime = RNA.fold_compound(
        five_prime_sense + five_prime_antisense
    ).mfe()[1]

    three_prime_sense = sense_rna[-4:]
    three_prime_antisense = antisense_rna[:4]
    delta_g_3prime = RNA.fold_compound(
        three_prime_sense + three_prime_antisense
    ).mfe()[1]

    diff = delta_g_3prime - delta_g_5prime

    if diff <= -3:
        return 100
    elif diff >= 0:
        return 0
    else:
        return (1 - (diff / -3)) * 100
```

Thermal asymmetry is an important feature of siRNA. It affects how the RNA is incorporated into the protein complex that performs gene silencing. This function first converts our input sequence to RNA, meaning that Uracil (U) replaces Thymidine (T). Next, it takes windows of the RNA sequence and computes the **Minimum Free Energy** (**MFE**) using the ViennaRNA package. MFE tells us how stable the RNA will be when it is folded in solution. When the 5' end of the RNA is less stable than the 3' end, the siRNA tends to work better. Finally, we perform a normalization on the score and return it. This is used in scoring candidate RNAs to find the best one.

6. Next, we need to check whether the target region of the gene we want to knock down will actually be accessible for the siRNA to pair with it. If the gene is folded in solution such that the target region is inaccessible, then the siRNA will not be effective.

```python
def analyze_target_accessibility(self, target_region):
    target_rna = target_region.replace('T', 'U')

    (structure, mfe) = RNA.fold(target_rna)
```

```
central_start = max(0, len(structure) // 2 - 10)
central_end = min(
    len(structure), len(structure) // 2 + 11)
central_region = structure[central_start:central_end]

unpaired_count = central_region.count('.')
unpaired_percentage = (
    unpaired_count / len(central_region)) * 100

return unpaired_percentage
```

This code first converts the sequence to RNA by replacing T with U. We then use the ViennaRNA `fold()` function to predict the secondary structure of the RNA. The `fold()` function will return both the structure of the RNA in a dot-notation structure (`https://www.tbi.univie.ac.at/RNA/ViennaRNA/refman/io/rna_structures.html`) and the minimum free energy. In dot notation, unpaired nucleotides between the sequence and the target are represented by dots.

We next focus on the central binding region of the RNA around a 21nt window. Then we calculate the number of unpaired nucleotides (represented by dots) and normalize as a percentage of the total length. We return the unpaired percentage.

Next, we will calculate the **Ui-Tei** value. This is based on a series of rules for siRNA design developed by Kumiko Ui-Tei (for more information, look in the *See also* section).

```
def calculate_ui_value(self, sequence):
    sequence = sequence.upper()
    score = 0

    if sequence[0] in 'AU':
        score += 33.3
    if len(sequence) > 9 and sequence[9] == 'A':
        score += 33.3
    if len(sequence) > 18 and sequence[18] == 'A':
        score += 33.3
    return score
```

This code examines our sequence and does the following:

- It gives it a higher score for having A or U in the first position

- It adds `score` if position 10 contains an A

- It adds `score` if position 19 contains an A

7. Now we will define one of our most important functions, which generates candidates based on the various scoring methods we've defined.

```python
def generate_candidates(self, mrna_sequence, length=21):
    candidates = []
    for i in range(len(mrna_sequence) - length + 1):
        target_start = max(0, i - 10)
        target_end = min(
            len(mrna_sequence), i + length + 10)
        target_region = mrna_sequence[
            target_start:target_end]

        sense = mrna_sequence[i:i+length]
        if not all(n in 'ACGT' for n in sense):
            continue
        antisense = str(Seq(sense).complement())[::-1]

        candidates.append({
            'position': i + 1,
            'sense': sense,
            'antisense': antisense,
            'target_region': target_region
        })

    return candidates
```

Here, we generate all the siRNA sequences we can from a given a target sequence. We use a default sliding window of 21nt and move across the target.

For each position, we expand the window around it by 10nt for accessibility analysis and create the antisense strand of the target, storing all these candidates in a dictionary.

Next, let's define a function to compute the **Reynolds** score! This score incorporates several sequence characteristics such as low GC content and absence of inverted repeats.

```python
def compute_reynolds_score(self, sequence):
    score = 0
    sequence = sequence.upper()
    gc_content = self.calculate_gc_content(sequence)
    if 30 <= gc_content <= 52:
        score += 1
    au_count = sum(
        1 for i in range(14, 19)
        if i < len(sequence) and sequence[i] in 'AT'
    )
```

So far, we have counted the number of As and Us. Let's now create a score based on our rules:

```
if au_count >= 3:
    score += 1
if len(sequence) > 2 and sequence[2] == 'A':
    score += 1
if len(sequence) > 9 and sequence[9] == 'A':
    score += 1
if len(sequence) > 12 and sequence[12] == 'T':
    score += 1
if len(sequence) > 12 and sequence[12] != 'G':
    score += 1
if len(sequence) > 18 and sequence[18] != 'G':
    score += 1
has_repeat = False
```

Next, we will check for repeats...

```
for i in range(len(sequence) - 7):
    if sequence.count(sequence[i:i+7]) > 1:
        has_repeat = True
        break
if not has_repeat:
    score += 1
if sequence[0] in 'AT':
    score += 1
```

Finally, we add one more scoring at the end of the sequence:

```
if len(sequence) > 18 and sequence[18] in 'AT':
    score += 1
return score
```

This function will score the sequence based on several criteria. We check whether the GC content is between 30 and 52, check for A and U at critical positions, lack of G at certain positions, and lack of internal repeats.

8. Great! Now let's bring it all together by creating our master function to score the siRNA candidates:

```
def score_candidates(self, candidates):
    scored_candidates = []

    for candidate in candidates:
        sense = candidate['sense']
        antisense = candidate['antisense']
        target_region = candidate['target_region']
```

9. We are now looping over the candidates. For each one, let's calculate the GC content:

```
gc_content = self.calculate_gc_content(sense)
seed_gc = self.calculate_seed_region_gc(antisense)

gc_score = 100 - abs((gc_content - 45) * 2)
gc_score = max(0, min(100, gc_score))

seed_gc_score = 100 - (
    (seed_gc / self.parameters['seed_max_gc']) * 100
)
seed_gc_score = max(0, min(100, seed_gc_score))
```

We have now obtained our GC scores.

10. Next, we need to create a motif score based on our forbidden patterns. We'll also calculate our additional scoring metrics:

```
motif_score = 100 if self.check_forbidden_patterns(sense) else 0
thermo_score = self.calculate_thermodynamic_asymmetry(
    sense, antisense
)
access_score = self.analyze_target_accessibility(
    target_region
)
reynolds_score = (
    self.compute_reynolds_score(sense) / 10
) * 100
ui_score = self.calculate_ui_value(sense)
```

11. Let us now add our scores together for a total score:

```
total_score = (
    gc_score * self.weights['gc_content'] +
    seed_gc_score * self.weights['seed_gc'] +
    thermo_score * self.weights['thermo_asymmetry'] +
    access_score * self.weights['secondary_structure'] +
    motif_score * self.weights['motif_penalty']
)
total_score = (
    0.7 * total_score +
    0.15 * reynolds_score +
    0.15 * ui_score
)
candidate_with_score = candidate.copy()
```

12. We'll update our candidates with their scores in a dictionary and return the sorted list:

```
candidate_with_score.update({
    'gc_content': gc_content,
    'seed_gc': seed_gc,
    'gc_score': gc_score,
    'seed_gc_score': seed_gc_score,
    'motif_score': motif_score,
    'thermo_score': thermo_score,
    'access_score': access_score,
    'reynolds_score': reynolds_score,
    'ui_score': ui_score,
    'total_score': total_score
})

scored_candidates.append(candidate_with_score)

return sorted(
    scored_candidates,
    key=lambda x: x['total_score'],
    reverse=True
)
```

To recap, this function loops over our siRNA candidates and applies our various scoring criteria. It checks the GC content, problematic sequence patterns, thermodynamic asymmetry, target accessibility, Reynolds score, and Ui-Tei score. We then calculate a composite score based on a weighted function. Recall that the weight parameters are defined in the init() function of the class. The final score consists of a weighted combination of 70% of the component scores, 15% of the Reynolds score, and 15% of the Ui-Tei score.

Finally, we sort and return the list of candidates with their scores. Note in the return statement, we use a lambda function. This is a short, unnamed inline function, which in this case, simply takes the name of the candidate (x) and returns total_score. This tells the sorting function to sort the candidates by total score, and we set reverse=True, indicating we want them sorted in descending order.

Note that our class contains a check_off_targets_blast() method, which is used to run BLAST against the target genome to check for additional places the siRNA might hit. It isn't directly used in this example, so we won't go over it further in the interest of space – but here's an AI tip you could use to try it out.

> **AI tip**
>
> **Prepare**: First, paste in the code for the `siRNADesigner` class
>
> **Prompt**: Write a code example that uses the `siRNADesigner` class to check for off-target effects
>
> **You should see**: Code to generate a list of top siRNA candidates using your class, and then run the off-target check routine

Next, you'll see another method, `predict_efficacy_with_thermocomposition()` – again, not used in this example. This method is complementary to the scoring methods we used here. If you want to try it out, you could explore an AI tip like the following (you may need to paste in your code first, so the tool knows what to refer to).

> **AI tip**
>
> **Prompt**: Enhance the siRNA off-target analysis example by also incorporating the `predict_efficacy_with_thermocomposition` method. Modify the example to apply this method to each candidate, include the resulting score in the evaluation process, and adjust the final ranking to account for all three scoring approaches: the original composite score, off-target analysis, and thermo-composition prediction. The final score should weigh these three components appropriately. Make sure to display and export the thermo-composition scores in the results.
>
> **You should see**: An updated `siRNADesigner()` example that now incorporates the additional scoring method

Okay, let's define a short function that retrieves our top candidates:

```
def get_top_candidates(self, scored_candidates, top_n=5):
    return scored_candidates[:top_n]
```

This function simply returns our top five scored candidates by default.

Now we can show some ways to visualize our candidates:

```
def visualize_candidates(self, candidates, top_n=10):
    df = pd.DataFrame(candidates[:top_n])

    plt.figure(figsize=(12, 6))
    ax = sns.barplot(x=df.index, y='total_score', data=df)

    plt.title('Top siRNA Candidates by Score')
    plt.xlabel('Candidate Index')
    plt.ylabel('Total Score')
```

```
for i, p in enumerate(ax.patches):
    ax.annotate(
        f"Pos: {df.iloc[i]['position']}",
        (p.get_x() + p.get_width() / 2., p.get_height()),
        ha='center', va='center',
        xytext=(0, 10),
        textcoords='offset points'
    )

plt.tight_layout()
plt.show()
```

We have now built a bar graph of the top candidates based on their score. Now let's make a graph of the components that make up our scores:

```
score_components = [
    'gc_score', 'seed_gc_score', 'thermo_score',
    'access_score', 'motif_score'
]
plt.figure(figsize=(14, 8))

df_melt = pd.melt(
    df,
    id_vars=['position'],
    value_vars=score_components,
    var_name='Score Component',
    value_name='Value'
)

sns.barplot(
    x='position', y='Value',
    hue='Score Component', data=df_melt
)

plt.title('Score Components by siRNA Candidate')
plt.xlabel('Position in Target mRNA')
plt.ylabel('Score Value')
plt.legend(title='Score Component')
plt.tight_layout()
plt.show()
```

Finally, we will build a scatter plot of the GC content of the top candidates:

```python
plt.figure(figsize=(10, 6))

sns.scatterplot(
    x='position', y='gc_content',
    size='total_score', hue='total_score',
    sizes=(50, 200), data=df
)

plt.axhline(y=30, color='r', linestyle='--', alpha=0.5)
plt.axhline(y=60, color='r', linestyle='--', alpha=0.5)

plt.title('GC Content Distribution of Top siRNA Candidates')
plt.xlabel('Position in Target mRNA')
plt.ylabel('GC Content (%)')
plt.tight_layout()
plt.show()
```

To recap, this method uses `pandas` to convert the data to a DataFrame. It then creates three key plots for us:

- Bar plot of top siRNA candidates by score

- Most important components of the score for each of the top candidates

- GC content distribution of the top candidates

Now we'll write a function to print a nice, formatted output of the top candidates:

```python
def format_output(self, candidates, top_n=5):
    for i, candidate in enumerate(candidates[:top_n], 1):
        print(
            f"Candidate #{i} (Score: {candidate['total_score']:.2f},"
            f"Position: {candidate['position']})"
        )
        print(f"Sense:     5'-{candidate['sense']}-3'")
        print(f"Antisense: 3'-{candidate['antisense']}-5'")
        print(f"GC Content: {candidate['gc_content']:.1f}%")
        print(f"Seed Region GC: {candidate['seed_gc']:.1f}%")
        print(f"Thermodynamic Asymmetry: "
              f"{candidate['thermo_score']:.1f}")
        print(f"Target Accessibility: "
              f"{candidate['access_score']:.1f}")
        print(f"Reynolds Score: {candidate['reynolds_score']:.1f}")
        print(f"Target Region: ...{candidate['target_region']}...")
        print("-" * 50)
```

This method is pretty straightforward. For each top candidate, it prints out the key information, including sequence, scores, GC content, and target region.

The final method in our class will be for exporting our candidates:

```python
def export_candidates(self, candidates, file_path):
    df = pd.DataFrame(candidates)
    df.to_csv(file_path, index=False)
    print(f"\nAll candidates exported to '{file_path}'")
```

That's it! Now that we've built our class, let's see how to run it:

```python
if __name__ == "__main__":
    try:
        import RNA
    except ImportError:
        print("ERROR: ViennaRNA Python package not found.")
        print("Please install it with: pip install ViennaRNA")
        print(
            "For installation instructions, visit: "
            "https://www.tbi.univie.ac.at/RNA/"
        )
        sys.exit(1)
    designer = SiRNADesigner()

    example_sequence = """
        ATGGGGAAGGTGAAGGTCGGAGTCAACGGATTTGGTCGTATTGGGCGCCTGGTCACC
        AGGGCTGCTTTTAACTCTGGTAAAGTGGATATTGTTGCCATCAATGACCCCTTCATT
        GACCTCAACTACATGGTTTACATGTTCCAATATGATTCCACCCATGGCAAATTCCATG
        GCACCGTCAAGGCTGAGAACGGGAAGCTTGTCATCAATGGAAATCCCATCACCATCTT
        CCAGGAGCGAGATCCCTCCAAAATCAAGTGGGGCGATGCTGGCGCTGAGTACGTCGTG
        GAGTCCACTGGCGTCTTCACCACCATGGAGAAGGCTG
    """
    example_sequence = re.sub(r'\s+', '', example_sequence)
    print("Designing siRNAs for target sequence...")
    candidates = designer.generate_candidates(example_sequence)
    print(f"Generated {len(candidates)} siRNA candidates")

    print("Scoring candidates based on design rules...")
    scored_candidates = designer.score_candidates(candidates)

    top_candidates = designer.get_top_candidates(
        scored_candidates)
    designer.format_output(top_candidates)
    designer.export_candidates(
        scored_candidates, 'sirna_candidates.csv')
    designer.visualize_candidates(scored_candidates)
```

The first line of this code checks whether the special `name` variable is set to `main`. Python does this when it is running a script directly, as opposed to importing a module. This ensures the code will execute.

Next, we check whether the ViennaRNA package is installed, and if not, provide instructions to install it.

We next instantiate an instance of our `siRNADesigner()` class. We set up a target sequence, which is, in this case, a partial sequence of the GAPDH gene, a Glyceraldehyde 3-phosphate Dehydrogenase (`https://www.genecards.org/cgi-bin/carddisp.pl?gene=GAPDH`).

Next, we will generate a list of candidates using the `generate_candidates()` method. We will then score the candidates and get our top N candidates. We will then format the output and print our top candidates. Finally, we export the candidates to an output CSV file and visualize our results.

Here are the RNAs designed:

```
Designing siRNAs for target sequence...
Generated 305 siRNA candidates
Scoring candidates based on design rules...


=== Top siRNA Candidates ===

Candidate #1 (Score: 66.88, Position: 116)
Sense:     5'-ACCTCAACTACATGGTTTACA-3'
Antisense: 3'-TGTAAACCATGTAGTTGAGGT-5'
GC Content: 38.1%
Seed Region GC: 42.9%
Thermodynamic Asymmetry: 0.0
Target Accessibility: 66.7
Reynolds Score: 90.0
Target Region: ...CCCTTCATTGACCTCAACTACATGGTTTACATGTTCCAATA...
-------------------------------------------------
Candidate #2 (Score: 65.72, Position: 70)
Sense:     5'-AACTCTGGTAAAGTGGATATT-3'
Antisense: 3'-AATATCCACTTTACCAGAGTT-5'
GC Content: 33.3%
Seed Region GC: 28.6%
Thermodynamic Asymmetry: 0.0
Target Accessibility: 71.4
Reynolds Score: 70.0
Target Region: ...GGCTGCTTTTAACTCTGGTAAAGTGGATATTGTTGCCATCA...
-------------------------------------------------
```

```
Candidate #3 (Score: 65.05, Position: 125)
Sense:      5'-ACATGGTTTACATGTTCCAAT-3'
Antisense: 3'-ATTGGAACATGTAAACCATGT-5'
GC Content: 33.3%
Seed Region GC: 42.9%
Thermodynamic Asymmetry: 0.0
Target Accessibility: 52.4
Reynolds Score: 100.0
Target Region: ...GACCTCAACTACATGGTTTACATGTTCCAATATGATTCCAC...
```

As you can see, we get our top siRNA candidates with their sense and antisense sequences. We can review their scores and their position within the target gene.

Our siRNA candidates have been saved in the `sirna_candidates.csv` file.

Now let's look at the visualizations. As you can see, we have our three key visualizations for our top siRNA candidates.

The following graph shows us the score for each of the candidates, and we use a label to show the position in the target sequence:

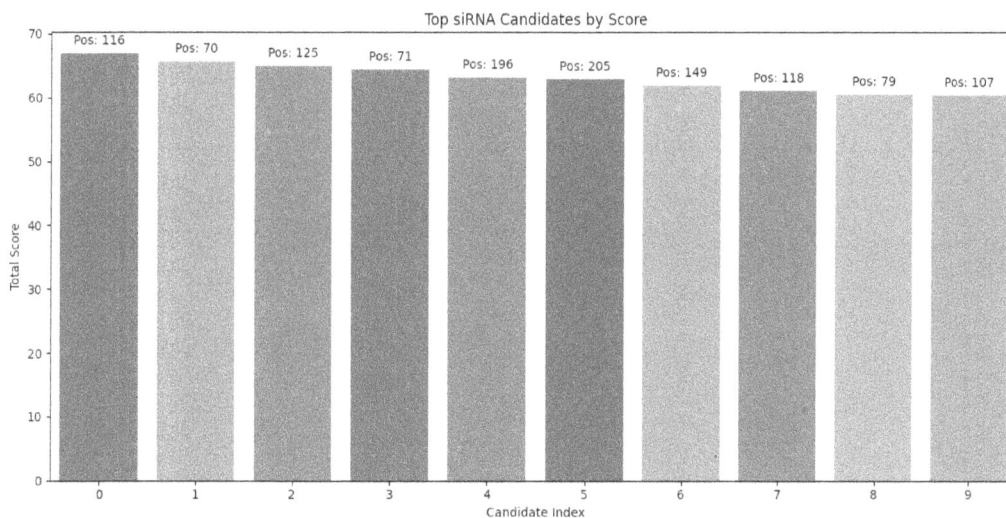

Figure 12.1 – Graph showing candidate scores and sequence positions

In the next figure, we see the relative importance of different scoring components for our top candidates. The importance of each scoring component is shown on the Y-axis, while the X-axis represents the position at which the candidate occurs within the mRNA. The bars in each group show the score components, from left to right, as follows: GC Score, Seed GC Score, Thermo Score, Access Score, and Motif Score.

You can see that motifs present tends to be one of the largest contributing factors, whereas seed GC content tends to be one of the least important.

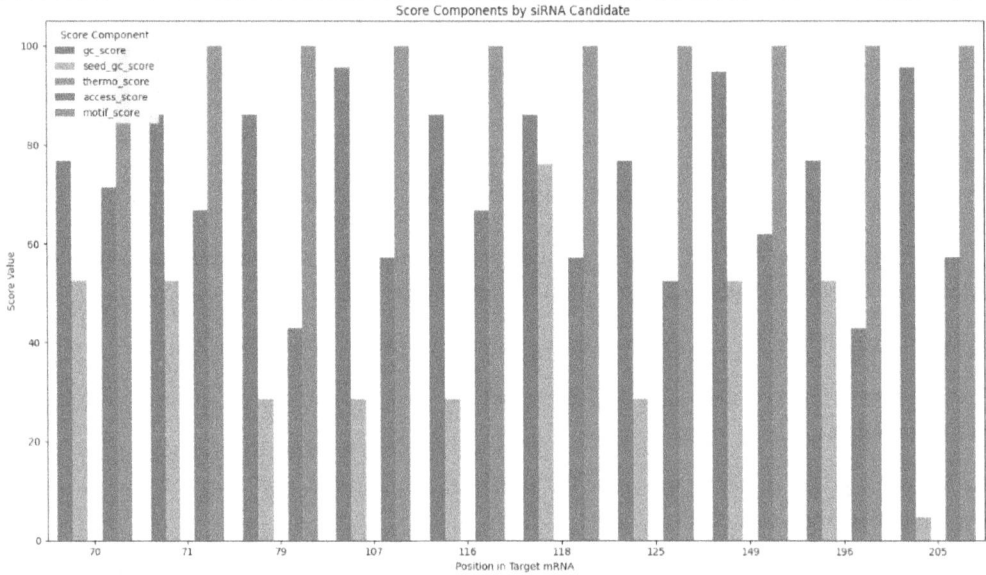

Figure 12.2 – Relative importance of scoring criteria for top siRNA candidates

The third graph shows the GC content of the candidates relative to their position within the target. We use color and size to represent the score as well.

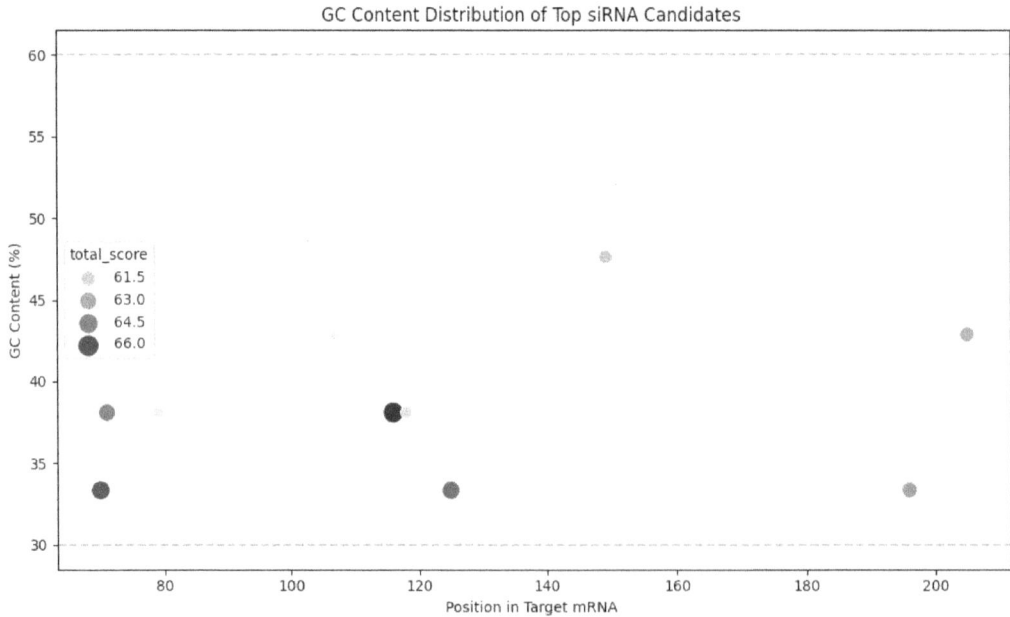

Figure 12.3 – GC content for siRNA candidates

This concludes our development of the siRNA system! You can see that you are able to use a variety of techniques, including motif finding and RNA folding, to find siRNAs that can knock down genes. This is a great real-world example of integrating multiple aspects of bioinformatics into a practical workflow.

There's more...

ViennaRNA can be used for many other purposes. RNA folding is important to study to understand the conservation of non-coding RNAs across species or to model how RNA interacts with other molecules. You can even design RNAs to fold into specific structures.

Machine learning is widely applied to the field of RNA design. For a good review on this topic, read Martinelli, *Machine learning for siRNA efficiency prediction: A systematic review*, Health Sciences Review, Jun 2024: `https://www.sciencedirect.com/science/article/pii/S2772632024000102`.

See also

- ViennaRNA is described here: Lorenz et al, *ViennaRNA Package 2.0*, Algorithms for Molecular Biology, Nov 2011: `https://almob.biomedcentral.com/articles/10.1186/1748-7188-6-26`

- For information on the Ui-Tei value, see: Ui-Tei et al, *Reduced base-base interactions between the DNA seed and RNA target are the major determinants of a significant reduction in the off-target effect due to DNA-seed-containing siRNA*: `https://ui-tei.rnai.jp/assets/files/pdf/Ui-TeiMHS2009_P2_26.pdf`

- For a good overview of siRNA design and discussion of the Reynolds score, read He et al, *Predicting siRNA efficacy based on multiple selective siRNA representations and the combination at score level*, Scientific Reports, Mar 2017: `https://pmc.ncbi.nlm.nih.gov/articles/PMC5357899/#:~:text=For%20example%2C%20Reynolds%20analyzed%20180,%2C%20and%20(8)%20position%2013`

- RNAxs integrates target site accessibility with siRNA scoring – Tafer et al, *The impact of Target site accessibility on the design of effective siRNAs*, Nat. Biotechnol., Apr 2008: `https://www.nature.com/articles/nbt1404`

- Read more about OOP here: `https://www.freecodecamp.org/news/how-to-use-oop-in-python/`

- You can learn more about Lambda functions in this article: `https://docs.aws.amazon.com/lambda/latest/dg/lambda-python.html`

Predicting food properties using bioinformatics

Bioinformatics is increasingly being used in a wide variety of areas. It is used extensively in the development of agricultural and food products. In this recipe, we are going to use bioinformatics to calculate some key food properties of an organism. Microorganisms are often used as **Single-Cell Protein (SCP)** by fermenting them and then adding the protein into various foods. We can actually tell a lot about the nutritional value of an organism by examining its proteome!

In this recipe, we will calculate many properties of the amino acids in an organism, using the E. coli proteome as an example. We'll use the properties of the proteome to calculate the number of essential amino acids, protein content, potential flavor profile, and other key characteristics of the organism.

By the end of this recipe, you'll see how bioinformatics can be applied to predict the food and nutritional properties of organisms used in food science.

Getting ready

First, we'll download the E. coli proteome from UniProt to analyze it:

```
! wget https://ftp.uniprot.org/pub/databases/uniprot/current_
release/knowledgebase/reference_proteomes/Bacteria/UP000000625/
UP000000625_83333.fasta.gz
```

Uncompress the file:

```
! gunzip UP000000625_83333.fasta.gz
```

Let's rename the file to keep it simple:

```
! mv UP000000625_83333.fasta example_organism.fasta
```

The code for this recipe can be found in the GitHub repository for this book, under `Ch12/Ch12-3-food-properties.ipynb`.

How to do it...

Here are the steps to try this recipe:

1. First, we will import our libraries:

```
import pandas as pd
import numpy as np
from Bio import SeqIO
import matplotlib.pyplot as plt
from collections import Counter
```

2. Now we will write a function to calculate food properties:

```python
def calculate_food_properties(proteome_file, output_file=None):
    essential_aas = {
        'F': 'Phenylalanine',
        'I': 'Isoleucine',
        'K': 'Lysine',
        'L': 'Leucine',
        'M': 'Methionine',
        'T': 'Threonine',
        'V': 'Valine',
        'W': 'Tryptophan',
        'H': 'Histidine'
    }

    non_essential_aas = {
        'A': 'Alanine',
        'R': 'Arginine',
        'N': 'Asparagine',
        'D': 'Aspartic acid',
        'C': 'Cysteine',
        'E': 'Glutamic acid',
        'Q': 'Glutamine',
        'G': 'Glycine',
        'P': 'Proline',
        'S': 'Serine',
        'Y': 'Tyrosine'
    }
```

We define amino acids as **essential** or **non-essential**. Essential amino acids cannot be synthesized by the body, so we must get them from food. As such, they are the most important food components. Non-essential amino acids can be biosynthesized by our metabolism.

3. Now we will define some key properties of amino acids:

```python
aa_properties = {
    'hydrophobic': ['A', 'I', 'L', 'M', 'F', 'V', 'P', 'G'],
    'hydrophilic': [
        'R', 'N', 'D', 'C', 'Q', 'E',
        'H', 'K', 'S', 'T', 'W', 'Y'
    ],
    'sulfur_containing': ['C', 'M'],
    'umami': ['E', 'G'],
    'sweet': ['A', 'G', 'S', 'T'],
    'bitter': ['I', 'L', 'V', 'F', 'Y', 'W', 'H']
}
```

Hydrophobic amino acids interact with lipid membranes, improving food absorption. **Hydrophilic** amino acids can bind water, which may enhance baking properties. Sulfur-containing amino acids undergo something called the Maillard reaction (https://www.sciencedirect.com/topics/agricultural-and-biological-sciences/maillard-reaction), which generates distinctive aromas and flavors. Finally, we define amino acids for the umami, sweet, and bitter tastes.

4.   Now we will read the E. coli proteome file using BioPython:

```
proteins = list(SeqIO.parse(proteome_file, "fasta"))
```

5.   Next, we will initialize key counters for our analysis:

```
total_aa_count = 0
aa_counter = Counter()
essential_aa_counter = Counter()
non_essential_aa_counter = Counter()
property_counter = {prop: 0 for prop in aa_properties}
```

These will keep track of the total amino acids, essential amino acids, and non-essential amino acids. The last line will create a dictionary with a counter for each of our properties.

6.   Next, we will process each protein and count up the relevant properties:

```
for protein in proteins:
    sequence = str(protein.seq)
    for aa in sequence:
        if aa in essential_aas or aa in non_essential_aas:
            aa_counter[aa] += 1
            total_aa_count += 1
            if aa in essential_aas:
                essential_aa_counter[aa] += 1
            elif aa in non_essential_aas:
                non_essential_aa_counter[aa] += 1
            for prop, aa_list in aa_properties.items():
                if aa in aa_list:
                    property_counter[prop] += 1
```

This code does the following:

I.   Loops over all the proteins in the organism

II.   Then loops over all the amino acids in that protein

III. For each amino acid, we will count whether it is essential or non-essential

IV.   We will also loop over each property, and if that amino acid is in the list for the property, we will count that as well.

7. Next, we will calculate the total protein content based on certain assumptions of the average weight of amino acids:

```
avg_aa_weight = 110  # g/mol
protein_content = (
    total_aa_count * avg_aa_weight / 6.022e23 * 100
)  # g/100g
```

Then, we calculate the ratio of essential amino acids:

```
essential_ratio = (
    sum(essential_aa_counter.values()) / total_aa_count
    if total_aa_count > 0 else 0
)
```

8. Now we will prepare our results in a dictionary:

```
results = {
    "protein_content": protein_content,
    "total_amino_acids": total_aa_count,
    "essential_amino_acid_ratio": essential_ratio,
    "amino_acid_composition": {
        aa: count/total_aa_count
        for aa, count in aa_counter.items()
    },
    "essential_amino_acids": {
        aa: count
        for aa, count in essential_aa_counter.items()
    },
    "non_essential_amino_acids": {
        aa: count
        for aa, count in non_essential_aa_counter.items()
    },
    "taste_properties": {
        prop: count/total_aa_count
        for prop, count in property_counter.items()
    }
}
```

This includes our protein content, total amino acids, and essential amino acid ratio. We also create a dictionary for amino acid composition, with the relative percentage of each amino acid. We provide counts for essential and non-essential amino acids. And finally, we create a dictionary of taste properties, with the relative frequency of each property based on the percentage of amino acids in the proteome possessing that property.

9. Now we are going to calculate a nutritional score based on the relative importance of the essential amino acids:

```python
essential_aa_importance = {
    'F': 0.8, 'I': 0.9, 'K': 1.0,
    'L': 0.9, 'M': 1.0, 'T': 0.8,
    'V': 0.9, 'W': 1.0, 'H': 0.8
}
nutritional_score = 0

for aa, count in essential_aa_counter.items():
    composition = (
        count / total_aa_count if total_aa_count > 0 else 0
    )
    nutritional_score += (
        composition * essential_aa_importance.get(aa, 0)
    )
results["nutritional_score"] = nutritional_score * 10
```

This code sets up the relative importance of each amino acid for nutrition, then loops over the essential amino acids. It then calculates the relative composition of each amino acid and multiplies it by the importance factor, scaling the nutritional score to a range of 0-10 at the end.

10. Now we will calculate a flavor profile:

```python
flavor_profile = {
    "umami": (
        sum(aa_counter[aa] for aa in aa_properties['umami'])
        / total_aa_count if total_aa_count > 0 else 0
    ),
    "sweet": (
        sum(aa_counter[aa] for aa in aa_properties['sweet'])
        / total_aa_count if total_aa_count > 0 else 0
    ),
    "bitter": (
        sum(aa_counter[aa] for aa in aa_properties['bitter'])
        / total_aa_count if total_aa_count > 0 else 0
    )
}
results["flavor_profile"] = flavor_profile
```

For each flavor, we will calculate the relative proportion of amino acids yielding that flavor and store it in a dictionary.

11. Finally, we will generate our analysis report and return the results:

```
if output_file:
    generate_analysis_report(results, output_file)
return results
```

12. Next, we will define a function for generating a report on the results:

```
def generate_analysis_report(results, output_file):
    aa_data = pd.DataFrame({
        'Amino Acid': list(
            results['amino_acid_composition'].keys()),
        'Frequency': list(
            results['amino_acid_composition'].values())
    })
    aa_data = aa_data.sort_values(
        'Frequency', ascending=False
    )
```

We set up a `pandas` dataframe with the amino acids and frequencies.

13. Next, we begin setting up our plots:

```
fig, axes = plt.subplots(2, 2, figsize=(15, 12))

axes[0, 0].bar(aa_data['Amino Acid'], aa_data['Frequency'])
axes[0, 0].set_title('Amino Acid Composition')
axes[0, 0].set_ylabel('Frequency')
axes[0, 0].tick_params(axis='x', rotation=45)
```

This will set up a plot that has four sections (subplots) and the first chart, which is a bar chart of amino acid frequency:

```
essential_count = sum(results['essential_amino_acids'].values())
non_essential_count = sum(
    results['non_essential_amino_acids'].values())
axes[0, 1].pie(
    [essential_count, non_essential_count],
    labels=['Essential', 'Non-essential'],
    autopct='%1.1f%%'
)
axes[0, 1].set_title('Essential vs Non-essential Amino Acids')
```

14. This code creates a pie chart showing the percentage of essential versus non-essential amino acids:

```
flavor_data = results['flavor_profile']
axes[1, 0].bar(flavor_data.keys(), flavor_data.values())
axes[1, 0].set_title('Flavor Profile')
axes[1, 0].set_ylabel('Proportion')
```

The preceding code produces a proportional bar chart of the flavor components:

```
prop_data = results['taste_properties']

axes[1, 1].bar(prop_data.keys(), prop_data.values())
axes[1, 1].set_title('Amino Acid Properties')
axes[1, 1].set_ylabel('Proportion')
axes[1, 1].tick_params(axis='x', rotation=45)
```

Our last chart is a bar chart of the amino acid properties and their proportion.

15. Finally, we adjust the padding of the plots and save the figure to a file:

```
plt.tight_layout()
plt.savefig(f"{output_file}_plots.png")
```

We then print out a summary report:

```
with open(f"{output_file}_summary.txt", 'w') as f:
    f.write("PROTEOME NUTRITIONAL ANALYSIS\n")
    f.write("=============================\n\n")
    f.write(
        f"Protein Content: "
        f"{results['protein_content']:.2f} g/100g\n"
    )
    f.write(
        f"Nutritional Score: "
        f"{results['nutritional_ score']:.2f}/10\n"
    )
    f.write(f"Essential Amino Acid Ratio: "
            f"{results['essential_amino_acid_ratio']:.2f}\n\n")
    f.write("FLAVOR PROFILE\n")
    f.write("=============\n")
    for flavor, value in results['flavor_profile'].items():
        f.write(f"{flavor.capitalize()}: {value:.2f}\n")

    f.write("\nESSENTIAL AMINO ACIDS\n")
    f.write("=====================\n")
    for aa, count in results['essential_amino_acids'].items():
        f.write(f"{aa}: {count}\n")
```

Awesome! Let's use our nutritional functions:

```python
if __name__ == "__main__":
    proteome_file = "example_organism.fasta"
    results = calculate_food_properties(
        proteome_file, "nutritional_analysis")

    print(
        f"Protein Content: "
        f"{results['protein_content']:.2f} g/100g"
    )
    print(
        f"Nutritional Score: "
        f"{results['nutritional_ score']:.2f}/10"
    )
    print(
        f"Essential Amino Acid Ratio: "
        f"{results['essential_amino_acid_ratio']:.2f}"
    )
    print("\nFlavor Profile:")
    for flavor, value in results['flavor_profile'].items():
        print(f"  {flavor.capitalize()}: {value:.2f}")
```

Here are our results!

```
Protein Content: 0.00 g/100g
Nutritional Score: 3.94/10
Essential Amino Acid Ratio: 0.44
Flavor Profile:
  Umami: 0.13
  Sweet: 0.28
  Bitter: 0.34
```

This is how the results look:

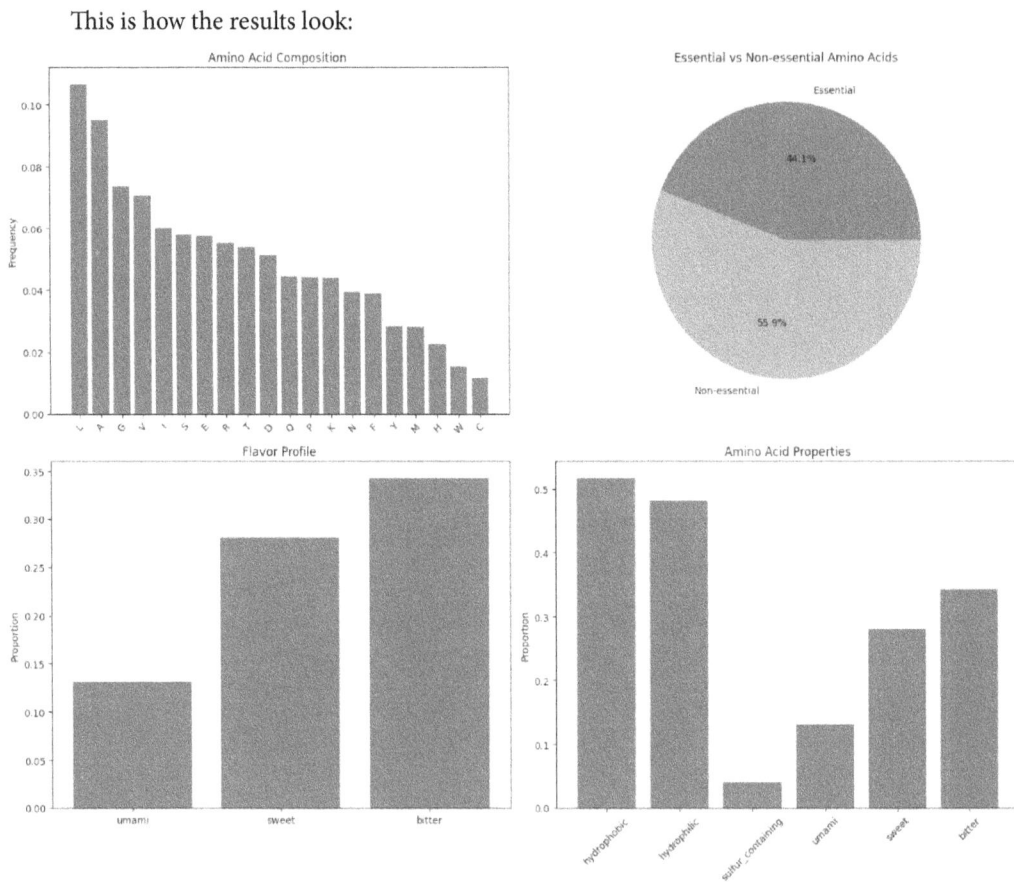

Figure 12.4 – Summary of nutritional properties for the E. coli proteome

You will also find a summary of the results in your working directory in the `nutritional_ analysis_summary.txt` file (as well as the plots).

That's it! In this recipe, you used bioinformatics to understand how an organism might be used to affect the nutrition, baking properties, and flavor of a food product. You could imagine many ways to take this further, by looking at how easily proteins can be digested or analyzing a proteome for substances that might be beneficial (or harmful)!

There's more...

Bioinformatics is increasingly used in many industries, such as food, agriculture, and others. For a recent review on bioinformatics in agriculture, read: Sahu et al, *Editorial: Bioinformatics, big data and agriculture: a challenge for the future,* Frontiers in Plant Science, Oct 2023: `https://www.frontiersin. org/journals/plant-science/articles/10.3389/fpls.2023.1271305/full.`

Bioinformatics can be used to develop insecticides – Da Costa et al, *Identification of Potential Insect Growth Inhibitor against Aedes aegypti: A Bioinformatics Approach*, IJMS, Jul 2022: `https://www.mdpi.com/1422-0067/23/15/8218`.

Bioinformatics can be used to discuss food digestibility – Jaeger et al, *In vitro digestibility of bioprocessed brewer's spent yeast: Demonstrating protein quality and gut microbiome modulation potential*, Food Research International, Feb 2025: `https://www.sciencedirect.com/science/article/pii/S0963996925000699`.

Bioinformatics can be used to study the microorganisms critical to successful food fermentation – Chelliah et al, *A review on the application of bioinformatics tools in food microbiome studies*, Briefings in Bioinformatics, Mar 2022: `https://academic.oup.com/bib/article/23/2/bbac007/6533500`.

Bioinformatics can also be used to discover **BioActive Peptides** (`https://www.sciencedirect.com/topics/food-science/bioactive-peptides`). These short peptides often come from food fermentation and can have many health benefits, such as antihypertension and immune stimulation – Du et al, *Bioinformatics approaches to discovering food-derived bioactive peptides: Reviews and perspectives*, TrAC, May 2023: `https://www.sciencedirect.com/science/article/pii/S0165993623001383?casa_token=wWdJZiBiRh8AAAAA:ZRRR166DG3dNG5e4H-ZtO99V71hWB9pD2hhRtK3cxO1R43nmwZje1B_jBrTcam5RLD-TcCFc`.

Discovering genes to make novel molecules

One of the most important tasks of bioinformatics is to characterize novel genes and pathways, and to discover new pathways that can make important molecules. These molecules might be used as industrial lubricants and biofuels or to cure cancer.

Comparative genomics is the analysis and comparison of multiple genomes, often to compare their gene content. It can be used to characterize the evolutionary history of organisms or to annotate genes based on their conservation across species. In this example, we will use it to find novel genes by comparing several organisms that are closely related. If one of those organisms is known to make a certain molecule and the others not, then we can subtract away all the genes in the organisms that are in common, leaving novel gene candidates.

Genome mining uses bioinformatics to discover and annotate genes in novel organisms or metagenomic samples. In particular, **extremophiles** can be very useful for mining as they contain genes to adapt to intense conditions such as cold or heat. Tools such as antiSMASH and DeepBGC are often used to annotate clusters of genes likely to be related to making a natural product (these are referenced in the *See also* section).

In this example, we will perform comparative genomics. We will imagine that we have discovered a new organism, a species of mycobacteria, called M. tuberculosis, in this case, that makes a valuable molecule – let's call it *YourRichium*. We know that there are three other closely related organisms that don't make this molecule. You'll discover which candidate genes might make the molecule!

Getting ready

First, let's install the packages we'll need (you may already have these installed):

```
! pip install biopython pandas
```

You will also need Prodigal, BLAST, and HMMER installed:

```
brew install brewsci/bio/prodigal
brew install blast
brew install hmmer

mkdir -p ~/pfam
cd ~/pfam
curl -O ftp://ftp.ebi.ac.uk/pub/databases/Pfam/current_release/Pfam-A.
hmm.gz
gunzip Pfam-A.hmm.gz
hmmpress Pfam-A.hmm
```

Add it to your path by placing in your `.zshrc` file:

```
export PFAM_DIR=~/pfam
source ~/.zshrc  # or source ~/.bash_profile
```

The code for this recipe can be found in `Ch12/Ch12-4-gene-discovery.ipynb`.

How to do it...

Here are the steps to perform this recipe:

1. First, we'll import our libraries:

   ```
   import os
   import subprocess
   from Bio import SeqIO
   from Bio.Blast.Applications import NcbiblastpCommandline
   import pandas as pd
   import argparse
   ```

2. Next, we set up Entrez:

   ```
   from Bio import Entrez
   Entrez.email = "your.email@example.com"
   ```

   You can replace the preceding email with your email.

3. Next, let's define a function to set up our directories:

```python
def setup_directories():
    directories = [
        'genomes', 'predictions', 'blast_results',
        'unique_genes', 'annotations'
    ]
    for directory in directories:
        os.makedirs(directory, exist_ok=True)
    return directories
```

4. Now we will define a function to download a genome:

```python
def download_genome(accession, output_dir):
    output_file = f"{output_dir}/{accession}.fasta"
    if not os.path.exists(output_file):
        print(f"Downloading {accession}...")
        handle = Entrez.efetch(
            db="nucleotide", id=accession,
            rettype="fasta", retmode="text"
        )
        with open(output_file, 'w') as out_f:
            out_f.write(handle.read())
        print(
            f"Downloaded {accession} to {output_file}")
    else:
        print(
            f"Genome {accession} already exists at "
            f"{output_ file}")
    return output_file
```

This function takes an NCBI accession number and a target directory as input and then uses Entrez.efetch() to download the file.

5. Let's create a function to predict genes from our genome file:

```python
def predict_genes(
    genome_file, output_dir, organism_type="prokaryote"
):
    genome_name = os.path.basename(genome_file).split('.')[0]
    output_prefix = f"{output_dir}/{genome_name}"
    protein_file = f"{output_prefix}_proteins.faa"
```

```
        if not os.path.exists(protein_file):
            print(f"Predicting genes for {genome_name}...")
            if organism_type == "prokaryote":
                cmd = (
                    f"prodigal -i {genome_file} -a {protein_file}"
                    f"-o {output_prefix}_genes.gff -f gff"
                )
                subprocess.run(cmd, shell=True, check=True)
            else:
                print(
                    "Eukaryotic gene prediction requires"
                    "tools like Augustus or MAKER"
                )
            print(f"Gene prediction complete for {genome_name}")
        else:
            print(f"Gene predictions already exist for"
                    f"{genome_ name}")
        return protein_file
```

This function will take in our genome file and run Prodigal on it.

6. The next function will create a BLAST database:

```
    def create_blast_database(protein_file):
        db_name = protein_file
        cmd = (
            f"makeblastdb -in {protein_file} "
            f"-dbtype prot -out {db_name}"
        )
        subprocess.run(cmd, shell=True, check=True)
        return db_name
```

This function will run our BLAST comparisons:

```
    def run_blast_comparison(
        query_proteins, subject_db, output_file, evalue=1e-5
    ):
        if not os.path.exists(output_file):
            print(
                f"Running BLAST comparison: "
                f"{query_proteins} vs {subject_db}"
            )
```

```
        blastp_cline = NcbiblastpCommandline(
            query=query_proteins,
            db=subject_db,
            out=output_file,
            outfmt=(
                "6 qseqid sseqid pident length mismatch gapopen"
                "qstart qend sstart send evalue bitscore"
            ),
            evalue=evalue,
            max_target_seqs=1
        )
        stdout, stderr = blastp_cline()
        print(
            f"BLAST comparison complete. "
            f"Results saved to {output_file}"
        )
    else:
        print(f"BLAST results already exist at {output_file}")
    return output_file
```

This function will take a set of query proteins and performs a BLAST search against a subject database. It constructs a BLAST command line with a standard, easy-to-use output format and saves the results to a file.

The next function is the core of our algorithm. It will use our BLAST results to identify genes in our organism of interest (target) that are not present in the reference genome(s) (reference gene sets):

```
def identify_unique_genes(
    target_protein_file, blast_results_files,
    output_dir, identity_threshold=30,
    coverage_threshold=50
):
    target_name = os.path.basename(
        target_protein_file
    ).split('_')[0]
    target_proteins = {}
```

We will loop over our target protein file:

```
    for record in SeqIO.parse(target_protein_file, "fasta"):
        target_proteins[record.id] = record
    proteins_with_hits = set()
```

Here, we loop over and parse each BLAST result file:

```
for blast_file in blast_results_files:
    with open(blast_file, 'r') as f:
        for line in f:
            parts = line.strip().split('\t')
            query_id = parts[0]
            identity = float(parts[2])
            alignment_length = float(parts[3])
            query_length = len(
                target_proteins[query_id].seq)
            coverage = (
                alignment_length / query_length) * 100
            if (
                identity >= identity_threshold
                and coverage >= coverage_threshold
            ):
                proteins_with_hits.add(query_id)
```

We have now built up a set of unique proteins for output:

```
unique_proteins = set(
    target_proteins.keys()) - proteins_with_hits
output_file = (
    f"{output_dir}/{target_name}_unique_proteins. faa")
with open(output_file, 'w') as out_f:
    for protein_id in unique_proteins:
        SeqIO.write(
            target_proteins[protein_id],
            out_f, "fasta"
        )
print(
    f"Identified {len(unique_proteins)} "
    f"unique proteins in {target_name}")
print(f"Unique proteins saved to {output_file}")
return output_file, unique_proteins
```

This function will identify a set of genes that are unique to our organism of interest.

We first get the name of our target organism and load in its proteins.

We then parse our BLAST results to find proteins with significant hits; our default criteria will be that we must have at least 30% sequence identity over at least 50% of the length of our query gene.

We next use a set () operator to perform a subtraction, to find the genes that are present only in the target organism and not in the reference.

We save and return the results.

7. For our final function, let's have a way to annotate our unique genes so we can understand them. We will not fully implement this function or utilize it in the recipe, but we'll leave it as an exercise for you to explore further.

```python
def annotate_unique_genes(unique_proteins_file, output_dir):
    output_name = os.path.basename(
        unique_proteins_file).split('_')[0]
    hmmer_output =(
        f"{output_dir}/{output_name}_pfam_annotations.txt"
    )
    cmd = (
        f"hmmscan --domtblout {hmmer_output} "
        f"pfam/Pfam-A.hmm {unique_proteins_file}"
    )
    # subprocess.run(cmd, shell=True, check=True)
    annotations = {}
    print(
        f"Functional annotation complete. "
        f"Results saved to {hmmer_output}"
    )
    return hmmer_output, annotations
```

This function would be used to annotate the genes further.

8. Great! Now we have all our functions defined. Let's use our code:

```python
def main():
    directories = setup_directories()
    target_accession = "NC_000962"
    reference_accessions = [
        "NC_002945",  # M. bovis
        "NC_008769",  # M. avium
        "NC_002677"   # M. leprae
    ]
    target_genome = download_genome(
        target_accession, directories[0])
    reference_genomes = [
        download_genome(acc, directories[0])
        for acc in reference_accessions
    ]
    target_proteins = predict_genes(
        target_genome, directories[1])
    reference_proteins = [
        predict_genes(genome, directories[1])
        for genome in reference_genomes
    ]
```

```
reference_dbs = [
    create_blast_database(proteins)
    for proteins in reference_proteins
]
```

We have now obtained target and reference proteins as well as our reference genomes. We are ready to run BLAST:

```
blast_results = []
for i, db in enumerate(reference_dbs):
    ref_name = os.path.basename(
        reference_genomes[i]).split('.')[0]
    target_name = os.path.basename(
        target_genome).split('.')[0]
    output_file = (
        f"{directories[2]}/"
        f"{target_name}_vs_{ref_name}. blast"
    )
    blast_results.append(
        run_blast_comparison(
            target_proteins, db, output_file
        )
    )
unique_genes_file, unique_gene_ids = identify_unique_genes(
    target_proteins,
    blast_results,
    directories[3]
)
annotation_file, annotations = annotate_unique_genes(
    unique_genes_file, directories[4]
)
print("\nSubtractive Comparative Genomics Results:")
print(f"Target organism: {target_accession}")
print(
    f"Reference organisms: "
    f"{', '.join(reference_ accessions)}")
print(
    "Total predicted proteins in target: "
    f"{len(list(SeqIO.parse(target_proteins, 'fasta')))}")
print(
    f"Number of unique proteins identified: "
    f"{len(unique_ gene_ids)}")
print(
    f"Unique protein sequences saved to: "
    f"{unique_genes_ file}")
```

```
        print(f"Functional annotations saved to: {annotation_file}")
if __name__ == "__main__":
    main()
```

For this example, we chose Mycobacterium and some related species. We do the following:

I.   Set up our directories

II.   Define our accessions; we will use M. tuberculosis H37Rv as our organism of interest (target), and the reference species will be M. bovis, M. avium, and M. leprae, respectively

III.   Download the genomes

IV.   Perform gene prediction

V.   Create BLAST databases

VI.   Run the BLAST comparisons

VII.   Identify unique genes

VIII.   Report our results

Here are our results:

```
Building a new DB, current time: 03/23/2025 13:50:48
New DB name:   ~/work/CookBook/Ch12/predictions/NC_002945_
proteins.faa
New DB title:  predictions/NC_002945_proteins.faa
Sequence type: Protein
Deleted existing Protein BLAST database named ~/work/CookBook/
Ch12/predictions/NC_002945_proteins.faa
Keep MBits: T
Maximum file size: 3000000000B
Adding sequences from FASTA; added 4013 sequences in 0.0374229
seconds.
Building a new DB, current time: 03/23/2025 13:50:48
New DB name:   ~/work/CookBook/Ch12/predictions/NC_008769_
proteins.faa
New DB title:  predictions/NC_008769_proteins.faa
Sequence type: Protein
Deleted existing Protein BLAST database named ~/work/CookBook/
Ch12/predictions/NC_008769_proteins.faa
Keep MBits: T
Maximum file size: 3000000000B
Adding sequences from FASTA; added 4027 sequences in 0.0403821
seconds.
Building a new DB, current time: 03/23/2025 13:50:49
New DB name:   ~/work/CookBook/Ch12/predictions/NC_002677_
proteins.faa
New DB title:  predictions/NC_002677_proteins.faa
Sequence type: Protein
```

```
Deleted existing Protein BLAST database named ~/work/CookBook/
Ch12/predictions/NC_002677_proteins.faa
Keep MBits: T
Maximum file size: 3000000000B
Adding sequences from FASTA; added 3999 sequences in 0.0365422
seconds.
Identified 62 unique proteins in NC
Unique proteins saved to unique_genes/NC_unique_proteins.faa
Functional annotation complete. Results saved to annotations/
NC_pfam_annotations.txt
Subtractive Comparative Genomics Results:
Target organism: NC_000962
Reference organisms: NC_002945, NC_008769, NC_002677
Total predicted proteins in target: 4085
Number of unique proteins identified: 62
Unique protein sequences saved to: unique_genes/NC_unique_
proteins.faa
Functional annotations saved to: annotations/NC_pfam_
annotations.txt
```

We have identified 62 unique genes! They will be saved in the `NC_unique_proteins.faa` file. This is a small enough list to begin further narrowing down your search. You could sift through the functional annotations and try to understand which genes might perform the reactions you are interested in. You could look for operons, which, in bacteria, are related groups of genes that are right next to each other. This generally indicates they are involved in the same pathway or biochemical process. Finally, you would express these genes, individually or in combination, in a host organism, such as yeast or E. coli, and then measure them for the product of interest.

There's more...

Gene-finding approaches such as the one presented here have been used to find many important natural products. For example, examining the genomes of cold-adapted extremophiles led to the discovery of **olefins** (`https://www.britannica.com/science/olefin`), an important hydrocarbon. These molecules help organisms to adapt to cold environments by making their membranes more fluid. But they can also be used to make biofuels – see: Sukovich et al, *Widespread Head-to-Head Hydrocarbon Biosynthesis in Bacteria and Role of OleA*, Applied and Environmental Microbiology, Jun 2010: `https://journals.asm.org/doi/full/10.1128/aem.00436-10`.

Alkanes represent another important biofuel molecule, being the primary component of gasoline. These genes were discovered using a comparative genomics approach like the one presented here. Different cyanobacteria, some of which produced alkanes and some which did not, were progressively compared and genes in common were subtracted away, until the genes for alkane biosynthesis were discovered – Schirmer et al, *Microbial Biosynthesis of Alkanes*, Science, Jul 2010: `https://www.science.org/doi/full/10.1126/science.1187936?casa_token=-etD39hGppQAAAAA%3AlirgJw4EIb_rCAMB3Ji7WlrQqG1qSYg_70qXrGv1pQJgQO5qCyG3VceUYPjHiYzipKXxCFa6s7Z9`.

Okay, let's try one last exercise! We can use AI to upgrade the preceding code so that it actually runs HMMER against a PFAM database to annotate the genes.

AI tip

Prepare: Make sure your code from this recipe has already been pasted into the AI.

Prompt: Update the preceding code to actually call HMMER.

Troubleshooting: During the writing of this code, I ran into several issues. I had to install the PFAM database and make sure the code pointed to my unique genes file from the previous work. I also had some issues with the summary parser not reading the file correctly. I also had the code updated to make it more robust by trying different e-values if no results are found on the first pass.

Here are some example **prompts** to fix these types of issues:

Write code to install the PFAM database for me

Update the above code so the input file is in `unique_genes/NC_unique_proteins.faa`

It looks like there is an issue with the parsing function

You can also paste in any errors you get to iterate and improve the code

You should see: Output code to run HMMER on your unique genes, parse the outputs, and report and summarize the results.

Here is what we get! The following is a partial output of the HMMER results:

```
Examining HMMER output file: annotation_results/NC_pfam_annotations.
txt
Total lines: 382
Data lines (non-comment): 369
First data line sample:
Abhydrolase_1         PF00561.26    245
NC_000962.3_145        -             301    5.7e-
34  118.3   0.2   1   1   1.7e-37   1.4e-
33  117.0   0.2    1   245    34   281    34   281 0.91 alpha/beta
hydrolase fold

Processed 369 data lines
Successfully parsed 144 domain hits
Found annotations for 40 proteins
Saved parsed results to annotation_results/NC_pfam_parsed.tsv

Examining raw HMMER output:
  Header lines: 13
  Data lines: 369
```

```
First few matches (if any):
  Abhydrolase_1         PF00561.26   245
NC_000962.3_145       -            301   5.7e-
34  118.3   0.2   1   1   1.7e-37  1.4e-
33  117.0   0.2    1   245    34   281   34   281 0.91 alpha/beta
hydrolase fold
  Abhydrolase_6         PF12697.13   217
NC_000962.3_145       -            301      3e-19   71.0  11.5   1   1
4.6e-23   3.7e-19   70.6  11.5    2   216    37   286   36   287
0.62 Alpha/beta hydrolase family
  Hydrolase_4           PF12146.15   239
NC_000962.3_145       -            301      6.9e-13  48.8   0.0   1   2
5.1e-13   4.1e-09   36.5   0.0    6   116    35   143   31   168
0.85 Serine aminopeptidase, S33
Annotation summary saved to annotation_results/NC_annotation_summary.
tsv
```

You can see that we have 144 protein domains annotated on our candidate genes. Each line includes details about the domain hit, its PFAM identifier, and the E-value of the hit.

Your output files will be in the `annotation_results` subdirectory.

The example for this code is provided in the notebook for this recipe.

We also create the following summary visualization:

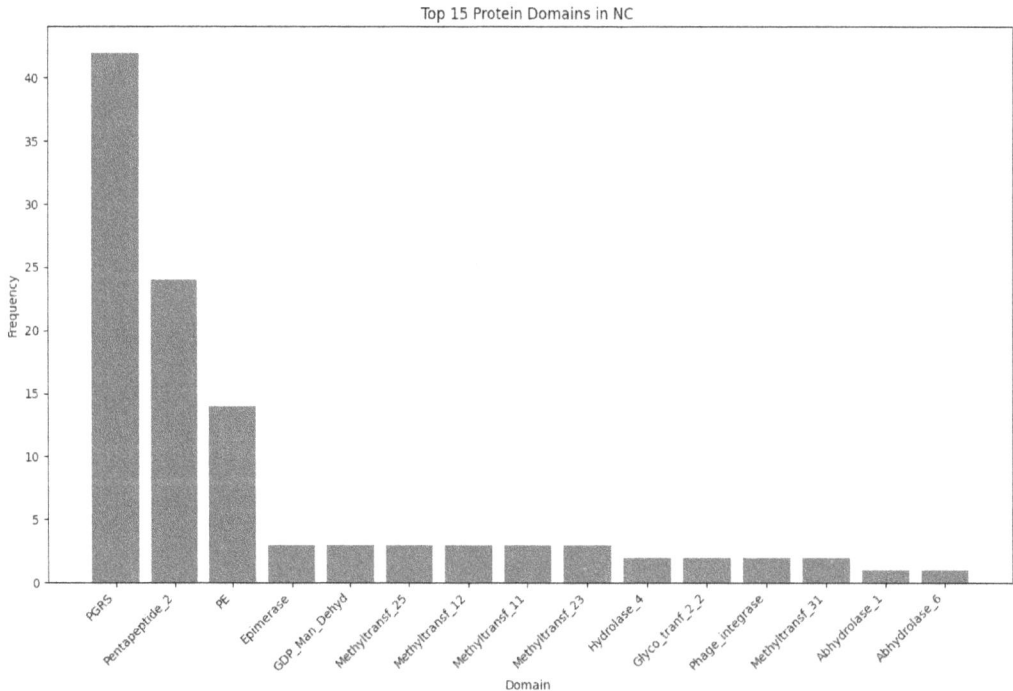

Figure 12.4 – Summary of protein domains for unique gene candidates

We see a variety of protein domains and their frequency graphed here. You could use this information to examine your candidate genes further. For example, let's say that you believe *YourRichium* synthesis involves a reaction catalyzed by a thioesterase (`https://www.sciencedirect.com/topics/biochemistry-genetics-and-molecular-biology/thioesterase`). You notice that there are A/B hydrolase domains on some of your proteins. You find that these are part of a superfamily of proteins called alpha/beta hydrolases, which include thioesterases. You could then use a database of these proteins, such as ESTHER, to explore their potential functions even further – Lenfant et al, *ESTHER, the database of the A/B hydrolase fold superfamily of proteins: tools to explore diversity of functions*, Nucleic Acids Research, Nov 2012: `https://pmc.ncbi.nlm.nih.gov/articles/PMC3531081/`.

You get the idea! You can see here how bioinformatics, AI, and literature research (and, of course, lab work!) create a virtuous cycle that can lead to significant and important discoveries. Scientists are even beginning to design entire synthetic genomes to enable optimized production of valuable products – James et al, *The design and engineering of synthetic genomes*, Nature Reviews Genetics, Nov 2024: `https://www.nature.com/articles/s41576-024-00786-y`.

Okay, let's clean up and close down our `conda` environment:

```
conda deactivate
```

See also

- antiSMASH is described in this paper: Blin et al, *antiSmash 6.0: improving cluster detection and comparison capabilities*, Nucleic Acids Research, May 2021: `https://academic.oup.com/nar/article/49/W1/W29/6274535`

- DeepBGC is discussed here: Liu et al, *Deep Learning to Predict the Biosynthetic Gene Clusters in Bacterial Genomes*, JMB, Aug 2022: `https://www.sciencedirect.com/science/article/pii/S0022283622001772?casa_token=VlnuGUa2uKoAAAAA:aPJLT01QE0_f95GL2HYTdKQc0XLkPjZ31ac--CsgB8BtcjhnhCYJMjzK-Xvg2tzg4TSEolja`

- Learn more about set operators in Python: `https://realpython.com/python-sets/`

Get This Book's PDF Version and Exclusive Extras

Scan the QR code (or go to `packtpub.com/unlock`). Search for this book by name, confirm the edition, and then follow the steps on the page.

Note: Keep your invoice handy. Purchases made directly from Packt don't require an invoice.

UNLOCK NOW

13

Genome Editing

In this chapter, we will learn how bioinformatics is used in genome editing and analysis. In the past several years, there have been tremendous advances in our ability to modify genomes. This has led to a revival in our ability to alter microbes to make biofuels, engineer food crops for greater yield or pest resistance, and even alter human cells to cure cancer.

The use of sequencing, especially long-read sequencing, has become critical in this area to support verification of edits. We'll see how the background we've built up so far is useful in applying bioinformatics to these emerging areas.

The ability to insert or attach barcodes, or short sequences, to cells or constructs within cells has also been revolutionary. By combining this approach with sequencing, we can design powerful and elegant experiments. We'll see how bioinformatics can contribute to this exciting area.

By the time you finish this chapter, you'll understand how to design strategies for editing genomes and resolve the accuracy of the edits. You'll also know how to use barcode counting algorithms to generate large-scale biological datasets.

In this chapter, we'll cover the following recipes:

- Designing guide RNAs for genome editing
- Counting barcodes in genomic libraries
- Using nanopore data to resolve genome edits

Technical requirements

In this chapter, we'll use the following tools and packages:

- Biopython
- NumPy and pandas
- Matplotlib and Seaborn

You will need a working directory called Ch13.

The code for this chapter is available in this book's GitHub repository: `https://github.com/PacktPublishing/Bioinformatics-with-Python-Cookbook-Fourth-Edition/tree/main/Ch13`.

Remember to activate your conda environment before beginning the recipes, like so:

```
conda activate bioinformatics_base
```

If you would like to set up a `conda` environment specific to this chapter, before activating `bioinformatics_base`, run the following code:

```
conda create -n ch13-genome-editing --clone bioinformatics_base
conda activate ch13-genome-editing
```

You can install the packages for this chapter as you go through the recipes. Alternatively, you can use the YAML file provided in this book's GitHub repository, which will install all the necessary packages immediately:

```
conda env update --file ch13-genome-editing.yml
```

Designing guide RNAs for genome editing

In this recipe, we will learn about the exciting world of genome editing! Early methods of genome editing included transposon mutagenesis, in which short "hopping" DNA sequences called **Transposable Elements (TEs)** are used to integrate into the genome and knock out genes (Liu et al., *Transposase-assisted target-site integration for efficient plant genome engineering*, Nature, Jun 2024, `https://www.nature.com/articles/s41586-024-07613-8`). But it was really the invention of **Clustered Regular Interspaced Short Palindromic Repeats (CRISPR)** (`https://www.broadinstitute.org/what-broad/areas-focus/project-spotlight/crispr-timeline`) that dramatically changed the field. These consist of sequences of DNA that are separated by spacers, and were first discovered in **Archaea** (`https://pmc.ncbi.nlm.nih.gov/articles/PMC7613921/`), a form of ancient life. CRISPR RNAs are transcribed and recognized by an enzyme known as **Cas9**. This enzyme recognizes the RNAs and the associated matching DNA in the genome and then cuts (edits) the genome. The initial applications involved knocking out genes, but researchers have now upgraded the CRISPR-Cas9 system to allow for gene knockup (increased gene expression) or knockdown (attenuating gene expression without a full knockout).

The RNAs used to perform CRISPR editing are called **guide RNAs** or **gRNAs**. To design appropriate guide RNAs, you need to understand where you are targeting in the genome, what off-target homologous regions might exist, whether the RNAs might fold back on themselves, and so forth.

When Cas9 binds to the target DNA, it looks for a special recognition site. This is the **Protospacer Adjacent Motif (PAM)** site (`https://www.idtdna.com/pages/support/faqs/what-is-a-pam-sequence-and-where-is-it-located`). As part of this recipe, we'll learn about identifying the PAM site to find places in the genome suitable for editing.

Cas9 identifies the target DNA and introduces a **Double-Strand Break** (**DSB**), in which the DNA is broken. There is a natural process in cells called **Homology-Directed Repair** (**HDR**) in which the cell can take a homologous sequence in and use it to help guide the repair of the region. We can take advantage of this by introducing a homologous repair sequence that is very similar on both ends but contains a point mutation in the middle, thus introducing a point mutation (SNV or Single Nucleotide Variant) in the sequence in a very precise way. Once the sequence has been repaired, it will return to normal except for this single mutation. We'll learn how to design a **repair template** in this recipe.

Here, we will learn about guide RNA design principles by designing a system to knock in a specific mutation in the BRCA1 gene (`https://www.cancer.gov/about-cancer/causes-prevention/genetics/brca-fact-sheet`). BRCA1 is a tumor suppressor gene involved in DNA repair and cell cycle regulation. As suggested by its name, mutations in this gene can lead to cancers such as breast cancer, which is where it got its name. Imagine that we have a patient with breast cancer, and it has been traced back to a single mutation (SNV) in the tumor. By editing cells to fix this mutation and introducing them into the patient, we can replace the tumorous cells with new, healthy cells! To do this, we would need to create a cell line of edited cells, so we'd need to design guide RNAs to introduce the right mutation in the right place.

Getting ready

First, you will want to get the human BRCA1 gene sequence. You can find it here: `https://www.ncbi.nlm.nih.gov/nuccore/NG_005905.2?report=fasta`.

You can download the file manually by choosing **Send to** | **Complete Record** | **File** from the menu and then choosing the **FASTA** format. You can leave **Show GI** unchecked.

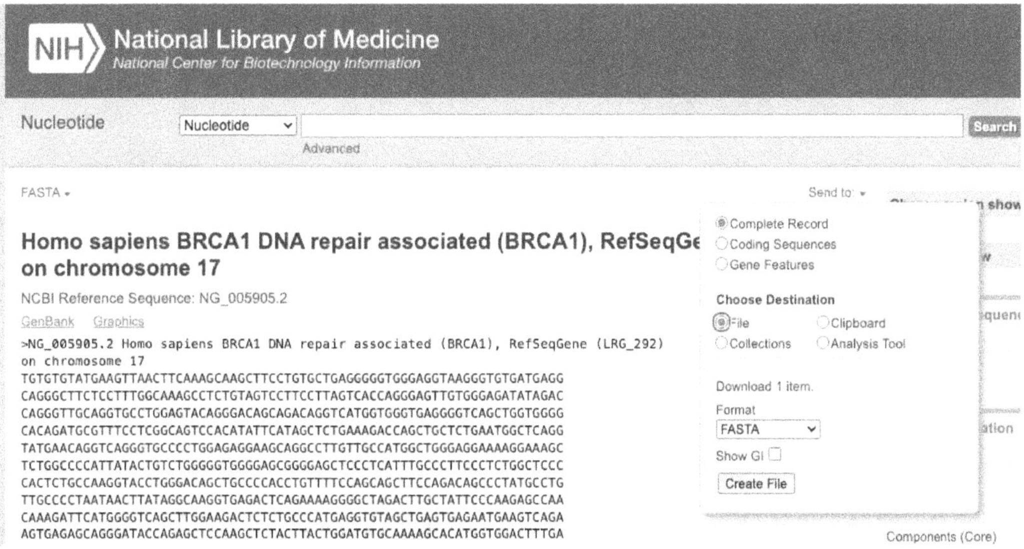

Figure 13.1 – Downloading the BRCA1 sequence from NCBI

Once you have your file, move it to your `Ch13` working directory and rename it `brca1_sequence.fasta`. Alternatively, you can download the file directly with this command:

```
! wget -O brca1_sequence.fasta "https://
eutils.ncbi.nlm.nih.gov/entrez/eutils/efetch.
fcgi?db=nucleotide&id=NG_005905.2&rettype=fasta&retmode=text"
```

This should place the `brca1_sequence.fasta` file in your working directory. You will use it in this recipe to design guide RNAs.

The code for this recipe can be found in `Ch13/Ch13-1-grna-design.ipynb`.

How to do it...

Let's design a system for finding guide RNAs that will introduce a mutation where we want it:

1. First, let's import our libraries:

    ```python
    from Bio import SeqIO
    import matplotlib.pyplot as plt
    import os
    ```

    We will use Biopython (`Bio`) and `os` for sequence file manipulation and `matplotlib` for plotting.

2. Next, we will write a simple function to load in our BRCA gene sequence:

    ```python
    def load_brca_sequence_from_file(filename, gene="BRCA1"):
        print(f"Loading {gene} sequence from file: {filename}")
        if not os.path.exists(filename):
            raise FileNotFoundError(
                f"File not found: {filename}")

        try:
            record = SeqIO.read(filename, "fasta")
            sequence = str(record.seq).upper()
        except Exception as e:
            raise Exception(
                f"Error reading FASTA file: {str(e)}")
        print(f"Sequence length: {len(sequence)} bp")
        return sequence, record.id
    ```

    This function takes in a filename and gene to parse out, with the default being BRCA1. It then checks whether the file exists.

    Next, it uses Biopython's `SeqIO` module to read in and parse the file into the `sequence` variable and sets the sequence to be uppercase. Finally, we print out the sequence length and return the sequence and record ID.

3. Next, we need to design a function to create our mutation guides. We will look in a region around the desired mutation position and attempt to find PAM sites. We will check both the forward and reverse strands:

```python
def design_mutation_guides(
    sequence, mutation_pos, mutation_from,
    mutation_to, window=100
):
    start = max(0, mutation_pos - window)
    end = min(len(sequence), mutation_pos + window)
    region_seq = sequence[start:end]
    if sequence[mutation_pos] != mutation_from:
        raise ValueError(
            f"Base at position {mutation_pos} is "
            f"{sequence[mutation_pos]}, "
            f"not {mutation_from}"
        )

    guides = []
    for i in range(
        max(0, mutation_pos - window),
        min(len(sequence), mutation_pos + window)
    ):
        if i + 2 < len(sequence) and sequence[i+1:i+3] == "GG":
            if i >= 20:
                guide_seq = sequence[i-20:i]
                cut_site = i - 3
                distance = mutation_pos - cut_site
                guides.append({
                    "sequence": guide_seq,
                    "pam": "NGG",
                    "strand": "+",
                    "start": i-20,
                    "end": i,
                    "mutation_distance": abs(distance),
                    "cut_site": cut_site
                })
        if i >= 1 and sequence[i-1:i+1] == "CC":
            if i + 20 < len(sequence):
                guide_seq = sequence[i+1:i+21]
                cut_site = i + 3
                distance = mutation_pos - cut_site
                guides.append({
                    "sequence": guide_seq,
```

```
                              "pam": "NGG",
                              "strand": "-",
                              "start": i+1,
                              "end": i+21,
                              "mutation_distance": abs(distance),
                              "cut_site": cut_site
                })
          return guides
```

This function takes in a sequence and a target position for introducing a mutation.

We will also specify the mutation from and to sequence – for example, "A" to "T" if we want to change an A to a T at that position. We will search within a certain radius around the target position, referred to as the **window size**. The default for this will be 100 base pairs.

First, we will extract the region around the mutation position (within our window size) with some simple string manipulation. Then, we will provide a simple check to make sure we don't try to change the mutation base to itself, as this wouldn't accomplish anything. We will also initialize our list of guides.

Next, we will search within the window for our guides, looking for a **PAM** site. Here, we are looking for an NGG PAM site, meaning any base followed by two Gs. We will walk across our sequence in the window area using a `for` loop. When we find a potential site, we will check that there is still enough room in the window to have a minimum **20-nucleotide** (20 nt) guide sequence. If there is a good candidate, we will find the **cut site**, which is going to be three base pairs upstream from the PAM site.

We are going to record the distance of the cut site within the guide from the mutation. This is important because CRISPR works best when the cut site occurs within 10-30 base pairs of the desired mutation.

The next step is to append the guide sequence information to our guides list. We will include our 20-nucleotide guide sequence, the PAM site used, the strand, the start and end position of the guide, the distance from the mutation, and the cut site.

We then do the same thing on the reverse strand, this time looking for "CC" in the reverse complement sequence. Finally, we return our list of candidate guides.

So, let's define a short function to calculate GC content:

```
def calculate_gc_content(sequence):
    gc_count = sequence.count("G") + sequence.count("C")
    return (gc_count / len(sequence)) * 100
```

This function simply counts the total number of Gs and Cs in the sequence and returns the percentage.

4. Now, let's define a function that will design the repair template:

```
def design_repair_template(
    sequence, mutation_pos, mutation_from,
    mutation_to, homology_arm_length=50
):
    left_arm_start = mutation_pos - homology_arm_length
    left_arm = sequence[left_arm_start:mutation_pos]
    right_arm_end = mutation_pos + 1 + homology_arm_length
    right_arm = sequence[mutation_pos+1:right_arm_end]
    repair_template = left_arm + mutation_to + right_arm
    return {
        "template": repair_template,
        "left_arm": left_arm,
        "right_arm": right_arm,
        "mutation": mutation_to,
        "length": len(repair_template)
    }
```

This function creates the repair template needed for HDR. It takes in the target sequence, the mutation position, and the desired mutation. We also provide it with the desired homology end length, which is set to 50 by default.

Next, we create the repair template by joining the homologous region from the left-hand side, the desired mutation, and the homologous region from the right-hand side.

Finally, we return the entire repair template, homology arms, desired mutation, and length of the repair template.

5. Now, let's define a function for scoring the different guide RNA candidates:

```
def score_guides(guides):
    scored_guides = []
    for guide in guides:
        seq = guide["sequence"]
        gc_content = calculate_gc_content(seq)
        gc_score = 1.0 - abs(gc_content - 50) / 50
        distance = guide["mutation_distance"]
        if 0 <= distance <= 10:
            distance_score = 1.0
        else:
            distance_score = max(
                0, 1.0 - (distance - 10) / 40)
        has_homopolymer = any(
            base * 4 in seq for base in "ATGC")
```

```
        has_self_comp = False
        overall_score = (0.3 * gc_score) + (
            0.7 * distance_score)
        if has_homopolymer or has_self_comp:
            overall_score *= 0.5
        guide_scored = guide.copy()
        guide_scored.update({
            "gc_content": gc_content,
            "gc_score": gc_score,
            "distance_score": distance_score,
            "has_homopolymer": has_homopolymer,
            "has_self_complementarity": has_self_comp,
            "overall_score": overall_score
        })
        scored_guides.append(guide_scored)
    return sorted(
        scored_guides,
        key=lambda x: x["overall_score"],
        reverse=True
    )
```

This function will do the following:

- Initialize a list of `scored_guides`.

- Calculate the GC content of the guide using the function we created previously. Guide RNAs with around 50% GC content tend to work best, so we penalize for deviations from this.

- Apply **proximity scoring**. This penalizes guides that are farther away from the mutation site.

- Penalize **homopolymers**, in which we see the same base in a row four or more times.

- Include a placeholder for checking **self-complementarity**. RNAs that can bind with themselves will be less effective – we discussed this in the *Designing siRNAs using Biopython and ViennaRNA* recipe of *Chapter 12*.

- Calculate the overall score, giving more weight to distance and less weight to GC content.

- Use `.copy()` to create a new list of scored guides and update it to include the relevant scoring criteria and overall score for each guide.

- Return the list of scored guides, sorted in descending order by overall score.

6. Now, we'll write a function for visualizing our guides:

```python
def visualize_guides(
    sequence, guides, mutation_pos,
    mutation_from, mutation_to, top_n=5
):
    plt.figure(figsize=(12, 6))
    region_start = max(0, mutation_pos - 100)
    region_end = min(len(sequence), mutation_pos + 100)
    plt.axvline(
        x=mutation_pos, color='red', linestyle='--',
        label=f'Mutation: {mutation_from} → {mutation_to}'
    )
    top_guides = guides[:top_n]
    for i, guide in enumerate(top_guides):
        if guide["strand"] == "+":
            plt.hlines(
                y=i+1, xmin=guide["start"],
                xmax=guide["end"], color='blue',
                alpha=0.7
            )
            plt.plot(guide["cut_site"], i+1, 'o',
                    color='green')
        else:
            plt.hlines(
                y=i+1, xmin=guide["start"],
                xmax=guide["end"], color='purple',
                alpha=0.7
            )
            plt.plot(guide["cut_site"], i+1, 'o',
                    color='green')

    plt.legend()
    plt.xlim(region_start, region_end)
    plt.ylim(0, top_n+1)
    plt.xlabel('Genomic Position')
    plt.yticks(
        range(1, top_n+1),
        [f"Guide {i+1}: Score={g['overall_score']:.2f}"
         for i, g in enumerate(top_guides)]
    )
    plt.title(
        f'Top {top_n} Guide RNAs for {mutation_from}'
        f'{mutation_pos}{mutation_to} Mutation'
```

```
        )
        plt.tight_layout()
        plt.savefig('guide_visualization.png')
        plt.close()
```

This function will create a genomic map for us with a 200-base-pair region where the mutation is in the center. Here, we will set up our plot and use Matplotlib's `axvline` function to draw a dashed vertical line at our mutation position. Then, we will extract our top guides (the default is 5) and plot them as colored lines against their genomic position.

7. Next, we need to iterate over the top guide sequences. For guides on the plus strand, we draw a blue line that covers the region of the guide sequence, and a green circle at the cut site. For guides on the minus strand, we use a purple line and a green circle.

Finally, we must add our legend, axis labels, and title. We will save our figure as a PNG file.

So, let's put it all together and use our functions!

The following code will run our core functions in the order needed. First, we must define our gene as BRCA1 and set up the position and desired mutation. Here, we load in our BRCA1 gene sequence from the FASTA file we created. Then, we design our guide candidates against the sequence.

Next, we must score the guides and then design our repair templates. Here, we print out our top five guide RNAs for review. Then, we call our visualization function to create an image of the genomic regions with our guide RNAs overlayed on it (saved as a PNG file). We print out the repair template information and then use pandas to write out the guide RNAs to a CSV file:

```
def main():
    gene = "BRCA1"
    mutation_pos = 42500
    mutation_from = "T"
    mutation_to = "C"
    brca_file = "brca1_sequence.fasta"

    try:
        sequence, gene_id = load_brca_sequence_from_file(
            brca_file, gene)
    except Exception as e:
        print(f"Error loading sequence: {str(e)}")
        print("Please ensure your FASTA file exists "
            "and is properly formatted.")
        print("You can download the BRCA1 sequence from "
            "NCBI and save it as brca1_sequence.fasta")
        return
    print(f"Designing guides for {gene} mutation "
        f"{mutation_from}{mutation_pos}{mutation_to}")
```

```
    try:
        guides = design_mutation_guides(
            sequence, mutation_pos,
            mutation_from, mutation_to
        )
        print(f"Found {len(guides)} potential guides")
    except ValueError as e:
        print(f"Error: {str(e)}")
        print("Check that the mutation position and "
              "reference base are correct.")
        return
    scored_guides = score_guides(guides)
    repair_template = design_repair_template(
        sequence, mutation_pos,
        mutation_from, mutation_to
    )
    print("\nTop 5 Guide RNAs:")
    for i, guide in enumerate(scored_guides[:5], 1):
        print(f"Guide {i}: {guide['sequence']} "
              f"(PAM: {guide['pam']}, "
              f"Strand: {guide['strand']})")
        print(f" Position: {guide['start']}-{guide['end']}, "
              f"Cut site: {guide['cut_site']}")
        print(f" Distance to mutation: "
              f"{guide['mutation_distance']} bp")
        print(f" GC content: {guide['gc_content']:.1f}%,"
              f"Score: {guide['overall_score']:.2f}")
    visualize_guides(
        sequence, scored_guides, mutation_pos,
        mutation_from, mutation_to
    )
    print("\nRepair Template Design:")
    print(f"Length: {repair_template['length']} bp")
    print(f"Left homology arm: {repair_template['left_arm']}")
    print(f"Mutation: {mutation_from} → "
          f"{repair_template['mutation']}")
    print(f"Right homology arm: {repair_template['right_arm']}")
    guides_df = pd.DataFrame(scored_guides)
    guides_df.to_csv(f"{gene}_mutation_guides.csv", index=False)
    print(f"\nResults saved to {gene}_mutation_guides.csv "
          "and guide_visualization.png")
if __name__ == "__main__":
    main()
```

That's it! We now have a set of candidate guide RNAs for introducing the desired mutation into the BRCA1 gene, as well as a repair template. These materials would be suitable for use in a CRISPR lab workflow to perform genome editing.

Here is our output:

```
Top 5 Guide RNAs:
Guide 1: TTCCACATGTTGGAAACATG (PAM: NGG, Strand: +)
  Position: 42466-42486, Cut site: 42483
  Distance to mutation: 17 bp
  GC content: 40.0%, Score: 0.82
Guide 2: ACATGTTGGAAACATGTGGA (PAM: NGG, Strand: -)
  Position: 42470-42490, Cut site: 42472
  Distance to mutation: 28 bp
  GC content: 40.0%, Score: 0.62
Guide 3: TATTAGGAACTTCCACATGT (PAM: NGG, Strand: +)
  Position: 42456-42476, Cut site: 42473
  Distance to mutation: 27 bp
  GC content: 35.0%, Score: 0.61
Guide 4: ACATTCCCAGAAATCATCTA (PAM: NGG, Strand: -)
  Position: 42525-42545, Cut site: 42527
  Distance to mutation: 27 bp
  GC content: 35.0%, Score: 0.61
Guide 5: CAGAAATCATCTAGCATTTG (PAM: NGG, Strand: -)
  Position: 42532-42552, Cut site: 42534
  Distance to mutation: 34 bp
  GC content: 35.0%, Score: 0.49

Repair Template Design:
Length: 101 bp
Left homology arm: TGAAGTTATTAGGAACTTCCACATGTTGGAAACATGTGGAAACAGAAGTA
Mutation: T → C
Right homology arm: GCTGAAAGTATTTAAATTTTGACCACATTCCCAGAAATCATCTAGCATTT

Results saved to BRCA1_mutation_guides.csv and guide_visualization.png
```

Figure 13.2 – Output for guide RNA design

We can see the top five guide RNAs, along with their details and scores, as well as the repair template design. Here is our output visualized as a graph:

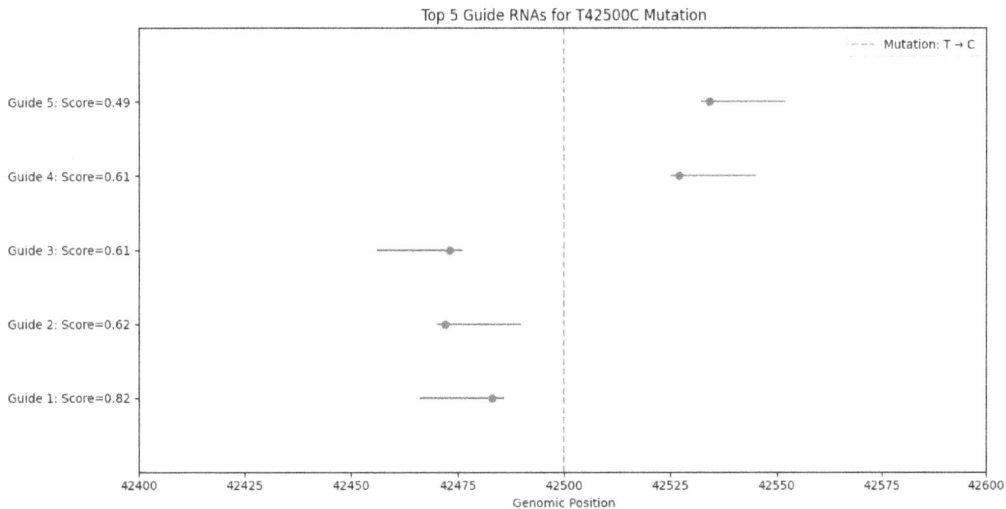

Figure 13.3 – Guide RNA design summary

This graph shows the top five guide RNAs and their positions along the sequence. We can see the different guides and their position along the genome. The cut site is marked by a circle. The mutation target position is highlighted in the center of the graph with a dashed line.

The guide RNAs will also be saved in the BRCA1_mutation_guides.csv file in your working directory.

There's more...

In this section, we'll go a little further and cover some of the latest advances in genome editing. But first, we have a tip on how to further upgrade the code we just wrote.

AI tip

You can update your scoring function so that it includes the actual implementation of self-complementarity by using the ViennaRNA package.

Preparation: Paste in the score_guides() function.

Prompt: Update the aforementioned scoring function to implement the self-complementarity score using ViennaRNA.

What you should see: An updated function that checks that the ViennaRNA package has been installed and uses the RNA.fold() function to calculate a self-complementarity score.

Genome editing is finding uses in many exciting areas. It is being used extensively in plant science to improve crop traits (Pan et al., *Guide RNA library-based CRISPR screens in plants: opportunities and challenges*, Current Opinion in Biotechnology, Feb 2023, `https://www.sciencedirect.com/science/article/abs/pii/S0958166922002178?casa_token=9Ng_0ter2oAAAAA:t-4lvycoGlSCxA4KqMijBFvfkfKPclCOqQCeMBt2-NUfCNE32DSM_DZtJQ2g4490JOwWzpDd`). CRISPR is also used in livestock editing to improve efficiency, farming practices, and animal welfare (Menchaca et al., *CRISPR in livestock: From editing to printing*, Theriogenology, Jul 2020, `https://www.sciencedirect.com/science/article/pii/S0093691X20300765?casa_token=lzXqSGW8_4oAAAAA:wiXa17FYxHQr3PEZta8_EXcX2NNHNHgkpgi7l-nOvBOPnJnfe82ZXJ84_yNOzu05eOpptgeO`). It is also widely used in medicine and cancer treatment, as we saw in this recipe.

Many exciting new applications of CRISPR are emerging. As mentioned previously, in addition to gene knockout, CRISPRa and CRISPRi have been developed to provide gene activation and inactivation, respectively (Bendixen et al., *CRISPR-Cas-mediated transcriptional modulation: The therapeutic promises of CRISPRa and CRISPRi*, Molecular Therapy, Jul 2023, `https://www.cell.com/molecular-therapy-family/molecular-therapy/fulltext/S1525-0016(23)00145-4`).

In addition to Cas9, other CRISPR enzymes have been discovered, and many have useful properties for different applications. For example, Cas12 recognizes a different PAM site and creates **sticky ends** instead of blunt ends, which means that when the enzyme cuts the DNA, it leaves overhangs of DNA on one strand, as opposed to having evenly matched strands. This can be useful for various applications, including increased precision. Cas13 can edit RNA instead of DNA!

Cas enzymes have also been a major target of protein engineering. For example, machine learning has been used to engineer around 1,000 unique Cas9 enzymes, each with their own PAM site recognition (Silverstein et al., *Custom CRISPR-Cas9 PAM variants via scalable engineering and machine learning*, Nature, Apr 2025, `https://www.nature.com/articles/s41586-025-09021-y`). Scientists are even working to engineer "PAM-free" Cas enzymes that can recognize any sequence (Collias & Beisel, *CRISPR technologies and the search for the PAM-free nuclease*, Nature Communications, Jan 2021, `https://www.nature.com/articles/s41467-020-20633-y`).

With that, we have seen how genome editing is transforming the world and how bioinformatics plays a huge role in this effort. The exciting advances in this space are only just beginning!

See also

- Read how a knock-in strategy similar to the one discussed here was used in BRCA1 to improve cancer treatment: Witz et al., *CRISPR/Cas9-mediated knock-in of BRCA1/2 mutations restores response to olaparib in pancreatic cancer cell lines*, Scientific Reports, Oct 2023, `https://www.nature.com/articles/s41598-023-45964-w`

- AwesomeCrispr provides a library of software tools and references: `https://github.com/davidliwei/awesome-crispr/blob/master/README.md`

- CRISPRware provides intelligent guide RNA design for library screening: Malekos et al., *CRISPRware: an efficient method for contextual gRNA library design*, bioRxiv, Jun 2024, `https://pmc.ncbi.nlm.nih.gov/articles/PMC11213142/`

- A good review on Cas protein structure: Wang et al., *Structural biology of CRISPR-Cas immunity and genome-editing enzymes*, Nature Reviews Microbiology, May 2022, `https://www.nature.com/articles/s41579-022-00739-4`

- CRISPR can even be used to image nucleic acids in real time: Chen et al., *CRIBAR: a fast and flexible sgRNA design tool for CRISPR imaging*, BioInformatics Advances, Feb 2025, `https://academic.oup.com/bioinformaticsadvances/article/5/1/vbaf022/8010465`

Counting barcodes in a genomic library

Another exciting application of genome editing is to create **barcoded genomic libraries**. In one typical example of this technique, CRISPR guide RNAs are introduced into cells so that each cell, on average, receives one guide RNA to knock out one gene, leaving behind a unique barcode sequence. The barcode and guide RNA from the CRISPR vector are integrated into the genome. This means that we can put many cells together into a single experiment (such as a fermentation tank) and see which ones out-compete the others, and then retrieve those cells and sequence them to identify the barcodes, a technique known as **bar-seq**. This will tell us which gene interventions worked the best. The identification of the individual genes or barcodes is referred to as bar-seq, and the overall process of running large strain libraries in fermentation settings is known as **pooled screening**.

Barcodes are also used in other applications, such as **demultiplexing**. Multiple samples are barcoded before being put on a sequencer so that they can be deconvoluted, or demultiplexed, before analysis. This allows for cost savings as many samples can be run on a sequencer to produce a tremendous amount of data per run. Illumina sequencers produce **Binary Base Call** (**BCL**) format files, which are then combined with the barcodes used for de-multiplexing using a program called BCLConvert, which then produces the FASTQ files for the individual samples. This recipe, however, will focus on the use of barcodes in screening genomic libraries.

Barcodes are typically 10-20 nucleotide-long sequences that are meant to be unique enough for the experiment. Given N barcodes, you will have 4^N potential unique barcodes, meaning you can design barcodes that are long enough for potentially very large experiments. You also need to consider GC content, homopolymer runs, self-similarity, and potential mutations in the barcode during construction.

To read out the barcodes, we typically use something like a **regular expression** so that we can perform "fuzzy matching" on the barcode. This helps ensure that if there are some accidental mutations, we can still recover the barcode. A regular expression encodes logic with some special characters to match strings. For example, consider the following expression:

```
\b[A-Z0-9._%+-]+@[A-Z0-9.-]+\.[A-Z]{2,}\b
```

This would use \b to define a word boundary. Here, the characters in the [] section say that we can match any one of several letters, numbers, and special characters. After that, we have to have an @ symbol. Next, we must match any number of characters again, after which we need a ., then two or more characters. This type of regular expression will match an email address – for example, `your.name@cookbook.com`.

To learn more about regular expressions, you should familiarize yourself with the Python `re` module: `https://docs.python.org/3/library/re.html`.

When analyzing barcodes, we need to know about the **Hamming distance**. This is the distance between any two barcodes – for example, `ACTG` and `AGCG` differ by two positions, so they have a Hamming distance of two. This is important because when designing barcode libraries, we want to design our set of barcodes so that all the barcodes have a minimum Hamming distance between them (typically, three is good). That way, if there are sequencing errors that erode one or two of the bases in the barcode, we can still distinguish them.

One popular tool for analyzing barcode sequences is **Bartender** (`https://github.com/LaoZZZZZ/bartender-1.1/blob/master/README.md`). It applies barcode patterns as regular expressions to cluster similar barcodes together. In many cases, barcodes are first generated randomly. The next task is to identify what they are. Bartender uses optimized algorithms to then cluster and identify the barcodes in the pool. One important class of random barcodes is **Unique Molecular Identifiers** (**UMIs**). These sequences are attached during the sequencing process to identify unique individual molecules. During sequencing, the molecules being sequenced are often amplified via PCR (**Polymerase Chain Reaction** - `https://en.wikipedia.org/wiki/Polymerase_chain_reaction`). During this process, duplicates of the same molecule can be amplified, leading to **PCR duplicates**. UMIs help filter out these duplicates (`https://www.ecseq.com/support/ngs/how-can-unique-molecular-identifiers-help-to-reduce-quantitative-biases`). UMIs also provide powerful error correction, allowing for highly sensitive variant detection. This is used in **Cell-Free DNA** (**cfDNA**) analysis to detect very low-level variants coming from tumor DNA circulating in your blood, enabling techniques such as **Molecular Residual Disease** (**MRD**) detection, in which a patient's cancer can be monitored to exquisite accuracy during their treatment (Hirotsu et al., *Dual-molecule barcode sequencing detects rare variants in tumor and cell-free DNA in plasma*, Scientific Reports, Feb 2020, `https://www.nature.com/articles/s41598-020-60361-3`). Bartender also understands that barcodes may change slightly during a **Time-Course Analysis** and can use the clustering approach to track the equivalent barcodes as they evolve in an experiment.

By the end of this recipe, you will understand how barcodes are extracted and read from sequencing data, and how barcoded libraries can be analyzed to gain biological insights from real-world experiments!

Getting ready

You don't need to install anything extra for this recipe. As this recipe is very long, we won't cover every single function. Instead, we will discuss each function and then, in some cases, refer you to the notebook for this recipe (available in this book's GitHub repository) so that you can review the actual code.

The code for this recipe can be found in `Ch13/Ch13-2-barcodes.ipynb`.

How to do it...

Here are the steps to complete this recipe:

1. First, we will import our libraries:

    ```
    import re
    import random
    from collections import Counter, defaultdict
    import matplotlib.pyplot as plt
    import seaborn as sns
    import pandas as pd
    import numpy as np
    ```

    Here, we are using `re` for regular expression manipulation. We have also included the `random` library so that we can generate pseudo-random numbers.

2. Now, we will set up a few basic items before going further:

    ```
    plt.style.use('ggplot')
    sns.set(font_scale=1.1)
    np.random.seed(42)
    random.seed(42)
    ```

    First, we set the plot style to the popular `ggplot` style. Next, we adjust the default font size by 10%. Then, we set random seeds for NumPy and Python's built-in random module, respectively. Setting the random seed to a number like this means the code will be reproducible when you run it.

3. Next, we will define a function for generating simulated FASTQ data:

    ```
    def generate_random_sequence(length):
        return ''.join(
            random.choice('ACGT') for _ in range(length))
    ```

    This function uses a **list comprehension** to make a random choice regarding our four bases (A, C, G, or T). The `for _` loop iterates `length` number of times, and finally, `.join()` concatenates the bases together. Recall that list comprehension is when we use a `for` loop directly inline to create the list.

4. Next, we need a function to map the barcodes to the genes. For each gene, there will be a barcode that represents its expression. We need to make up a series of gene names and then assign this series a random barcode. This will simulate what would happen in a real gene expression library:

```python
def create_barcode_gene_map(num_barcodes=50):
    genes = [
        f"Gene_{chr(65+i//4)}{i%4+1}"
        for i in range(num_barcodes)]
    barcodes = [
        generate_random_sequence(12)
        for _ in range(num_barcodes)]
    return dict(zip(barcodes, genes))
```

This function will give us gene names such as A1, B2, and so on. It will also give us 12 base pair barcodes via our `generate_random_sequence()` function. Here, we return the paired encoding by using the `zip()` function to combine or *zip up* the two lists, and then turn this result into a dictionary. Note that with four positions and 12 letters, we can generate 4^12 or 16,777,216 barcodes. That is plenty to cover all the genes in the genome!

5. Next, we need a function that will generate a simulated FASTQ file as if it had come from a real experiment. Imagine that we were measuring the numerous barcodes in a cell from a real experiment by sequencing it. We would find a variety of genes that had been up or down-regulated based on certain conditions. This would be reflected in the relative abundance of the barcodes for those genes. This function will simulate what we would get from a real sequencing experiment:

```python
def generate_fastq_content(
    num_reads, barcode_gene_map, barcode_distribution
):
    barcodes = list(barcode_gene_map.keys())
    lines = []

    for i in range(num_reads):
        barcode = np.random.choice(
            barcodes, p=barcode_distribution)
        pre_seq = generate_random_sequence(
            random.randint(15, 25))
        post_seq = generate_random_sequence(
            random.randint(30, 60))
        full_seq = pre_seq + barcode + post_seq
        quality = ''.join(
            chr(random.randint(33, 73))
            for _ in range(len(full_seq))
        )
        lines.append(f"@Read_{i+1} barcode={barcode}")
        lines.append(full_seq)
```

```
        lines.append("+")
        lines.append(quality)
    return lines
```

Let's review this code:

I.   First, we initialize our list of barcodes and the lines of our file.

II. Next, we use the NumPy `random.choice()` method to choose a set of barcodes based on `barcode_distribution` supplied to the function. The distribution is a set of probabilities used to skew the barcode choices – if none is given, a uniform distribution will be used.

III. Then, we generate a pre- and a post-sequence. We join the barcode in between the pre- and post-sequences to create a simulated read containing the barcode.

IV. Since FASTQ files require a quality line, we also generate a quality string with random Q scores from 33 to 73.

V.   Finally, we construct the FASTQ lines so that they include the header, sequence, and quality scores.

6. Now, we need a function to create different probability distributions for the barcodes. This will allow us to see a different result for each sample in the experiment:

```
def create_sample_distributions(num_barcodes, num_samples=3):
    distributions = []
    for sample in range(num_samples):
        dist = np.random.dirichlet(
            np.ones(num_barcodes) * 0.5)
        boost_indices = np.random.choice(
            num_barcodes, size=5, replace=False)
        for idx in boost_indices:
            dist[idx] *= random.uniform(3, 8)
        dist = dist / dist.sum()
        distributions.append(dist)

    return distributions
```

This code takes in the desired number of barcodes and samples and then uses `np.random` to create a **Dirichlet** distribution, a common parameterized probability distribution that is often used in cases like this (`https://www.sciencedirect.com/topics/mathematics/dirichlet-distribution`).

This code also selects five barcodes and puts them in `boost_indices` before randomly boosting their value by a factor between 3 and 8. This will ensure that each sample has a few overrepresented genes. Finally, the code appends the distributions to our list and returns it.

7. Next, we need a simple `parse_fastq()` function to parse out our faux FASTQ reads. Please refer to this book's notebook if you wish to review this function. It should be straightforward for you by this point, but if you want more details, you can paste it into your AI tool.

Now, we will define a function that will extract the barcodes from the simulated reads:

```
def extract_barcodes(reads, barcode_pattern):
    barcode_dict = {}
    for header, sequence in reads:
        read_id = header.split()[0][1:]
        header_match = re.search(r'barcode=(\w+)', header)
        if header_match:
            barcode_dict[read_id] = header_match.group(1)
            continue
        match = barcode_pattern.search(sequence)
        if match:
            barcode_dict[read_id] = match.group(1)
    return barcode_dict
```

This function takes in the reads and a barcode regular expression pattern. It then searches either the header or the sequence for the barcode and adds it to a dictionary.

Here, we are using Python's `regex` (regular expression) module. In this case, `.match()` looks for a pattern at the beginning of a string, `.search()` searches the entire string, and `.group()` is used to cleanly return the found pattern.

8. Next, we need to map the barcodes to the genes and count the number of barcodes for each gene:

```
def map_to_genes_and_count(barcode_dict, barcode_gene_map):
    barcode_counts = Counter(barcode_dict.values())
    gene_counts = defaultdict(int)
    for barcode, count in barcode_counts.items():
        if barcode in barcode_gene_map:
            gene = barcode_gene_map[barcode]
            gene_counts[gene] += count
    return Counter(gene_counts)
```

First, this function creates a `Counter` object from the Python `Collections` module to keep track of the counts for each barcode. Then, it creates `defaultdict` for the gene counts, a dictionary that provides default values for any non-existing keys.

After, it loops over the barcodes; for each barcode that is mapped to a gene, we increment the count for that gene.

9. Next, we will define some graphing functions to visualize our data. The `plot_gene_heatmap()` function will build a heat map showing the expression of various genes over the different samples. The `plot_gene_proportions()` function will make a stacked bar chart showing the relative proportion of each gene by sample, based on its barcode abundance. Finally, the `plot_top_genes_barplot()` function will create a bar chart of the top 10 genes and their relative abundance in each sample. You can review these functions in this book's notebook.

OK, let's set up our simulation and run the analysis!

10. First, we will create sample distributions for each of the three samples so that they will have contrasting barcode profiles:

```
sample_distributions = create_sample_distributions(
    num_barcodes, num_samples=3)
```

11. Next, we will generate the sample sequencing data for our three samples:

```
sample_fastq_data = {}
for i, dist in enumerate(sample_distributions):
    sample_name = f"Sample_{i+1}"
    sample_fastq_data[sample_name] = generate_fastq_content(
        num_reads=3000,
        barcode_gene_map=barcode_gene_map,
        barcode_distribution=dist
    )
    print(f"Generated "
            f"{len(sample_fastq_data[sample_name])//4} "
            f"reads for {sample_name}")
```

12. Now, we can set up our regex pattern. For illustration purposes, this pattern is simple and picks up any 12-base-pair sequence of ACGT:

```
barcode_pattern = re.compile(r'([ACGT]{12})')
```

In this example, we associated each gene and barcode in the header of the simulated FASTQ data to keep things simple. We used a very simple regex pattern. However, in real life, we might use a more sophisticated regular expression and barcode pattern. The following tip provides an example of how we could upgrade our code in such a case.

> **AI tip**
>
> **Background**: Let's imagine we want to work with 10x Genomics single-cell data (`https://www.10xgenomics.com/blog/single-cell-rna-seq-an-introductory-overview-and-tools-for-getting-started`), which uses a 16bp cellular barcode at the beginning of the read, followed by a UMI, which is used to identify reads that came from the same molecule. See `https://www.illumina.com/techniques/sequencing/ngs-library-prep/multiplexing/unique-molecular-identifiers.html` for more details. There will also be a potential Poly-T sequence after the UMI. We could specify this type of barcode with this pattern:
>
> ```
> barcode_pattern = re.compile(r'^([ACGT]{16})([ACGT]{12})
> (T{5,})?(.*)')
> ```
>
> **Preparation**: Paste in the code that you used to create distributions for three different samples.
>
> **Prompt**: Upgrade this code so that it uses a barcode pattern suitable for 10x Genomics single cell sequencing data, which uses a 16bp cell barcode, followed by a 12bp UMI, followed by a potential Poly-T sequence.
>
> **What you should see**: You should see code that can use the new barcode pattern and properly capture the different elements of the barcode. It will also include UMI deduplication to improve read counting for greater accuracy.

OK, let's continue with our example!

13. Next, we will loop over the samples and analyze them:

```python
sample_results = {}
for sample_name, fastq_lines in sample_fastq_data.items():
    reads = parse_fastq(fastq_lines)
    barcode_dict = extract_barcodes(
        reads, barcode_pattern)
    gene_counts = map_to_genes_and_count(
        barcode_dict, barcode_gene_map)

    sample_results[sample_name] = gene_counts
    print(f"\nAnalysis of {sample_name}:")
    print(f"  Total reads: {len(reads)}")
    print(f"  Reads with identified barcodes: "
          f"{len(barcode_dict)} "
          f"({len(barcode_dict)/len(reads)*100:.1f}%)")
    print(f"  Unique genes detected: {len(gene_counts)}")
    print(f"  Top 5 most abundant genes:")
    for gene, count in gene_counts.most_common(5):
        print(f"    {gene}: {count} reads "
              f"({count/len(reads)*100:.1f}%)")
print("\nGenerating visualizations...")
plot_gene_heatmap(sample_results)
```

```
plot_gene_proportions(sample_results)
plot_top_genes_barplot(sample_results)
print("Analysis complete!")
```

This code will do the following:

- Initialize a dictionary to keep track of the results for the samples

- Loop over the samples and their corresponding FASTQ data:

- Parse the FASTQ lines and extract the barcodes

- Map the barcodes to the genes and count them

- Store the gene counts in the `sample_results` dictionary for each sample

- Print out summary statistics

- Create our three key visualizations

That's it! We now have a functioning example of a barcode analysis system.

Here is the output we get:

```
Analysis of Sample_1:
  Total reads: 3000
  Reads with identified barcodes: 3000 (100.0%)
  Unique genes detected: 42
  Top 5 most abundant genes:
    Gene_E1: 1092 reads (36.4%)
    Gene_G1: 349 reads (11.6%)
    Gene_G2: 190 reads (6.3%)
    Gene_M2: 142 reads (4.7%)
    Gene_K2: 125 reads (4.2%)

Analysis of Sample_2:
  Total reads: 3000
  Reads with identified barcodes: 3000 (100.0%)
  Unique genes detected: 46
  Top 5 most abundant genes:
    Gene_E3: 469 reads (15.6%)
    Gene_L2: 262 reads (8.7%)
    Gene_K2: 215 reads (7.2%)
    Gene_H2: 176 reads (5.9%)
    Gene_M2: 172 reads (5.7%)

Analysis of Sample_3:
  Total reads: 3000
  Reads with identified barcodes: 3000 (100.0%)
  Unique genes detected: 45
  Top 5 most abundant genes:
```

```
Gene_A1: 377 reads (12.6%)
Gene_H1: 352 reads (11.7%)
Gene_J4: 281 reads (9.4%)
Gene_B1: 256 reads (8.5%)
Gene_F1: 216 reads (7.2%)
```

The following is its Heat Map:

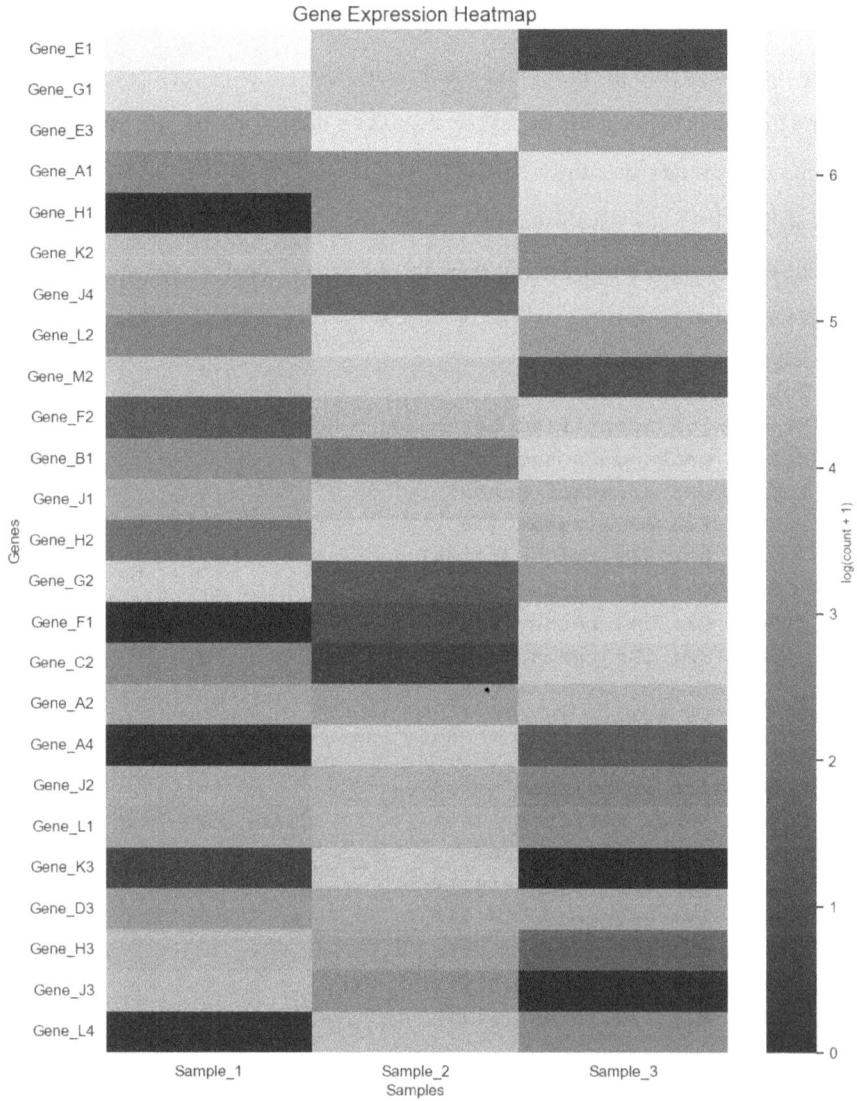

Figure 13.4 – Heat map from barcode analysis

Next, we will produce a bar chart for each sample. We will use different bar sections to show the proportion of each gene found in the barcode counting. Note that the bar sections can be read in reverse order of the legend - Gene E1 is the bottom section and "Other genes" is the top section:

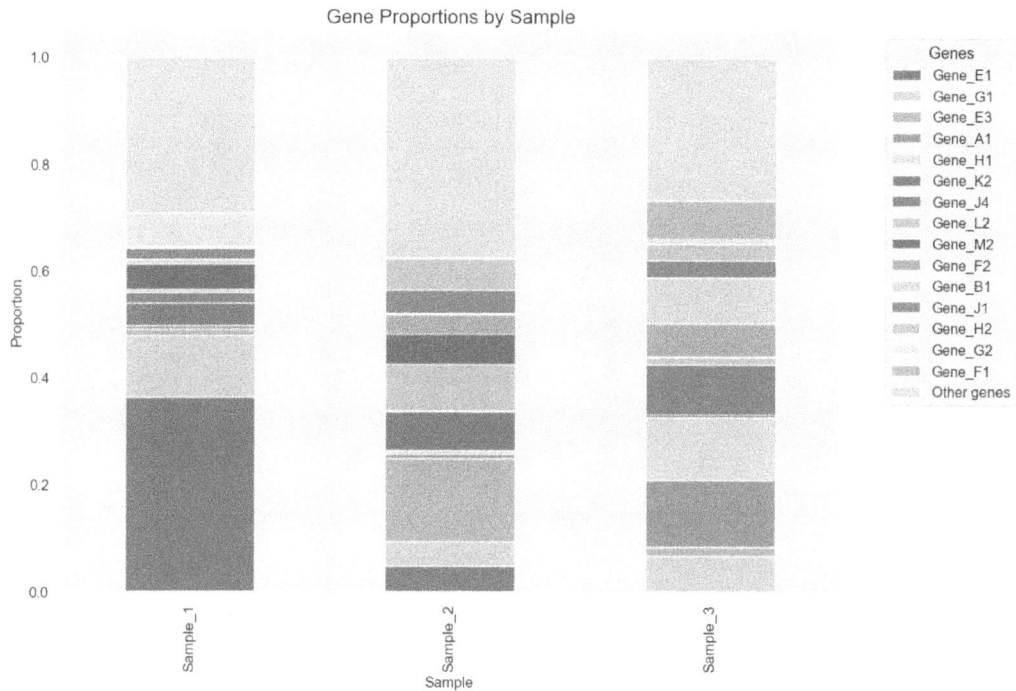

Figure 13.5 – Proportions of each gene by sample, based on barcode abundance

Here, we have a side-by-side bar chart showing the top genes found in the analysis along the X-axis, and their relative count for each sample shown in three different bars in each grouping. From left to right the bars in each group are: Sample 1, Sample 2, and Sample 3:

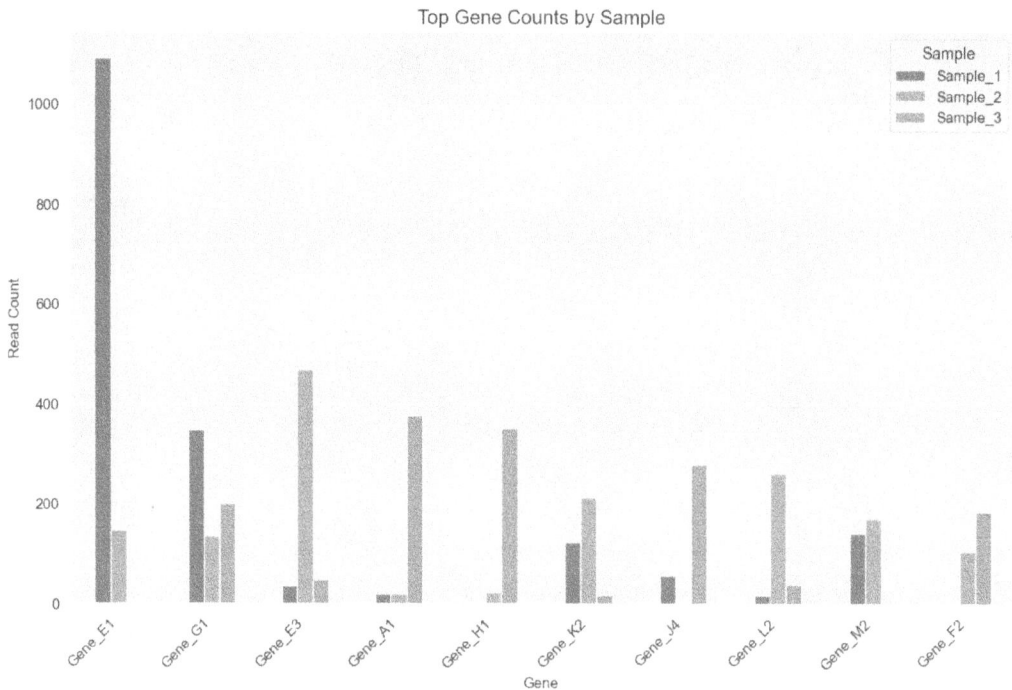

Figure 13.6 – Barcode counts for each of the top genes, by sample

This summarizes our results. Here, we can see that each sample has a different set of top genes and that each sample has a unique gene expression pattern. In a real experiment, these samples could be different strains of an organism that make varying levels of a product, or cells from three different tissues. We would be able to quickly gain insights into the biology going on in the samples. We could review the gene annotations for top up or down-regulated genes or use **pathway analysis** to see whether the genes worked together in common biochemical systems.

There's more...

Barcode sequencing technology is a transformative method for generating big datasets in biology. For example, **Perturb-seq** generates CRISPRi guide RNAs with barcodes for all major transcriptional regulators in a cell, and then exposes the cells to a variety of conditions. The readout comes from single-cell RNAseq. This provides a rich dataset that allows gene expression changes to be tied back to their genetic regulators, along with an analysis of how those regulators are important under different cellular conditions. This technology uses three levels of combinatorial barcode indexing (Jiang et al., *Systematic reconstruction of molecular pathway signatures using scalable single-cell perturbation screens*, Nature Cell Biology, Feb 2025, https://www.nature.com/articles/s41556-025-01622-z).

Barcode technology is used to perform powerful pooled screening in fermentation tanks (Wehrs et al., *Investigation of Bar-seq as a method to study population dynamics of Saccharomyces cerevisiae deletion library during bioreactor cultivation*, Microbial Cell Factories, Aug 2020, `https://link.springer.com/article/10.1186/s12934-020-01423-z`). It can be combined with imaging (Walton et al., Pooled genetic screens with image-based profiling, Molecular Systems Biology, Nov 2022, `https://www.embopress.org/doi/full/10.15252/msb.202110768`) and has been used in drug development (Wong et al., Multiplex barcoded CRISPR-Cas9 screening enabled by CombiGEM, PNAS, Feb 2016, `https://www.pnas.org/doi/10.1073/pnas.1517883113`) and organoid development (Mukhare et al., Integration of Organoids with CRISPR Screens: A Narrative Review, Biology of the Cell, Apr 2025, `https://onlinelibrary.wiley.com/doi/full/10.1111/boc.70006`). Barcode screening can also be combined with microfluidics for incredibly high-throughput experiments (Li et al., UDA-seq: universal droplet microfluidics-based combinatorial indexing for massive-scale multimodal single-cell sequencing, Nature Methods, Jan 2025, `https://www.nature.com/articles/s41592-024-02586-y`).

This powerful tool generates large datasets that directly interrogate complex biology, and bioinformatics is poised to have a huge impact on the analysis and interpretation of this data!

See also

- Here is a tutorial on regular expressions: `https://www.geeksforgeeks.org/write-regular-expressions/`

- The Bartender paper: Zhao et al., *Bartender: a fast and accurate clustering algorithm to count barcode reads*, BioInformatics, Mar 2018, `https://academic.oup.com/bioinformatics/article/34/5/739/4562326`

- mBARq is another tool that can be used for barcode analysis: Sintsova et al., *mBARq: a versatile and user-friendly framework for the analysis of DNA barcodes from transposon insertion libraries, knockout mutants, and isogenic strain populations*, BioInformatics, Feb 2024, `https://academic.oup.com/bioinformatics/article/40/2/btae078/7606336`

- SeqWalk can help you design robust libraries with good separation between barcodes: Gowri et al., *Scalable design of orthogonal DNA barcode libraries*, Nature Computational Science, Jun 2024, `https://www.nature.com/articles/s43588-024-00646-z`

- A tutorial on designing CRISPR screens is provided by Braun et al.: Braun et al., *Tutorial: design and execution of CRISPR in vivo screens*, Nature Protocols, Jul 2022, `https://www.nature.com/articles/s41596-022-00700-y`

- Dive deeper into the science of MRD: Kurtz et al., *Enhanced detection of minimal residual disease by targeted sequencing of phased variants in circulating tumor DNA*, Nature Biotechnology, Jul 2021, `https://www.nature.com/articles/s41587-021-00981-w`

Using nanopore data to resolve genome edits

In this recipe, we'll learn how to determine whether the position where we've edited a genome is correct. For example, we may wish to insert a foreign gene into a strain. In this case, we might want to ensure that the gene landed in the correct region and did not interfere with other genes or their promoters. We may also want to insert a barcode, a promoter library, or other entities. To check the accuracy of such edits, we could use PCR or short reads, but to be absolutely sure, long-read sequencing is recommended.

By the end of this recipe, you'll understand how to simulate long read data and apply it to a genomic edit and verify whether or not the edit has been performed successfully.

Getting ready

No major packages need to be installed for this recipe. The libraries we'll need have been outlined as required.

The code for this recipe can be found in Ch13/Ch13-3-genome-editing.ipynb.

How to do it...

Here are the steps to complete this recipe:

1. First, we need to import our libraries:

    ```
    import random
    import matplotlib.pyplot as plt
    import numpy as np
    import pandas as pd
    import seaborn as sns
    from Bio.Seq import Seq
    from Bio.SeqRecord import SeqRecord
    from collections import Counter
    from IPython.display import display, HTML
    ```

    Here, we are using Biopython and some of our other standard libraries. We are also bringing in the iPython display module (https://ipython.readthedocs.io/en/stable/api/generated/IPython.display.html), which provides methods for displaying rich media, such as HTML or audio, in a notebook.

2. Next, we will set our random seeds so that our results will be reproducible:

    ```
    np.random.seed(42)
    random.seed(42)
    ```

3. Now, let's set up a function that will generate a simulated DNA sequence:

```
def generate_random_sequence(length, gc_content=0.5):
    p_g = p_c = gc_content / 2
    p_a = p_t = (1 - gc_content) / 2
    nucleotides = []
    for _ in range(length):
        nuc = np.random.choice(
            ['A', 'C', 'G', 'T'],
            p=[p_a, p_c, p_g, p_t]
        )
        nucleotides.append(nuc)
    return ''.join(nucleotides)
```

This function will assign probabilities to the bases (A, C, G, T) depending on the average GC content provided. It will then make a random choice based on those probabilities and construct a sequence of the desired length.

4. Now, we will simulate the process of inserting a gene into a genome. First, we will construct the reference genome:

```
print("Creating reference genome...")
reference_genome = generate_random_sequence(10000)
print(f"Reference genome length: {len(reference_genome)}")
print(f"Reference sample: {reference_genome[:50]}...")
```

This uses our random sequence generator to set up a reference genome (reference_genome) of 10,000 base pairs.

5. Then, we will print out the length and a subset of the sequence:

```
print("\nCreating gene insertion...")
insertion_sequence = generate_random_sequence(
    1500, gc_content=0.6)
print(f"Insertion length: {len(insertion_sequence)}")
print(f"Insertion sample: {insertion_sequence[:50]}...")
```

This creates a simulated insertion sequence of 1,500 base pairs with 60% average GC content.

6. Now, we just need to insert our simulated "gene" into the reference sequence:

```
insertion_position = 4000
modified_genome = (
    reference_genome[:insertion_position] +
    insertion_sequence +
    reference_genome[insertion_position:]
)
print(f"\nModified genome length: {len(modified_genome)}")
print(f"Insertion position: {insertion_position}")
```

Here, we added `insertion_sequence` at position 4,000 in the reference genome by adding the left-hand side, insertion, and right-hand side together.

Here is our output:

```
Creating reference genome...
Reference genome length: 10000
Reference sample:
CTGGAAATGGATTAAACGCCGACCCTAGGAGAATTTCAGCACATCGCGGA...


Creating gene insertion...
Insertion length: 1500
Insertion sample:
GGCCCGTGCGATCCAGGCGAGCGTACATCCGGTGTAATTCCCGGCCCAAA...


Modified genome length: 11500
Insertion position: 4000
```

This shows us how we created our reference genome, the insertion, the length of the new "genome," and the position where we inserted the gene.

Nanopore reads are inherently noisy and can contain base errors (substitutions of the wrong base) and insertions and deletions, particularly around homopolymer runs.

7. We'll need a function to simulate the errors in the reads:

```
def apply_nanopore_errors(sequence):
    result = ''
    for base in sequence:
        p_error = 0.05
        if random.random() < p_error:
            error_type = random.choice(
                ['substitution', 'deletion', 'insertion']
            )
            if error_type == 'substitution':
                possible_bases = [
                    b for b in 'ACGT' if b != base]
                result += random.choice(possible_bases)
            elif error_type == 'deletion':
                continue
            else:
                result += random.choice('ACGT') + base
        else:
            result += base
    return result
```

This function loops over the sequence and assigns a 5% error rate. So, 5% of the time, we choose to introduce an error, and we pick between substitution, insertion, and deletion.

If we are introducing a substitution, we pick a base other than the current base and replace it with that base. If we are introducing a deletion, we simply skip over the base. Finally, if we are introducing an insertion (the final `else` condition), we pick a random base and add it in front of our main base. If we are not in the error condition, we just add the current base to the result.

In the end, we return our new sequence with errors introduced.

8. Next, we will define a function that will generate a set of nanopore reads.

   Our function, `generate_nanopore_reads()`, will take in a genome, a desired number of reads, and a length range for the reads. You can review the code for this in this recipe's notebook.

   This function creates several reads, with a random length between the minimum and maximum desired length. It picks a random position in our genome and then extracts a sequence for a read of that length. Then, we call our `apply_nanopore_errors()` function to add errors that are consistent with typical sequencing results. Finally, we assign the reads an ID, store the information, and return the read set.

   Let's use the function `generate_nanopore_reads()` to generate a nanopore read set for our genome:

```
print("\nGenerating nanopore reads...")
reads, read_info = generate_nanopore_reads(
    modified_genome, n_reads=1000)
print(f"Generated {len(reads)} nanopore reads")
print("\nSample of read information:")
display(read_info.head())
```

Here, we have generated 1,000 reads from the genome. Here is the output:

```
Generating nanopore reads...
Generated 1000 nanopore reads

Sample of read information:
```

	read_id	start	end	length
0	Read_1	204	6111	5907
1	Read_2	4150	10443	6293
2	Read_3	548	7735	7187
3	Read_4	2069	10806	8737
4	Read_5	3181	8655	5474

Figure 13.7 – Sample output for the generated reads

This shows a sample overview of the first five reads, including the read ID, start and end positions, and length of the read.

9. Next, we will define some functions that will align the reads with the genome. We will use a **k-mer-based alignment** strategy for this. In this approach, we break a read into k-length subsets called k-mers, and find the maximum number of k-mers in the read that match adjacent locations in the genome. This approach can be more efficient than traditional approaches such as Smith-Waterman.

10. First, we'll write a short function that will find all the k-mers in the sequence:

```
def find_kmers(sequence, k=15):
    kmers = []
    for i in range(len(sequence) - k + 1):
        kmers.append(sequence[i:i+k])
    return kmers
```

This function simply loops over the sequence and pulls out subsets (k-mers) with a default length of 15.

11. Now, we will define a function for mapping reads to the genome:

```
def map_read_to_genome(
    read, reference, k=15, mismatch_penalty=0.2
):
    read_kmers = find_kmers(read, k)
    kmer_positions = {}
    for i in range(len(reference) - k + 1):
        kmer = reference[i:i+k]
        if kmer not in kmer_positions:
            kmer_positions[kmer] = []
        kmer_positions[kmer].append(i)
    position_counts = Counter()
    for i, kmer in enumerate(read_kmers):
        if kmer in kmer_positions:
            for pos in kmer_positions[kmer]:
                expected_offset = i
                actual_offset = pos
                position_counts[pos - expected_offset] += 1
    if position_counts:
        best_offset = position_counts.most_common(1)[0][0]
        confidence = (
            position_counts[best_offset] / len(read_kmers)
        )
        if len(position_counts) > 1:
            second_best = (
                position_counts.most_common(2)[1][1]
                if len(position_counts) > 1 else 0
            )
```

```
        ratio = (second_best /
            position_counts.most_common(1)[0][1]
            if position_counts.most_common(1)[0][1] >0 else 0
        )
        confidence *= (1 - ratio * mismatch_penalty)
        return best_offset, confidence
else:
    return -1, 0.0
```

First, this function obtains the k-mers for the read. Then, it loops over the reference genome and all k-mers for it. Next, it initializes a `Counter()` object to count the number of read k-mers that match k-mer positions in the genome. After, we use the `most_common()` function (`https://www.geeksforgeeks.org/python-most_common-function/`) to find the best matching position in the genome with the highest number of k-mer matches coming from the read, which is therefore the most likely position for that read to align in the genome. We calculate a confidence score by looking at the ratio of k-mers that align with the genome versus the total number of k-mers in the read. We also apply a penalty for reads that map equally to both locations. Finally, we return the best match position and the confidence score. We'll use this function to align reads with our genome.

12. Now, we will map our simulated reads to the reference genome:

```
print("\nMapping reads to reference and modified genomes...")
mapping_results = []
k_size = 21
for read in reads:
    read_seq = str(read.seq)
    ref_pos, ref_conf = map_read_to_genome(
        read_seq, reference_genome, k=k_size)
    mod_pos, mod_conf = map_read_to_genome(
        read_seq, modified_genome, k=k_size)
    read_len = len(read_seq)

    if read_len > 0:
        normalized_ref_conf = ref_conf * (read_len / 1000)
        normalized_mod_conf = mod_conf * (read_len / 1000)
        conf_diff = normalized_mod_conf - normalized_ref_conf
    else:
        conf_diff = 0
    threshold = max(0.05, 0.2 - (read_len / 10000))
    read_type = 'ambiguous'
    if conf_diff > threshold:
        if (mod_pos < insertion_position and
            mod_pos + read_len > insertion_position +
```

```
            len(insertion_sequence)):
            read_type = 'spanning'
        elif ((mod_pos <= insertion_position and
                mod_pos + read_len > insertion_position) or
               (mod_pos < insertion_position +
                len(insertion_sequence) and mod_pos +
                read_len >= insertion_position +
                len(insertion_sequence))):
            read_type = 'supporting'
        else:
            read_type = 'non-insertion'
    elif conf_diff < -threshold:
        read_type = 'reference'

    mapping_results.append({
        'read_id': read.id,
        'read_length': read_len,
        'ref_position': ref_pos,
        'ref_confidence': ref_conf,
        'mod_position': mod_pos,
        'mod_confidence': mod_conf,
        'confidence_diff': conf_diff,
        'read_type': read_type
    })
```

This code will take each read and map it to both our "normal" reference genome and the "modified" reference genome with the insertion. By default, we will assume that our read mapping is ambiguous. We calculate a confidence difference, looking at the number of reads being mapped to the normal versus the modified genome. If the confidence score is too close to call, we'll leave this as an ambiguous mapping call. Note that we pass through a k-mer size of 21 here (as opposed to the function's default of 15) to get fewer ambiguous matches. Optimizing the k-mer size is an important factor to understand in k-mer-based alignment. Shorter k-mers lead to a higher number of ambiguous matches, especially in complex genomic regions. They also are less tolerant to errors seen in long-read technologies, as the longer the k-mer, the greater the chance it will contain an error. Conversely, longer k-mers improve the chance of finding unambiguous matches, especially in repetitive regions of the genome. We also apply an adjustment to our threshold based on read_length, as longer reads should have higher confidence.

If the reads map better to our modified genome, we'll examine the following cases:

- If the read can completely span the insertion and cross into the genomic regions on the left and right, we will call this a **spanning read**. This is an ideal situation in which we have significant evidence not only of the insertion but also that it has a flanking region on both the left and right-hand sides. This helps us be sure it was really inserted in the right place and wasn't inserted somewhere else, or even worse, caused some sort of recombination event in which it picked up a portion of the target genome but also rearranged itself somewhere else, which would be an **off-target effect**.

- If the read crosses one end of the genome into the insertion, we call this a **supporting read**. This provides good evidence of correct insertion, if combined with supporting reads coming from both sides of the insertion.

- If the read maps to the modified genome, but not at the insertion location, we call it a **non-insertion read**.

- If a read is better mapped to the unmodified genome, we call it a **reference read**.

13. Finally, we will create all the details for our analysis and append them to the `mapping_results` list.

Now, let's summarize our analysis:

```
alignment_df = pd.DataFrame(mapping_results)
display(alignment_df.head())
read_type_counts = alignment_df['read_type'].value_counts()
print("\nRead type distribution:")
display(read_type_counts)
```

Here, we are converting the data into a pandas DataFrame. We are displaying the header and then summarizing counts for the different read types.

Here is what we get:

```
Alignment results summary:
```

	read_id	read_length	ref_position	ref_confidence	mod_position	mod_confidence	confidence_diff	read_type
0	Read_1	5907	203	0.058026	204	0.068422	0.061408	spanning
1	Read_2	6293	2637	0.027674	4137	0.027292	-0.002408	ambiguous
2	Read_3	7187	541	0.045123	541	0.075038	0.214998	spanning
3	Read_4	8737	567	0.036182	2067	0.036022	-0.001403	ambiguous
4	Read_5	5474	1690	0.032710	3185	0.038651	0.032519	ambiguous

```
Read type distribution:
ambiguous     496
spanning      488
supporting     16
Name: read_type, dtype: int64
```

Figure 13.8 – Alignment results summary

We can definitely see some reads spanning or supporting our insertion, so our strategy appears able to detect the inserted sequence. We are not really getting reference reads because our confidence threshold doesn't distinguish between the normal and reference genome enough to cross the threshold. You can play around with tuning that if you want, but it will not impact this exercise.

14. Now, we will analyze and summarize the evidence for our gene insertion:

```
spanning_reads = len(alignment_df[
    alignment_df['read_type'] == 'spanning'])
supporting_reads = len(alignment_df[
    alignment_df['read_type'] == 'supporting'])
total_reads = len(alignment_df)

print(f"\nInsertion evidence summary:")
print(f"Total reads: {total_reads}")
print(f"Reads spanning the insertion: {spanning_reads}"
      f"({spanning_reads/total_reads*100:.1f}%)")
print(f"Reads supporting the insertion: {supporting_reads}"
      f"({supporting_reads/total_reads*100:.1f}%)")
```

This gives us some totals for the reads that support the insertion event:

```
Insertion evidence summary:
Total reads: 1000
Reads spanning the insertion: 488 (48.8%)
Reads supporting the insertion: 16 (1.6%)
```

Good! We have a fair number of reads that support our insertion or even span it completely. In a real-world example, our genome would be much larger, and the percentages would be lower – but so long as you have something in the range of 20-30x coverage supporting both ends of your insertion, you would have a strong case for proper gene insertion.

15. Next, we'll visualize our results. The first graph is for read type distribution. You can review this code under *Graph 1 – Read Type Distribution* in this recipe's notebook. Here is the resulting graph:

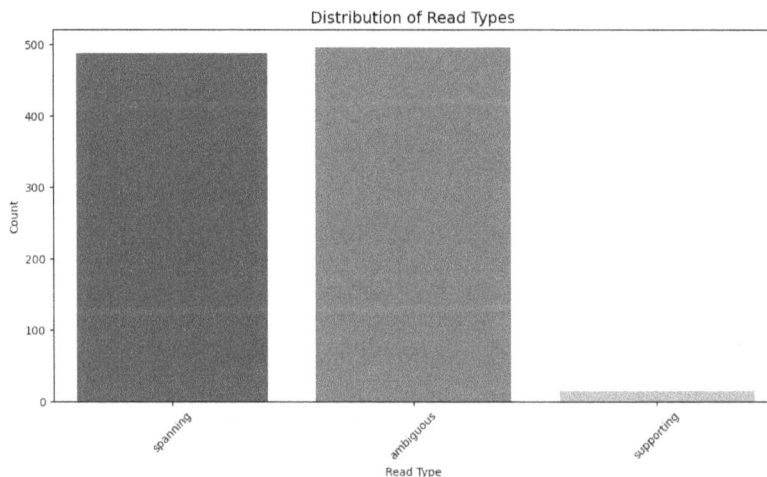

Figure 13.9 – Distribution of read types from insertion analysis

This figure provides a summary of the number of different read types that have been found.

Now, let's create a graph to help us analyze the confidence scores of the reads. This will help us see how much we are disambiguating a putative normal genome from a modified genome. Review the code for *Graph 2 – Confidence Difference Histogram* in this recipe's notebook. Here is the output:

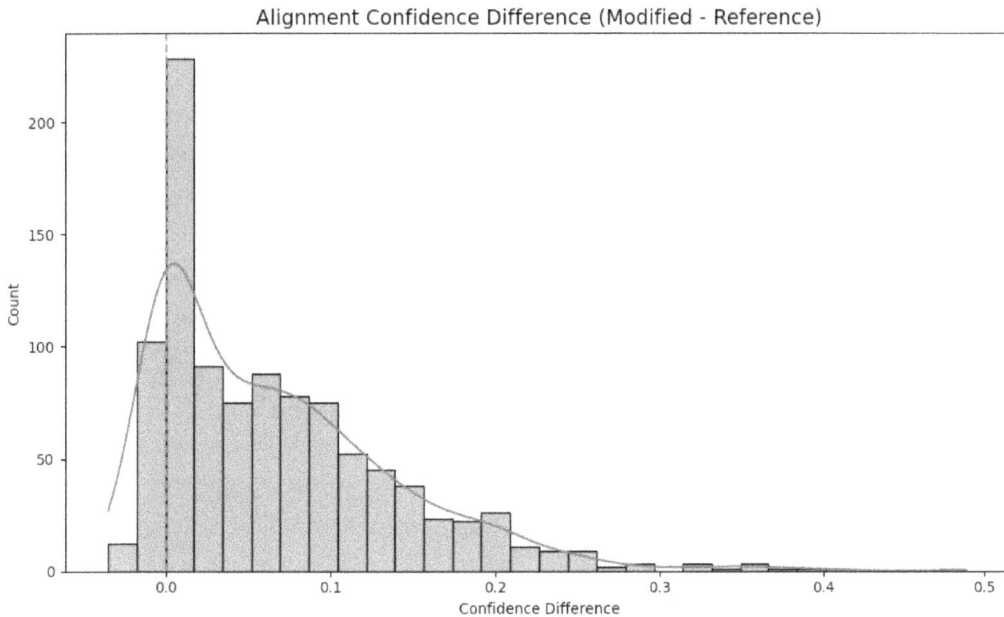

Figure 13.10 – Histogram of confidence differences in insertion analysis

This histogram shows a peak near 0. It represents our ambiguous reads, which are equivalent to reference reads. This is expected since much of the genome is normal and does not contain the insertion. However, on the right-hand side, we can see a long tail, indicating quite a few reads that have high confidence of separation between the normal and modified genomes. These are the reads that provide evidence for the insertion.

16. Now, check out the code for *Graph 3 – Read Alignment Visualization*. This graph uses line plotting in Matplotlib to show the reads lined up against the insertion region. Here is what we get:

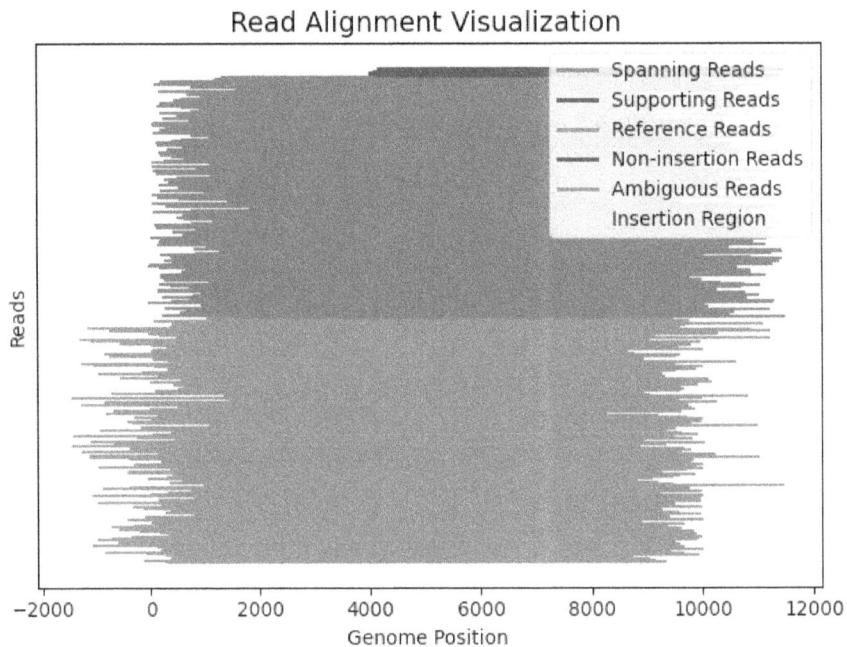

Figure 13.11 – Read alignment visualization for insertion analysis

The very top small group of lines are supporting reads - they only cover the right side of the insertion. The second large section of reads are spanning reads. The reads on the bottom half are ambiguous reads. Overall, this shows a large number of spanning reads crossing the insertion region, suggesting good evidence for insertion.

Now, let's define a function that verifies the insertion and summarizes our results. Review the code for the `verify_insertion()` function in this recipe's notebook.

This code calculates percentages for reads that support the insertion and applies a threshold, requiring at least five fully spanning reads and 15 supporting reads. This provides an example of calculating a simple confidence score based on the evidence.

At this point, we can calculate the verification result and display it using HTML:

Insertion Verification Result
Status: VERIFIED
Confidence: 0.05
Details: Insertion VERIFIED with moderate confidence (0.05). Found 488 spanning reads and 16 supporting reads.

Figure 13.12 – Results of insertion verification analysis

Congratulations! You have verified the insertion. Your strain engineers will be very happy!

Let's create another nice figure for them to summarize the evidence. Review *Step 6 – Coverage Analysis Across Insertion Site* in this recipe's notebook.

This code calculates a coverage depth by looking at the number of reads covering each position in the sequence and then adds dashed lines to highlight the region of the expected insertion. Here is what we'll see:

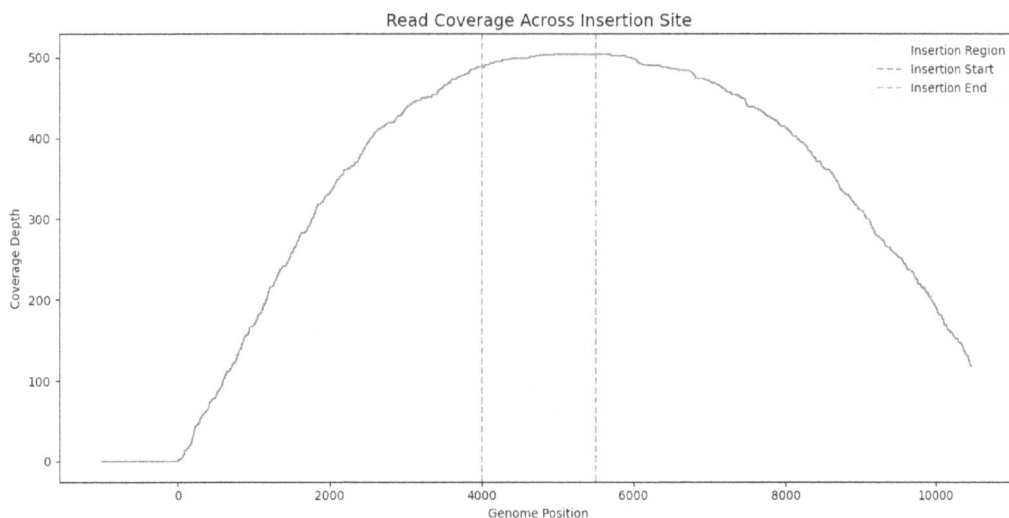

Figure 13.13 – Read coverage across the insertion site

The final piece will be a textual summary of the overall results and some guidance for the scientist regarding potential next steps. Review *Step 7 – Conclusion and Next Steps* in this recipe's notebook. Here is the output:

```
--- Conclusion ---
Insertion VERIFIED with moderate confidence (0.05). Found 488
spanning reads and 16 supporting reads.

Recommended next steps:
1. Validate insertion boundaries with targeted PCR
2. Confirm insertion sequence with Sanger sequencing
3. Verify expression of inserted gene (if applicable)

This analysis demonstrates how nanopore sequencing can be used
to verify gene insertions
through read mapping and coverage analysis. The long reads
provided by nanopore technology
are particularly valuable for spanning entire insertion regions,
enabling confident verification.
```

This explains the result and level of confidence. It suggests some next steps for potential further lab validation of the result if desired.

That concludes this recipe. As you can see, sequencing data, particularly long-read data, can be immensely valuable for confirming genome edits. It's a powerful tool that has greatly improved our ability to verify strain construction efforts and analyze off-target effects when they occur.

There's more...

You can play around with the preceding example more if you want by altering the number of reads, the length of insertion, and the k-mer length. You can also try out a sequence without any insertion to confirm that it will not be found. In general, you will see that having too few reads will lead to "under sequencing," and you will not be able to detect insertions. The k-mer length will impact the ambiguity of your mappings as well, and hence your ability to detect insertions.

Long-read sequencing is critical for verifying genomic insertions. In this recipe, we used an **alignment-based** approach, in which we aligned reads around the region of the insertion to verify that it was present, and that it contained the correct genomic region on the left-hand side (**left flank**) and the right-hand side (**right flank**). Another approach that can be very helpful is to take reads that contain all or part of the insertion and perform **de novo assembly** on them. This can be especially helpful when the insertion is incomplete or imperfect and can reveal the off-target region of the insertion. Another interesting approach is to annotate the reads themselves directly. Typically, this is done with a de novo annotation strategy or by using a **lift-over** strategy to bring over annotations (from your reference genome plus your genetic parts library). This approach can be very helpful for providing direct read-level (single-molecule) evidence for the insertion and neighboring region.

Now that we've come to the end of this chapter, let's clean up and close down our conda environment:

```
conda deactivate
```

See also

- SeqVerify is a software package that can be used for interrogating genomic edits: Smela et al., *SeqVerify: An accessible analysis tool for cell line genomic integrity, contamination, and gene editing outcomes*, Stem Cell Reports, Oct 2024, https://www.cell.com/stem-cell-reports/fulltext/S2213-6711(24)00242-X

- SuperDecode is another tool that can be used for verifying edits that supports multiple sequencing technologies: Li et al., *SuperDecode: An integrated toolkit for analyzing mutations induced by genome editing*, Molecular Plant, Apr 2025, https://www.cell.com/molecular-plant/fulltext/S1674-2052(25)00092-9?uuid=uuid%3A637b4f6c-bd3e-4714-b986-b15f29bacd53

- Read more about genome editing for plant improvements: Mall et al., *CRISPR/Cas-Mediated Genome Editing for Sugarcane Improvement*, Sugar Tech, Jan 2024, https://link.springer.com/article/10.1007/s12355-023-01352-2

Get This Book's PDF Version and Exclusive Extras

UNLOCK NOW

Scan the QR code (or go to packtpub.com/unlock). Search for this book by name, confirm the edition, and then follow the steps on the page.

Note: Keep your invoice handy. Purchases made directly from Packt don't require an invoice.

14

Cloud Basics

In this chapter, we'll go over some of the basics of cloud computing and get you prepared for a deeper discussion on workflow systems in *Chapter 15*.

Cloud computing has transformed the world of business, and bioinformatics in particular. Not too long ago, most compute servers were kept "on-prem," or on-premises, meaning that physical hardware was kept at a business center or in a data center. Nowadays, most businesses are moving to virtual servers that are provided on demand, commonly known as "cloud computing." The most widely used cloud providers include:

Provider	URL
Amazon Web Services (AWS)	`https://aws.amazon.com/free/`
Google Cloud Platform (GCP)	`https://cloud.google.com/`
Microsoft Azure	`https://azure.microsoft.com/`
Oracle Cloud Infrastructure (OCI)	`https://www.oracle.com/cloud/`

Table 14.1 – Major cloud providers

Cloud computing allows you to provision or "spin up" significant resources rapidly. You can make a huge database server or 100 large compute nodes in minutes. You can then get rid of these resources the moment you no longer need them, thereby maximizing the efficient use of your capital. You can specify your resources using code to keep things highly reproducible, and you can keep versioned copies of your systems in online repositories, to ensure accuracy and reproducibility. In short, the cloud offers amazing and transformative capabilities.

In this chapter, we'll be focusing on AWS as it is one of the biggest and most commonly used platforms. But other platforms also have their advantages, and the concepts are very similar. For instance, Google's platform, GCP, offers amazing services in the area of incredibly scalable data lakes and databases – `https://cloud.google.com/spanner`.

Microsoft Azure offers deep integration with Microsoft products such as SQL Server, and has a unique Azure OpenAI offering – `https://azure.microsoft.com/en-us/products/ai-foundry/models/openai`.

Oracle's OCI cloud offers features such as direct integration with Microsoft Azure, cloud services within your own data center, low egress costs, and self-healing databases – `https://www.oracle.com/autonomous-database/`.

Before choosing a cloud platform for your company, you should carefully research the options and weigh the pros and cons, including on-prem as one of the options for comparison. You should take time to understand the existing software stack within your company and the options available in each cloud provider, and perform a **Total Cost of Ownership** (**TCO**) analysis before deciding – `https://www.nops.io/blog/cloud-total-cost-of-ownership/`.

By the end of this chapter, you'll know how to set up a basic cloud account and start becoming familiar with the many resources available there. We'll see how to create and destroy cloud resources programmatically, and we'll learn how to set up applications on the cloud. This will provide a solid foundation to discuss how bioinformatics workflows can be launched on the cloud in the next chapter.

In this chapter, we will cover the following recipes:

- Setting up an AWS account
- Building cloud components using `boto3`
- Building a container in AWS ECR

Technical requirements

In this chapter, we will use the following tools and packages:

- AWS
- `boto3`
- Docker

You will want to make a working directory called `Ch14`.

The code for this chapter is available in this book's GitHub repository at `https://github.com/PacktPublishing/Bioinformatics-with-Python-Cookbook-Fourth-Edition/tree/main/Ch14`.

Remember to activate your `conda` environment before beginning the recipes, like this:

```
conda activate bioinformatics_base
```

Alternatively, if you would like to set up a `conda` environment specific to this chapter, before activating `bioinformatics_base`, run the following:

```
conda create -n ch14-cloud-computing --clone bioinformatics_base
conda activate ch14-cloud-computing
```

You will be able to install the packages for the chapter as you go, or you can use the YAML file provided in the repository:

```
conda env update --file ch14-cloud-computing.yml
```

Setting up an AWS account

AWS is an amazing system for running computations and storing data on the cloud. You can build virtual servers, run websites, and more.

One of the key uses of the cloud is bioinformatics data storage. Sequencing and other types of experiments generate massive amounts of data these days, and cloud storage provides a reliable, expandable, and cost-effective method for storing this data.

On AWS, the main storage is called **Simple Storage Service**, or **S3**. It can grow to store your data as needed, and you pay only for what you use. It can store exabyte levels of data and has incredibly high reliability. You can set up **archiving** rules to send the data to **Glacier** or **Glacier Deep Archive**, which hold the data in slower-access but lower-cost states to save money (`https://aws.amazon.com/pm/s3-glacier/`). This means Glacier and Glacier Deep Archive take longer to retrieve your data – you have to request to **unarchive** your files before using them. However, data in Glacier or Glacier Deep Archive costs significantly less to store. You can also set up **replication** to make copies of your data in multiple regions of the world for maximum reliability.

AWS operates in multiple geographic **Regions** subdivided into **Availability Zones** (we will use the terms interchangeably here). These represent concentrations of data centers in major regions of the world. You will want to keep in mind in which zone your cloud resources are being created. This is because cloud components are tied to a zone, so you'll need to keep track of where you created them. The zone will also impact the pricing of data that you download or transfer from one zone to another. Placing cloud resources in various zones also helps maximize the availability of your systems by introducing redundancy, and by placing services as close as possible geographically to your customers.

Here is a list of the Availability Zones – `https://docs.aws.amazon.com/global-infrastructure/latest/regions/aws-availability-zones.html`.

> Note
> There are various AWS Certification courses you can take to show that you have mastered cloud computing skills. I highly recommend that you consider getting a certification; it can be a powerful way to improve your resume – `https://aws.amazon.com/training/`.

In this recipe, we are going to set up a basic free AWS account for you so that you can play around with basic concepts in cloud computing. You'll learn about the web interface and terminal-based access to the system. You can continue to use this account and take additional courses to go deeper. The concepts presented here are similar for other cloud providers if you decide to use them in your future work.

Getting ready

First, we will set up our AWS account. Go to `https://aws.amazon.com/free/`.

This will take you to a page to set up a free AWS account. Be sure to write down the information you will be providing in a secure location. You may want to use a password manager such as LastPass (`https://www.lastpass.com/`). You will set up your **root user email** and **account name**. Be sure to use a secure, personal email account that you have access to. You will also set a **root user password**. You will then fill out some additional contact information. You will also be asked to provide a credit card. This card will only be used if you go beyond the free tier usage. But you must take care to monitor any costs you may incur on this account (see the billing discussion that follows).

> **Note**
>
> If you are uncomfortable with setting up an account and providing credit card information right now, you can just read through the examples in this chapter and familiarize yourself with AWS and how it works.

You will need access to your mobile phone to receive SMS confirmation messages. You can also choose the Basic (free) support plan. When you are done, you will be taken to the AWS Management Console:

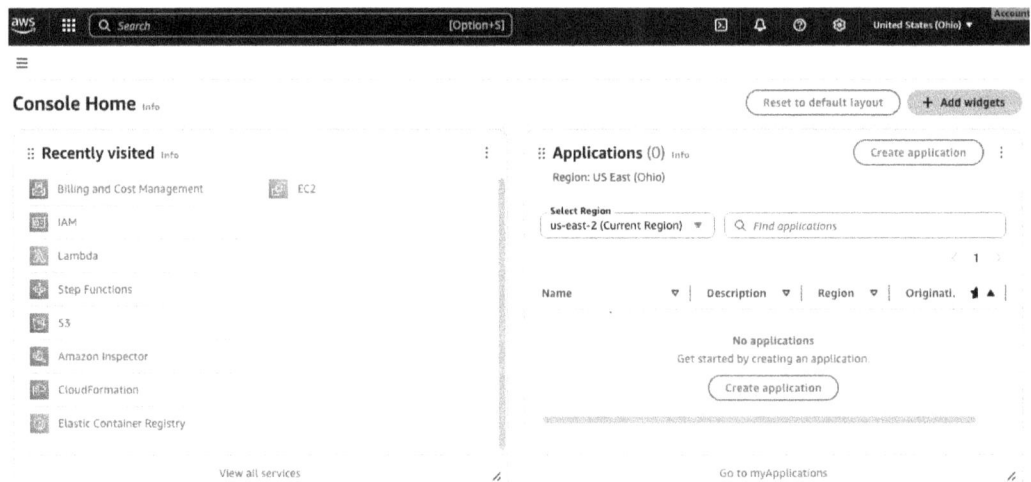

Figure 14.1 – AWS Management Console

On this page, you can manage your account and search for resources. In the upper right, you will see the Availability Zone you are currently set to. In the preceding figure, it is set to **Ohio**, which is in the **US-East-2** zone. It's important to know about Availability Zones because they determine the rough geographical region in which your compute resources are located. This can matter for performance if you are trying to optimize for users in a certain Region. It also impacts **inter-Region transfer** costs (`https://aws.amazon.com/blogs/architecture/overview-of-data-transfer-costs-for-common-architectures/`). When you transfer data between Regions, there will be a charge. There are Regions for the East and West coasts of the US and multiple Regions around the globe.

You may want to secure your root user further by adding **Multi-Factor Authentication (MFA)**. You might want to create a separate **Identity and Access Management (IAM)** user as well so that you are not always logging in as the root user – see `https://docs.aws.amazon.com/IAM/latest/UserGuide/id_users_create.html`.

Great, you now have an AWS account! Next, we will perform a few simple operations in the account to get more familiar with it. We will use the Free Tier account, which provides a set of free services to use to try out AWS, so it should not incur any costs.

Before going further, you may want to learn more about navigating the AWS Management Console:

`https://aws.amazon.com/getting-started/hands-on/getting-started-with-aws-management-console/`

`https://www.youtube.com/watch?v=i331jNgsL_4`

There is no notebook for this section.

How to do it...

> **Note**
>
> Learning AWS could easily be a topic for another (large) book, so we won't go into all the details here. But we encourage you to study on your own, and several key resources are provided in the *See also* section.

Let's go over a few basics in AWS. First, let's build a simple virtual server. In AWS, compute servers are called **Elastic Compute Cloud instances**, or **EC2 instances** for short. Let's build a small EC2 instance using the AWS Management Console – `https://docs.aws.amazon.com/AWSEC2/latest/UserGuide/EC2_GetStarted.html`:

1. Go to the EC2 console dashboard. One of the easiest ways to navigate in AWS is simply to type the name of the resource you are looking for in the search bar. So, you can just type EC2 if you want, and you will be taken to the EC2 console page:

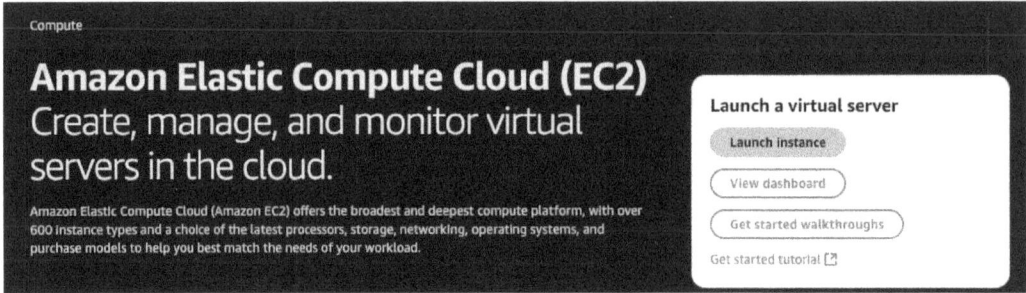

Figure 14.2 – EC2 console page

2. From here, click **Launch instance**. Give your instance a name (record this information) and choose the instance type. You will want to choose `t2.micro` as this is part of the Free Tier. You can also choose an **Amazon Machine Image** (**AMI**) from here. AMIs can come pre-configured with a wide variety of software. You can also use clean base images of popular **operating systems** (OSs) such as Red Hat or Ubuntu. Let's choose the default Amazon Linux AMI.

3. We will also need a **key pair**. This gives you a pair of SSH keys for logging in to your instance. Choose **Create a new key pair**. Give your key pair a name (record this information). Choose the defaults of **RSA** and **.pem**. Download your key pair and put it in an AWS directory (e.g., `~/key-pairs`).

4. You can also set up network information from here. Notice that your default is set to **Allow SSH traffic from anywhere – 0.0.0.0/0**. This will allow anyone to potentially connect to your EC2 instance if they have the security information, so it is not really secure. This will be fine for playing around with, but in a real situation, you would want to coordinate with your IT department to set up appropriate security rules.

5. Now, click **Launch Instance**.

   You will see a **Success** message and a page with the next steps:

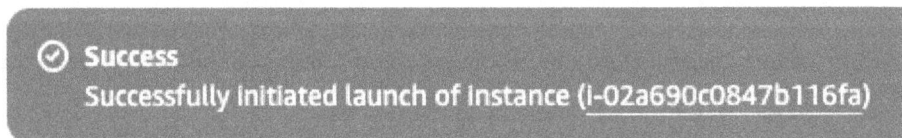

Figure 14.3 – Result of creating an EC2 instance

You may want to copy the preceding link and save it. The link is in parentheses just after the word "instance," you can Right-Click on it and choose **Copy Link Address** to get the link. This link will be the URL to your EC2 instance.

6. Next, we can connect to the instance using the GUI. Click **Connect to Instance** and then **Connect using EC2 Instance Connect**. The default username is **ec2-user**.

Here it is! You are now logged in to your EC2 instance:

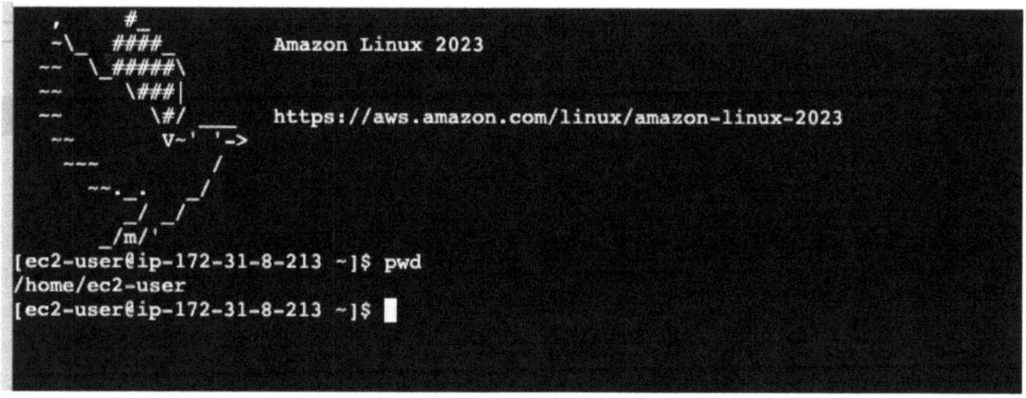

```
       #_
  ~\_  ####_                 Amazon Linux 2023
 ~~  \_#####\
 ~~     \###|
 ~~       \#/ ___            https://aws.amazon.com/linux/amazon-linux-2023
  ~~       V~' '->
   ~~~         /
     ~~._.   _/
        _/ _/
      _/m/'
[ec2-user@ip-172-31-8-213 ~]$ pwd
/home/ec2-user
[ec2-user@ip-172-31-8-213 ~]$ █
```

Figure 14.4 – The terminal of the EC2 instance

Now, let's make sure our key pair is set up so that we can connect from our Mac terminal.

7. Go back to the **EC2 Connect to Instance** page and then go to the **SSH Client** tab. You will first copy the chmod command and run it from your terminal, like this:

```
chmod 400 ~/key-pairs/shanebrubaker-key-pair-1.pem
```

Make sure to replace the preceding path and filename with yours.

8. Next, copy the IP address of your EC2 instance using the example ssh command that is given to you:

```
ssh -i ~/key-pairs/shanebrubaker-key-pair-1.pem ec2-user@ec2-54-
219-204-167.us-west-1.compute.amazonaws.com
```

Again, make sure you replace the preceding command with your information.

You will see the terminal screen of your instance just like when we connected through the console GUI.

That's it! You have now connected to your instance directly from your Mac terminal.

OK, let's shut down our instance. This is important to remember, even though this particular instance might be free for a while, you want to shut it down, so you don't have to clean it up later. If this were a bigger instance, it could cost a significant amount of money per hour. So, you always want to be mindful of shutting down instances and turning off resources being consumed in AWS to reduce costs.

9. Go back to your EC2 **Instances** panel.

You will see your instance running there. Select it using the checkbox, and then click **Instance state** in the upper right:

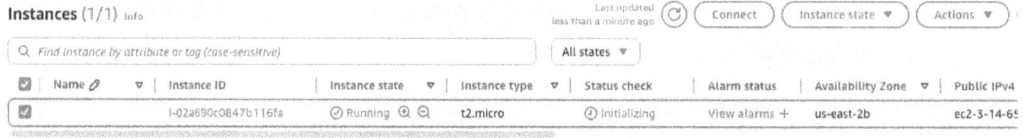

Figure 14.5 – Terminating the instance

Click **Terminate (delete) instance**. This will shut down your instance, and you will no longer be charged for it. Note that this deletes your instance completely and any data on it – it cannot be retrieved or turned back on later. Within a few seconds, you should see **Terminated** as the instance state, and if you go back to the EC2 **Instances** page and refresh it, you will see zero instances running.

Great job! We have now set up an AWS account and gained some basic familiarity with it. You should consider taking a more extensive course on AWS development and using some of the resources in the *See also* section to learn more.

There's more...

Let's review a few basics on billing in AWS so you are familiar with it. Remember, as long as there are no services running within your account, you should not be billed.

You can search for `Billing and Cost Management` in the search bar to take you to a page with information on your current and estimated charges. You can also set various alerts about usage charges. You will probably want to go into **Billing Preferences** and check the box for **AWS Free Tier alerts**. This will send you an email if you are close to exceeding your free usage limits. You can also set up a variety of billing alerts using **CloudWatch** (`https://docs.aws.amazon.com/AmazonCloudWatch/latest/monitoring/monitor_estimated_charges_with_cloudwatch.html`).

You may also want to set up a **budget**. This will alert you if you go over certain spending limits. Go to **Billing and Cost Management** and then choose the **Zero Spend Budget** template and add your email address. This will alert you of any level of costs being incurred in your account.

You will want to become familiar with the following major types of charges within AWS:

- **Free Tier**: You can spin up resources that fall within the Free Tier without any charge. For example, a small EC2 instance could be created and used completely for free. There are many services within AWS that fall within this Free Tier. You will also have expanded access to free services via the Free Tier trial period for 12 months. You can find a list of Free Tier services here: `https://aws.amazon.com/free/?p=ft&z=subnav&loc=1&all-free-`

```
tier.sort-by=item.additionalFields.SortRank&all-free-tier.sort-
order=asc&awsf.Free%20Tier%20Types=*all&awsf.Free%20Tier%20
Categories=*all.
```

- **Transfer fees**: There is no transfer fee when uploading data into AWS. However, there are fees for taking the data back out (downloading it). There are also fees for transferring data between Regions. You can read more here: `https://aws.amazon.com/blogs/architecture/overview-of-data-transfer-costs-for-common-architectures/`.

- **Storage fees**: S3 storage incurs charges (`https://aws.amazon.com/s3/pricing/`). You will at some point want to learn more about archiving data using Glacier and Glacier Deep Archive (`https://aws.amazon.com/s3/storage-classes/glacier/`). These archiving tiers cost significantly less money than regular live storage. **Lifecycle rules** are used to define policies for when data is archived (`https://docs.aws.amazon.com/AmazonS3/latest/userguide/object-lifecycle-mgmt.html`). Keep in mind that there is also an extra charge for unarchiving items too early.

- **Compute fees**: Computational tasks, instances, lambdas, and other services run on AWS will also incur charges (`https://aws.amazon.com/ec2/pricing/on-demand/`).

There are many other services and capabilities within AWS, and we will only touch on a small fraction of them here. One service to know about in the field of bioinformatics is **AWS HealthOmics** (`https://aws.amazon.com/healthomics/`). This service is specifically for storing and working with genomics and health care data. It provides metadata capabilities for tracking data and offers a unique storage model that charges by the base pair instead of the byte. It includes genomic data compression capabilities, Ready2Run workflows for common bioinformatics pipelines, and HIPAA compliance for working with patient data.

See also

- You can find an AWS tutorial here: `https://aws.amazon.com/getting-started/`

- For more background on S3, check out `https://www.geeksforgeeks.org/introduction-to-aws-simple-storage-service-aws-s3/`

- To develop your AWS skills further, check out these Udemy courses:

  - AWS for Beginners: `https://www.udemy.com/course/aws-for-beginners-the-ultimate-foundational-course/`

  - AWS Certified Developer: `https://www.udemy.com/course/aws-certified-developer-associate-dva-c01/`

  - AWS Certified Solutions Architect: `https://www.udemy.com/course/aws-solutions-architect-professional/`

- Find an AWS tutorial video here: `https://www.youtube.com/watch?v=UmQnenLf_Cs`

Building cloud components using boto3

As we saw in the *Setting up an AWS account* recipe, it is possible to do many things in AWS through the Management Console. You could build your own systems of cloud components, servers, and databases, all through a point-and-click interface.

However, today's cloud practitioners prefer to build systems through scripts that can be tracked, checked in, and reproduced. You can use a library from AWS called `boto3` to programmatically create, manipulate, and destroy resources – `https://boto3.amazonaws.com/v1/documentation/api/latest/index.html`. You can also define templates that specify components and their parameters. This is known as **Infrastructure as Code** (**IaC**).

By using IaC, we can make sure that our components are made in a way that is documented. We can then check our component definitions into a code repository such as GitHub and take advantage of version control. We can also make the generation of cloud components automated. In AWS, we use a service called **CloudFormation** to provision resources using IaC (`https://aws.amazon.com/cloudformation/`).

In this recipe, we will learn how to build AWS components using the `boto3` library and take a quick look at a simple CloudFormation template.

For this recipe, we'll begin working in the `Ch14-2-boto3.ipynb` notebook.

Here is an overview of some of the many AWS components you can build programmatically and their main uses.

The following are the types of AWS components:

Component	Description	Usage
Aurora	Cloud database	Powerful, highly scalable database with serverless option
AWS Batch	Virtual cluster	Launch large arrays of jobs
Bedrock	Generative AI	Build LLM applications
DynamoDB	NoSQL database	Document (non-relational) database
EC2	Elastic compute	Virtual compute servers
ECS	Stands for Elastic Container Service	Deploy, manage, and scale containerized applications
EKS	Stands for Elastic Kubernetes Service	Deploy containerized applications

Component	Description	Usage
EventBridge	Event handling	Build event-driven architectures
Fargate	Serverless compute	Deploy scalable systems
IAM	Stands for Identity Access Management	Control security and permissions
Lambda	Serverless code	Run small serverless jobs
RDS	Relational database servers	Run Postregs, MySQL, SQL Server, and other popular databases
Route 53	DNS service	Register domain names and route traffic to your services
S3	Stands for Simple Storage Service	File storage and backups
SageMaker	Machine learning	Train and deploy machine learning models
State machines	Execute step functions	Build orchestration systems
VPC	Stands for Virtual Private Cloud	Isolate your own virtual network

Table 14.2 – Major AWS components and services

With EC2, you can access many different types of instances, including the latest GPUs. You can run powerful workflows such as NVIDIA Parabricks on these GPU instances to greatly accelerate your bioinformatics workflows (`https://www.nvidia.com/en-us/clara/genomics/`). You can also use **Field-Programmable Gate Array** (**FPGA**) instances to power tools such as DRAGEN.

You can set up **serverless applications** in AWS. This amazing innovation allows you to spin up resources when needed, without even creating or destroying them. These tools, such as **lambdas**, allow you to build highly scalable and very efficient applications – `https://aws.amazon.com/pm/lambda/`. Many popular services, such as the **Aurora** database, offer both `fully-fledged server` resources and serverless versions.

Event-triggering services such as **EventBridge** are often combined with lambdas and other AWS components to build entire systems that respond to events, orchestrate computations, and store results.

In this recipe, we'll learn how to create some basic cloud components programmatically. This will give you a good introduction and prepare you to take things even further by creating multiple components and even entire cloud systems using what you learn here.

Getting ready

First, we need to install the AWS client, often referred to as the AWS CLI – `https://aws.amazon.com/cli/`:

```
! pip install awscli
```

Next, we will install `boto3`:

```
! pip install boto3
```

To let our client easily access our AWS account, we want to set up a security key.

To get the **AWS access key ID**, go to your AWS console and look in the upper right for your Account Name. Click on this and then choose **Security Credentials**. Then, click on **Create Access Key**. Now, copy the ID and the key to your notepad.

Go into the terminal of your local machine and type the following:

```
aws configure
```

You will be prompted to enter the access ID and key you copied in the preceding step . Choose your default Region (eg. us-west-2). Set the default output format to JSON. Recall that you can see your default Region in the upper right of the AWS Management Console when you first log in; see *Figure 14.1*.

You will now be configured to connect automatically to your account using the client. Here are a few commands you can try to test it out:

```
aws help
```

This will give you an overview of the commands available.

For each service in AWS, there are also a series of subcommands in the form of the following:

```
aws [service] help
```

For example, to list S3 buckets, you would use the following:

```
aws s3 ls
```

You can read through the documentation and familiarize yourself with the various AWS CLI commands.

How to do it...

Here are the steps to try this recipe:

1. First, let's build an S3 bucket using `boto3`. We will import the libraries we need:

    ```
    import boto3
    from botocore.exceptions import ClientError
    ```

The botocore exceptions module will help us handle any AWS service errors (https://boto3.amazonaws.com/v1/documentation/api/latest/guide/error-handling.html).

2. Now, let's define a function to create an S3 bucket using boto3:

```python
def create_bucket(bucket_name, region=None):
    try:
        s3_client = boto3.client(
            's3', region_name=region)
        if region is None:
            s3_client.create_bucket(
                Bucket=bucket_name)
        else:
            location = {'LocationConstraint': region}
            s3_client.create_bucket(
                Bucket=bucket_name,
                CreateBucketConfiguration=location
            )
        print(
            f"Bucket '{bucket_name}' created successfully."
        )

    except ClientError as e:
        print(f"Error: {e}")
        return False
    return True
```

This function will take the name of the bucket to create and the AWS Region as parameters. We wrap our bucket creation code in a try..except block to catch the ClientError exception object. We then create an s3_client object using the boto3.client() method - this will return an instance of a low-level service client class, using our current AWS session. If the Region is not set, we will just create the S3 bucket in the default Region. For buckets in US-East-1, you are not supposed to set a location constraint. For any other Region, we will create a location dictionary.

3. We will then use the create_bucket() method of our s3_client to instantiate the bucket. We will also print out the bucket name if created successfully.

OK, great! Let's build a function to enable **versioning** on our bucket:

```python
def enable_versioning(bucket_name):
    s3 = boto3.client('s3')
    versioning_config = { 'Status': 'Enabled' }
    s3.put_bucket_versioning(
        Bucket=bucket_name,
        VersioningConfiguration=versioning_config
```

```
    )
    print(
        f"Versioning enabled on bucket "
        f"'{bucket_name}'."
    )
```

Versioning is a powerful tool for ensuring the safety of data (`https://docs.aws.amazon.com/AmazonS3/latest/userguide/Versioning.html`).When this is enabled, every version of an object is saved. So, when you delete an object, it is just *marked* as deleted and not really removed. Likewise, a new version does not truly overwrite the current version but is simply marked as *latest*. You need to be careful, though, as versioning will cost you extra money, and you must delete multiple versions to truly erase a file.

4. Now, we will define a function to upload a file to our S3 bucket:

```
def upload_sample_file(
    bucket_name, file_name, object_name=None
):
    if object_name is None:
        object_name = file_name
    s3 = boto3.client('s3')
    try:
        s3.upload_file(
            file_name, bucket_name, object_name)

        print(
            f"File '{file_name}' uploaded to "
            f"'{bucket_name}/{object_name}'.")
    except ClientError as e:
        print(f"Upload error: {e}")
```

This code will use the boto3 `upload_file()` method to copy a given file into your bucket. If the object name is provided, it will use that as the name we end up with in S3 – otherwise, it will use the provided filename.

We are ready to make our bucket!

```
bucket = "sab-bucket-4"
region = "us-west-2"
if create_bucket(bucket, region):
    enable_versioning(bucket)
    with open(
        "sample.txt", "w"
    ) as f:
        f.write("This is a sample file.")
        upload_sample_file(bucket, "sample.txt")
```

Our main code here first sets a bucket name. You can replace this with any name you want, but remember, it must be *globally unique*. This means the name must be unique across your entire AWS account; you cannot have two buckets with the same name, even if they are in two different Regions. In fact, bucket names must be globally unique across ALL AWS accounts, see: `https://docs.aws.amazon.com/AmazonS3/latest/userguide/bucketnamingrules.html`

5. Next, the preceding code will set the Region and create the bucket with our function. Then, we enable versioning and upload our sample file. You can make any simple text file to test the upload; alternatively, we have provided a `sample.txt` file in the GitHub repo.

OK, you have now created the bucket! You can check this by going into your AWS Management Console and looking in S3 (make sure you are in the right Region). Alternatively, you can check from the terminal like this:

```
aws s3 ls
```

This will list out all your S3 buckets. You can also list the contents of the bucket and see the file like this:

```
aws s3 ls s3://sab-bucket-4
```

OK, let's clean up our resources now:

```
def empty_bucket(bucket_name):
    s3 = boto3.resource('s3')
    bucket = s3.Bucket(bucket_name)
    try:
        bucket.object_versions.delete()
        print(
            f"All objects and versions deleted from "
            f"'{bucket_name}'.")
    except ClientError as e:
        print(f"Error emptying bucket: {e}")
        return False
    return True
```

This function will use the `object_versions.delete()` method to empty out the bucket of any files first. The S3 bucket needs to be empty of any files before we can remove it:

```
def delete_bucket(bucket_name):
s3 = boto3.client('s3')
    try:
        s3.delete_bucket(Bucket=bucket_name)
        print(f"Bucket '{bucket_name}' deleted successfully.")
    except ClientError as e:
        print(f"Error deleting bucket: {e}")
        return False
    return True
```

This function deletes the bucket using the boto3 `delete_bucket()` method.

Now, we will run the code:

```
bucket_to_delete = "sab-bucket-4"
if empty_bucket(bucket_to_delete):
    delete_bucket(bucket_to_delete)
```

Remember to replace the `bucket_to_delete` variable in the preceding code with your bucket name.

Great! We have now cleaned up our resources. You can double-check this in the console or by running `aws ls` again.

AI tip

We can use `boto3` to manipulate many other components in AWS. Let's look at an example of using it to deal with EC2 instances.

Prepare: Type `Review this code:` and then paste in your code for this recipe.

Prompt: Now write an example to create and then remove an EC2 instance using `boto3`, similar to the preceding code.

You should see: Code to set up an EC2 instance, launch it, and then terminate the instance and clean it up.

Let's now look at an example of using CloudFormation to deploy an S3 bucket using IaC.

AI tip

Prepare: Enter the prompt, "Write code to create a simple CloudFormation template to deploy an S3 bucket."

You should see: A template (`.yaml`) file for the infrastructure, and a sample command to deploy it.

The preceding AI prompt will write for you a simple CloudFormation template file for an S3 bucket, in the form of a `.yaml` file. Use a text editor to make a file for this called `s3-bucket.yaml`. It will look like this:

```
AWSTemplateFormatVersion: '2010-09-09'
Description: CloudFormation template to create an S3 bucket

Resources:
  MyS3Bucket:
    Type: AWS::S3::Bucket
    Properties:
      BucketName: my-simple-cf-bucket-123456  # Must be globally unique

Outputs:
  BucketName:
    Description: The name of the created S3 bucket
    Value: !Ref MyS3Bucket
```

Figure 14.6 – YAML file for S3 bucket

Now use the command provided in the terminal:

```
aws cloudformation create-stack --stack-name my-s3-stack
--template-body file://s3-bucket.yaml --capabilities CAPABILITY_
NAMED_IAM
```

This uses the AWS CLI to deploy the CloudFormation stack. Run the command and then go to the AWS Management Console, into CloudFormation, and you will see the stack created (make sure you have the right Region selected):

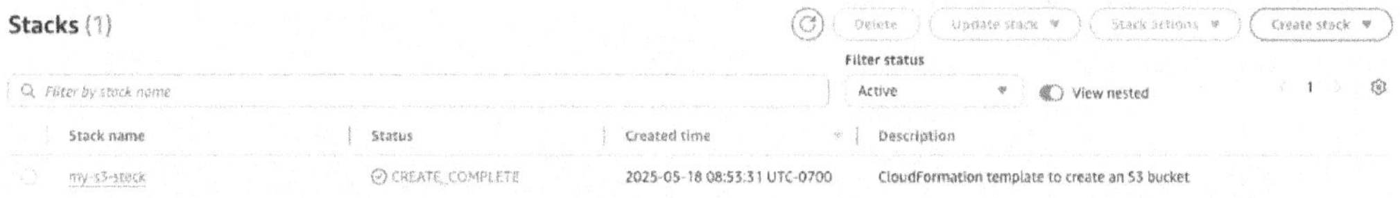

Figure 14.7 – The CloudFormation stack

Now, click on the stack name and click on the **Resources** tab. You will see your S3 bucket:

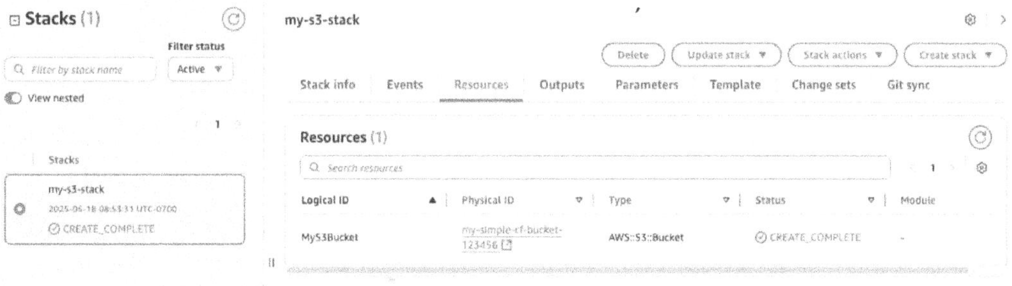

Figure 14.8 – The S3 bucket resource within the CloudFormation stack

Great! You have now deployed an S3 bucket using IaC.

Let's clean up our resources:

```
aws cloudformation delete-stack --stack-name my-s3-stack
```

This will remove the stack and the S3 bucket. Note that this will work because the S3 bucket is already empty, as we just created it (normally, S3 buckets have to be empty before they can be removed).

We have now learned how to set up, inspect, and remove components from AWS. This enables you to programmatically set up and control entire cloud infrastructures.

There's more...

There are other popular IaC systems. One is Terraform – `https://developer.hashicorp.com/terraform`. Unlike CloudFormation, which is specific to AWS, Terraform works with all the major cloud providers. Other related tools include Azure Resource Manager (**ARM**) for Microsoft's platform (`https://azure.microsoft.com/en-us/get-started/azure-portal/resource-manager`) and Google Cloud Deployment Manager for GCP (`https://docs.cloud.google.com/deployment-manager/docs/`). Another popular tool is Pulumi (`https://www.pulumi.com/`), which lets you author using many programming languages.

You should take the time to learn to secure your AWS account and the components within it. In particular, it is useful to learn more about **EC2 security groups**. Security is a big topic and beyond the scope of this book, but here are a few key resources:

- **EC2 security groups**: `https://docs.aws.amazon.com/AWSEC2/latest/UserGuide/ec2-security-groups.html`

- **Free Tier security**: `https://aws.amazon.com/free/security`

- **Udemy course**: `https://www.udemy.com/course/aws-certified-security-specialty/`

See also

- Udemy CloudFormation masterclass: `https://www.udemy.com/course/aws-cloudformation-master-class/`

- **Udemy Terraform course**: `https://www.udemy.com/course/mastering-terraform-beginner-to-expert/`

Building a container in AWS ECR

Containers are compartmentalized systems for storing code and associated dependencies. With a container, we can set up just the bioinformatics code and libraries we need to run a particular module. This allows us to make our pipelines very modular and reproducible. They are lightweight because they do not contain an entire OS and can be spun up quickly. This happens because they share the underlying resources of the machine they run on, as compared to virtual machines, which contain an entire OS of their own. They are highly portable and help to eliminate system-specific issues. Once built, a container does not change. In this way, you can keep track of different versions of a module by making multiple containers, tagging them with the version, and then storing them in a **container repository**.

Using containers, you can build a bioinformatics pipeline or workflow system, which stores different pipeline definitions as sets of versioned containers. This allows you to create a highly reproducible system and check changes from one version of the pipeline to the next. Comparing a version to a previous version is called a regression. A **regression** is the process of comparing two versions and their outputs carefully to see whether anything has changed. Some changes might be considered minor, such as a small change in the value of a metric. Other changes could be major, such as an alteration in a patient variant call. We often speak of **scientific changes** versus **technical changes** when doing regressions. A technical, or engineering, change is a modification to how a system is implemented, intended to speed it up, improve data tracking, or do anything else that is expected not to change the outputs. For these types of changes, we expect no differences in the regression. Scientific changes occur when we are actually trying to update the output to make it better, and so we expect changes in the outputs – these are carefully reviewed by a computational biologist.

Popular container platforms such as Docker help you store, manage, and manipulate containers (`https://docs.docker.com/`). Containers are run on **orchestration platforms** such as Kubernetes, ECS, or OpenShift.

In this recipe, we will build a simple Docker container and push it to the AWS repository, known as Elastic Container Registry, or **ECR** – `https://aws.amazon.com/ecr/`.

Getting ready

First, we need to install Docker Desktop. Follow the instructions here to install it on your machine – `https://docs.docker.com/desktop/setup/install/mac-install/`.

Also, make sure your Docker daemon is running – `https://docs.docker.com/engine/daemon/start/`.

How to do it...

Here are the steps to try this recipe:

1. First, we will import our libraries:

    ```
    import os
    import boto3
    import subprocess
    from botocore.exceptions import ClientError
    import logging
    ```

    We will use the `boto3` library for interaction with AWS.

2. Next, we will configure logging:

```
logging.basicConfig(
    level=logging.INFO,
    format='%(asctime)s - %(levelname)s - %(message)s'
)
logger = logging.getLogger(__name__)
```

3. Now, let's set some basic AWS parameters:

```
AWS_REGION = 'us-east-1'
REPOSITORY_NAME = 'my-simple-app'
IMAGE_TAG = 'latest'
```

This will set our Region and give a name for our repository. We also *tag* the image with a version. In this case, we will set it to latest.

4. OK, now let's define a function to create a simple Docker file:

```
def create_dockerfile():
    logger.info(
        "Creating Dockerfile and application files...")
    os.makedirs('app', exist_ok=True)
    with open('app/app.py', 'w') as f:
        f.write(
            '''
            from flask import Flask
            app = Flask(__name__)
            @app.route('/')
            def hello():
                return "Hello from the Docker container!"
            if __name__ == "__main__":
                app.run(host='0.0.0.0', port=5000)
            '''
        )
    with open('app/requirements.txt', 'w') as f:
        f.write('flask==2.2.3\n')
    with open('Dockerfile', 'w') as f:
        f.write(
            '''
            FROM python:3.9-slim
            WORKDIR /app
            COPY app/requirements.txt .
            RUN pip install --no-cache-dir -r requirements.txt
            COPY app/ .
            EXPOSE 5000
            CMD ["python", "app.py"]
```

```
        ' ' '
    )
logger.info(
    "Dockerfile and application files created successfully")
```

This code creates an app subdirectory for our simple application. It will be a very simple **Flask** application. Flask is a lightweight framework for running web applications on AWS. We will write out a simple Python script to give a hello message from the Flask application, as well as a requirements.txt file, which provides the libraries needed for the application (in this case, it is just Flask). Note the use of the triple-quotes (' ' '), which allow for multiple lines of text to be printed out.

Here is what our app.py file will look like:

```
(base) shanebrubaker@SHANEs-MacBook-Pro app % cat app.py

from flask import Flask
app = Flask(__name__)

@app.route('/')
def hello():
    return "Hello from the Docker container!"

if __name__ == "__main__":
    app.run(host='0.0.0.0', port=5000)
(base) shanebrubaker@SHANEs-MacBook-Pro app %
```

Figure 14.9 – The app.py file

The requirements.txt file will look like this:

```
(base) shanebrubaker@SHANEs-MacBook-Pro app % cat requirements.txt
flask==2.2.3
(base) shanebrubaker@SHANEs-MacBook-Pro app %
```

Figure 14.10 – The requirements.txt file

A requirements.txt file is used to list all the libraries and dependencies for a Python program. This one simply installs flask 2.2.3.

5. OK, now we will define a function to build the Docker image:

```
def build_docker_image():
    image_name = f"{REPOSITORY_NAME}:{IMAGE_TAG}"
    logger.info(f"Building Docker image: {image_name}")

    try:
        subprocess.run(
            ["docker", "build", "-t", image_name, "."],
            check=True,
            stdout=subprocess.PIPE,
            text=True
        )
```

```
        logger.info("Docker image built successfully")
        return image_name
    except subprocess.CalledProcessError as e:
        logger.error(f"Failed to build Docker image: {e}")
        raise
```

We create an image name of the form `repository name : tag`. We then run `docker build` with our image as the target.

This function will create the ECR repository:

```
def create_ecr_repository(ecr_client):
    try:
        logger.info(
            f"Checking if repository "
            f"{REPOSITORY_NAME} exists...")
        ecr_client.describe_repositories(
            repositoryNames=[REPOSITORY_NAME])
        logger.info(
            f"Repository {REPOSITORY_NAME} already exists")

    except ecr_client.exceptions.RepositoryNotFoundException:
        logger.info(
            f"Creating repository {REPOSITORY_NAME}...")
        response = ecr_client.create_repository(
            repositoryName=REPOSITORY_NAME,
            imageScanningConfiguration={'scanOnPush': True},
            encryptionConfiguration={'encryptionType':'AES256'}
        )
        logger.info(
            f"Repository created: "
            f"{response['repository']['repositoryUri']}")
```

The preceding function first checks to see whether our repository already exists using the `describe_repositories()` method. If it needs to be created, we make it using `create_repository()`. We provide the repository name, and we will scan the image for security vulnerabilities when we push it to the repository. We also set encryption for the repository.

6. Next up is a function to get our ECR login:

```
def get_ecr_login_command(ecr_client):
    response = ecr_client.get_authorization_token()

    auth_data = response['authorizationData'][0]
    token = auth_data['authorizationToken']
    endpoint = auth_data['proxyEndpoint']

    return token, endpoint
```

This function gets the authentication information needed for ECR using get_authorization_token().

7. Next, we will write a function to push our image to ECR:

```python
def tag_and_push_image(local_image_name, repository_uri):
    logger.info(f"Tagging image as {repository_uri}")
    try:
        subprocess.run(
            ["docker", "tag",
              local_image_name, repository_uri],
            check=True,
            stdout=subprocess.PIPE,
            text=True
        )
        logger.info("Image tagged successfully")
    except subprocess.CalledProcessError as e:
        logger.error(f"Failed to tag image: {e}")
        raise

    try:
        logger.info(f"Pushing image to {repository_uri}")
        subprocess.run(
            ["docker", "push", repository_uri],
            check=True,
            stdout=subprocess.PIPE,
            text=True
        )
        logger.info("Image pushed successfully")
    except subprocess.CalledProcessError as e:
        logger.error(f"Failed to push image: {e}")
        raise
```

The preceding function will make Docker command calls to tag the image and push it to the repository.

Here is a function to log in to ECR:

```python
def login_to_ecr(region, endpoint):
    try:
        print(f"Logging into ECR at {endpoint}...")
        subprocess.run(
            f"aws ecr get-login-password "
            f"--region {region} | docker login "
            f"--username AWS --password-stdin "
            f"{endpoint.replace('https://', '')}",
```

```
        shell=True,
        check=True
    )
    print("Logged in to ECR successfully.")
except subprocess.CalledProcessError as e:
    print(f"Failed to log in to ECR: {e}")
    raise
```

This function authenticates Docker with ECR using the AWS CLI, so we can push the image.

With that, we now have all our functions defined!

8. Now, we can run our code:

```
def main():
    try:
        create_dockerfile()
        local_image_name = build_docker_image()

        logger.info(
            f"Connecting to AWS ECR in region {AWS_REGION}..."
        )
        ecr_client = boto3.client(
            'ecr', region_name=AWS_REGION)
        create_ecr_repository(ecr_client)
        response = ecr_client.describe_repositories(
            repositoryNames=[REPOSITORY_NAME])
        repository_uri = response[
            'repositories'][0]['repositoryUri']
        repository_uri_with_tag = (
            f"{repository_uri}:{IMAGE_TAG}")
        token, endpoint = get_ecr_login_command(
            ecr_client)
        logger.info("Logging in to ECR...")
        login_to_ecr(
            AWS_REGION, endpoint.replace("https://", ""))
        tag_and_push_image(
            local_image_name, repository_uri_with_tag)

        logger.info(
            f"Successfully built and pushed Docker "
            f"image to: {repository_uri_with_tag}")
        logger.info("You can pull this image using:")
        logger.info(
            f"docker pull {repository_uri_with_tag}")
```

```
        return repository_uri_with_tag

    except Exception as e:
        logger.error(f"An error occurred: {e}")
        raise

if __name__ == "__main__":
main()
```

This function creates our Docker image and builds it. It then uses `boto3` to establish an ECR client and create the repository. We retrieve the repository **Uniform Resource Identifier** (**URI**), which uniquely identifies the ECR repository.

9. Next, we retrieve our ECR login credentials. We log in to ECR and then tag and push the image.

That's it! You have now built a simple container and pushed it to ECR.

Next, we will look at our container in Docker Desktop, which we covered in *Chapter 1, Computer Specifications and Python Setup* in the *Installing the required software with Docker* recipe. You can learn more about Docker Desktop here: `https://www.docker.com/blog/getting-started-with-docker-desktop/`.

Here is what our container looks like in Docker Desktop (the second and third items in the following screenshot):

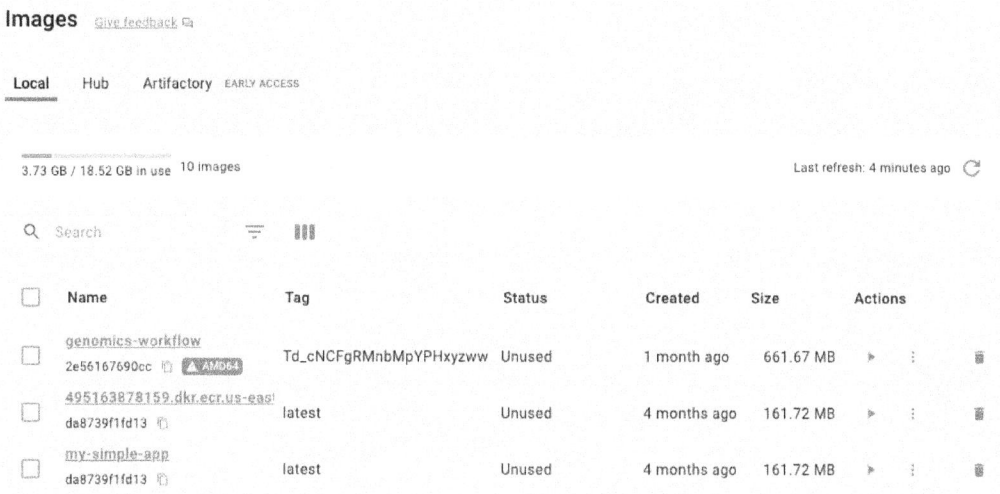

Images Give feedback

| Local | Hub | Artifactory EARLY ACCESS |

3.73 GB / 18.52 GB in use   10 images   Last refresh: 4 minutes ago

	Name	Tag	Status	Created	Size	Actions
	genomics-workflow 2e56167690cc ⬛ AMD64	Td_cNCFgRMnbMpYPHxyzww	Unused	1 month ago	661.67 MB	▶ ⋮ 🗑
	495163878159.dkr.ecr.us-eas da8739f1fd13	latest	Unused	4 months ago	161.72 MB	▶ ⋮ 🗑
	my-simple-app da8739f1fd13	latest	Unused	4 months ago	161.72 MB	▶ ⋮ 🗑

Figure 14.11 – Docker Desktop shows the local and ECR versions of the container

Here is what your repository looks like in ECR:

Private repositories (1)

Q	Search by repository substring

	Repository name ▲	URI
○	my-simple-app	⎘ 495163878159.dkr.ecr.us-east-1.amazonaws.com/my-simple-app

Figure 14.12 – The repository in ECR

Finally, let's clean up our components. Review the code in the notebook for the `delete_ecr_image()` and `delete_ecr_repository()` functions. Once you run this code, you will have cleaned up your image and the ECR repository. You can double-check this in the AWS console interface by setting it to the Region used here (US-East-1) and searching for `Elastic Container Registry`.

Once you have your Docker container in ECR, you may want to install it and run it on an EC2 instance – `https://plainenglish.io/blog/how-to-deploy-a-docker-image-to-amazon-ecr-and-run-it-on-amazon-ec2-3a8445`.

You should also learn about the `docker exec` command and how to run code from within your Docker container – `https://docs.docker.com/reference/cli/docker/container/exec/`.

There's more...

Singularity containers are another popular tool used in bioinformatics. They are especially useful for their reproducibility and security model (`https://docs.sylabs.io/guides/3.5/user-guide/introduction.html`).

BioContainers is an amazing repository of standard containers used in Nextflow, Galaxy, and other workflow management systems (`https://github.com/biocontainers`).

We have seen how to use cloud resources to build systems and containerize applications. In *Chapter 15, Workflow Systems*, we'll take a look at how to bring these pieces together to build complex, accurate bioinformatics workflows for analyzing and interpreting data.

OK, let's clean up and close down our `conda` environment:

```
conda deactivate
```

See also

- The Celeste system from the *All of Us* program provides an excellent example of a containerized infrastructure that includes the use of AWS Batch and other major AWS components, as well as DRAGEN – Siddiqui et al., *Celeste: a cloud-based genomics infrastructure with variant-calling pipeline suited for population-scale sequencing projects*, MedRxiv, May 2025 – `https://www.medrxiv.org/content/10.1101/2025.04.29.25326690v1`

- CREDO is a tool for creating reproducible Docker containers for bioinformatics – Alessandri et al., *CREDO: a friendly Customizable, REproducible DOcker file generator for bioinformatics applications*, BMC Bioinformatics, March 2024 – `https://link.springer.com/article/10.1186/s12859-024-05695-9`

- Learn more about containers in Kadri et al, *Containers in Bioinformatics: Applications, Practical Considerations, and Best Practices in Molecular Pathology*, The Journal of Molecular Diagnostics, May 2022 – `https://www.sciencedirect.com/science/article/pii/S1525157822000381`

- Read about how Singularity containers can be used to improve reproducibility in Mitra-Behura et al., *Singularity Containers Improve Reproducibility and Ease of Use in Computational Image Analysis Workflows*, Frontiers in Bioinformatics, January 2022 – `https://www.frontiersin.org/journals/bioinformatics/articles/10.3389/fbinf.2021.757291/full`

Get This Book's PDF Version and Exclusive Extras

UNLOCK NOW

Scan the QR code (or go to `packtpub.com/unlock`). Search for this book by name, confirm the edition, and then follow the steps on the page.

Note: Keep your invoice handy. Purchases made directly from Packt don't require an invoice.

15

Workflow Systems

One of the core technologies used in bioinformatics is **pipelines**, also known as **workflow systems**. Almost all bioinformatics tasks require more than one step. In many cases, certain steps must also be parallelized, and their results gathered together. These sets of ordered tasks are referred to as bioinformatics pipelines or workflows. In bioinformatics, you can find three main types of workflow system:

- Frameworks such as Galaxy (`https://usegalaxy.org`) or DNAnexus (`https://www.dnanexus.com/`), which are geared toward end users.

- Programmatic workflows, which are geared toward code interfaces that, while generic, often originate from the bioinformatics or machine learning space. Two examples are Snakemake (`https://snakemake.readthedocs.io/`) and Nextflow (`https://www.nextflow.io/`).

- Totally generic workflow systems such as Apache Airflow (`https://airflow.incubator.apache.org/`), which take a less domain-specific approach to workflow management. These are especially well suited to data management tasks, often called **Extract-Transform-Load** (ETL). We won't cover Airflow directly in this chapter but will provide some references for comparison.

In this chapter, we will cover the following recipes:

- Introducing Galaxy
- Writing a bioinformatics workflow with Snakemake
- Writing a bioinformatics workflow with Nextflow

Technical requirements

The code for this chapter can be found at `https://github.com/PacktPublishing/Bioinformatics-with-Python-Cookbook-Fourth-Edition/tree/main/Ch15`.

You will want to create a `Ch15` folder and set up your notebooks there. In this chapter, we will work within three sub-folders for the respective tools we'll be learning about: Galaxy, Snakemake, and Nextflow.

Remember to activate your `conda` environment before beginning the recipes, like this:

```
conda activate bioinformatics_base
```

Or, if you would like to set up a `conda` environment specific to this chapter, before activating `bioinformatics_base`, run the following:

```
conda create -n ch15-workflows --clone bioinformatics_base
conda activate ch15-workflows
```

You will be able to install the packages for the chapter as you go, or you can use the YAML file provided in the repository:

```
conda env update --file ch15-workflows.yml
```

Introducing Galaxy

In this recipe, we'll be learning about a popular public workflow system called Galaxy that can even be used by non-technical users. But before we dive into Galaxy, let's go over some basic concepts and terminology relating to workflow systems that we'll be using throughout this chapter.

A workflow is typically defined as a series of **tasks**, which are jobs that need to be run. When one task has to finish before another can begin, we call this a **dependency**. In many cases, a task needs to be parallelized. For example, we may want to break a FASTA file into many pieces, annotate each piece, and then combine the results. This type of process is typically called a **scatter-gather**.

We typically represent these dependencies in a workflow as a **Directed Acyclic Graph (DAG)**. Take a look at this illustrative DAG:

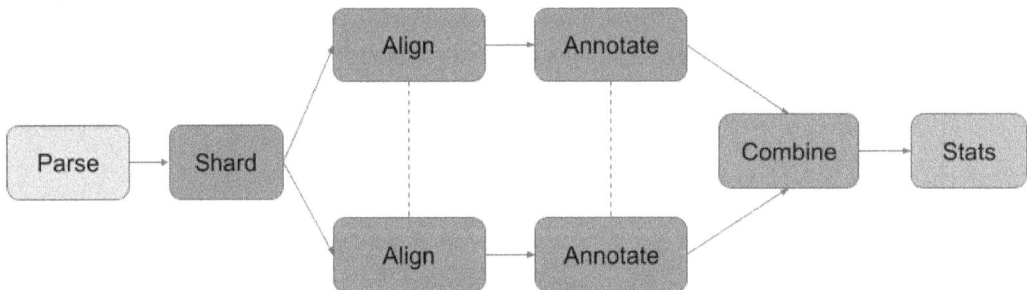

Figure 15.1 – Example of a directed acyclic graph in bioinformatics

Let's imagine that we want to parse a multi-FASTA file of gene sequences, align them to a reference genome, annotate each gene, and then combine and output the results. The DAG starts with a **parse** step, which opens and processes the FASTA file. We then **shard** the file into many pieces, creating one sub-file for each gene (sharding is a term for breaking up or shredding a file, database, or sequence into pieces). Note that the shard step has a dependency on the parse step, so it cannot begin until parsing is completed. We then take each gene file and run many parallel tasks on each one. We align each gene to the reference genome and then run an annotation program (for example, we could translate the gene and then run InterProScan on it). These types of parallel tasks are where the scalability of bioinformatics workflows really comes into play. In modern cloud computing environments, such as AWS Batch, thousands or even tens of thousands of these tasks can be run at once. These types of tasks are also often called **parallel tasks**, or **array jobs**. Next, we **combine** the results from each parallel task into a single file. Note that the shard, parallel tasks, and combine segments here are just another name for doing a scatter-gather. Finally, we run a **stats** module to summarize the output with some basic statistics. These steps taken together represent a fairly typical workflow in bioinformatics.

Workflow systems allow us to standardize the language in which pipelines are represented and make them more portable. One important feature that workflow systems assist us with is **reproducibility**. It is important that bioinformatics workflows are reproducible – given the same inputs and the same versions of all underlying software, they should produce identical results. A workflow system can be used to automatically run the same pipeline more than once and then systematically compare the outputs. This is called a **regression**. In a regression, we compare outputs of multiple runs of the software. If the results are the same, we say that we have "passed the regression." In many runs of the exact same workflow, we should normally get the same result. Most bioinformatics software is actually **deterministic**, meaning that you get the same result on the same inputs each time. However, if some of the software components in your workflow are **non-deterministic**, you won't even be able to get the same workflow to produce identical results on repeated runs! Some of the most common workflow languages are **Workflow Definition Language** (WDL), **Common Workflow Language** (CWL), and **Domain-Specific Language** (DSL). WDL is a popular workflow language that is popular on platforms such as DNAnexus. You can use a local runner called miniWDL (`https://github.com/chanzuckerberg/miniwdl`) to try it out. CWL (`https://www.commonwl.org/`) is another important open source standard. Finally, DSL is the latest workflow language from Nextflow. It contains many powerful tools that we will cover in the *Writing a bioinformatics workflow with Nextflow* recipe. In this recipe, we will discuss Galaxy, which is especially important for bioinformaticians supporting users who are less inclined to code their own solutions. While you may not be a typical user of these pipeline systems, you might still have to support them.

Galaxy is an open-source system that empowers non-computational users to do computational biology. It is the most widely used, user-friendly pipeline system available. Galaxy can be installed on any server by any user, but there are also plenty of other servers on the web with public access, the flagship being `http://usegalaxy.org`.

We will start by creating a free account at `http://usegalaxy.org` and playing around with it a bit. Reaching a level of understanding that includes knowledge of basic workflows is recommended.

Here are some tutorials on Galaxy that you can review:

- Galaxy Training: `https://training.galaxyproject.org/`

- Galaxy Tutorials for BioInformatics YouTube collection: `https://www.youtube.com/watch?v=zIEmgzOTOUk&list=PLe1-kjuYBZ07QXg-rokpjbcqYJrgvEIHE`

- *Galaxy Basics for genomics*: `https://training.galaxyproject.org/topics/introduction/tutorials/galaxy-intro-101/tutorial.html`

- Galaxy training materials: `https://github.com/galaxyproject/training-material`

In this recipe, we will learn about a couple of ways to access Galaxy:

- By using a Docker container
- By creating an account on a public server called `usegalaxy.org`

This will give you a general sense of how to play around with Galaxy and try it out using the tutorials available. We'll also use one or both of these options in the *Using the Galaxy API* bonus recipe. The bonus recipe can be found in the GitHub repository for this book under `Chapter 15` in: `galaxy/Ch15-1-bonus-using-galaxy-apis.pdf`.

Getting ready

In this recipe, we will carry out a local installation of a Galaxy server using Docker. As such, a local Docker installation is required. This is covered in *Chapter 1*, in the *Installing the required software with Docker* recipe.

You may also find it helpful to register an account with Docker here: `https://app.docker.com/signup`.

You should then test that you can log in to Docker from the terminal like this:

```
docker login
```

Before we go over using Galaxy within Docker, we will try it through the public website. Let's get you signed up for a free Galaxy account!

Go to `http://usegalaxy.org`; this is the page that will be displayed:

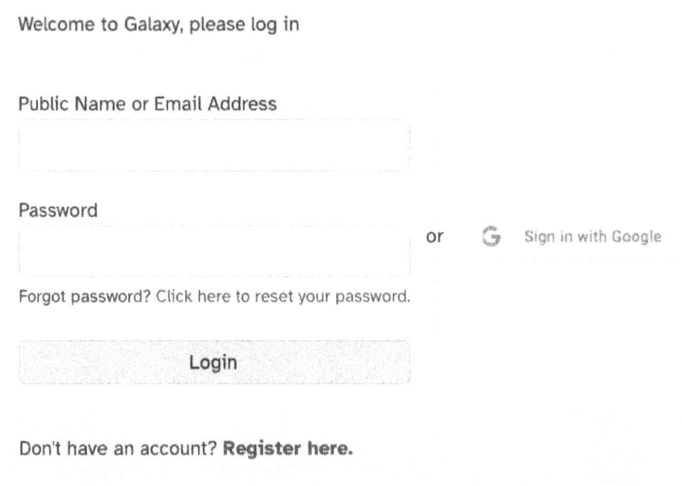

Welcome to Galaxy, please log in

Public Name or Email Address

Password

or   G   Sign in with Google

Forgot password? Click here to reset your password.

Login

Don't have an account? **Register here.**

Figure 15.2 – The Galaxy login and registration page

Click on `Register here` and then enter your email address and desired password.

Once that is done, you will be able to log in and use this public Galaxy server. You are encouraged to go through some of the provided tutorials at this point and familiarize yourself with Galaxy. You can find them here: `https://training.galaxyproject.org/`.

The code for this recipe can be found in `Ch15/galaxy/Ch15-1-introducing-galaxy. ipynb`.

How to do it...

OK, next let's pull our Galaxy Docker image and get it up and running!

Make sure you already have Docker Desktop installed. This is covered in *Chapter 1*, in the *Installing the required software with Docker* recipe.

Go ahead and make sure Docker Desktop is up and running as well. It should appear as an icon in the upper-right toolbar. You can make sure it is started by opening it. On a Mac, you can hit *Command + spacebar*, start typing `Docker`, and find **Application**.

From the terminal, run the following:

```
docker pull bgruening/galaxy-stable
```

This will retrieve the latest Docker image for Galaxy.

> **Note**
> Docker containers are transient with regard to disk space. This means that when you stop the container, all disk changes will be lost. This can be solved by mounting volumes from the host on Docker, as in the next step. All content in the mounted volumes will persist.

Now that we have our Galaxy Docker container, we are going to connect it to our filesystem and run it:

1. Let's create a directory to connect data on the container to our filesystem:

   ```
   mkdir /tmp/galaxy_data
   ```

   It is possible you will need to use the `sudo` command in front of this if you need admin permissions to make this folder on your system.

2. Now we need to run our Docker container:

   ```
   docker run -d -p 8080:80 --platform linux/amd64 -v /tmp/galaxy_
   data:/export --name galaxy bgruening/galaxy-stable
   ```

   This will start the container in "detached mode" (the `-d` option), meaning it will run in the background. It will map port `80` in the container to port `8080` on your machine, meaning that the URL for the container will be `http://localhost:8080`.

> **Note**
>
> In some cases, port `8080` on your system might be taken by another application. If you need to find out the port your container is running on, you can type the following (from the terminal):
>
> `docker ps`
>
> You can also take the container ID (which you can find in the Docker Desktop GUI or in the output of the `docker ps` command) and type the following:
>
> `docker port <container-id>`

Note that we also specify the `platform` option. This will force the container to run as if it were on a Linux AMD64 architecture, since that is what it is meant to run on, and you are most likely running this on macOS or a similar type of architecture.

We also specify a mountpoint for the data in the container (the `-v` option). As mentioned in the preceding info box, this means that persistent data from your Galaxy instance will be accessible from the `/tmp/galaxy_data` directory on your native operating system via the terminal.

Finally, we give the container an easy name to reference (the `--name` option) and provide the actual image to pull.

3. If you want to see your container running from the terminal, you can list Docker containers with the following:

```
docker ps -a
```

> **Troubleshooting tip**
>
> If you find that you need to get rid of the container and recreate it, you can remove the container like this: `docker rm galaxy`

Great! Now we can inspect our Docker container in Docker Desktop. Go into Docker Desktop (it should be in your toolbar, or on a Mac, you can hit *Command* + spacebar, `start typing Docker`, and find **Application**). You should see something like this:

Containers Give feedback

Container CPU usage   Container memory usage

No containers are running.   *No containers are running.*

	Name	Image	Status	CPU (%)	Port(s)
	galaxy b6ad072c3c69 AMD64	bgruening/galaxy-stable	Exited (143)	N/A	8080:80

Figure 15.3 – Galaxy container running in Docker Desktop

4. Note that if you want to remove the container later, you can go into Docker Desktop and remove it by using the trashcan icon on the right side. You can click on the container named **galaxy** to enter the container management interface:

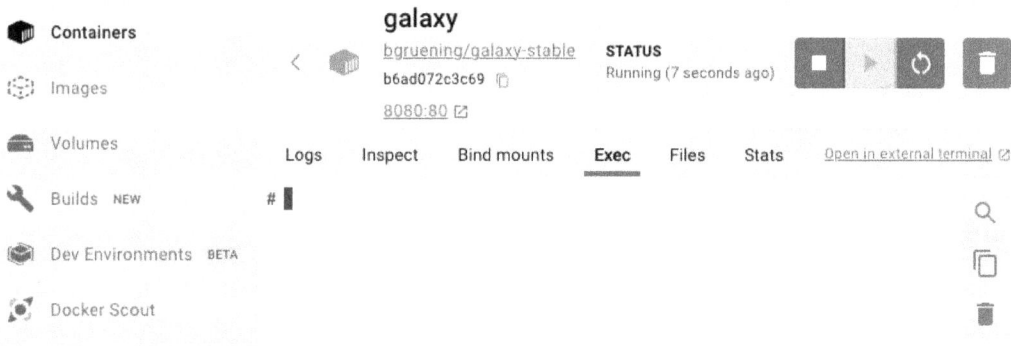

Figure 15.4 – Docker Desktop container management interface

From here, you can view logs of what is happening on the server under **Logs**, or see statistics on CPU, disk, and network performance under **Stats**. In the **Exec** window, you get a handy terminal (shown in *Figure 15.4*) that allows you to interact with the container (note that the container needs to be running to see the **Exec** menu. This is done by either using the docker run command that we executed earlier or hitting the **Play** button next to the container in the GUI).

OK, great! Now let's go to our Docker container. In a browser, go to http://localhost:8080. Here is what you should see:

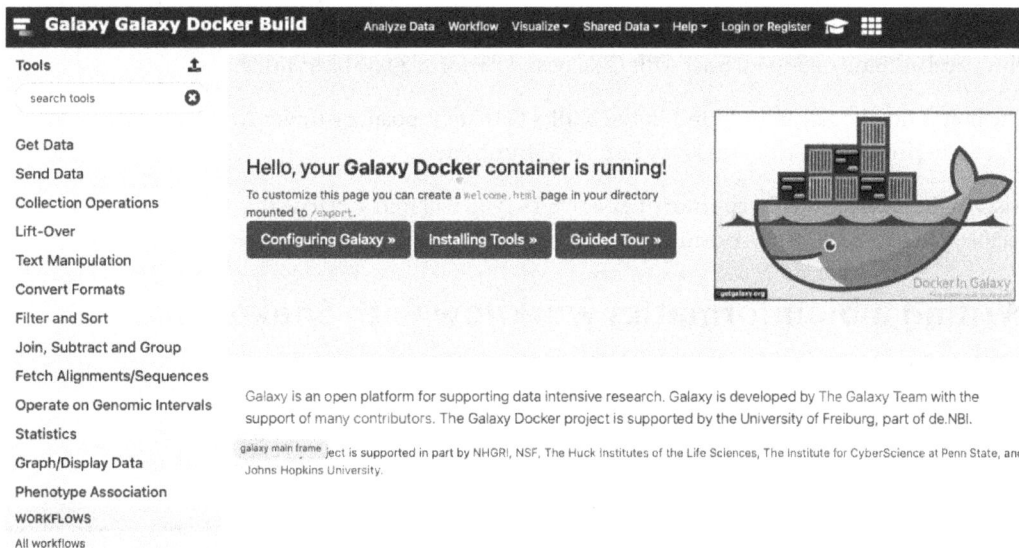

Figure 15.5 – Galaxy running in Docker

From here, you will see that you have the same functionality that you would expect on any public Galaxy server. You can upload data, install tools, and manipulate file formats.

> **Note**
>
> If you wanted to log in to this Docker image, you would use the following:
>
> **Username**: `admin`
>
> **Password**: `password`

There's more...

Galaxy is used for many purposes and by many communities. It is an excellent teaching tool for bioinformatics; see Hiltemann et al., *Galaxy Training: A powerful framework for teaching!*, PLOS Computational Biology, Jan 2023 – `https://journals.plos.org/ploscompbiol/article?id=10.1371/journal.pcbi.1010752`.

Galaxy is great for RNA-Seq analysis: Batut et al., *RNA-Seq Data Analysis in Galaxy*, RNA BioInformatics Protocol, Apr 2021 – `https://link.springer.com/protocol/10.1007/978-1-0716-1307-8_20`.

Other bioinformatics workflow sites you should check out include DNAnexus (`https://www.dnanexus.com/`), Terra (`https://app.terra.bio/`), Velsera (`https://velsera.com/`), Basepair (`https://www.basepairtech.com/`), Seqera (`https://seqera.io/`), and AWS HealthOmics (`https://aws.amazon.com/healthomics/`). Now that you have had an introduction to Galaxy, the next step is to learn how to interact with it programmatically! For this, we have provided a bonus recipe. In it, you will explore using the Galaxy REST API and a Python library called BioBlend to interact with Galaxy and perform a few basic tasks.

The bonus notebook can be found in the book's GitHub repository under `Ch15` in the `galaxy/Ch15-1-bonus-using-galaxy-apis.ipynb` file.

Also, in the book's GitHub repository under `Ch15`, you will find a PDF with a full explanation of the recipe: `galaxy/Ch15-1-bonus-using-galaxy-apis.pdf`.

Writing a bioinformatics workflow with Snakemake

Here, we will learn about a popular bioinformatics workflow management tool called Snakemake (`https://snakemake.readthedocs.io/en/stable/`).

Snakemake is implemented in Python and shares many traits with it. That being said, its fundamental inspiration is a Makefile, the framework used by the venerable "make" building system - `https://en.wikipedia.org/wiki/Make_(software)`. Snakemake has a few advantages over other workflow systems: first, it is **Pythonic** (easy to develop if you know Python); second, its Makefile-based approach is straightforward. What makes Snakemake unique is that it works in reverse order to interpret **rules** in Makefiles. Snakemake will not rerun a rule if the output file is already present unless you force it to do so (also called **computational reuse**). This is a big advantage if you are using **EC2 Spot Instances** in AWS, which are "spare" instances obtained at a fraction of the normal cost – if a job fails, it can restart from where it left off. Lastly, Snakemake is highly scalable and portable – you can develop on your laptop and then move to HPC or cloud environments easily.

In this recipe, we will develop a simple Snakemake pipeline that covers the outline of a typical bioinformatics workflow that would do quality control, alignment, and variant calling. We are only going to stub out mock examples for the alignment and variant calling steps, and we will fully implement the quality control step using FastQC. We'll also implement a cool interactive dashboard in the Jupyter notebook.

Getting ready

You will want to move to the `snakemake` subdirectory under your `Ch15` working directory.

For this exercise, let's use a `conda` environment provided by Snakemake:

```
conda deactivate
conda create -n snakemake -c conda-forge -c bioconda snakemake
conda activate snakemake
```

Make sure FastQC is installed in the Snakemake environment:

```
conda install -c bioconda fastqc
```

Check that FastQC is installed:

```
fastqc -h
```

> **Note**
> This example will fall back to generating a mock FastQC result if it cannot find FastQC installed.

> **Tip**
> You may get a notice that you need to update `conda` to use Snakemake, or you may see an error message like this: `Conda must be version 24.7.1 or later, found version 23.5.2`.
>
> If so, try updating your `conda` like this:
>
> **`conda update --name snakemake conda`**
>
> Or you can run the following:
>
> **`conda install conda=25.5.1`**
>
> After doing this, restart your terminal.

There are two main ways you can explore this recipe:

- **Running Snakemake from the notebook**: If you prefer working mostly in the notebook
- **Running Snakemake from the terminal**: The most standard way of running Snakemake

You will build the Snakefile and required components through the Jupyter notebook provided. At that point, you can jump out of the notebook and run Snakemake from the command line, which is the most typical way it is used. We also provide a handy dashboard in the notebook in case you want to run it from there.

The code for this recipe can be found in `Ch15/snakemake/Ch15-2-snakemake.ipynb`.

How to do it...

OK, let's get started on our Snakemake pipeline!

1. First, we will import our libraries. Review the *Import Libraries* section of the code in the notebook.

   We make use of several standard libraries, such as `os` for making system calls. We use `ipywidgets` to display the interactive dashboard.

2. Next is a function to install required packages. Review the `install_packages()` function in the notebook.

   This is a handy function to install the packages if needed. We've already installed these in the *Getting ready* section, so you probably will not need to run this function . But if you would like this function to execute, you will uncomment the following line in the notebook:

   ```
   # install_packages()
   ```

3. Next, we will enable our Jupyter widgets. Review Section 3 in the notebook, *Enable widgets in Jupyter*.

   This code checks to see whether we are in Jupyter with the `get_ipython()` function and then enables inline plotting so that we can use Matplotlib rendering inside notebook cells. This lets us make handy and visually appealing widgets.

4. Review Section 4 in the notebook, *Create directory structure and simulated FASTQ data*:

   I. We introduce a `FastqSimulator` class. This class will help us generate some simulated data for FastQC. It includes a `generate_sequence()` function to generate a random DNA sequence of a defined length.

   II. We then use `generate_quality()` to make quality lines and `generate_fastq_file()` to put it all together and make the FASTQ file.

5. Next, we will set up a project structure with directories for each area:

    I.   Review the `setup_project()` function. We will create directories for our raw and processed data, for results from different analyses, and for logs and configuration files.

    II.   We will also go ahead and run `FastqSimulator` to make a set of sample files.

6. Now we arrive at a central feature of the recipe, creating the Snakemake workflow:

    I.   Review the *Create Snakemake workflow* section of the notebook. It mostly consists of a large text definition for what our Snakefile will look like surrounded by triple-quotes. In just a bit, we will review the contents of the Snakefile.

7. Next, we are going to create some mock scripts for parts of the workflow that we don't want to fully implement yet. This will help you understand the process of building a larger and more complex workflow in Snakemake but without making the exercise so long that it becomes onerous.

    We create a mock FastQC script by creating a simple HTML output file with some basic statistics. Similarly, we make mock alignment and variant calling output files with faux content. We do the same for a MultiQC output and mock pipeline summary file.

8. Next, we are going to define our pipeline controller class:

    I.   Review the `BioinformaticsPipelineController` class in the notebook. When created, this class will run its `__init__()` function. This function will initialize basic parameters and then create the interactive dashboard. It does this using the `create_dashboard()` function. This function creates headers and buttons using the `widgets` library (`https://ipywidgets.readthedocs.io/en/stable/`).

    II.   We create start and stop buttons, a dropdown to select particular steps, and a progress bar. There will also be a window for output and results.

    III. We next **bind** the buttons to functions. In this way, each button is connected to one of the functions in our code and is activated by an action, in this case, the `on_click()` action.

    IV.   Next, we define an `update_status()` function, which we will use to update the status of the pipeline and set the progress bar to a value as it progresses. We also define a function for logging called `log_output()`.

    V.   We define `run_snakemake_command()` to put together the various parameters for running Snakemake. We define some functions for running the full pipeline or various individual steps in the pipeline. We also need some functions to stop the pipeline while it is running or clear out the existing results.

9. OK, let's run the interactive dashboard!

    Section 9 of the notebook, *Run the Interactive Dashboard*, prints out some header information and initializes our `BioinformaticsPipelineController()`. We then display our dashboard and a few more **Quick Action** buttons for convenience.

Running Snakemake from the notebook

You can skip this approach if you want and run Snakemake from the terminal in the upcoming section. If you want to skip running from the notebook and focus on running Snakemake from the terminal, proceed to *Reviewing the Snakemake workflow structure* in this recipe and then continue with the *Running Snakemake from the terminal* section of the chapter.

Here is our interactive dashboard:

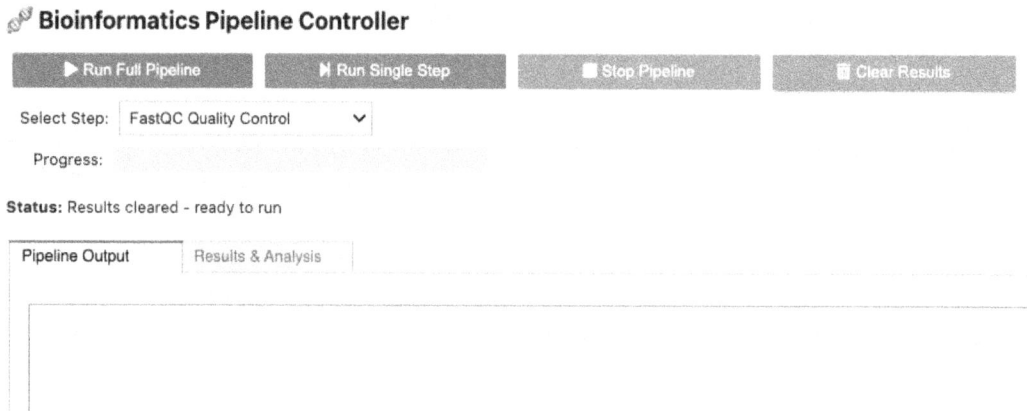

Figure 15.6 – Interactive dashboard for Snakemake

Choose **FastQC Quality Control** under the **Select Step** dropdown.

Now click **Run Single Step**.

You will see your pipeline running in the **Pipeline Output** window. You will see **Analysis complete for sample …** in the window and the progress bar should go to full completion, indicating that the run is complete. In a moment, we'll take a look at the result files that are generated.

Reviewing the Snakemake workflow structure

Let's go back over what happened here and review the components of the Snakemake configuration files and results directories. You will want to go into the terminal for this part.Go to your Ch15/snakemake working directory and take a look. Here is an overview of what you will see:

```
-rw-r--r--    1 shanebrubaker  staff    139 Jun  7  2025 config.yaml
drwxr-xr-x    5 shanebrubaker  staff    160 Jun  7  2025 envs
drwxr-xr-x    4 shanebrubaker  staff    128 Jun  7  2025 reference
drwxr-xr-x    6 shanebrubaker  staff    192 Jun  7  2025 bioinformatics_pipeline
drwxr-xr-x    6 shanebrubaker  staff    192 Jun  7  2025 fastq_pipeline
drwxr-xr-x   10 shanebrubaker  staff    320 Jun  7  2025 data
drwxr-xr-x    2 shanebrubaker  staff     64 Jun  7  2025 config
drwxr-xr-x    7 shanebrubaker  staff    224 Jun  7  2025 scripts
-rw-r--r--@   1 shanebrubaker  staff  14889 Jun 15 08:28 dag.pdf
drwxr-xr-x   17 shanebrubaker  staff    544 Aug 30 18:15 logs
-rw-r--r--    1 shanebrubaker  staff   3356 Sep 13 11:38 Snakefile
drwxr-xr-x    5 shanebrubaker  staff    160 Sep 13 11:38 results
-rw-r--r--    1 shanebrubaker  staff  70471 Dec  4 21:02 Ch15-2-snakemake.ipynb
```

Figure 15.7 – The Snakemake working directory

Take a look at the Snakefile:

```
less snakefile
```

Here is a portion of the contents:

```python
import os
from pathlib import Path

# Configuration
SAMPLES = ["sample1", "sample2", "sample3"]
DATA_DIR = "data"
RESULTS_DIR = "results"

# Target rule - what we want to produce
rule all:
    input:
        # FastQC reports (real)
        expand(f"{RESULTS_DIR}/fastqc/{{sample}}_R1_fastqc.html", sample=SAMPLES),
        expand(f"{RESULTS_DIR}/fastqc/{{sample}}_R2_fastqc.html", sample=SAMPLES),
        # Mock outputs
        expand(f"{RESULTS_DIR}/alignment/{{sample}}.bam", sample=SAMPLES),
        expand(f"{RESULTS_DIR}/variants/{{sample}}.vcf", sample=SAMPLES),
        f"{RESULTS_DIR}/multiqc_report.html",
        f"{RESULTS_DIR}/pipeline_summary.json"

# Real FastQC rule
rule fastqc:
    input:
        fastq=f"{DATA_DIR}/raw/{{sample}}_{{read}}.fastq.gz"
    output:
        html=f"{RESULTS_DIR}/fastqc/{{sample}}_{{read}}_fastqc.html",
        zip=f"{RESULTS_DIR}/fastqc/{{sample}}_{{read}}_fastqc.zip"
    params:
        outdir=f"{RESULTS_DIR}/fastqc"
    log:
        "logs/fastqc_{sample}_{read}.log"
    shell:
        """
        # Check if fastqc is available, if not use mock
        if command -v fastqc >/dev/null 2>&1; then
            fastqc {input.fastq} -o {params.outdir} --extract 2> {log}
        else
            echo "FastQC not found, creating mock output..." > {log}
            python scripts/mock_fastqc.py {input.fastq} {params.outdir} {wildcards.sample} {wildcards.read}
        fi
        """
```

Figure 15.8 – The Snakefile defines the Snakemake pipeline

The Snakefile is the core of the Snakemake system and defines the workflow. It contains the rules for carrying out tasks and specifies the dependencies between them. We start out by defining important configuration information and listing out our samples and key directories.

The first rule is the **all** rule, which defines the final deliverables for the pipeline. It uses the expand() helper function to take all the samples and specify an expected result for each sample in the results directory. Snakemake uses **backward chaining** to work backward from the desired outputs of the pipeline, recursively unpacking additional rules until it has built the job DAG needed to execute all required steps to get those outputs.

The next rule is for the FastQC results. It says that for each FASTQ input file, we expect to get an HTML and ZIP output file. It provides an output directory parameter. We also specify a location for log files. We can then see the shell command, which provides the actual runtime call. This command checks to see whether FastQC is available. If so, it runs a FastQC command with the input FASTQ files and output directory parameters and redirects STDERR to the logs directory. This last part, with 2>, means redirect STDERR to a location – note that 1> is STDOUT. STDOUT is for the primary output and STDERR is for error messages.

If FastQC is not available, we run our mock function script.

You can use the logs directory to help debug your pipeline.

Tip: Snakemake plugin

Before continuing, it is worth recommending the SnakeCharm plugin for the PyCharm **Integrated Development Environment** (**IDE**): https://github.com/JetBrains-Research/snakecharm.

Instead of just using less to review the Snakefile, we can use this plugin to review our Snakemake files in the IDE. This is much more visually appealing, providing color highlighting specific to Snakemake and other tools.

It is recommended that you create a new project in PyCharm for your Snakemake work. Make sure to follow these instructions for enabling SnakeCharm:

1. Go to **Settings** | **Plugins** and search for SnakeCharm. Hit **Install**.

2. Go to **Settings** | **Languages & Frameworks** | **Snakemake**. Choose **Enable Snakemake Support**.

3. Go to **Settings** | **Project** (*the name of your project*) **Python Interpreter** | **Add Interpreter** | **Add Local Interpreter** | **Conda Environment** | **Use existing environment**. Choose the Snakemake conda environment that you set up in the *Getting ready* section. It should be called snakemake.

The setup instructions can be found here: https://github.com/JetBrains-Research/snakecharm/wiki#setup-snakemake-support.

Now use **File** | **Open** to open your Snakefile.

This is what our Snakefile looks like in the PyCharm IDE with SnakeCharm:

```
main.py        Snakefile  ×
1
2    import os
3    from pathlib import Path
4
5    # Configuration
6    SAMPLES = ["sample1", "sample2", "sample3"]
7    DATA_DIR = "data"
8    RESULTS_DIR = "results"
9
10   # Target rule - what we want to produce
11   rule all:
12       input:
13           # FastQC reports (real)
14           expand(f"{RESULTS_DIR}/fastqc/{{sample}}_R1_fastqc.html", sample=SAMPLES),
15           expand(f"{RESULTS_DIR}/fastqc/{{sample}}_R2_fastqc.html", sample=SAMPLES),
16           # Mock outputs
17           expand(f"{RESULTS_DIR}/alignment/{{sample}}.bam", sample=SAMPLES),
18           expand(f"{RESULTS_DIR}/variants/{{sample}}.vcf", sample=SAMPLES),
19           f"{RESULTS_DIR}/multiqc_report.html",
20           f"{RESULTS_DIR}/pipeline_summary.json"
```

Figure 15.9 – The Snakefile in the PyCharm SnakeCharm IDE

Running Snakemake from the terminal

Let's practice running Snakemake from the command line. Go to your terminal and type the following:

```
snakemake --use-conda --cores 4 &
```

This will run your Snakemake pipeline with four CPU cores. By default, it will look at your Snakefile and run the rules specified there. You will see output reflecting the run. Take a moment to review the output and check that your pipeline ran successfully. Again, if you run into issues, you can review the `logs` directory.

Let's next visualize the DAG for our pipeline. For this, we'll need to make sure Graphviz is installed:

```
brew install graphviz
```

Graphviz contains the `dot` tool, which is great for visualizing DAGs (`https://graphviz.org/docs/layouts/dot/`).

Now run this Snakemake command:

```
snakemake --dag | dot -Tpdf > dag.pdf
```

This will generate the DAG and place it in `dag.pdf`. Now run the following:

```
open dag.pdf
```

Here is what we see:

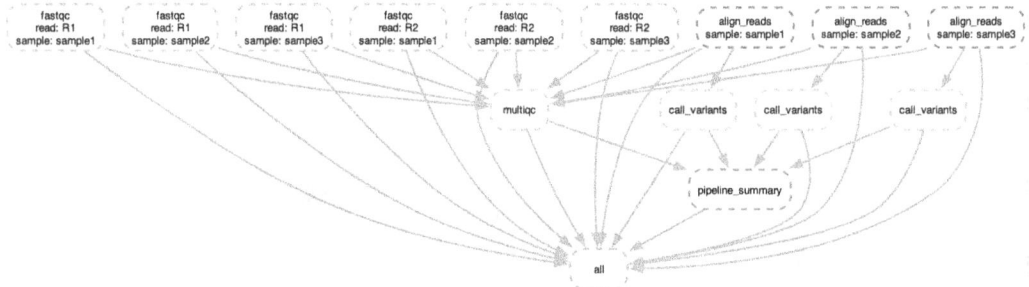

Figure 15.10 – The DAG for the Snakemake pipeline

We can see the structure of the pipeline. FastQC and alignment jobs feed into MultiQC. Alignments feed into variant calling, which then goes into a pipeline summary rule. All of these tasks together make up the full pipeline represented by the `all` rule.

Now, let's review what we see in our results directories. Take a look at one of the `.html` files in the `results/fastqc` directory.

Here is a sample output from a FastQC analysis:

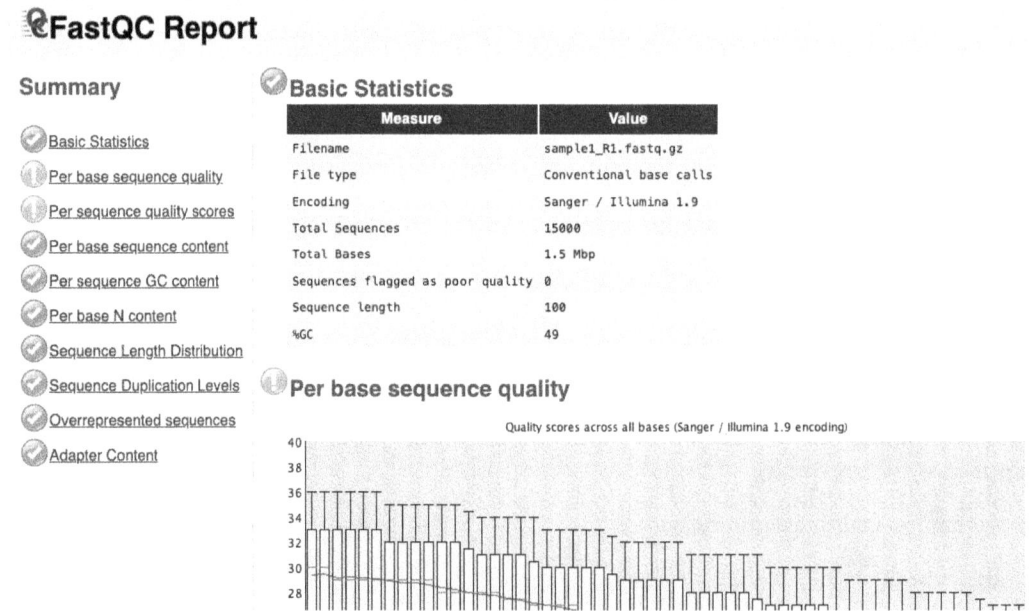

Figure 15.11 – FastQC output file from the Snakemake pipeline

> **Tip**
>
> In some cases, when you try to run Snakemake, it may tell you that a pipeline is already running and your directory is locked. To unlock your directory, run the following:
>
> ```
> snakemake --unlock
> ```

There's more...

If you want, you can continue to develop your Snakemake pipeline further to include real implementations of alignment and variant calling. You can work on this yourself or upload the entire notebook into an AI tool and ask it to help you build out real functions for the alignment and variant calling steps. You'll need to identify and install suitable tools for these steps – you can get some ideas from the previous chapters!

Here is a good example of a basic genomics workflow in Snakemake to review: `https://snakemake.readthedocs.io/en/stable/tutorial/basics.html`.

You should definitely take some time to check out the other things you can do with iPython widgets! You can make sliders to control interactive plots, radio buttons, multi-select boxes, and much more. You can find a more in-depth tutorial here: `https://saturncloud.io/blog/understanding-jupyter-notebook-widgets/`.

See also

- There is a great slide tutorial for Snakemake here: `https://slides.com/johanneskoester/snakemake-tutorial#/27`

- Snakemaker transforms your work in the terminal or notebook into a Snakemake pipeline: Masera et al., *Snakemaker: Seamlessly transforming ad-hoc analyses into sustainable Snakemake workflows with generative AI*, arXiv, Apr 2025 – `https://arxiv.org/abs/2505.02841`

- The original Snakemake paper can be found here: Koster & Rahmann, *Snakemake—a scalable bioinformatics workflow engine*, Bioinformatics, Aug 2012 – `https://academic.oup.com/bioinformatics/article/28/19/2520/290322`

Writing a bioinformatics workflow with Nextflow

In this recipe, we will learn about one of the most powerful, modern workflow management systems, Nextflow (`https://www.nextflow.io/`). We are going to implement a simple pipeline in Nextflow that does some basic bioinformatics tasks, such as trimming and quality control, as a way to learn more about it. We'll implement FastQC and MultiQC, but we'll just mock out the trimming step. We'll also build an interactive dashboard to manage it in our notebook, just like the last example.

Unlike Snakemake, which is Pythonic, Nextflow is based on Java. This means you need a current version of Java installed so that you can run a **Java Virtual Machine (JVM)**. It is based on **Groovy**, a Java-based programming language (`https://groovy-lang.org/`).

This means that Nextflow can use Java and Groovy libraries directly. Nextflow also has features for **computational reuse** and can scale easily from laptop to HPC to cloud.

By the end of this recipe, you will have a solid understanding of workflow management in Nextflow.

Getting ready

Move to your `Ch15/nextflow` directory.

Remember to reactivate the `Ch15-workflows` conda environment (or use your `bioinformatics-base` environment if you were using that):

```
conda deactivate
conda activate Ch15-workflows
```

Install Nextflow, FastQC, MultiQC, and Trimmomatic:

```
conda install -c bioconda nextflow fastqc multiqc trimmomatic
```

Note that Trimmomatic is only needed if you want to upgrade the exercise later to include a real implementation of trimming.

We also want to make sure Seaborn is installed:

```
pip install seaborn
```

There are two main ways you can explore this recipe:

- **Running Nextflow in the notebook**: A Jupyter widget dashboard is provided if you are more comfortable working out of the notebook
- **Running Nextflow from the terminal**: This is the more standard way of running Nextflow

You will first build the core Nextflow files and necessary components using the notebook. You will then have the option to try running Nextflow from the notebook dashboard, or you can proceed to focus on running it from the terminal.

The code for this recipe can be found in `Ch15/nextflow/Ch15-3-nextflow.ipynb`.

How to do it...

OK, let's check out Nextflow!

1. First, let's import our libraries. Review the *Import Libraries* section in the notebook.

2. Let's define our pipeline class.

   Review this code in the notebook. We define a `SimpleFastQCPipeline()` class. The main workflow definition file in Nextflow is called `main.nf`. This class is going to help us define that file, as well as setting up the necessary directories and configuration files. We'll review the contents of the `main.nf` file in just a moment.

   We define a `setup_pipeline()` function to create our directories for data, reference files, results, and reports. This function creates our workflow and configuration files and our sample data and checks for the installation of required components.

Note

You will want to run the code in the Jupyter notebook at this point so that your pipeline files will be created. You will then examine the files from the terminal using `less` or your favorite IDE.

Tip: Nextflow plugin

Before going further, we should also mention that you can view Nextflow files through the Nextflow plugin of the **Visual Studio Code (VSCode)** IDE. Here is an overview:

1. Download and install VSCode: `https://code.visualstudio.com/download`.

2. Remember to pick the appropriate operating system.

3. For a Mac, be sure to check **Apple Silicon** if needed.

4. On a Mac, you will want to unzip the download and then drag the VSCode application icon into your **Applications** directory in Finder. This will make it so you can use the *Command + spacebar* shortcut to find the application.

5. Install the Nextflow plugin for VSCode. Go into VSCode and then to **Settings | Extensions**. Search for `Nextflow`. Press **Install**. The extension is also located here: `https://marketplace.visualstudio.com/items?itemName=nextflow.nextflow`.

6. Go to **File | Open** and open your `main.nf` file.

Here is what our Nextflow pipeline definition looks like in VSCode:

```
Extension: Nextflow       main.nf    ×
Users > shanebrubaker > work > CookBook > Ch15 > nextflow > main.nf
  1    #!/usr/bin/env nextflow
  2    nextflow.enable.dsl=2
  3
  4    // Parameters
  5    params.reads = "data/*_{R1,R2}.fastq.gz"
  6    params.reference = "reference/genome.fa"
  7    params.outdir = "results"
  8    params.test_mode = false
  9
 10    log.info """
 11    ===============================================
 12    Simplified FastQC Pipeline
 13    ===============================================
 14    reads        : ${params.reads}
 15    reference    : ${params.reference}
 16    outdir       : ${params.outdir}
 17    test_mode    : ${params.test_mode}
 18    ===============================================
 19    """
 20
 21    // Main workflow
 22    workflow {
 23        // Create input channels
 24        reads_ch = Channel.fromFilePairs(params.reads, checkIfExists: true)
 25
 26        // Run FastQC
 27        fastqc(reads_ch)
 28
 29        // Trim reads (simplified)
 30        trimmomatic(reads_ch)
 31
 32        // Generate final report
 33        multiqc(fastqc.out.zip.collect())
 34    }
```

Figure 15.12 – The main.nf file in the VSCode editor with the Nextflow extension

Reviewing the Nextflow workflow structure

Let's take a quick look at our directory structure (use `ls -ltr` from your terminal):

```
drwxr-xr-x   6 shanebrubaker  staff     192 Jun 15 12:50 data
drwxr-xr-x   3 shanebrubaker  staff      96 Jun 15 12:50 reference
drwxr-xr-x   5 shanebrubaker  staff     160 Jun 15 13:15 results
drwxr-xr-x   4 shanebrubaker  staff     128 Jun 15 13:15 reports
drwxr-xr-x  12 shanebrubaker  staff     384 Aug 30 19:05 work
-rw-r--r--@  1 shanebrubaker  staff    8363 Sep 13 12:48 main.nf
-rw-r--r--   1 shanebrubaker  staff     251 Sep 13 12:48 nextflow.config
-rw-r--r--   1 shanebrubaker  staff   44814 Dec  4 21:32 Ch15-3-nextflow.ipynb
```

Figure 15.13 – Nextflow example directory structure

The data directory will hold our randomly generated sample FASTQ files. We also create a directory for our reference genome. Our results directory will contain the output files for each of our main steps: Trimmomatic, FastQC, and MultiQC. The reports directory will contain a summary of our pipeline run.

Let's examine the contents of `nextflow.config`:

```
process {
    cpus = 1
    memory = '2 GB'
    time = '30m'
}

executor {
    name = 'local'
    cpus = 4
}

report {
    enabled = true
    file = 'reports/execution_report.html'
}

timeline {
    enabled = true
    file = 'reports/timeline.html'
}
```

Figure 15.14 – Nextflow configuration file

This file contains some generic high-level parameters for the pipeline. The first section, `process`, contains high-level compute guidelines. We set it to use a certain amount of CPU and memory. Each process will also be allowed up to 30 minutes of **wall clock** time to run. Note that these are global default settings; you can override them at a task level in `main.nf`.

The next section, executor, specifies where and how to run the pipeline. In this case, we will run it on the local machine with four CPUs.

The report section specifies where to place pipeline summary reports. The last section, timeline, tells Nextflow to generate a plot of the timeline of the processes run in the pipeline. We'll review these reports later in the recipe. Note that by separating the config file from main.nf, we can decouple our application logic from our resources – we can later increase the size of execution nodes, for example, simply by changing the config file.

Let's next turn our attention to the contents of the main.nf file:

```
nextflow.enable.dsl=2

// Parameters
params.reads = "data/*_{R1,R2}.fastq.gz"
params.reference = "reference/genome.fa"
params.outdir = "results"
params.test_mode = false

log.info """
=============================================
Simplified FastQC Pipeline
=============================================
reads        : ${params.reads}
reference    : ${params.reference}
outdir       : ${params.outdir}
test_mode    : ${params.test_mode}
=============================================
"""

// Main workflow
workflow {
    // Create input channels
    reads_ch = Channel.fromFilePairs(params.reads, checkIfExists: true)

    // Run FastQC
    fastqc(reads_ch)

    // Trim reads (simplified)
    trimmomatic(reads_ch)

    // Generate final report
    multiqc(fastqc.out.zip.collect())
}
```

Figure 15.15 – Part of the Nextflow main.nf file

In the preceding figure, we see a portion of the `main.nf` file. Let's review this file carefully:

1. We first enable the Nextflow DSL2 standard, which is the latest version of their workflow specification language.

2. Next, we set some parameters for inputs and outputs.

3. We then use the `log.info` directive to print some logging information at the beginning about our key parameters.

4. We then create our primary workflow definition. Review the `workflow` section of the file.

This section first creates a Nextflow **Channel** (`https://www.nextflow.io/docs/latest/channel.html`), which is how Nextflow controls the flow of data. We then define the key steps of the pipeline, which include FastQC, trimming, and MultiQC.

Unlike the **backward chaining** we saw in Snakemake, Nextflow is **forward driven**. This means it starts with our source files and pushes data through **channels** where the data is processed by **consumer** functions until we complete the run.

Next, we'll look in depth at the `fastqc()` process.

This process first adds a tag to make log output more descriptive and then sets a `publishDir` output directory. It defines the inputs as a tuple of the sample names and input read paths. Our outputs will include the FastQC HTML and `.zip` reports.

We next define a `script` block. If the pipeline is in test mode, we will generate a fake report using some inline HTML. If not, we will check to see whether FastQC is installed. If it is, we will run FastQC on our FASTQ files; otherwise, we will fall back to running a mock report.

The next sections of `main.nf` specify a mocked process for running read trimming with Trimmomatic, and a full implementation of a MultiQC process.

Finally, at the end, we will see a `workflow.onComplete` section, which prints out a final log message with some pipeline execution statistics.

OK, great, we have set up our Nextflow pipeline and reviewed the core contents of our `main.nf` file. You can now choose whether you want to run Nextflow in your notebook with the interactive dashboard or run Nextflow from the terminal.

Running Nextflow in the notebook

You can skip this section and go straight on to *Running Nextflow from the terminal* if you wish to take that approach instead.

Next, we will go over how to run our Nextflow pipeline from within the Jupyter notebook using a handy dashboard:

1. Now we are going to set up our interactive dashboard.

2. Review Section 3, *Interactive Controller*, in the notebook. We will define a class, `SimpleFastQCController()`. This class uses iPython widgets as before to set up an interactive dashboard. It includes a loop to monitor the progress of the pipeline and provide an updated progress bar.

3. Next, we will initialize our dashboard. Review the *4. Initialize Display* section in the notebook.

    This code will create an instance of our `SimpleFastQCPipeline()` class and then call its `setup_pipeline()` method. We then create a controller from the `SimpleFastQCController()` class, passing our pipeline into it. We then create and display the control interface. Here is our dashboard:

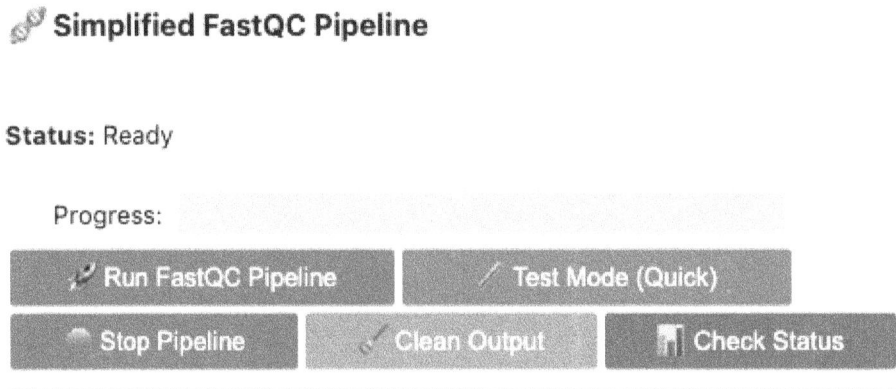

    ### 🧬 Simplified FastQC Pipeline

    Status: Ready

    Progress:

🔬 Run FastQC Pipeline	✏️ Test Mode (Quick)
🛑 Stop Pipeline	✅ Clean Output	🧾 Check Status

    Pipeline Output:

    Figure 15.16 – Interactive dashboard for Nextflow pipeline

    We have buttons to run our pipeline and perform in **Test Mode**, which we saw when we reviewed the `main.nf` file. You can also stop the pipeline, clean up the output, or check in on the status of the pipeline. Just below the buttons, you will see the **Pipeline Output** window, which will show the Nextflow command line used to execute the pipeline.

    > **Note**
    >
    > If you want to restart the pipeline, you may need to exit your Jupyter notebook and restart it. You typically do this by closing your Jupyter notebook tabs, going back to the terminal where you started it, hitting *Ctrl* + *Z* to stop the process, and then typing `jupyter notebook` again to restart it.

4. Great! Let's run our pipeline. Go ahead and click **Run FastQC pipeline**.

5. Once completed, you will see the message **Pipeline Execution Complete!**.

> **Note**
>
> You may see this warning: **WARN: Task runtime metrics are not reported when using macOS without a container engine!**.
>
> This is a harmless warning that can be ignored. It just means that when Nextflow is executing tasks natively on macOS, instead of inside a container, it cannot gather runtime metrics such as CPU usage; this is why you won't see that information in your pipeline reports.

Running Nextflow from the terminal

Let's practice running Nextflow from the terminal:

```
nextflow run main.nf --reads data/*_{R1,R2}.fastq.gz --reference
reference/genome.fa --outdir results
```

This will run the Nextflow pipeline based on our `main.nf` file, just as we ran it in the notebook. We will supply the input reads data as anything matching R1 or R2 and ending in `.fastq.gz` in our data directory. We also provide the reference genome and output directory.

> **Troubleshooting**
> If you run into any problems in Nextflow, you can review the log:
>
> `cat .nextflow.log`

Reviewing the results of the Nextflow pipeline

Now we can check out our results:

1. First, let's check out our pipeline reports. Go into your terminal and run the following:

    ```
    open reports/execution_report.html
    ```

This is an overall summary report on pipeline execution. Here is what we see:

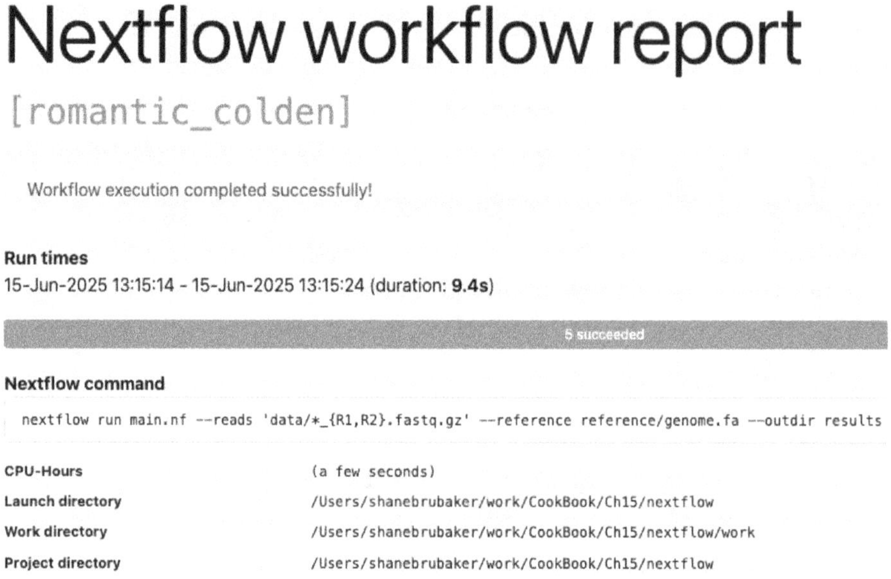

Nextflow workflow report

`[romantic_colden]`

Workflow execution completed successfully!

Run times
15-Jun-2025 13:15:14 - 15-Jun-2025 13:15:24 (duration: **9.4s**)

```
                                                      5 succeeded
```

Nextflow command

```
nextflow run main.nf --reads 'data/*_{R1,R2}.fastq.gz' --reference reference/genome.fa --outdir results
```

CPU-Hours	(a few seconds)
Launch directory	/Users/shanebrubaker/work/CookBook/Ch15/nextflow
Work directory	/Users/shanebrubaker/work/CookBook/Ch15/nextflow/work
Project directory	/Users/shanebrubaker/work/CookBook/Ch15/nextflow

Figure 15.17 – Nextflow workflow summary report

This report includes a summary of overall execution time, working directories, and hashes that represent unique IDs for the pipeline run. You can also see information about resource utilization (if it had been run in a container) and a detailed report of task status. In a real scenario, you could use this report to help you optimize compute performance and overall cost for your pipeline.

2. Next, let's look at the timeline report:

    ```
    open reports/timeline.html
    ```

Here is what it looks like:

Processes execution timeline

Launch time: 15 Jun 2025 13:15
Elapsed time: 10.2s
Legend: job wall time / memory usage (RAM)

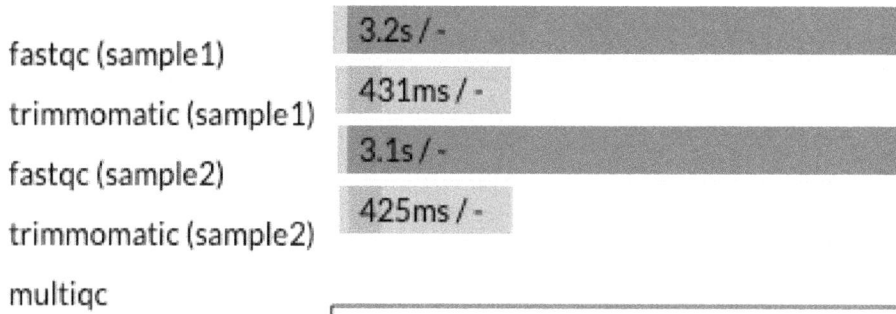

Figure 15.18 – Nextflow timeline report

We can see the order of execution and time taken by each task in the pipeline.

3. Now let's review our results:

```
open results/fastqc/sample1_R1_fastqc.html
```

We see the FastQC report for this sample:

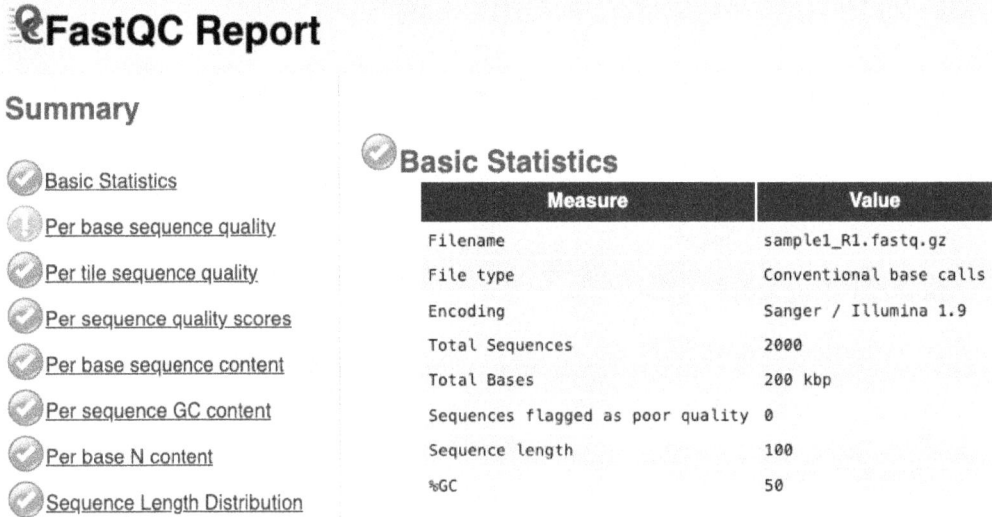

Figure 15.19 – FastQC report for sample 1

4. Now, check out the MultiQC report. From the terminal, type the following:

```
open results/multiqc/multiqc_report.html
```

Here is what we see:

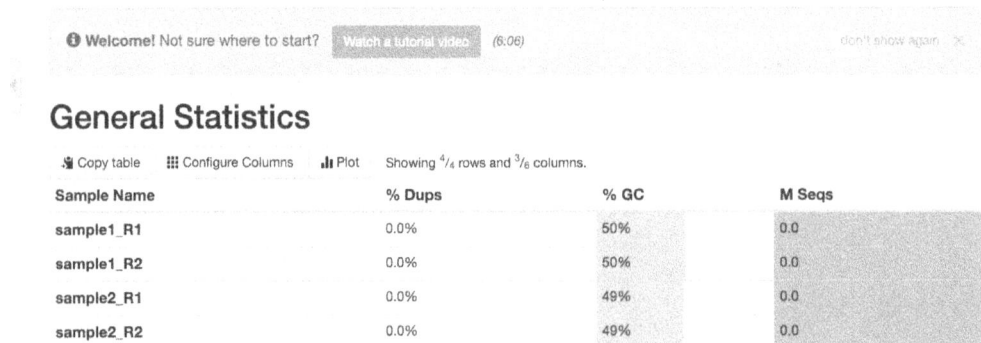

Figure 15.20 – MultiQC report from Nextflow pipeline

This provides the sequencing metrics from the FastQC report aggregated into this MultiQC page. If you had more QC tools in the pipeline, you could potentially aggregate more reports here as well.

There's more...

You can take this example further by implementing trimming with Trimmomatic (`https://github.com/timflutre/trimmomatic`). You could even extend the pipeline to perform alignment, variant calling, or other major bioinformatics tasks if you like.

Nextflow is becoming the predominant workflow management language. It is supported by the powerful **nf-core** (`https://nf-co.re/`) community, which provides standardized containers for bioinformatics tools and Nextflow pipelines for hundreds of common analyses.

With the introduction of DSL 2, Nextflow has many powerful features (`https://seqera.io/blog/dsl2-is-here/`). **Nextflow modules** allow you to modularize and compose your workflows for powerful reusability. Nextflow can run on almost any environment, including HPC and cloud platforms.

Seqera offers a commercial platform for running Nextflow. They have integrated Parabricks and Fusion into their platform to optimize compute and storage efficiency (`https://seqera.io/blog/nvidia-parabricks-fusion/`).

Airflow (`https://airflow.apache.org/`) is typically thought of as a data loading and transformation platform but can also be used for bioinformatics workflows. SciDAP (`https://airflowsummit.org/sessions/2021/scidap/`) provides a CWL extension for Airflow.

You've now learned a great deal about bioinformatics workflow management and the tools and platforms used to implement it. We have covered some of the most common workflow platforms and the key workflow definition languages they use. In the next chapter, we are going to look at two new and interesting approaches to workflow management: Flyte and AWS state machines!

OK, let's clean up and close down our `conda` environment:

```
conda deactivate
```

See also

- Check out this Nextflow tutorial, which uses the `uv` package manager: `https://github.com/sebrauschert/nextflow-example`
- Read more about nf-core here: Langer et al., *Empowering bioinformatics communities with Nextflow and nf-core*, bioRxiv, May 2024 – `https://www.biorxiv.org/content/10.1101/2024.05.10.592912v1.abstract`

- There is a tool for automating tests in Nextflow: Patel et al., *NFTest: automated testing of Nextflow pipelines*, BioInformatics, February 2024 – `https://academic.oup.com/bioinformatics/article/40/2/btae081/7606335`

- FLOWViZ is an excellent example of a pipeline built in Airflow: Luis et al., *FLOWViZ: An Airflow Based Workflow Middleware for Computational Phylogenetics*, Oct 2023 – `https://www.preprints.org/frontend/manuscript/58cd5a927889b582e8fab1902ca1758e/download_pub`

Get This Book's PDF Version and Exclusive Extras

UNLOCK NOW

Scan the QR code (or go to `packtpub.com/unlock`). Search for this book by name, confirm the edition, and then follow the steps on the page.

Note: Keep your invoice handy. Purchases made directly from Packt don't require an invoice.

16

More Workflow Systems

In this chapter, we will learn about two more interesting workflow systems. **Flyte** is a more recent workflow system developed at Lyft. It is Pythonic and runs on Kubernetes. We'll cover it first. Then we'll learn how to build a workflow system using AWS Step Functions and State Machines. This can be a great approach when you want an AWS-native solution that integrates well with other AWS services, such as EventBridge, SQS, and SNS.

We will cover the following recipes in this chapter:

- Building a bioinformatics workflow with Flyte
- Launching a bioinformatics orchestrator with AWS Step Functions

Technical requirements

The code for this chapter can be found at `https://github.com/PacktPublishing/Bioinformatics-with-Python-Cookbook-Fourth-Edition/tree/main/Ch16`.

You will want to create a `Ch16` folder and set up your notebooks there. In this exercise, we will work within two sub-folders for the respective tools we'll be learning about: `flyte` and `aws`.

Remember to activate your `conda` environment before beginning the recipes, like this:

```
conda activate bioinformatics_base
```

Or, if you would like to set up a `conda` environment specific to this chapter, before activating `bioinformatics_base`, run the following:

```
conda create -n ch16-more-workflows --clone bioinformatics_base
conda activate ch16-more-workflows
```

You will be able to install the packages for the chapter as you go, or you can use the YAML file provided in the repository:

```
conda env update --file ch16-more-workflows.yml
```

Building a bioinformatics workflow with Flyte

Flyte (`https://flyte.org/`) is a powerful modern workflow system that was originally developed at Lyft. It is now available as an open source platform for the community. Flyte can run a wide variety of workflow types, including those for machine learning and bioinformatics.

Flyte runs on **Kubernetes** (`https://kubernetes.io/`). Flyte can be run in a local environment (on your laptop) or in a remote environment. The remote environment could be on a Docker cluster running on your laptop, on a **High Performance Computing** (**HPC**) cluster at your company, or in the cloud on a Kubernetes cluster such as **OpenShift** (`https://developers.redhat.com/products/openshift/`).

Flyte defines **tasks**, which are the fundamental units of computation, implemented as containerized applications. It builds a **workflow** as a series of tasks that can have **dependencies**, meaning, for example, that **Task B** cannot be run until **Task A** is completed. Flyte entities are broadly kept under **projects**, which help you organize your work. Flyte **domains** may point to staging or production environments to help you maintain different deployments. Flyte also **versions** your workflows for careful tracking of changes. Flyte **launch plans** define the overall configuration for workflows and allow you to schedule regular runs.

Some of the key features of Flyte include the following:

- Automatic determination of dependent tasks; independent tasks are parallelized automatically
- Strong typing ensures accurate data types; data classes are used for data handling: `https://www.geeksforgeeks.org/python/data-classes-in-python-an-introduction/`
- Workflows are modular and can be composed from sub-workflows
- Universal file handling allows seamless usage of workflows across local and cluster-based environments
- Granular control over resource usage (CPU, RAM)
- Caching allows for **computational reuse** when nothing has changed
- **Map** tasks allow for the implementation of scatter-gather tasks for highly parallel implementation of computations
- **Dynamic workflows** can programmatically alter their parameters at runtime
- **Branching** in workflows allows for conditional logic to execute different paths in a pipeline
- **Human-in-the-loop** execution allows you to pause your workflow and wait for human interaction
- **Decks** allow you to integrate visualizations into your workflows: `https://docs-legacy.flyte.org/en/v1.15.0/user_guide/development_lifecycle/decks.html`

We will be performing this recipe from the terminal. In this recipe, we will create and run a basic Flyte workflow showing an overview of a typical bioinformatics pipeline. We will mock most of the internal functions so that we can focus on the functionality of Flyte. By the end of this recipe, you should have an understanding of how to install Flyte, define a basic workflow, and run it.

Getting ready

You will be working out of the `Ch16/flyte` subdirectory.

Make sure your `conda` environment is activated:

```
conda activate ch16-more-workflows
```

Let's install Flyte. From the terminal, type the following:

```
pip install "flytekit[all]"
```

Alternatively, you can install it using conda:

```
conda install -c conda-forge flytekit
```

We also want the local Flyte cluster installed (this will be used to run Flyte workflows on your local computer):

```
brew tap flyteorg/homebrew-tap
brew install flytectl
```

The Python toolkit for interacting with Flyte is called `flytekit`. Another tool called `flytectl` helps you set up and manage a local Docker cluster for use with Flyte. We'll discuss these more later. Note the use of the `brew tap` command: this is used to add an external repository to brew: `https://docs.brew.sh/Taps`.

We will be performing this recipe from the terminal. The code for the exercise can be found in the `Ch16_1_flyte.py` file:

> **Note**
>
> Note that the filename here has underscores instead of dashes. When Flyte imports your code, it imports it as a module in Python, and dashes are not allowed in these types of identifiers. So, we need the filename to contain underscores, not dashes.

How to do it...

Let's begin by reviewing the code in the Flyte workflow. Take a look at the `Ch16_1_flyte.py` file. Remember, you can use `less` or a Python **IDE** (short for **Integrated Development Environment**) such as **PyCharm** (`https://www.jetbrains.com/pycharm/`) to review the file.

1. The first section of the file provides documentation of the workflow in a docstring. This is the section in triple-quotes. It includes some information on installing dependencies and running the pipeline.

   Next, we will import our libraries. Review the *Import Libraries* section in the file.

   We will use some standard libraries and include the `Tuple` module (`https://typing.python.org/en/latest/spec/tuples.html`). Tuples are used to provide type hints for functions that return multiple values. We will also use the `dataclasses` library for storing our data (`https://docs.python.org/3/library/dataclasses.html`). We next suppress some warnings.

   > **Note**
   > The reason we suppress the preceding warnings is that when running the Flyte workflow, you may get a couple of warnings saying `parameter -i is used more than once`. This is simply a minor issue with a duplicate parameter in the Flyte client interface and can be safely ignored.

   We next import the task, workflow, and resources modules from `flytekit`. **FlyteKit** (`https://github.com/flyteorg/flytekit`) is the core library for developing and interacting with Flyte. These modules perform the following:

   - **Task**: Forms the core unit of execution in a Flyte workflow. The `@task` decorator turns a regular Python function into a Flyte task.

   - **Workflow**: The `@worfklow` decorator defines the orchestration of Flyte tasks. It creates the workflow DAG and manages the dependencies of tasks and the flow of data from one task to another.

   - **Resources**: This module controls the amount of resources assigned to each task, such as CPU and memory.

   We bring in the `FlyteFile` module (`https://www.union.ai/docs/flyte/user-guide/data-input-output/flyte-file-and-flyte-directory/`). This handles files across multiple environments, so your workflows can run either locally or on the cloud. It can intelligently use local file pointers or point to S3 buckets when run on the cloud, downloading files into containers as needed. We also use `FlyteDirectory`, which serves a similar function for entire collections of files.

2. The next section defines key Python data classes we will use. Review the *Dataclass Definitions* section of the file.

    These data classes (`https://www.datacamp.com/tutorial/python-data-classes`) hold information on our QC metrics and alignment results. They store key metrics such as the total number of reads, mean read quality, GC content, and so on. By using data classes, we provide built-in type checking for Flyte. They are also **serializable** – this means Flyte can easily transmit them between tasks. When run in containers, Flyte can turn the data into a JSON object, store it in S3, and then unpack it in the next task to facilitate the easy transfer of data between containerized tasks.

3. Now we will define our first task. Take a look at the *Task Definitions* section of the file. This first task will perform quality control on our data. This is the `quality_control()` task. Review it in the file.

    Let's focus on the first section of the task. The `@task` decorator tells Flyte that this function will be used as a unit of work in Flyte. The resources allowed for the task are defined using the resources module: in this case, we will use 1 CPU and memory of up to 1 GB. The `cache_version` line tells Flyte that if this task has been run before and we are using the same inputs, we can use the cached version of the inputs.

4. Next, we provide our function definition. The inputs for Flyte are typed so that they can be checked for accuracy as the workflow proceeds. The first input, the FASTQ file, is a `FlyteFile` class. The `->` operator tells us that we are next going to define the outputs of the task. The outputs will be a **tuple** consisting of a `FlyteDirectory` class, which will return our `reports` folder, and the `QCMetrics` data class, which will contain the metrics from the QC analysis.

    Review the rest of the code for the `quality_control()` task now. This function does the following:

    *   Defines an output directory using the `tempfile` module (`https://www.geeksforgeeks.org/python/create-temporary-files-and-directories-using-python-tempfile/`)

    *   Downloads the input FASTQ file for processing

    *   Loops over the FASTQ file and parses each line to get basic quality metrics such as the total number of reads and bases, mean quality score, and GC content

    *   Loads the results into the `qc_metrics` data class

    *   Creates report outputs and returns the metrics

5. The next task is for read alignment. Review the `align_reads()` task in the code.

    This code is given 2CPUs and 2 GB of RAM. It is set to utilize cached outputs. It downloads the input FASTQ file and reference genome and then creates a mock alignment BAM file. It creates simulated metrics and returns them in the `alignment_metrics` data class.

    Later, you can take a look at the `sample_sorted.bam` file to see the output of this step (it will be in our working directory after you run the Flyte workflow).

6. Our next task will simulate calling variants. Take a look at the `call_variants()` task in the code.

    This task gets 1 CPU and 1 GB of RAM, and will utilize caching. It will bring in the reference file and BAM alignment as inputs. It will then generate a mock VCF file using some random effects to make it look more realistic. It returns the VCF file.

    You can check out the `sample_variants.vcf` file after running the workflow to see what the result of this step looks like.

7. The next task will generate a report and some visualizations. Review the code for `generate_report()`.

    This code takes 1 CPU and 1 GB of RAM and is not set for caching. It will take in the metrics files and produce visualizations and summary reports. We'll review those in a moment.

8. Now we come to the workflow definition. This is at the heart of our recipe, defining the core series of tasks we will run. Review the *Workflow Definition* section of the code.

    The `@workflow` decorator is used to define our workflow, which will be called `genomics_pipeline`. It will take in our FASTQ file, reference genome, and the name of our sample. It will return a `FlyteDirectory` class of outputs and three `FlyteFile` objects containing output files and a summary report. The workflow consists of our core steps of quality control, alignment, variant calling, and summary reporting. Because Flyte knows the first two steps are independent, it can run quality control and alignment in parallel. Since variant calling relies on the BAM file, Flyte knows that it must wait for the alignment task to complete first, creating a **dependency**. At the end, we generate the analysis report.

9. Okay, now we have defined the main Flyte workflow for this recipe. The next few functions will provide some additional functionality for running in standard Python mode. First, we are going to define a function to create our sample files. Take a look at the `create_sample_data()` function.

    This function simply creates a mock FASTQ file and reference genome file.

10. Finally, we have our `main()` function. This will be used in standard Python mode to print out instructions and create our sample files.

    This function will generate our sample input files if they do not already exist. It then prints out usage instructions, including how to run in local and remote mode, information on requirements, and details on output files.

11. That's it! We have now set up our Flyte workflow and ancillary functions. Next, we will run the workflow and learn more about the Flyte ecosystem.

Before we start, let's review our Flyte workflow. We will use `pyflyte`, which is the **command-line interface (CLI)** for Flyte. From the terminal, type the following:

```
pyflyte run Ch16_1_flyte.py --help
```

If you encounter issues with this command or other Flyte commands, it may be due to the flytekit version installed or your Python version. Check that you're using a recent Python version compatible with your flytekit version. If needed, try setting up a fresh `conda` environment. You may also need to ensure you're pointing to the Flyte installation in your new environment, as Flyte may sometimes be installed globally. For example, if your `conda` environment is called flyte-fresh, you would run: `~/anaconda3/envs/flyte-fresh/bin/pyflyte run Ch16_1_flyte.py --help`

This command provides you with an overview of the workflow and tasks defined within:

```
Usage: pyflyte run Ch16_1_flyte.py [OPTIONS] COMMAND [ARGS]...

Run a [workflow|task|launch plan] from Ch16_1_flyte.py

─ Options ──────────────────────────────────────────────────────────
  --help  Show this message and exit.

─ Commands ─────────────────────────────────────────────────────────
  genomics_pipeline  workflow (Ch16_1_flyte.genomics_pipeline)  Complete genomics workflow: QC -> Alignment -> Variant Calling -> Report
  align_reads        task (Ch16_1_flyte.align_reads)  Align reads to reference genome (simulated for local execution)
  call_variants      task (Ch16_1_flyte.call_variants)  Call variants from aligned reads (simulated)
  generate_report    task (Ch16_1_flyte.generate_report)  Generate comprehensive analysis report with visualizations
  quality_control    task (Ch16_1_flyte.quality_control)  Perform quality control analysis on FASTQ files
```

Figure 16.1 – Output of workflow overview

Okay, let's try out our Flyte workflow! First, we are going to run the code in standard Python mode to generate the sample files:

```
python Ch16_1_flyte.py
```

Now, run the workflow:

```
pyflyte run Ch16_1_flyte.py genomics_pipeline --fastq_file sample.
fastq --reference_genome reference.fa
```

> **Note**
>
> Keep in mind that copying directly from the book may not always work due to spaces or other special characters. You may want to copy from the instructions in the terminal or paste your command into a Notepad tool first for inspection.

You will now see the output of the run. Once completed, you can check your working directory for the output files. This will include the BAM file, VCF file, and summary plots (PNG files). You can review `sample_analysis_report.md` and `sample_summary.csv` using `less`.

Let's take a look at our analysis summary visualization:

```
open sample_analysis_plots.png
```

Here is what you will see:

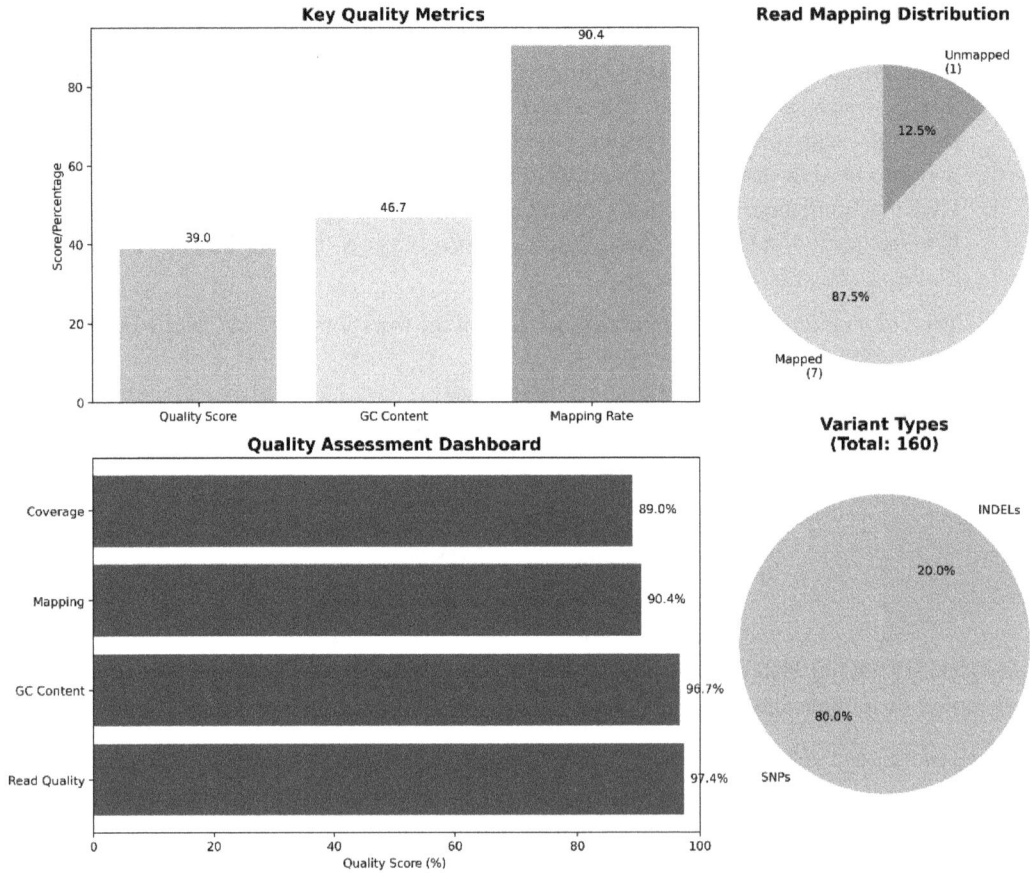

Figure 16.2 – Flyte workflow analysis summary plot

This plot shows quality scores and alignment rates at the upper left. The **mapping rate** represents the number of reads successfully utilized during alignment. At the bottom left, we see the coverage, mapping, GC content, and read quality represented as a horizontal bar chart. On the right, we see a pie chart of the mapping rate and the types of variants found, respectively. Let's check out the other visualization we made:

```
open sample_quality_summary.png
```

This is what we see:

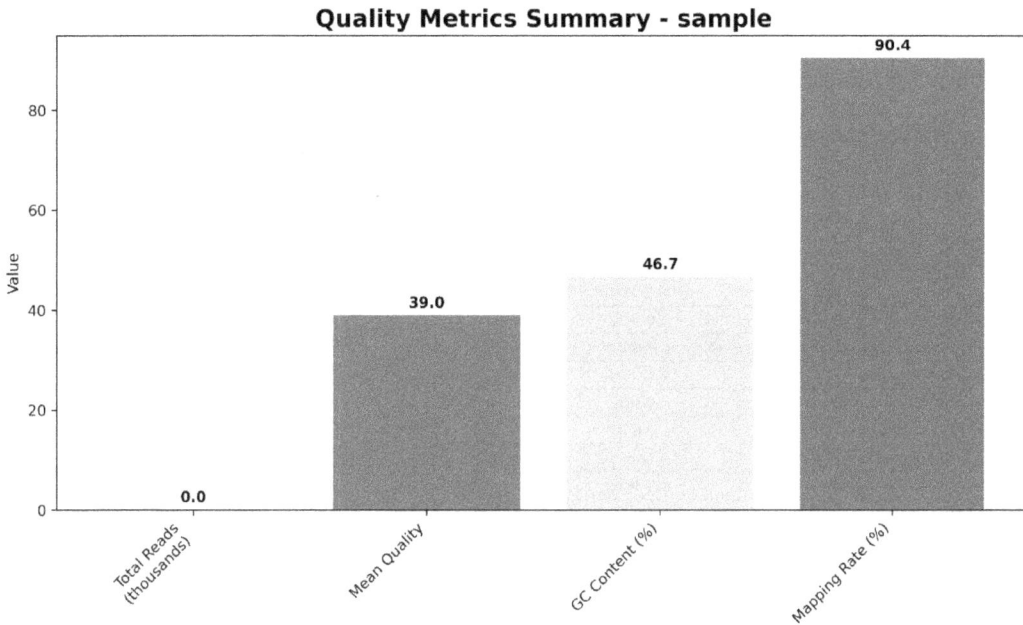

Figure 16.3 – Flyte workflow sample quality report

This bar chart summarizes the total reads, mean read quality, GC content of reads, and their mapping rate.

Great job! You have learned the basics of defining tasks and workflows in Flyte. In the example so far, we ran the workflow locally. In the next section, we will see how to spin up a Docker-based system and practice running Flyte workflows on it.

There's more...

We've run our Flyte system locally. Next, let's run our Flyte workflow using a local cluster with Docker. This approach will be a little closer to a full production architecture where we would run Flyte on a Kubernetes cluster in an HPC or cloud environment.

Make sure Docker is running. On a Mac, you can launch Docker by clicking *Command + Space* and then typing `Docker` and hitting *Enter*. You should see the Docker icon in your upper-right taskbar, and you can check whether it is running. You can also check whether Docker is running by typing this in your terminal:

```
docker ps
```

You should see a header with container IDs and information, either with currently running containers or just the header if nothing is running. We are going to use `flytectl`, which is Flyte's cluster infrastructure management tool (`https://www.union.ai/docs/flyte/api-reference/flytectl-cli/`). Now run the following:

```
flytectl demo start
```

This will spin up a local Flyte cluster. Here is what you should see:

```
  Starting container...
  Waiting for cluster to come up...
  Activated context "flyte-sandbox"!
  +------------------------------------------------------+----------+-----------+
  |                        SERVICE                       |  STATUS  | NAMESPACE |
  +------------------------------------------------------+----------+-----------+
  | flyte-sandbox-kubernetes-dashboard-5b4465fcfb-hk4cb  | Running  | flyte     |
  +------------------------------------------------------+----------+-----------+
  | flyte-sandbox-docker-registry-648bd974fc-p5khd       | Running  | flyte     |
  +------------------------------------------------------+----------+-----------+
  | flyte-sandbox-postgresql-0                           | Running  | flyte     |
  +------------------------------------------------------+----------+-----------+
  | flyte-sandbox-minio-797448f46d-n6r4x                 | Running  | flyte     |
  +------------------------------------------------------+----------+-----------+
  | flyteagent-bd4bcf75d-czbck                           | Running  | flyte     |
  +------------------------------------------------------+----------+-----------+
  | flyte-sandbox-buildkit-55b956cfbb-w8mlk              | Running  | flyte     |
  +------------------------------------------------------+----------+-----------+
  | flyte-sandbox-proxy-74bcd9c78f-1f84w                 | Running  | flyte     |
  +------------------------------------------------------+----------+-----------+
  | flyte-sandbox-5fd66d7fc4-d52zq                       | Running  | flyte     |
  +------------------------------------------------------+----------+-----------+
  Flyte is ready! Flyte UI is available at http://localhost:30080/console
  Run the following command to export demo environment variables for accessing flytectl
      export FLYTECTL_CONFIG=/Users/shanebrubaker/.flyte/config-sandbox.yaml
  Flyte sandbox ships with a Docker registry. Tag and push custom workflow images to localhost:30000
  The Minio API is hosted on localhost:30002. Use http://localhost:30080/minio/login for Minio console, default credentials - username: minio, password
  : miniostorage
```

Figure 16.4 – Local Flyte cluster details

Next, we run the following:

```
pyflyte --config ~/.flyte/config-sandbox.yaml register Ch16_1_flyte.py
```

This will register your workflow with Flyte. Here is a portion of the output:

```
Successfully serialized 6 flyte objects
   Registration Ch16_1_flyte.call_variants type TASK successful with version pv8atVJ0527kp5qzrILYjQ
   Registration Ch16_1_flyte.generate_report type TASK successful with version pv8atVJ0527kp5qzrILYjQ
   Registration Ch16_1_flyte.align_reads type TASK successful with version pv8atVJ0527kp5qzrILYjQ
   Registration Ch16_1_flyte.quality_control type TASK successful with version pv8atVJ0527kp5qzrILYjQ
   Registration Ch16_1_flyte.genomics_pipeline type WORKFLOW successful with version pv8atVJ0527kp5qzrILYjQ
   Registration Ch16_1_flyte.genomics_pipeline type LAUNCH_PLAN successful with version pv8atVJ0527kp5qzrILYjQ
Successfully registered 6 entities
```

Figure 16.5 – Output of Flyte workflow registration

This is the next command:

```
pyflyte --config ~/.flyte/config-sandbox.yaml run --remote Ch16_1_
flute.py genomics_pipeline --fastq_file sample.fastq --reference_
genome reference.fa
```

This will run your workflow on your local cluster. We provide the location of our sandbox configuration, workflow name, input FASTQ file, and the input reference genome.

> **Note**
>
> If you need to update and rerun a workflow remotely, you will follow the steps for registering and running your workflow again, as previously.

You will be given a URL link to an HTML dashboard to review the run. Copy and paste the URL into your browser.

You can also open your Flyte local cluster dashboard by going to `http://localhost:30080/`.

Here is what it looks like:

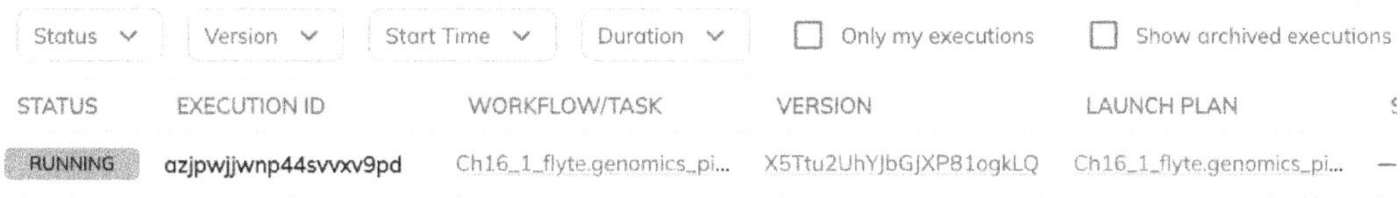

Figure 16.6 – Genomics workflow execution displayed in the Flyte dashboard

This is the Flyte workflow and cluster management interface. You will see here an overview of your workflow and the tasks within it. You can see the status of each task as either **SUCCEEDED** or **FAILED**. While your workflow is still running, you'll see the jobs changing status. During the run, you will see a **Terminate** button at the upper right, which will allow you to stop the workflow execution. When the workflow is done, you will see the **Relaunch** button, which lets you run the workflow again. You can control many aspects of your workflow through this GUI interface.

Whenever you want to see the details of your workflow run, you can go to **Workflows** (in the left taskbar), click on your workflow name, and then click on the execution ID. This will take you to a detailed page showing your tasks, their execution status, and timestamps. If a task has errored, you can click on the **I** button (information) for more details on the problem.

> **Note**
>
> You may get different errors when you run locally versus remotely. That's because, when running remotely, you will be launching jobs inside a container. There may be different software environments or dependencies that are not installed in the container that you will need to watch out for.

When defining a workflow that uses a container registry, you will use the `@ImageSpec` decorator to define the requirements for the container (`https://www.union.ai/docs/flyte/user-guide/core-concepts/image-spec/`). It might look something like this:

```
cluster_image = ImageSpec(
    name="genomics-production",
    packages=["pandas", "numpy", "scipy", "matplotlib",
              "seaborn", "biopython", "pysam",
              "scikit-learn", "tensorflow" ],
    registry="your-company-registry.com"
)
```

This image specification would name your image and provide information on required packages to install, such as pandas, NumPy, and so on. You would point it at your desired container registry.

Okay, let's go back to discussing our workflow.

From the workflow execution page, you can also see a DAG representing your workflow. Click on the **Graph** tab:

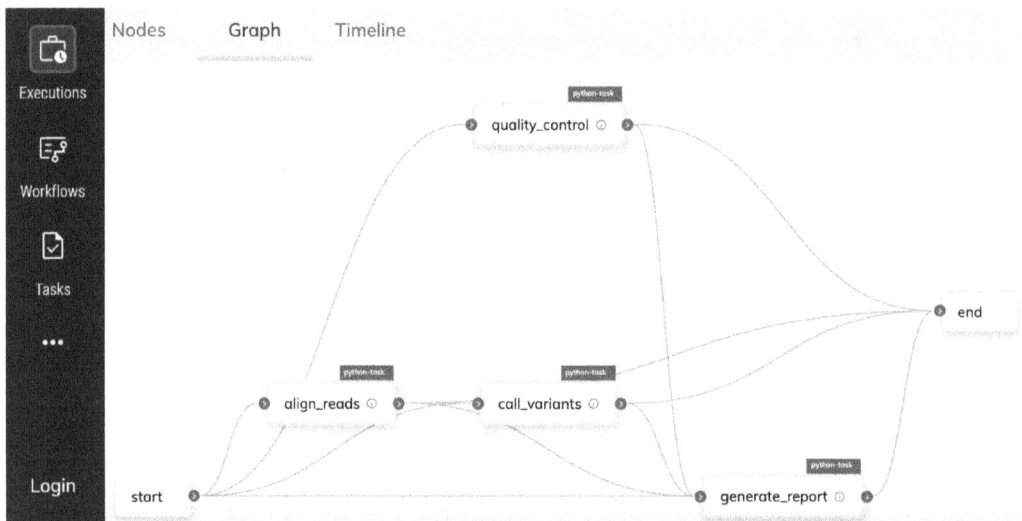

Figure 16.7 – The genomics pipeline workflow shown in the Flyte Graph interface

You can see that your tasks are outlined in green, showing that they were successfully completed.

Now click on the **Timeline** tab:

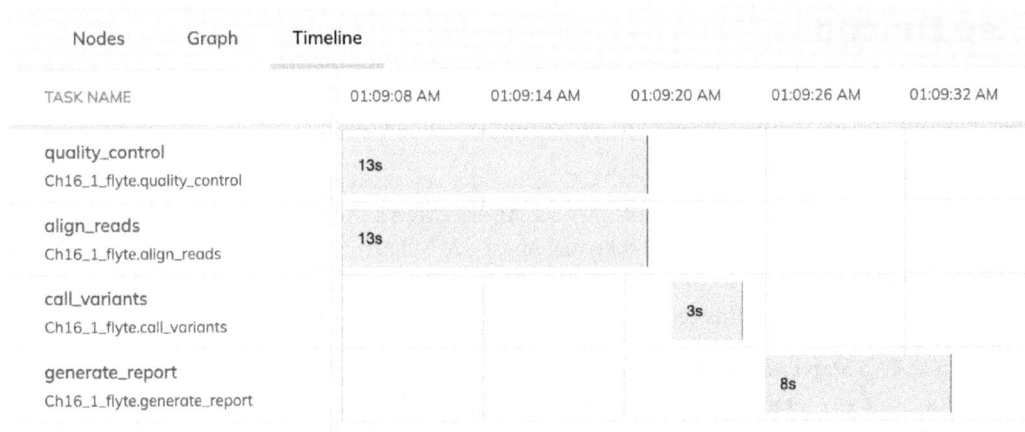

Figure 16.8 – Flyte Timeline display for the genomics workflow execution

This displays a timeline of your run showing the order and execution of runs.

If you were running this workflow in a real production scenario, you would be launching it on a real cluster such as OpenShift. You would be managing your containers and workflows much as we've discussed here.

You can see how Flyte can easily be used to develop and prototype workflows on your local system and then productionize these workflows on bigger clusters in the cloud. This is a great way to develop and a big strength of Flyte!

See also

- You can find a gentle introduction to Flyte here: `https://medium.com/@rahulmadan_18191/master-local-workflows-with-flyte-a-beginners-guide-using-python-0f33115e6c1a`

- This article provides a great tutorial on performing read alignment using Flyte: `https://flyte.org/blog/bioinformatics-on-flyte-read-alignment`

- Dive deeper into FlyteKit: `https://patford12.medium.com/getting-started-with-flytekit-python-sdk-d9614224d639`

- Check out Flyte School: `https://www.youtube.com/watch?v=0cP9pLLeqT4&list=PL-OJo2SeWc8I6PgnrTCAl-JdWaqV-MJSN`

- This article provides an interesting comparison of Flyte and Airflow: `https://flyte.org/blog/orchestrating-data-pipelines-at-lyft-comparing-flyte-and-airflow`

Launching a bioinformatics orchestrator with AWS Step Functions

You could also build your own workflow management system using cloud computing platforms and the tools available on them. In this recipe, we'll show you how to do just that!

You should familiarize yourself with the AWS genomics reference architecture (`https://aws.amazon.com/blogs/architecture/automated-launch-of-genomics-workflows/`). This 7-part series describes how AWS state machines, AWS Batch, ECR, and other key components can be combined to provide a comprehensive workflow management system. This will help orient you to what we are doing in this recipe.

You can use AWS **Step Functions** to build an orchestration system: `https://aws.amazon.com/step-functions/`. **State Machines** are a computational model in which a system transitions between different states based on its inputs: `https://www.youtube.com/watch?v=gv5fQrD8XUo`. AWS implements State Machines in the form of Step Functions to create visual workflows that can orchestrate and interact with AWS services. The terms *State Machine* and *Step Function* within AWS are often used interchangeably. They are **serverless**, meaning that they do not require a dedicated server to be running all the time, and are instead spun up on demand.

Step Functions typically calls **Lambdas**, which are serverless functions that can run code: `https://aws.amazon.com/lambda/`. It can also call other AWS services, such as **Simple Notification Service** (**SNS**), to send emails or perform many other operations. Lambdas provide high scalability and excellent cost management as they only run when needed.

AWS state machines are implemented using **Amazon States Language** (**ASL**): `https://docs.aws.amazon.com/step-functions/latest/dg/concepts-amazon-states-language.html`

Step Functions tasks can be executed in **synchronous** or **asynchronous** mode. Synchronous tasks wait for the execution of their predecessor task. Asynchronous tasks are launched immediately and do not wait for the previous task.

In this recipe, we'll construct a simple State Machine orchestrator using AWS Step Functions. We'll mock a basic FASTQ analysis so that we can focus on the mechanics of Step Functions. By the end of this example, you'll have a basic understanding of Step Functions and how to implement it with `boto3`.

Getting ready

We covered setting up your own AWS account and installing the required client software in *Chapter 14, Cloud Basics*. We'll be using your AWS account for this recipe. If you have not set one up, you can simply read the chapter and follow along with the concepts presented.

Let's make sure our `conda` environment is activated:

```
conda activate ch16-more-workflows
```

Make sure `boto3` is installed:

```
pip install boto3
```

If you have not already done so, you may need to set up your AWS client and default regions using the following:

```
aws configure
```

See *Chapter 14* if you need more details on configuring your account. For this recipe, we will be working in the `aws` subdirectory. The code for this recipe is found in `deploy_bioinformatics_workflow.py`. We'll be working from the terminal for this example.

How to do it...

Let's review the code.

1. Look over the *Import Libraries* section in the file.

    We will use the `boto3` client for interacting with AWS.

2. Next, we configure key AWS parameters:

    ```
    REGION = 'us-west-2'
    LAMBDA_FUNCTION_NAME = 'fastq-quality-analysis'
    STATE_MACHINE_NAME = 'simple-bioinformatics-pipeline'
    IAM_ROLE_NAME = 'BioinformaticsWorkflowRole'
    ```

    Remember to change your region name if you prefer working in a different AWS region. This is the region where any resources created by the script will be deployed.

3. Now review the `create_lambda_function_code()` and `lambda_handler()` sections.

    This code is essentially a function that creates another function. When executed, it will create the function inside the triple-quotes, and that portion will become the actual Lambda function when launched on AWS.

    This lambda function will implement a mock QC step that does the following:

    I.   Obtains the input sample ID and FASTQ file.

    II.   Simulates processing time and generates mock quality metrics using the `random` module.

    III.   Uses some logic to determine whether the quality of the sample is PASS, WARNING, or FAIL.

    IV.   Package the return values into the `results` dictionary.

    V.   Add recommendations based on the metrics.

    VI.   Return the results.

VII. Finally, the function will create a ZIP file of the lambda function in memory. This is used to programmatically deploy the lambda function using the ZIP file. Here is an overview of the process:

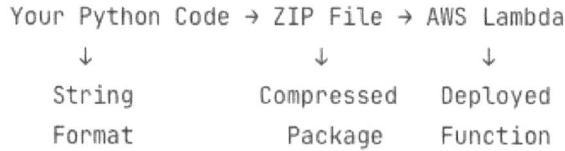

```
Your Python Code → ZIP File → AWS Lambda
        ↓                ↓            ↓
     String         Compressed    Deployed
     Format          Package      Function
```

Figure 16.9 – Overview of the AWS lambda deployment process

4. Next, we have `create_state_machine_definition()`.

This function is at the heart of our orchestrator. It defines the state machine workflow using **ASL**. The workflow has the following steps:

5. `ValidateInput` will bring in the sample ID and FASTQ file. It will add a timestamp using the special variable `$$.Execution.StartTime`. It is a **PASS** state, which means that it passes along its input to an output: `https://docs.aws.amazon.com/step-functions/latest/dg/state-pass.html`.

6. `FastqQualityAnalysis` is a **TASK** state: `https://docs.aws.amazon.com/step-functions/latest/dg/state-task.html`. It runs our lambda function, which is passed in as `LAMBDA_FUNCTION_NAME`. It defines a path to capture the results. It defines a `Retry` parameter in which it will wait two seconds and then try the task again, up to two times. Error handling is also defined, in which any error will return the `AnalysisFailure` state and the error message. This task implements **exponential backoff**, in which the wait time between each retry increases exponentially – for example, if the `BackOffRate` parameter is 2, the first wait is multiplied by 2^0, the next by 2^1, and so on. In this way, services that are overloaded will be given an increased time to recover, increasing the stability of the system. Here is an overview of the retry process:

```
Time →     0s     2s     6s    10s
           |      |      |      |
Initial   ▮      |      |      |    ← Lambda fails
Attempt    |      |      |      |

Wait      |░░░░░|      |      |    ← Wait 2 seconds (2 × 2.0⁰ = 2)
Period 1   |      |      |      |

Retry #1  |     ▮|      |      |    ← First retry fails
           |      |      |      |

Wait      |      |░░░░░|      |    ← Wait 4 seconds (2 × 2.0¹ = 4)
Period 2   |      |      |      |

Retry #2  |      |     ▮|      |    ← Final attempt
           |      |      |      |

Result    |      |      | ☑/✗  ← Success or give up
```

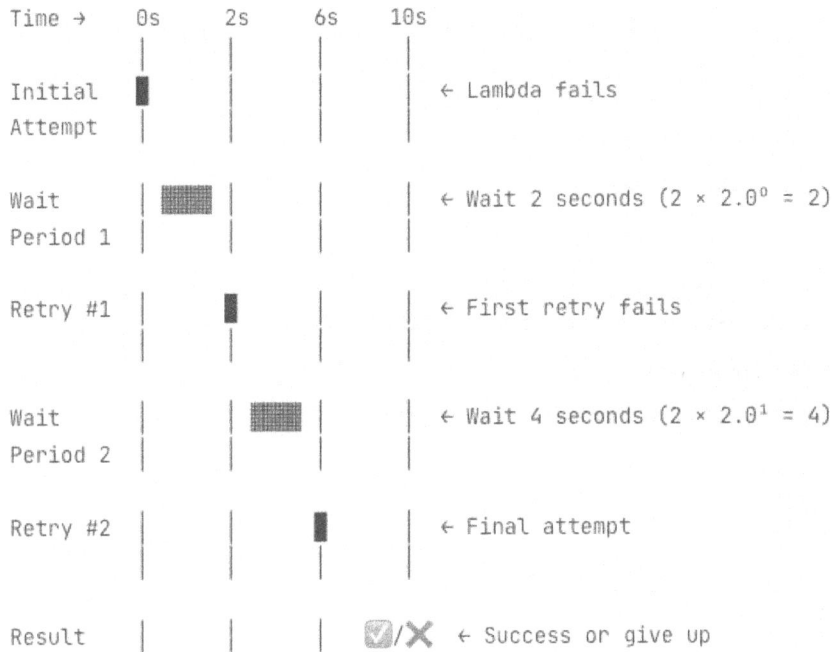

Figure 16.10 – Illustration of exponential backoff in AWS State Machines

7. Next, we have CheckQualityResults. This is a **CHOICE** state, which determines whether the analysis gave a **PASS**, **WARNING**, or **FAILURE** result (defaults to FAILURE): https://docs.aws.amazon.com/step-functions/latest/dg/state-choice.html.

8. Finally, we have the three **terminal states**. These define how the state machine ends. It will either have **Success**, **Warning**, or **Failure**. In each case, we will provide a message and some return values.

9. Our next function is create_iam_role(). It will create the IAM role for us using the boto3 client. The role will have permission to execute our lambda functions.

10. Next, we have create_lambda_function(). This will use boto3 to create the lambda function.

11. Now review create_step_function(). This function will create our State Machine by obtaining the definition of the workflow and using the boto3 Step Functions tools.

12. The next function is called run_workflow_example(). This will actually trigger our state machine. It will provide test input and then call the start_execution() function and wait for the completion of the workflow.

13. Next, we come to our `main()` function. This will do the following:

    I.   Deploy the IAM role

    II.   Create our Lambda function

    III.   Build the State Machine

    IV.   Run the workflow with sample data

14. The final function is called `cleanup()`. We will run this at the end to remove any resources we created, so that we don't incur additional costs.

Let's run our workflow:

```
python deploy_bioinformatics_workflow.py
```

Here is what we see:

```
 Deploying Simple Bioinformatics Workflow...
 Region: us-west-2

Creating IAM role...
 Created IAM role: BioinformaticsWorkflowRole
 Waiting for IAM role propagation...

Creating Lambda function...
 Created Lambda function: fastq-quality-analysis

Creating Step Functions state machine...
 Created State Machine: simple-bioinformatics-pipeline

Running example workflow...
 Started workflow execution: arn:aws:states:us-west-2:495163878159:execution:simple-bioinformatics-pipeline:test-execution-1754011332
 Waiting for execution to complete...
 Execution completed with status: SUCCEEDED
 Execution Results:
{
  "result": "WARNING",
  "message": "FASTQ quality analysis completed with warnings",
  "qualityMetrics": {
    "totalReads": 2321782,
    "averageQualityScore": 26.81,
    "gcContent": 54.43,
    "duplicateRate": 6.39,
    "processingTimeSeconds": 4.19
  },
  "sampleId": "SAMPLE_001",
  "recommendations": [
    "Consider quality trimming"
  ]
}

 Deployment completed successfully!
 State Machine ARN: arn:aws:states:us-west-2:495163878159:stateMachine:simple-bioinformatics-pipeline
 Lambda Function ARN: arn:aws:lambda:us-west-2:495163878159:function:fastq-quality-analysis

 To run the workflow manually:
aws stepfunctions start-execution \
  --state-machine-arn arn:aws:states:us-west-2:495163878159:stateMachine:simple-bioinformatics-pipeline \
  --input '{"sampleId": "TEST_SAMPLE", "fastqFile": "s3://bucket/file.fastq.gz"}'
```

Figure 16.11 – Output of the State Machine deployment

This output shows the deployment of the various components and the creation of the State Machine. We run a test and see the example output returned. We are given the **ARN** of the State Machine and the lambda. This is the **Amazon Resource Name**, which uniquely specifies resources in AWS. These identifiers are unique across all of AWS, even beyond your own account. The ARN can be used to execute the State Machine or monitor the workflow. You could even use it to create a **CloudWatch** alarm for failed executions.

These are the resources created:

- **State machine**: The overall workflow orchestrator, which manages the inputs, lambdas, and outputs
- **Lambda function**: The function that runs our mock FastQC analysis
- **IAM role**: The identity role needed for deploying the resources

Remember that this example will create the AWS resources in the region you specified (in this case, it was **US-West-2**). If you are looking for the resources in the AWS Management Console, make sure to set it to the right region using the dropdown in the upper-right corner.

To see your Step Functions, look for **Step Functions** in the AWS search bar in your GUI console. This is what your Step Functions looks like in the interface:

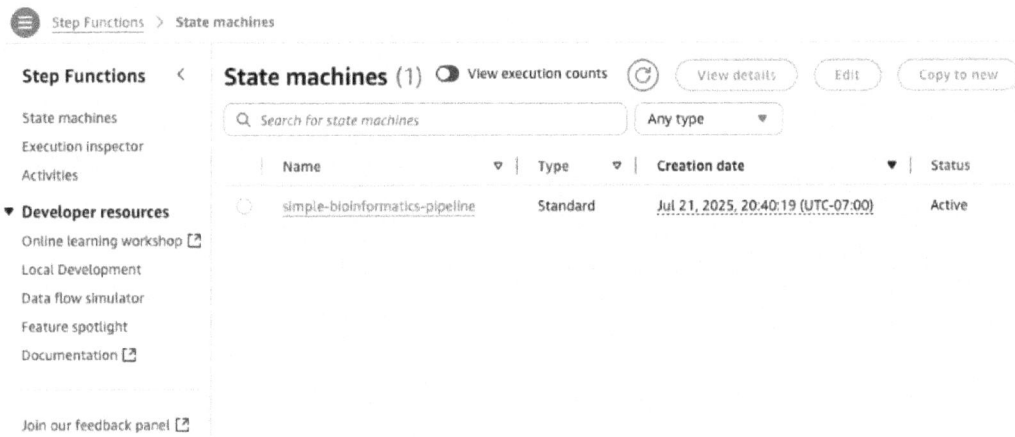

Figure 16.12 – AWS console display of the step function

If you drill down by clicking on the step machine named `simple-bioinformatics-pipeline`, you can see details of the execution:

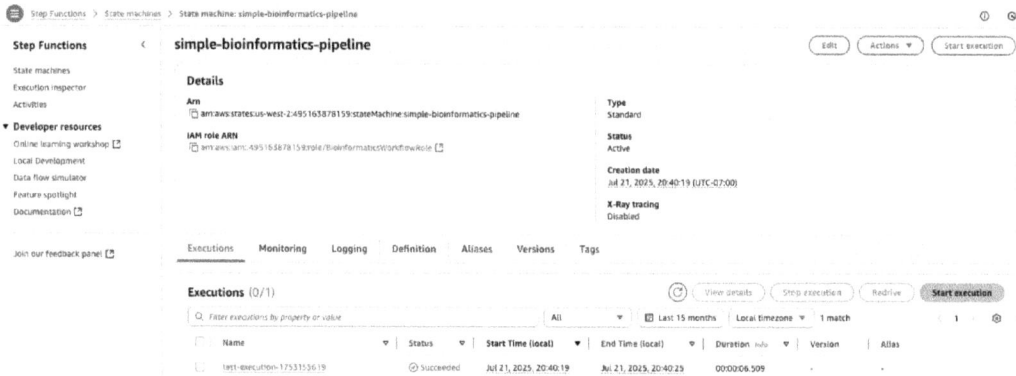

Figure 16.13 – Successful execution of the Step Functions

To see your Lambda function, search for `Lambda` in the AWS search bar. Then click on the name of the lambda function you created: `fastq-quality-analysis`

Here is what you will see:

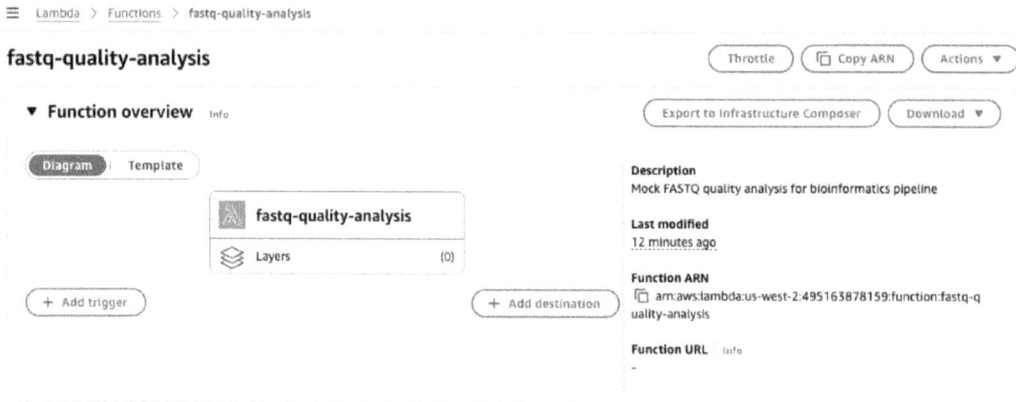

Figure 16.14 – Lambda function for FASTQ quality analysis

The function, its code, and other details about it are shown in the interface.

To see the IAM role created by the deployment, search for **IAM** in the AWS Management Console, and then click on **Roles**. Here is what is shown:

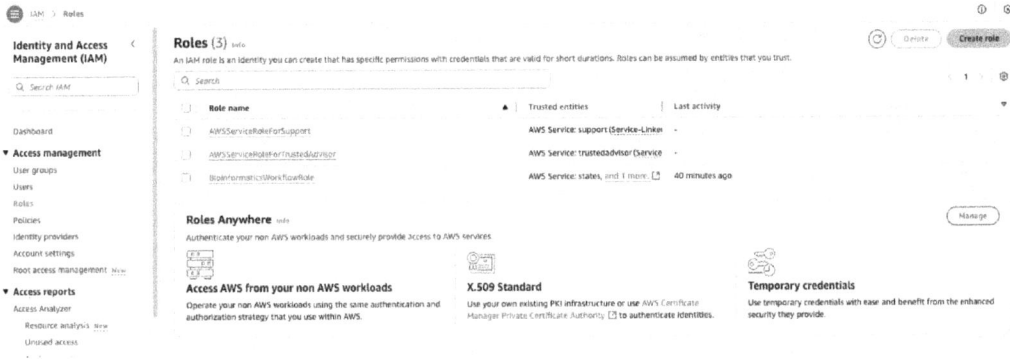

Figure 16.15 – IAM roles shown in the AWS Management Console

This shows `BioinformaticsWorkflowRole`, which was created to run the Step Functions.

Okay, great – we have run our workflow and reviewed the resources deployed using the AWS Management Console. Now let's clean up the resources so that we don't incur additional costs:

```
python deploy_bioinformatics_workflow.py cleanup
```

You can double-check the areas where we created resources using the AWS Console again to make sure that you've removed them (be sure you are set to the right AWS Region). You may also want to review your AWS cost center and budget at this time, just to make sure there are no costs being incurred in your account.

There's more...

Lambdas have some key limitations to be aware of. They can only run for up to 15 minutes. They also have limitations in the memory and other resources they can use. For larger jobs, you will want to look at using **AWS Batch**.

A lambda can trigger an AWS Batch job. It might look something like this:

```
import boto3
import json
def lambda_handler(event, context):
    batch_client = boto3.client('batch')
    response = batch_client.submit_job(
        jobName='genomics-analysis-' +
        context.aws_request_id,
```

```
        jobQueue='genomics-processing-queue',
        jobDefinition='genomics-pipeline-job',
        parameters={
            'inputFile': event['s3_object_key'],
            'outputBucket': 'processed-genomics-data',
            'sampleName': event.get(
                'sample_name', 'unknown'
            )
        }
    )
    return {
        'statusCode': 200,
        'body': json.dumps({
            'jobId': response['jobId'],
            'jobName': response['jobName']
        })
    }
```

The preceding code defines a lambda that uses the boto3 `batch_client.submit_job()` function to launch an AWS Batch job.

Lambda layers provide a way to use shared libraries across your functions: `https://docs.aws.amazon.com/lambda/latest/dg/chapter-layers.html`. This can be very helpful for reducing the size of your package deployments.

> **AI tip**
>
> You can try building a more complex State Machine with multiple dependent functions:
>
> **Prepare**: Paste or upload the `deploy_bioinformatics_workflow.py` code into your AI chat window.
>
> **Prompt**: Update this example to include three lambdas that execute sequentially.
>
> **You should see**: Code to build a state machine with three :Lambdas that execute synchronously. The system will pass through data from one Lambda to the next. IAM roles and deployment scripts with instructions will be included.

Step Functions and Lambdas can also be deployed using **CloudFormation**, `https://docs.aws.amazon.com/step-functions/latest/dg/tutorial-lambda-state-machine-cloudformation.html`.

> **AI tip**
>
> **Prompt**: Write code to deploy a three-state workflow using AWS Step Functions for bioinformatics analysis using CloudFormation.
>
> **You Should See**: CloudFormation templates for deploying a State Machine with three steps, along with deployment instructions.

You have now gotten a basic introduction to orchestration using AWS Step Functions. It is possible to build complex, production-ready bioinformatics workflows using these approaches. Such systems can be integrated with AWS services such as **EventBridge** and **DynamoDB** to provide event-driven executions, configuration databases, and other capabilities to build full-fledged workflow management systems, just like the others we have seen in this chapter.

You've now learned a great deal about bioinformatics workflow management and the tools and platforms used to implement them. You are poised to utilize all the skills and tools you've learned in previous chapters to build complex, scalable workflows!

Okay, let's clean up and close down our `conda` environment:

```
conda deactivate
```

See also

- This AWS tutorial shows you how to create a State Machine: `https://docs.aws.amazon.com/step-functions/latest/dg/tutorial-creating-lambda-state-machine.html`

- Watch this tutorial on Step Functions: `https://www.youtube.com/watch?v=GVpmVu8vcNQ`

- This course covers Lambdas, Step Functions, and other serverless capabilities: `https://www.udemy.com/course/aws-lambda-serverless-developer-guide-with-hands-on-labs/`

Get This Book's PDF Version and Exclusive Extras

UNLOCK NOW

Scan the QR code (or go to `packtpub.com/unlock`). Search for this book by name, confirm the edition, and then follow the steps on the page.

Note: Keep your invoice handy. Purchases made directly from Packt don't require an invoice.

## 17

Deep Learning and LLMs for Nucleic Acid and Protein Design

In this chapter, we will cover one of the most important topics of our time. The rise of **Machine Learning** (**ML**), powered by exponential advances in computing technology, has transformed many industries, including bioinformatics. ML is the use of computation to learn and infer results from data. It includes a variety of algorithms from the simple to the complex, all the way from traditional data science techniques that we learned about in *Chapter 4*, such as K-means, PCA, and decision trees, to **Neural Networks** (**NNs**), which we will cover in this chapter. **Deep Learning** (**DL**) is a subset of ML that uses NNs with many layers. One of the key reasons DL has become so powerful in the last decade or so is simply that computers have reached a level of performance at which they could simulate more than a handful of deep layers in NNs, leading to a breakout in the capability of ML models.

ML has begun to transform bioinformatics in numerous ways and at every level of analysis. It impacts data generation through algorithms embedded into sequencing equipment. For example, ML is typically integrated into nanopore devices to enhance signal detection; see Wan et al., *Beyond sequencing: machine learning algorithms extract biology hidden in Nanopore signal data*, Trends in Genetics, Mar 2022 – `https://www.cell.com/trends/genetics/fulltext/S0168-9525(21)00257-2`. It is used in sequence alignment; see Dotan et al., *BetaAlign: a deep learning approach for multiple sequence alignment*, Bioinformatics, Jan 2025 – `https://academic.oup.com/bioinformatics/article/41/1/btaf009/7945664`. ML is used in variant calling tools such as DRAGEN and DeepVariant; see Asri et al., *Pangenome-aware DeepVariant*, bioRxiv, Jun 2025 – `https://pubmed.ncbi.nlm.nih.gov/40501862/`. We have already seen that ML is used extensively in protein structure prediction using tools such as AlphaFold. ML is driving rapid advances in protein design; see Albanese et al., *Computational protein design*, Nature Reviews Methods Primers, Feb 2025 – `https://www.nature.com/articles/s43586-025-00383-1`. Scientists are even using it to design proteins that go beyond nature; see Li et al., *RareFold: Structure prediction and design of proteins with noncanonical amino acids*, bioRxiv, May 2025 – `https://www.biorxiv.org/content/10.1101/2025.05.19.654846v1`.

AI is being used extensively in synthetic biology; see Groff-Vindman et al., *The convergence of AI and synthetic biology: the looming deluge*, NPJ Biomedical Innovations, Jul 2025 – `https://www.nature.com/articles/s44385-025-00021-1`.

We are also witnessing rapid applications in cancer research and many other areas; see Yang et al., *MLOmics: Cancer Multi-Omics Database for Machine Learning*, Scientific Data, May 2025 – `https://www.nature.com/articles/s41597-025-05235-x`.

Because ML is such a large topic, we will not be able to cover everything in this chapter. We will provide you with extensive resources here to deepen your studies and explore further. By the end of this chapter, you should have a good sense of the basic concepts of ML and how it is impacting bioinformatics by looking at a few key examples.

In this chapter, we will cover the following recipes:

- Machine learning with PyTorch
- Designing proteins with LLMs
- Designing DNA with LLMs

Technical requirements

The code for this chapter can be found at `https://github.com/PacktPublishing/Bioinformatics-with-Python-Cookbook-Fourth-Edition/tree/main/Ch17`.

Create a `Ch17` folder and set up your notebooks there.

Remember to activate your `conda` environment before beginning the recipes, like this:

```
conda activate bioinformatics_base
```

Or, if you would like to set up a `conda` environment specific to this chapter, before activating `bioinformatics_base`, run the following:

```
conda create -n ch17-machine-learning --clone bioinformatics_base
conda activate ch17-machine-learning
```

You will be able to install the packages for the chapter as you go, or you can use the YAML file provided in the repository:

```
conda env update --file ch17-machine-learning.yml
```

Machine learning with PyTorch

It is first important to understand the basics of NNs. NNs originated from early efforts to understand how the human brain functions. However, the first NNs were not all that brain-like in that they tended to contain linear components (the brain's neurons are highly non-linear) and they were trained using backpropagation – we'll cover this in a moment and discuss why it is not like human learning. Despite that, increased computational power has somewhat compensated for this and led to the breakthroughs we see today. NNs contain nodes, like neurons, that are connected to each other by weights, similar to dendrites in the human brain, and each node implements some transform called an **activation function** on the signals coming into it.

Here is an overview of a simple NN:

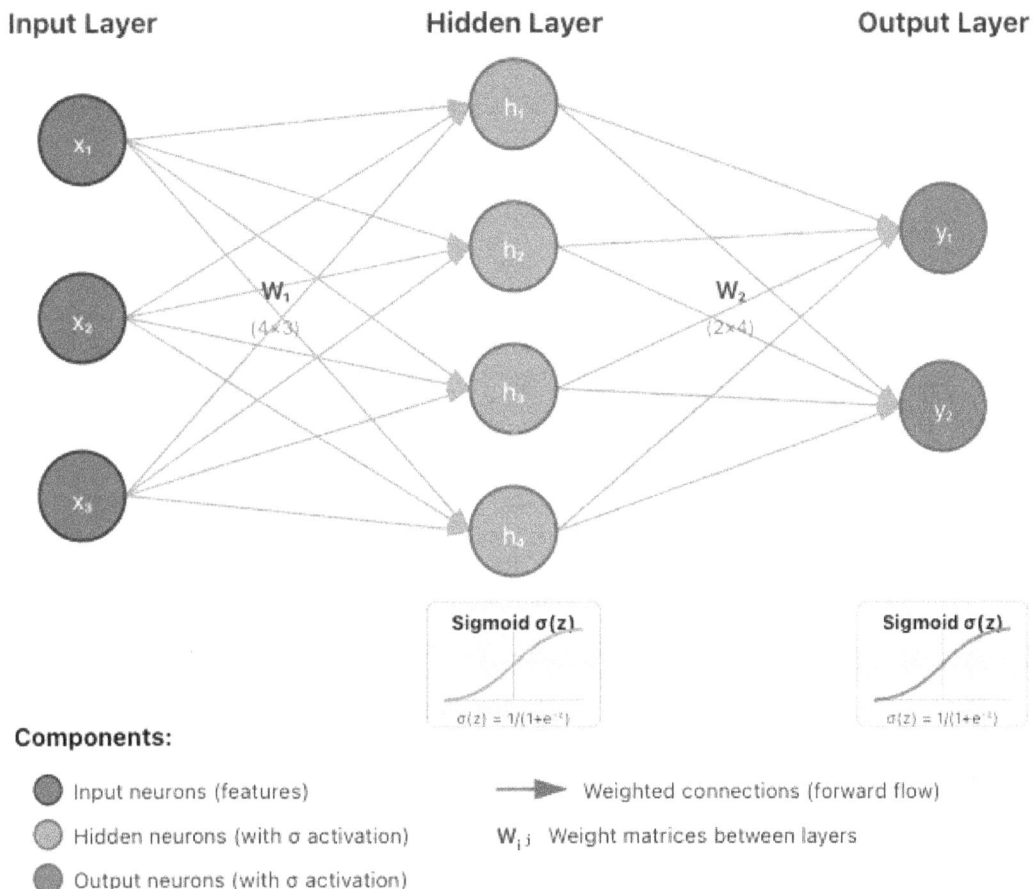

Figure 17.1 – A standard NN

We first have an **input layer**, which takes in the input data to be processed. Next, there are one or more **hidden layers**, which perform the computations. The inputs from the first layer are summed up as they come into the next layer, and then the activation function is applied. In this sense, we can think about whether there is enough signal coming into any given node to cause it to "fire" like a neuron and pass its signal on to the next layer. Finally, there is an **output layer**, which translates our signals into the final predictions. The weights coming into each node are stored in a matrix, *Wij*. The activation function is typically something like a sigmoid (`https://en.wikipedia.org/wiki/Sigmoid_function`). This is useful because it smooths out non-linear effects at the extreme highs and lows of activation, while behaving in a mostly linear fashion in the primary range. Other common activation functions include tanh (hyperbolic tangent) and ReLU (`https://www.geeksforgeeks.org/deep-learning/relu-activation-function-in-deep-learning/`).

This type of NN is called a **feed-forward NN** because signals flow from left to right. There are other types of NN architectures, such as **Recurrent Neural Networks** (**RNNs**), which allow for loops in which signals can travel backward. RNNs are especially well suited to time-series or linguistic analysis. The preceding network is also called **fully connected** (or *dense*) because each neuron in a layer is connected to every other neuron in the next layer. By contrast, the human brain is **sparsely connected**.

The most important part of an NN comes next: instead of trying to write an explicit program to solve a problem (as we've been doing throughout this book), we are going to *train* the NN to perform a task. In this way, it can potentially learn patterns that we might not easily be able to understand or program. This is the power of ML. We train the network by first performing a forward pass with our data and then comparing the results from the output layer with the results we wanted. For example, let's say we want to recognize pictures of cats and distinguish them from other animals. The picture of a cat would be transformed into a numerical representation and fed into the input layer. It would then go through the NN and be read out in the output layer. We would compare it to known pictures of cats. This is known as **labeling** and is critical for ML. Most ML models require a large corpus of labeled data containing both correct and incorrect examples for training.

Let's imagine we put in a picture and get an incorrect result. The next thing we do is work backward by looking at the weights in the network and penalizing them by adjusting the weights. This process is known as **backpropagation** and is the most common training technique for NNs. Over the course of many training runs, called **epochs**, we will eventually arrive at a network that is relatively well trained to recognize cats. Backpropagation suffers from numerous limitations – it can be computationally expensive, prone to getting stuck in local minima, and, in deep networks, may suffer from issues such as exploding gradients, which can make training unstable; see `https://www.geeksforgeeks.org/deep-learning/vanishing-and-exploding-gradients-problems-in-deep-learning/`. Despite these limitations, given sufficient computational power and large datasets, it has been able to achieve amazing things. It is also not very biological, in that it would seem hard for neurons to change their synaptic strength by sending an error signal backward. You can read more about this in Lillicrap et al., *Backpropagation and the brain*, Nature Reviews Neuroscience, Apr 2020 – `https://www.nature.com/articles/s41583-020-0277-3`. NNs do, however, exhibit many brain-like behaviors, including aspects of **Hebbian learning**, first identified in the brain, in which *neurons that fire together wire together*.

These are the core concepts of NNs. Before going further, you may wish to explore some background material here:

- **NN basics**: `https://www.geeksforgeeks.org/machine-learning/neural-networks-a-beginners-guide/`

- **ML for beginners**: `https://victorzhou.com/blog/intro-to-neural-networks/`

- **Udemy course on NNs**: `https://www.udemy.com/course/fundamentals-in-neural-networks/`

One further advancement to discuss is the **Convolutional Neural Network** (**CNN**), which we will use in this recipe. CNNs introduce the idea of a *convolution* in which a **kernel** or filter is slid over the input data. The kernel is a small two-dimensional array of weights that can recognize simple **features**. For example, a feature might be a diagonal line or maybe something that looks like the outline of a cat's ear. In this sense, it is easy to think of a CNN as being like the human eye, building up an image from a series of smaller features.

CNNs are ideal for image recognition, and this is the easiest way to think about them. But you can transform many types of data into the equivalent of an image – for example, coverage data along every position in a genome. That's exactly what we are going to do in this recipe: we will use coverage data to train a CNN to recognize large changes in coverage that may represent a **duplication** of a gene (2x coverage) or a **deletion** of a gene (0 coverage). This is known as **Copy Number Variation** (**CNV**) analysis and is critical in genetic testing; see Bowman et al., *Whole genome sequencing for copy number variant detection to improve diagnosis and management of rare diseases*, Developmental Medicine & Child Neurology, Jun 2024 – `https://onlinelibrary.wiley.com/doi/full/10.1111/dmcn.15985`.

Here is an illustration of how CNNs work:

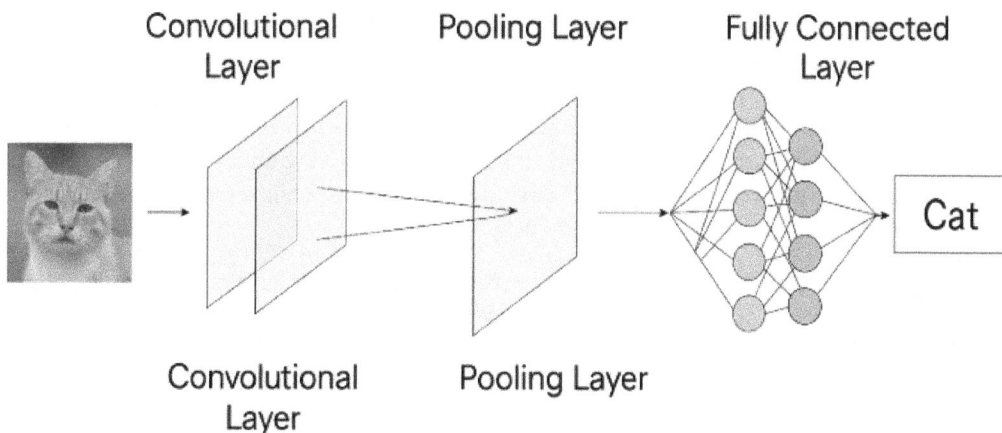

Figure 17.2 – Overview of CNN architecture

Before continuing, you may want to learn more about CNNs with these resources:

- **Introduction to CNNs**: `https://www.datacamp.com/tutorial/introduction-to-convolutional-neural-networks-cnns`

- **Video on CNNs**: `https://www.youtube.com/watch?v=m0fmBY4t5BI`

- **Udemy course on CNNs**: `https://www.udemy.com/course/convolutional-neural-networks-deep-learning/`

- **Review article**: Alzubaidi et al., *Review of deep learning: concepts, CNN, architectures, challenges, applications, future directions*, Journal of Big Data, Mar 2021 – `https://link.springer.com/article/10.1186/s40537-021-00444-8`

CNNs have different layers that are alternated to produce the overall architecture of the NN model. The choice and order of the layers can be very important for building a strong model. In this recipe, we'll build a CNN model and visualize its architecture. The key layers are as follows:

- **Input layer**: Formats the raw data

- **Convolutional layer**: Applies the kernels to the input data

- **Activation layer**: Applies non-linear functions to the data

- **Pooling layer**: Sub-samples and simplifies the data to avoid over-fitting

- **Fully connected layer**: Similar to a traditional NN layer, makes the final classification

- **Output layer**: Makes the classification and formats it to the expected output

In this recipe, we will build a CNN to perform genomic copy number analysis. We will use **PyTorch** (`https://pytorch.org/`), one of the most popular libraries for ML in Python. It supports the easy construction of a wide variety of NN types. It supports **GPU** (short for **graphical processing unit**) acceleration using **CUDA** (`https://developer.nvidia.com/cuda-toolkit`), which is a framework developed by NVIDIA for using the GPUs available on your system for NN models. If CUDA is not available on your system, the code in this recipe will default to using your CPU. This simply means that the recipe may run more slowly as it will use your CPU instead of a GPU – but for the purposes of this recipe, which is small, it should not have a big impact. PyTorch interfaces with **Hugging Face** (`https://huggingface.co/`), a popular public repository for AI models. Hugging Face provides many **Transformer** models.

Transformer models convert words or **tokens** into vectors that represent the data, called **embeddings**. Tokens are typically words or sub-words in a sentence – they are essentially the smallest unit of encoding in a ML model. Transformers use a mechanism called **attention** to look at other words or tokens in the data, even over a long range, to find relationships in the data. They "mask" some of the tokens and then try to reconstruct the sentence, iterating in this way over and over until they learn the material deeply. The most common example usage of Transformers is on words and sentences written in English or other languages. Familiar examples in this space include chat platforms such as ChatGPT and Claude. However, the same methods can be applied to strings of DNA or protein sequences. These biological sequences can also be thought of as languages that contain words and sentences. Unlike RNNs, Transformers can work in parallel on large sequences of text or data. This has made them a transformative tool. Transformers were first described in Vaswani et al., *Attention is all you need*, NIPS 2017 – `https://proceedings.neurips.cc/paper/2017/hash/3f5ee243547dee91fbd053c1c4a845aa-Abstract.html`.

One popular early Transformer was **Bidirectional Encoder Representations from Transformers** (**BERT**), which analyzes sentences in both directions; see Devlin et al., *BERT: Pre-training of Deep Bidirectional Transformers for Language Understanding*, arXiv, Oct 2018 – `https://arxiv.org/abs/1810.04805`.

Transformers are used as encoder-decoders. Once they have learned the representation of a corpus of text, they can also generate an equivalent response from a section of that text. This gave rise to **generative AI**, in which an AI model can respond to questions, leading to programs such as **ChatGPT** and **Claude**, which we have covered in this book.

Besides PyTorch, other important Python libraries to know about for ML are **TensorFlow** (`https://www.tensorflow.org/`) and **Keras** (`https://keras.io/`).

It is also important to understand **cross-validation**. In this method, the training data is split into *k* subsets or **folds**. During each round of training, the model is trained on all but one of the subsets of data, and then tested against the **held-out** dataset. This is repeated until all subsets have been used. This method reduces overfitting, provides for more robust training, and gives a fairer test of model performance (`https://scikit-learn.org/stable/modules/cross_validation.html`).

Large, pre-trained models are now available for many tasks, including Natural Language Processing (NLP), DNA sequence analysis, and protein analysis and design. These are called **foundation models**. You can use them to avoid incurring massive training costs and extend them for your own use. You can fine-tune them or extract embeddings from them. Another key concept to know about is **Retrieval-Augmented Generation** (**RAG**). In this method, models are allowed to retrieve external documents during searches in order to enhance results and obtain up-to-date information.

Another important advancement in AI has been the rise of **end-to-end learning**. This innovation meant that instead of doing a lot of preprocessing and filtering before training a model, training occurs directly from raw input to output. An example would be trying to parse French phrases first to categorize them to train a language model, versus simply giving raw French phrases and their English translations to a model and having it learn from that directly. This method has turned out to be very powerful for training NNs. The recipe presented here is an example of end-to-end learning; we will be going directly from raw sequence coverage data to calls about gene deletions or duplications.

Getting ready

Let's install PyTorch!

```
! pip install torch
```

You should have the other libraries you need (scikit-learn, numpy, pandas, matplotlib, and seaborn) from previous chapters already.

The code for this recipe can be found in `Ch17/Ch17-1-machine-learning.ipynb`.

How to do it...

1. Let's begin by importing the libraries:

```
import torch
import torch.nn as nn
import torch.optim as optim
import torch.nn.functional as F
from torch.utils.data import Dataset, DataLoader
import numpy as np
import pandas as pd
import matplotlib.pyplot as plt
import seaborn as sns
from sklearn.metrics import (
    classification_report, confusion_ matrix)
import warnings
warnings.filterwarnings('ignore')
```

We will bring in the `torch` library to utilize PyTorch. This is our core ML library. From `torch` (PyTorch), we will bring in the nn library. This module greatly simplifies the construction of NNs. The `optim` library provides the optimization functions for training the NN. The next library, `torch.nn.functional`, will provide our convolution functions. The functions in this module are **stateless**, meaning they will not store and update their parameters. This can provide added flexibility for us in constructing our network. We also bring in `torch` utilities for data handling and some of our standard libraries for data handling and visualization. From scikit-learn, we will import `classification_report` and `confusion_matrix`. The classification report will provide us with a handy summary of the precision, recall, and other key characteristics of our model. The confusion matrix will be used to summarize our predicted versus actual performance. Finally, we bring in the `warnings` module and suppress minor warnings in our notebook.

2. Next, we will set the random seeds:

```
torch.manual_seed(42)
np.random.seed(42)
```

This will help make our work more reproducible.

3. Now we are going to define a class for simulating CNV data. Review the code for the CNVDataGenerator class in the notebook. This class will help us create simulated CNV data by making bins of coverage data and then simulating increases or decreases in this data, much like real variations in copy number. This class will implement the following features:

    ▪ Take in a number of samples to generate (default is 1000). We also set a number of bins and the size of each bin.

    ▪ Define a generate_normal_coverage() function, which provides a realistic baseline of sequencing coverage. It defines a mean coverage of 30x depth and a standard deviation, and then uses a normal distribution to set a base coverage. It then adds a simulated GC bias (some regions tend to sequence more poorly when the percentage of G and C bases is higher) and a random noise component.

    ▪ Define an introduce_cnv() function, which will simulate a CNV by either introducing a **deletion** by multiplying the coverage by a reduction factor or simulating a **duplication** event by multiplying the coverage by an amplification factor.

    ▪ Build a function to generate a complete CNV dataset called generate_dataset(). This function creates a number of samples as defined by the input. It will introduce copy number changes into the underlying data and label them with a 0 for no change, a 1 for a deletion (both copies of the gene or region have gone away) or a 2 for a duplication (there are now two copies of a gene where there used to be one). It first sets a baseline coverage with labels as normal (zero). It then randomly adds one or more CNVs of a random length and labels them as a deletion (1) or a duplication (2). It outputs an array, X, of coverage data, an array of labels, y, and cnv_positions containing metadata for each sample about the type and position of each CNV in the sample.

4. We next define a simple class to hold CNV datasets:

```
class CNVDataset(Dataset):
    def __init__(self, X, y):
        self.X = torch.FloatTensor(X)
        self.y = torch.LongTensor(y)
    def __len__(self):
        return len(self.X)
    def __getitem__(self, idx):
        return self.X[idx], self.y[idx]
```

This class is a wrapper around PyTorch datasets (https://docs.pytorch.org/ tutorials/beginner/basics/data_tutorial.html). The dataset provides a convenient wrapper around data for PyTorch that stores the data and corresponding labels in a way that is optimized for NN training. There are three key methods that are defined. The init method is the constructor for the class and stores the data as PyTorch tensors. The len method simply returns the total number of samples in the dataset. The `getitem` method returns a sample based on an index. The datasets will be used by the PyTorch `DataLoader` class to batch creation during training.

5. Our next class is the CNV detector. This is the CNN at the heart of our recipe, which detects and classifies CNVs. Review this code in the notebook. It is in the `class CNVDetectorCNN` section in the `Ch17-1` notebook:

    *   This class implements a CNN that will output three states for copy number: normal (no change), deletion, or duplication. It takes sequence coverage as input and outputs a copy number prediction for each position in the sequence.

    *   The first function performs class initialization. Note that we first call the `super()` method. This is a key part of **Object-Oriented Programming** (**OOP**) in which we call the parent class constructor from this inherited class. We then set the sequence length, which defaults to `200`, and the number of classes to predict, which will be three here.

    *   Next, we set up our convolutional layers. We will first set up three convolutional layers with different kernel sizes to capture different levels of coverage variation. For example, small kernels tend to capture noisy fluctuations in coverage, whereas larger kernels will capture broader CNV signatures. This is known as multi-scale design and is designed to capture features at various levels to ensure a robust and accurate model.

    *   Next, we set up two layers designed to combine the multi-scale features.

    *   We next create batch normalization layers and layers for dropout regularization. These will help us provide more stable training and reduce overfitting.

    *   We extract global average behavior features into the `global_pool` variable.

    *   The classification layers are set up, which provide the final feature classification decision (normal, deletion, or duplication).

    *   Finally, we set up the dropout layer.

6. The next key function we will set up in this class is the `forward()` method, which will provide forward pass analysis. This will pass data through our CNN and perform the analysis:

    *   First, perform a dimensional transformation to use one-dimensional convolutional analysis.

    *   Set up convolution layers using ReLU units.

    *   Combine the features into a single tensor using the `torch.cat()` method.

    *   Perform batch normalization and dropout.

    *   Add the second and third convolutional layers.

    *   Extract global features.

    *   Combine the original signal and original features.

- Set up the fully connected layers.

- Reshape the output to match the expected dimensions.

That concludes our detector function, which implements the CNN layers and analyzes the data for CNV calls.

7. This next class provides training and evaluation of our CNN models. Review the `CNVTrainer` class in the notebook.

This class performs the training for our CNN. It will also help us evaluate the accuracy of our model:

- First, define the `init()` method. The device will default to CPU here if CUDA is not available. We will also pass in our CNN model. We will track training loss and accuracy, as well as validation results, to see how well our model generalizes.

- The next function is `train_epoch()`. This will perform the training on our model. We will pass in `train_loader`, which is the dataset with our training batches. This function will also be passed an **optimizer function**. Optimizers calculate a **loss function** during each training round based on how close or far away the NN model is from performing well. The optimizer seeks to maximize an **objective function**, which defines how well the NN output matches the labeled training data. With each round of training, the optimizer calculates a **gradient**, which determines how much and in which direction to update the weights to (hopefully) make the NN perform better in the next round of training. One popular optimization method you should learn about is **Stochastic Gradient Descent** (**SGD**); see `https://en.wikipedia.org/wiki/Stochastic_gradient_descent`. In this recipe, we will be using the highly popular **Adaptive Moment Estimation** (**Adam**) optimizer; see `https://www.geeksforgeeks.org/deep-learning/adam-optimizer/`.

- We call `self.model.train()` to set our model to training mode.

- Next, iterate through the training batches. Our coverage data will be in `batch_x`, and our labels will be in `batch_y`. We will zero out our gradients before each pass, so they do not accumulate.

- During each pass, we perform a forward calculation to make predictions and then calculate a loss using our criterion.

- Then, a backward pass occurs. During this step, gradients (differences between the network weights and the desired outputs) are computed via backpropagation, and then network weights are adjusted using our optimizer to improve performance.

- Lastly, calculate and return summary information for this epoch of training.

- The next function is `validate()`, which will help us to validate our network architecture against a new set of (non-training) data. We will pass in the `DataLoader` class containing our validation batches and the criterion we plan to use:

- Set our model to evaluation mode using `self.model.eval()`.

- Loop over the batches with gradients off.

- Run the model for each batch and calculate the loss using our criterion.

- Compute and return accuracy metrics.

- The final piece of this class is the `train()` function. This is the master function that will train and validate our CNN model:

- Set our optimizer as the Adam optimizer.

- Use **cross-entropy loss** as the loss criterion (`https://en.wikipedia.org/wiki/Cross-entropy`).

- Set up a scheduler to adjust the learning rate based on training. We want our model to learn more quickly at the beginning and then slow down changes as we refine toward the end.

- Next, iterate through a specific number of training **epochs**. Each epoch will represent one pass through the entire training dataset. During each training phase, we will call `train_epoch()` to set the model to training mode, run forward passes on our training data, calculate gradients and losses, and update our model parameters based on the optimizer.

- Then, validate the model. This is performed after each epoch to help monitor for overfitting of the model. We will also practice **early stopping**, which means we will stop training early if we are failing to improve the validation of the model (`https://en.wikipedia.org/wiki/Early_stopping`).

- Adjust the learning rate based on validation loss.

- Store the metrics (this will be used for plotting later on).

- Print a progress message every 10 epochs.

- This wraps up our training class! We now have the core elements of our CNN model defined. The next few functions will be focused on visualizing and interpreting our results.

8. Now we are going to add a function to visualize our network architecture. Review the code for the `visualize_network_architecture()` function in the notebook.

    This function will help us visualize the architecture of our CNN. We first set up our layers as follows:

    - **Layer 0**: Input coverage sequences

    - **Layer 1**: Multi-scale convolutions

    - **Layer 2**: Concatenate multi-scale features + batch normalization

    - **Layer 3**: Second convolutional layer with feature integration

    - **Layer 4**: Third convolutional layer with deeper feature learning

    - **Layer 5**: Global pooling combined with original signal

- **Layer 6**: First fully connected layer

- **Layer 7**: Second fully connected layer

- **Layer 8**: Output layer (classify as **normal**, **deletion**, or **duplication**)

Then, loop through and draw a box for each layer. We add arrows between each box and annotation as to the purpose of each layer on the side. We also add a legend in the upper right.

The plot is displayed. It will also be saved to the `ch17-1-network-architecture.png` file in your working directory.

9. The next function will be used for visualizing training results. Review the `visualize_training_results()` function in your notebook:

   - Pass in the `CNVTrainer` instance and set up our plots. The x axis will be based on the number of epochs of training.

   - Create a loss plot to show the loss in training, and an accuracy curve to show the change in accuracy.

   - Display the plot. It will also be saved to `ch16-1-training-results.png`.

10. The following function helps you visualize CNV predictions. Review the `visualize_cnv_predictions()` function in your notebook:

    - The function takes as inputs the model, test data, true CNV labels, and number of samples to visualize. We also set our model to evaluation mode.

    - We define the class names for normal, deletion, and duplication.

    - Next, with gradients disabled, we process the samples. The sample input is retrieved, and we run our model to get the prediction.

    - The coverage data is plotted. The true CNVs are plotted by filling in a background showing their true location.

    - The predicted CNVs are plotted with triangular markers.

    - The plot is displayed and saved to the `ch17-1-cnv-results.png` file.

11. Now review the `main()` code used to run the CNN model. This code does the following:

    - Checks whether CUDA is available. If not, it sets the device to CPU.

    - Generates our synthetic data using the `CNVDataGenerator` class and prints a summary.

    - Splits our data into training and validation sets.

    - Creates the `DataSet` and `DataLoader classes` using the `CNVDataSet` class.

    - Builds the model using the `CNVDetectorCNN` class and shows its parameters.

- Visualizes the CNN model architecture.

- Trains the model using the `CNVTrainer` class.

- Visualizes the training results.

- Evaluates the model on the test set.

- Generates the classification report.

- Visualizes the predictions.

12. OK, that's it! Let's utilize our system to train and run an ML model for CNV prediction. Go ahead and run the code and review the results.

Here is our output:

```
=== CNV Detection with PyTorch ===

Using device: cpu

1. Generating synthetic CNV data...
Generated 800 samples with 200 genomic bins each
Class distribution:
  Normal: 130,661 bins (81.7%)
  Deletion: 14,893 bins (9.3%)
  Duplication: 14,446 bins (9.0%)

Dataset splits:
  Training: 560 samples
  Validation: 120 samples
  Test: 120 samples

2. Creating neural network model...
Model created with 694,232 total parameters (694,232 trainable)
```

The first part of the output goes over the basic setup. We generated 800 samples with 200 genomic regions (bins) each. The majority of these bins are made to be normal, and some percentage of deletion and duplication events are introduced into the data.

Next, the datasets are created. We will have 560 samples for training. 120 samples are reserved for validation, and 120 samples for final testing.

Then, the CNN model is initialized, and we see the number of parameters it will have. These parameters will be trained by the optimizer during our training passes.

Then, we have the next section of output:

```
3. Visualizing network architecture...
```

Here is an overview of our CNN architecture:

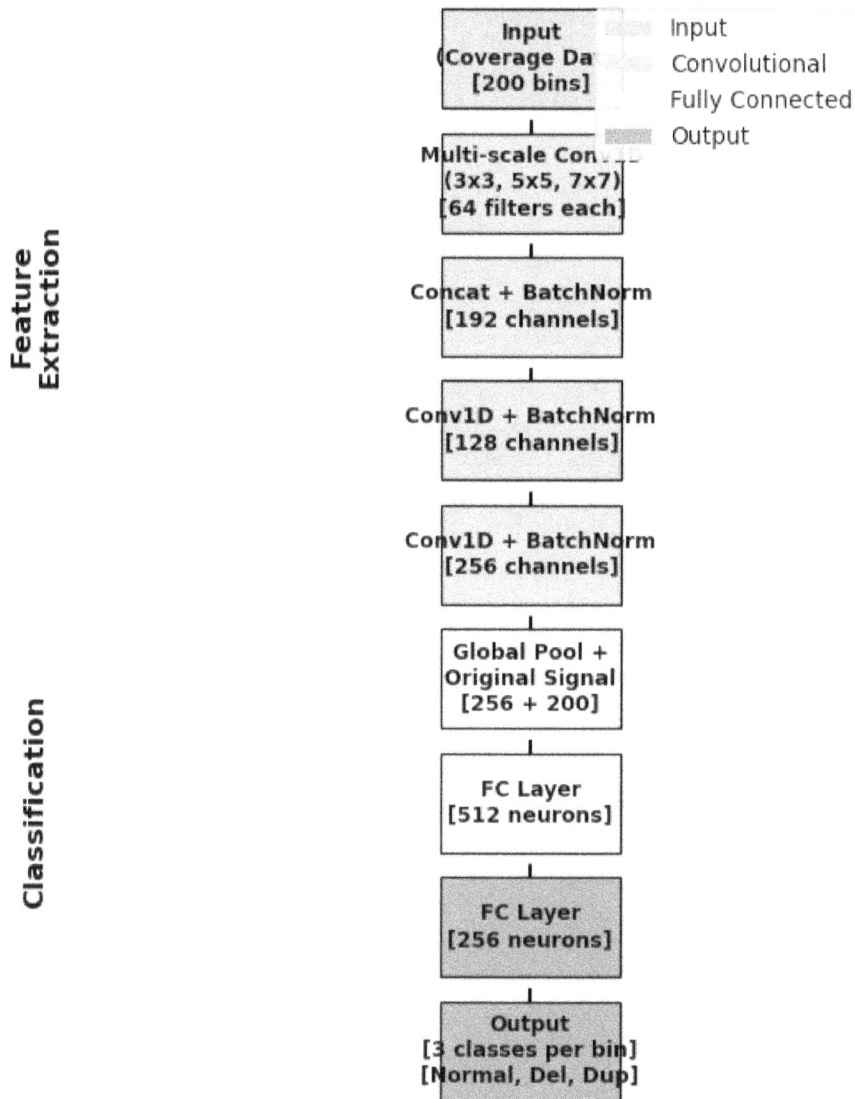

Figure 17.3 – CNN model architecture

The preceding diagram shows our CNN model with the layers we set up as discussed previously.

Here is the output from training the model:

```
4. Training the model...
Starting training...
Epoch    0: Train Loss: 1.1350, Train Acc: 60.16%, Val Loss:
0.6938, Val Acc: 80.95%
Epoch   10: Train Loss: 0.4850, Train Acc: 82.04%, Val Loss:
0.4667, Val Acc: 82.43%
Epoch   20: Train Loss: 0.4073, Train Acc: 85.59%, Val Loss:
0.3924, Val Acc: 86.36%
Training completed!
```

Next, we will visualize our training results:

```
5. Visualizing training results...
```

The following chart shows the results of the training in graphical format. The chart on the left shows the training loss, which is the difference between the model's outputs and the expected outputs, and the same calculation for the differences during validation. On the right, we see the converse of this, in which accuracy represents the match of model outputs to expected results for training and validation phases, respectively.

Here are the results of the training analysis:

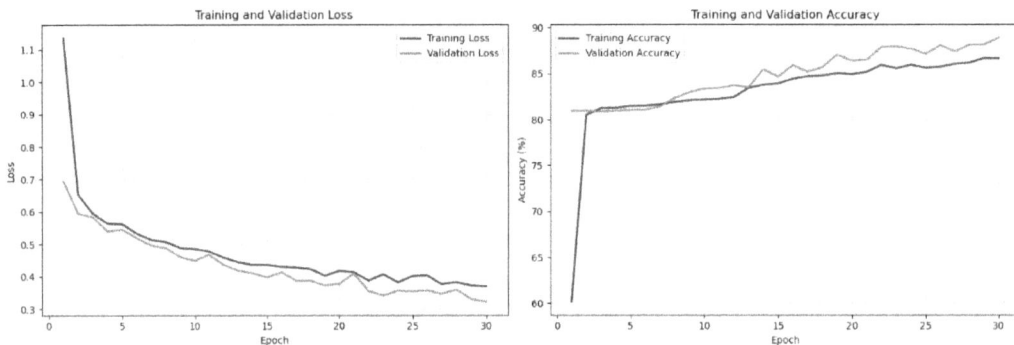

Figure 17.4 – CNN model training results

In the preceding visualization, we see that we train for multiple epochs with our training loss going down over time. At the same time, the accuracy is going up and the validation loss is going down. Once we achieve acceptable validation accuracy, we have finished training!

Now we will evaluate our model:

```
6. Evaluating on test set...
Test Loss: 0.2953
Test Accuracy: 89.50%

Detailed Classification Report:
              precision     recall   f1-score    support

      Normal       0.91       0.98       0.94      19902
    Deletion       0.78       0.47       0.59       2151
 Duplication       0.80       0.54       0.64       1947

    accuracy                             0.89      24000
   macro avg       0.83       0.66       0.72      24000
weighted avg       0.89       0.89       0.88      24000
```

We can see the output of the evaluation, including the overall accuracy and these key characteristics for each CNV category:

- **Precision**: Measures the accuracy of positive predictions:

  - Precision is defined by this equation: **True Positives (TPs)** / (True Positives + False Positives)

  - It tells us the following: Of the CNVs called, how many were actually correct?

- **Recall**: Measures the completeness of our CNV calls:

  - Recall is defined by the following equation: TP / (True Positives + False Negatives)

  - **It tells us the following**: Of the CVNs known to exist, how many were actually found?

- **F1 score**: A balanced combination of the precision and recall:

  - **It is defined by the following equation**: 2 × (Precision × Recall) / (Precision + Recall)

  - **It gives us**: The harmonic mean of the precision and recall scores

- **Support**:

  - Raw number of instances of each class

  - For example, we found 19,902 "normal" cases

The following chart shows a graphical overview of the detection results for three samples:

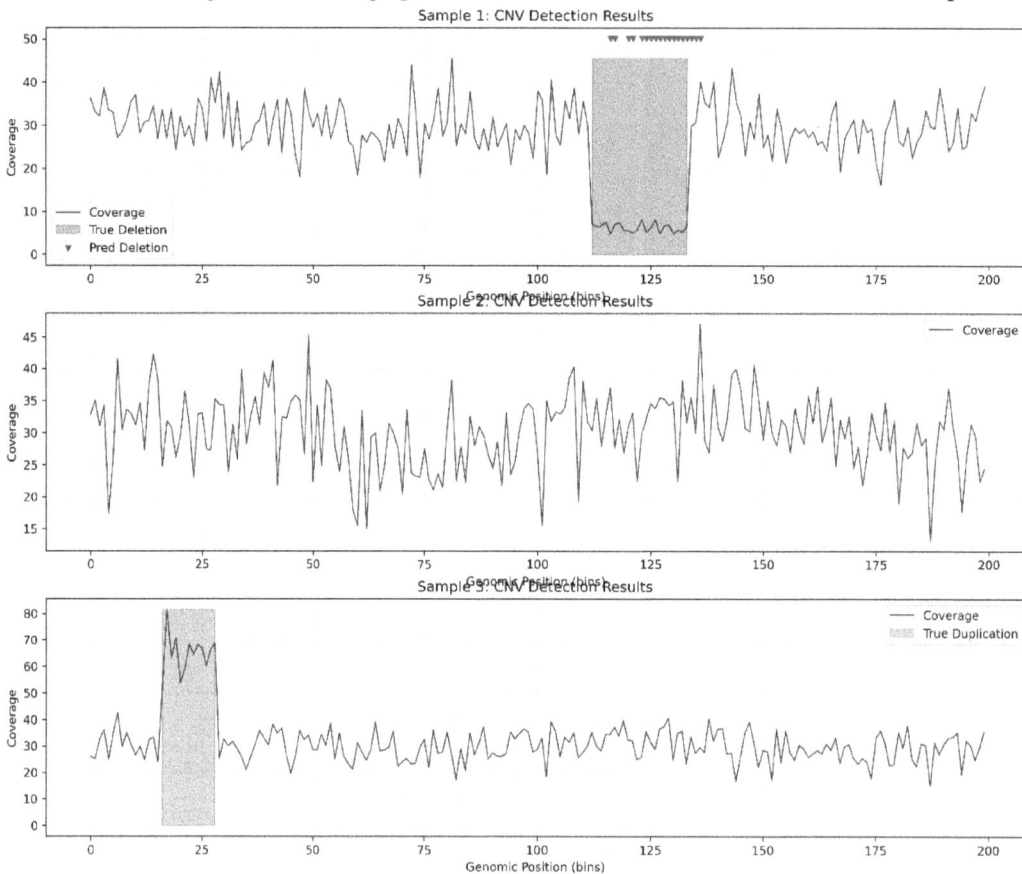

Figure 17.5 – CNV analysis results

Here we have our genomic CNV visualization. It shows the results for a few samples. In the first sample, we have a large dip in coverage representing a known deletion, which was correctly analyzed and called by our CNN. The next sample is a normal sample, and the third sample shows a potential duplication event, marked by an increase in coverage.

> **Note**
>
> You may get different results and CNV calls from those shown here. This is because the coverage graph we are generating has randomness in it and therefore may lead to different coverage levels and hence different calls.

That's it! We have completed our analysis. As you can see, the model is not perfect. It can be trained further by potentially giving it more training data or training it for longer or with different optimization strategies. You could also try different CNN architectures by playing around with the number, order, and type of layers. With proper architecture and tuning, CNNs can perform amazing feats of classification with very high accuracy.

There's more...

NNs and PyTorch are being used widely throughout the field of bioinformatics. For example, OpenDock uses PyTorch to perform modeling of chemical ligands binding to proteins; see Hu et al., *OpenDock: a pytorch-based open-source framework for protein-ligand docking and modelling*, Bioinformatics, Nov 2024 – `https://academic.oup.com/bioinformatics/article/40/11/btae628/7829142`.

CNNs are being used in many aspects of bioinformatics. For example, a CNN has been used in the prediction of schizophrenia; Henriques et al., *Integrating AI and genomics: predictive CNN models for schizophrenia phenotypes*, Journal of Integrative Bioinformatics, Jun 2025 – `https://pmc.ncbi.nlm.nih.gov/articles/PMC12569582/`.

As we mentioned, current implementations of NNs are very powerful but not terribly brain-like. However, researchers are exploring many avenues to make even more realistic and powerful models. For example, there are efforts to go beyond backpropagation for training; see Huang et al., *Brain-like training of a pre-sensor optical neural network with a backpropagation-free algorithm*, Photonics Research, Vol 13, 2025 – `https://opg.optica.org/prj/fulltext.cfm?uri=prj-13-4-915&id=569722`.

As you can see, ML is a huge area with many facets. Advances in this area are rapidly impacting bioinformatics and related fields, leading to exciting new possibilities!

In the next recipe, we will learn about an important emerging area of ML, **Large Language Models (LLMs)**!

See also

- Learn about the fundamental work by Rosenblatt (perceptrons) that led to NNs: `https://www.youtube.com/watch?v=i1G7PXZMnSc`
- This video is the first in a series that provides a great visual introduction to NNs: `https://www.youtube.com/watch?v=aircAruvnKk`
- Here is a Udemy course on ML: `https://www.udemy.com/course/machine-learning-course-with-python/`
- A course on TensorFlow can be found here: `https://www.udemy.com/course/tensorflow-developer-certificate-machine-learning-zero-to-mastery/`

- This course covers NNs using PyTorch: `https://www.udemy.com/course/the-complete-neural-networks-bootcamp-theory-applications/`

- This course goes deeply into PyTorch: `https://www.udemy.com/course/pytorch-ultimate/`

- This model uses a CNN to perform variant calling: Khazeeva et al., *DeNovoCNN: a deep learning approach to de novo variant calling in next generation sequencing data*, Nucleic Acids Research, Sep 2022 – `https://academic.oup.com/nar/article/50/17/e97/6609811`

- Here is a good review on foundation models: Guo et al., *Foundation models in bioinformatics*, National Science Review, Apr 2025 – `https://academic.oup.com/nsr/article/12/4/nwaf028/7979309`

Designing proteins with LLMs

In this recipe, we will learn about another exciting development in ML that is impacting bioinformatics: LLMs. LLMs are based on Transformers and are large because they have a huge number of parameters. They have learned language and information from the internet to power tools such as ChatGPT and Claude. But they have now also been unleashed on DNA and proteins to give us powerful tools for biomedicine and synthetic biology.

Protein engineering has a long history, beginning with random mutagenesis of proteins followed by screening to find new variants with desired properties. Scientists also create new proteins using rational design in which they select certain variants based on studies of the active site or review of evolutionary orthologs. **Focused libraries** can also be created in which all amino acids, or combinations of amino acids, are scanned in critical sections of the protein using custom DNA libraries that produce many defined variants of the protein.

With the emergence of ML, there has been an explosion of capabilities in protein design, as AI models can now design completely new enzymes from scratch, in ways that would be nearly impossible for human scientists. These techniques use **generative AI** to come up with novel protein sequences. Companies such as **Arzeda** (`https://www.arzeda.com/`) offer comprehensive services for new protein design using ML.

In this recipe, we will learn how to design a protein with LLMs, using a popular tool called **ProtGPT2**; see Ferruz et al., *ProtGPT2 is a deep unsupervised language model for protein design*, Nature Communication, Jul 2022 – `https://www.nature.com/articles/s41467-022-32007-7`.

ProtGPT2 is a language model trained on a large number of protein sequences. It can generate realistic yet novel protein sequences that can be effectively folded by AlphaFold.

By the end of this recipe, you will have learned how to install and run Transformers using models from Hugging Face to run the ProtGPT2 designer. You will have seen how to design your own proteins using LLMs and investigate their properties. This should give you a good sense of how powerful this technique can be!

Getting ready

Let's make sure we have all our packages installed:

```
! pip install torch transformers numpy matplotlib seaborn pandas
```

The code for this recipe can be found in `Ch17/Ch17-2-protein-design.ipynb`.

How to do it...

1. We begin by importing our libraries:

    ```python
    import torch
    from transformers import GPT2LMHeadModel, GPT2Tokenizer
    import re
    import numpy as np
    import pandas as pd
    import matplotlib.pyplot as plt
    import seaborn as sns
    from typing import List, Optional
    import warnings
    from tqdm.auto import tqdm
    import logging
    import os
    import sys
    ```

    We will use the PyTorch library as before. We will also bring in the `transformers` library (`https://pytorch.org/hub/huggingface_pytorch-transformers/`). This library contains pre-trained models for NLP. We will import `GPT2LMHeadModel`, which provides an interface to the Hugging Face GPT-2 model from OpenAI. We will use this for language modeling to implement our protein sequence generation. We also import `GPT2Tokenizer`, which converts text sequences to tokenizers that the model understands.

    We will also use the Python `typing` library, which provides support for type hints (`https://docs.python.org/3/library/typing.html`). This will help us provide data type checking in our code.

    We are going to use the `tqdm` library (`https://tqdm.github.io/`) to make a progress bar.

2. The next section is used to suppress some warnings. Review the section entitled `Comprehensive Warning Suppression` in the notebook.

    This section uses **monkey patching** (`https://dev.to/karishmashukla/monkeying-around-with-python-a-guide-to-monkey-patching-obc`) to suppress extra warnings coming from the transformer model that we don't want to see in our notebook.

3. Next, we set our plotting style and print a short header for our analysis.

4. Great! Now we are ready to define the core class of our system, which will use ProtGPT2 to design proteins. Review the code for the `ProtGPTDesigner` class in the notebook. This class will do the following:

    - Wrap the ProtGPT2 protein language model. First, initialize and download the ProtGPT2 model from Hugging Face. Choose a CUDA or CPU implementation.

    - Configure the model. We will use a **decoder-only** architecture, which means we will only use the decoder block of the transformer (no encoder). We will use this to generate our output sequences.

    - Define the `generate_protein()` function. The `endoftext` prompt simply tells ProtGPT2 to start generating proteins from scratch. We also supply the number of sequences to generate, the maximum length, and other key parameters. The `temperature` parameter controls how conservative or creative the model is in designing the protein sequence.

    - Define a `validate_sequence()` function. This function will check the accuracy of the designed protein sequences. It checks for valid amino acids to be returned (no unusual characters). It also calculates for us the amino acid composition and key biochemical properties, such as hydrophobicity (recall we discussed how important these were in *Chapter 12, Metabolic Modeling and Other Applications*).

    - The next function is `batch_generate_and_validate()`. This function loops through and generates and validates our proteins while showing a progress bar. It will handle any errors and return the results as a pandas DataFrame.

    That concludes our core protein designer class!

5. Now we will initialize our model:

    ```
    designer = ProtGPT2Designer()
    ```

    This simply instantiates a `ProtGPTDesigner` object.

6. Next, let's look at an example of *de novo* protein generation. Review the *Example 1 – de novo Protein Generation* section in the notebook.

    Here, we will generate three proteins from scratch, with no particular guidelines on what to generate. We will then display and validate the results.

7. The next example looks at motif-based protein generation. *Review Example 2 – Motif-based Generation* in the notebook:

    - This example generates proteins based on a motif. The motif is provided via the `prompt` parameter.

    - The protein designer `generate_protein()` function is again used to create protein sequence examples. The proteins are then validated and displayed.

8. Next, we will explore generating a whole batch of proteins. Review *Example 3 – Batch generation and analysis* in the notebook.

This example generates and validates 20 proteins and then displays a summary of their properties.

9. Let's visualize our results! Review the next section, *Visualization*, in your notebook.

10. This code creates figures for length distribution, validity ratio, amino acid composition, and length versus validity with hydrophobic content overlaid.

11. The plot will be displayed and saved to a file called `ch17-2-protein-design.png`.

12. Now we will display our sample sequences:

```
print("\n  Sample Generated Sequences:")
print("=" * 50)
valid_proteins = df_proteins[
    df_proteins['is_valid'] == True].head(5)
for idx, row in valid_proteins.iterrows():
    print(f"\n{row['protein_id'].upper()}:")
    print(f"Sequence: {row['sequence']}")
    print(f"Length: {row['length']} | "
          f"Validity: {row['validity_ratio']:.3f} | "
          f"Hydrophobic: {row['hydrophobic_content']:.3f}")
```

The preceding code picks out the top five valid proteins and then loops over them. It prints the protein ID, sequence, length, validity ratio, and hydrophobic content.

13. In the next example, we will perform a temperature comparison. We will use a few different temperatures and compare the resulting proteins. Review the `Temperature Comparison` section in your notebook.

    Recall that lower temperatures are more conservative and higher temperatures are more creative. What you should see is that lower-temperature protein designs tend to have higher validity and are closer to the training data. At higher temperatures, the proteins are less predictable and tend to have lower validity.

14. In the last cell of the notebook, you will see a section that says **Optional: Save results**. This is some extra code that will let you save your proteins as a CSV file. If you like, you can uncomment this code and run it to save your results to a file in your working directory.

Great work! We have now set up an LLM to design proteins using ProtGPT2. We have gone through several different examples to explore potential protein sequences.

Here is a portion of the output:

```
Loading ProtGPT2 model and tokenizer...
    Using device: cpu
  Model loaded successfully!
  Model parameters: 774,030,080
  Ready to generate proteins!
```

Note that this model has ~774 million parameters! The parameters are the weights in the model that are tuned during optimization, as we discussed in the *Machine learning with PyTorch* recipe. The more parameters a model has, the "smarter" it will be. This is quite impressive, but modern LLMs such as Claude are getting into the hundreds of billions of parameters, and ChatGPT-4 is believed to be in the trillions!

OK, let's proceed with our first example. We will generate some proteins *de novo*, meaning from scratch, with no special instructions:

🔖 EXAMPLE 1: De novo protein generation
===
🧬 Generating 3 protein sequence(s)...

🧬 Generated Protein 1:
Sequence: MEVRMNTYIYALFVLALLSPAYAQETTAAPADNTTTAAPATTTAAPANTTAAPANTTAAP
ANTTAAPANTTAAPANTTA...
✏️ Length: 300 amino acids
☑️ Validity: 1.000
🔬 Valid: Yes
⬤ Hydrophobic: 0.567
⚡ Charged: 0.013

  Generated Protein 2:
Sequence: MTMEQIMRVMDTMSREEQEAYLRSVAARKAAAAARAAAERAAAAAAAQRAAAAAAAQRAA
AAAAAQRAAAAAAAAAARA...
  Length: 296 amino acids
☑️ Validity: 1.000
🔬 Valid: Yes
  Hydrophobic: 0.764
⚡ Charged: 0.057

  Generated Protein 3:
Sequence: MKKITLLLLTTIAFSFASYAQTATPVVSKTVDGATYTTTVSTNTYNSTVTINGVKFTATT
GNVTISANGDITINPTATG...
✏️ Length: 103 amino acids
☑️ Validity: 1.000
  Valid: Yes
  Hydrophobic: 0.379
⚡ Charged: 0.068

You can see the proteins and a portion of their sequence, along with some key characteristics. Validity is the percentage of real, standard amino acids in the protein. Hydrophobicity represents the percentage of hydrophobic amino acids, and the charged characteristic does the same for charged amino acids.

Let's now generate some proteins to which we will attach a defined motif, which could represent a signal peptide:

```
☞ EXAMPLE 2: Protein generation with starting motif
=====================================================
🧬 Generating 2 protein sequence(s)...

🧬 Motif-based Protein 1:
Sequence: MKKLLFLLLMLFTVVSCSNDDDDNTTPTPTPNPPVEPTQDFTATLVGTWQLNTYEYSDGN
LVESQDDGDWCEVNNTLVF...
🏷 Length: 155 amino acids
☑ Validity: 1.000

🧬 Motif-based Protein 2:
Sequence: MKKLLFLLLLFFSINFCFSQENSNKFGYVSVEELLTKYPDYKTAQDELDSYTQDIQNEYNN
KLQAFQTEYQAYMAKANNM...
🏷 Length: 166 amino acids
☑ Validity: 1.000
```

Here, we have generated some proteins and then appended our signal peptide to them. We then validate the full protein sequence (with the signal motif attached) and display the results and summary characteristics.

In the next example, we will generate a larger batch of proteins to look at their average characteristics:

```
Generated 20 proteins
☑ Valid proteins: 20
🏷 Average length: 345.8 ± 181.6
☞ Validity ratio: 1.000 ± 0.000

📊 Summary Statistics:
      length    validity_ratio   hydrophobic_content   charged_content  \
count
      20.000000           20.0            20.000000         20.000000
mean  345.800000            1.0             0.338504          0.147171
std   181.612543            0.0             0.207078          0.167409
min    72.000000            1.0             0.009766          0.000000
25%   258.500000            1.0             0.216557          0.013504
50%   376.500000            1.0             0.357911          0.085878
75%   417.000000            1.0             0.428004          0.228253
max   898.000000            1.0             0.772414          0.552699


        polar_content   aromatic_content
count      20.000000          20.000000
mean        0.279123           0.061040
```

```
std        0.157838        0.097172
min        0.003341        0.000000
25%        0.189612        0.000000
50%        0.249613        0.022032
75%        0.380644        0.079791
max        0.588076        0.405573
```

We see the average sequence length, plus or minus the standard deviation of the length. We also get information on the validity ratio. We then see a table of summary statistics, including validity ratio, hydrophobicity and charge, polarity, and aromaticity.

Next, we will look at the plot we generated of protein analysis visualization:

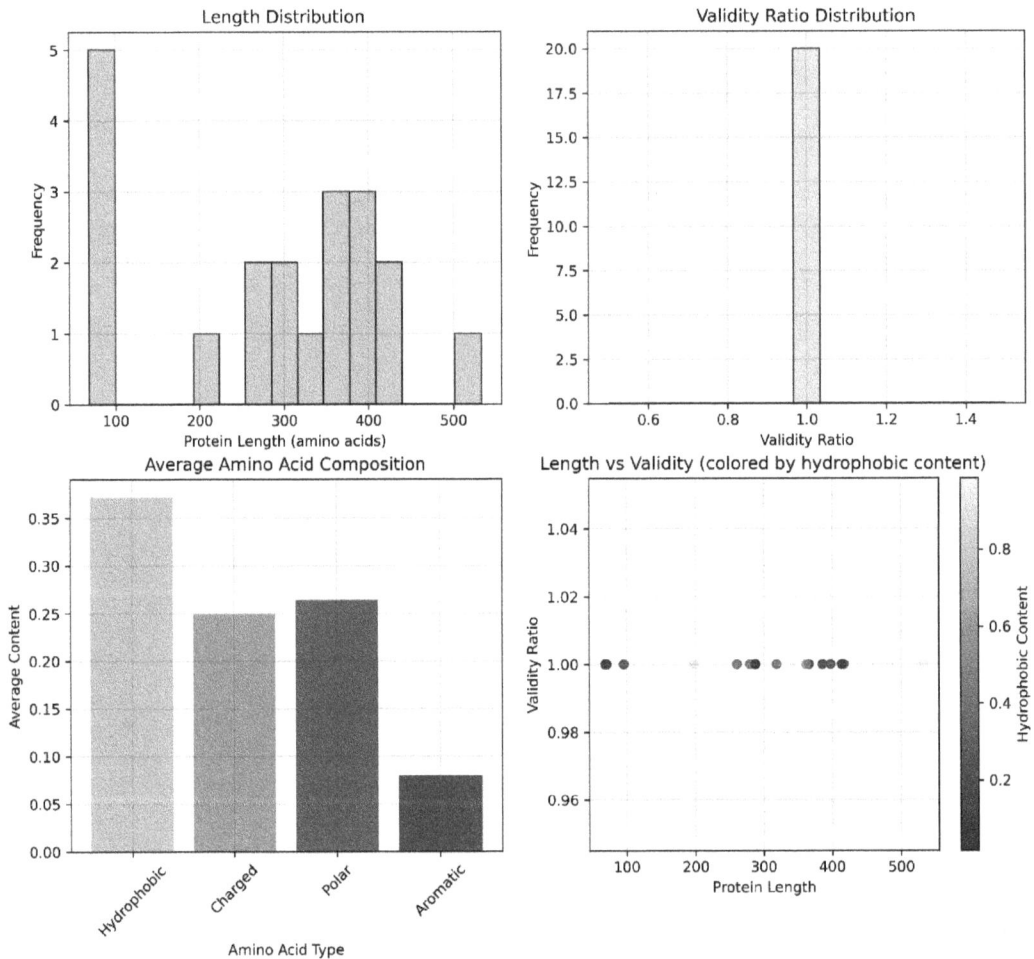

Figure 17.6 – Analysis of ProtGPT2 protein design results

This plot shows a length distribution histogram in the upper left. The next plot, in the upper right, shows the distribution of validity ratios. The plot on the lower left shows the average content of hydrophobic, charged, polar, and aromatic amino acids. Finally, the plot on the lower right shows protein length versus validity ratio, with hydrophobicity overlaid.

Here is the next part of the output:

```
Sample Generated Sequences:
========================================================


PROTEIN_001:
Sequence:
METIMKLLLTTTLLATAASAFAAGDTTTTGTGTTTGTTTTDTTTTTTDTTTTTGTTTTTTDTTTTTGS
Length: 66 | Validity: 1.000 | Hydrophobic: 0.227


PROTEIN_002:
Sequence: MKKILFIIIILIFSFSCERNKYYNYLYYNGKITEIVIKNDNSYIYKNIIN-
NDTLLKNINKNNTYYYKINNNKIIEINNKYEYKYIDNINNNIISKEFYFNYNKYKINDKIINYYYYD-
SLNNKLKTIIDYNYNNNLIKETYNYIDNKYKKKIIYYYNNKIIKEYIYNKKNKKIYFEYYNDKDNN-
LIKEYYYNSDNNLIKKIEYNNKNQIIEKKIFNKK
Length: 217 | Validity: 1.000 | Hydrophobic: 0.272


PROTEIN_003:
Sequence: MSSTTPHRRRRGRTALLAVLTALVALALAGPAAAAPADAAPAAQAAPAAPAAPAA
PAAPAAPAAPAAPAAPAAPAAPAAPAAPAAPAAPAAPAAPAAPAAPAAPAAPAAPAAPAAPAAPAA-
PAEPAAPAAPAAPAAPAAPAAAAAPAAPAAPAAPAAPAAAAAPAAPAAPAAPAAPAAPAAAAAPAAA
AAPAAAPAAPAAPAAPAAPAAPAAPAAPAAPAASAAPAAPAAAAAPAAAAPAAAAPAAA
APAAPAAPAAPAAPAAPAAPAAPAAPAAPAAAAAPAAAAPAAPAAAAAPAAAAAPAAAAAPAA
PAAAAPAAPAAPAAPAAPAAPAAPAAPAAPAAPAAPAAPAAPAAPAAPAAPAAPAAPAAPAAPA
APAAPAAPAAPAAPAAPAAPAAPAAPAAPAAPAAPAA
Length: 447 | Validity: 1.000 | Hydrophobic: 0.960


PROTEIN_004:
Sequence: MSQTALLIGNGLSQAFHPSFSYQKLFDACVKAGLDGNQKYFHNLFSLSGFEDFETVL-
KALDHNRQIQAALDLPEGKEAELKQKIRDYLLNILMDIHVEYPQDQDQDQDQDQDQDQDQDQDQDQD-
QDQDQDQDQDQDQDQDQDQDQDQDQDQDQDQDQDQDQDQDQDQDQDQDQDQDQDQDQDQDQDQDQDQD-
QDQDQDQDQDQDQDQDQDQDQDQDQDQDQDQDQDQDQDQDQDQDQDQDQDQDQDQDQDQDQDQDQDQD-
QDQDQDQDQDQDQDQDQDQDQDQDQDQDQDQDQDQD
Length: 300 | Validity: 1.000 | Hydrophobic: 0.143


PROTEIN_005:
Sequence: MRKYTSILLTLILLVGCASAPKPEPEPTPEPTPEPTPTPTPTPTPTPEPKPTE-
PEKPEEPKPEKPVEPKPEKPVEPKPEKPVEPKPEKPVEPKPEKPVEPKPEKPVEPKPEKPVEPKPEK-
PVEPTPKPEPEPAKEEPKKPEEPKKETPAPSEPKKEEPKKEEPKKEEPKKEEPKKEEPKKEEPKKEEP-
KAEEPKKEEPKAEEPKKEEPKKEEPKAEEPKKEEPKKEEPKAEEPKKEEPKKEEPKAEEPKKEEPK-
KEEPKAEEPKKEEPKKEEPKKEEPKAEEPKKEEPKKEEPKKEEPKAEEPKKEEPKKEEPKKEEP-
KKEEPKAEEPKKEEPKKEEPKAEEPKKEEPKKEEPKKEEPKAEEPKKEEPKKEEPKAEEPKK
Length: 389 | Validity: 1.000 | Hydrophobic: 0.344
```

This gives us the first five valid proteins and their basic characteristics. You may actually notice that some of the proteins are highly repetitive. This is a common problem in which LLMs can become stuck in **local minima**. Higher temperatures, which are less conservative, tend to overcome this problem. In the next example, we explore varying the temperature:

```
✦ ADVANCED EXAMPLE: Temperature comparison
=======================================================

✎   Temperature: 0.5
  Seq 1: MDAPDHDPAAPRRRGLFGGLGRRPAAAPPVAAPAPAPAPAPAPAPAPA... (L=259,
V=1.00)
  Seq 2: MSIIMNFFRLILLILFSLISFSSFSQELESNLKSLENSLKNLEEKIESLK... (L=210,
V=1.00)

    Temperature: 0.8
  Seq 1: MSDHNSPHAGHDHGEHASHDHGSHDGHDHGSHSGHDHGSHSGHDHGSHDG... (L=152,
V=1.00)
  Seq 2: MFYLPRNGALVTLALLIGVAAWLVWAPQRAWSQVPASSPPTASAPATSTP... (L=207,
V=1.00)

    Temperature: 1.2
  Seq 1: MEPDQDPPPLPSGWLPCWSQDHRRPYYFHPATNISQWDPPAPPTPPVAPP... (L=229,
V=1.00)
  Seq 2: MLIYITFLLFLFFLINIILNLYIYFNFFLIYTFFFLFFLFYFILLYLYLF... (L=193,
V=1.00)

🌿 Analysis complete! You can now:
    • Modify generation parameters in any cell
    • Generate more proteins with different settings
    • Analyze specific protein properties
    • Export sequences for further analysis
```

You can see some improvement in variety as we increase the temperature.

As you can see, protein design is very powerful. As larger LLMs are created with more parameters, we will see increasing sophistication in the types of proteins we can design!

There's more...

Now we have seen how to design our own proteins using AI! Wouldn't it be cool if we could go even further and model the structure of one of our own proteins? We would like to create a structural model like those we discussed in *Chapter 9, Protein Structure and Proteomics*. It turns out that using today's powerful modern structural bioinformatics and AI tools, we can!

One powerful technique for looking at protein structures using bioinformatics is **homology modeling**. In this technique, a protein structure similar to the sequence you supply is identified, and your input sequence is threaded through the template to obtain a putative structure. SWISS-MODEL (`https://swissmodel.expasy.org/`) is a powerful site for performing homology modeling. You can go to the SWISS-MODEL website and input one of the protein sequences from your LLM modeling as an example.

Here is an example using one of the generated sequences:

```
Target Protein Sequence: MKKLLFLLLFFSINFCFSQENSNKFGYVSVEELLTKYPDYK-
TAQDELDSYTQDIQNEYNNKLQAFQTEYQAYMAKANNM
```

Here is the result from SWISS-MODEL:

Figure 17.7 – Result of SWISS-MODEL protein analysis

The preceding results show the output from SWISS-MODEL, including an overview of the protein properties, the template used, and an interactive three-dimensional model of the protein.

If you want to go further and model your protein *de novo*, you could use a system such as **ColabFold** (`https://github.com/sokrypton/ColabFold`), which provides notebooks that act as interfaces to protein folding tools; see Mirdita et al., *ColabFold: making protein folding accessible to all*, Nature Methods, Jun 2022 – `https://pubmed.ncbi.nlm.nih.gov/35637307/`. You can also use the **AlphaFold 3** server (`https://alphafoldserver.com/welcome`). Try going to this server and inputting your test sequence. Here is what you will see:

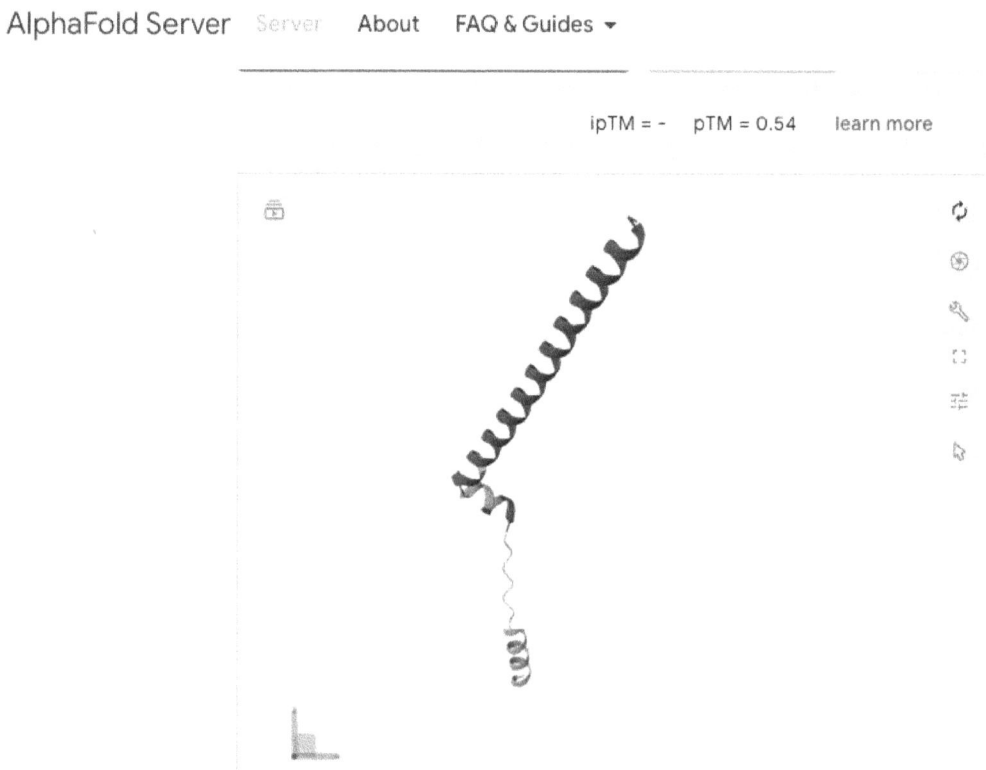

Figure 17.8 – Output from the AlphaFold 3 server

In the preceding output, you can see that AlphaFold has folded your protein and rendered it into a three-dimensional model. Of course, this protein was a fairly simple example and so does not have a lot of three-dimensional structure. But you can see now how we can construct a complete workflow that goes from concept through protein design using LLMs, through folding and reviewing the protein.

You have now seen that advances in AI have given us powerful abilities to design proteins, the building blocks of life itself. In the next recipe, we'll look at how to design genes and even entire genomes using similar principles!

See also

- This extensive course includes sections on transformers and LLMs: `https://www.udemy.com/course/deep-learning-masterclass-with-tensorflow-2-over-15-projects/`

- Here is more background on LLMs for protein design: Valentini et al., *The promises of large language models for protein design and modeling*, Frontiers in Bioinformatics, Nov 2023 – `https://www.frontiersin.org/journals/bioinformatics/articles/10.3389/fbinf.2023.1304099/full`

- BAGEL (`https://github.com/softnanolab/bagel`) is a Python framework for protein design: Lala et al., *BAGEL: Protein Engineering via Exploration of an Energy Landscape*, bioRxiv, Jul 2025 – `https://www.biorxiv.org/content/10.1101/2025.07.05.663138v1`

- DeepBLAST can find remote homology in protein searches using three-dimensional alignment: Hamamsy et al., *Protein remote homology detection and structural alignment using deep learning*, Nature Biotechnology, Sep 2023 – `https://www.nature.com/articles/s41587-023-01917-2`

- A good review on homology modeling can be found here: Hasani & Barakat, *Homology Modeling: an Overview of Fundamentals and Tools*, International Review on Modeling and Simulations, Apr 2017 – `https://www.researchgate.net/profile/Horia-Jalily-Hasani/publication/317713260_Homology_Modeling_an_Overview_of_Fundamentals_and_Tools/`

- Prostruc provides a Python tool for homology modeling: Pawar et al., *Prostruc: an open-source tool for 3D structure prediction using homology modeling*, Frontiers in Chemistry, Nov 2024 – `https://www.frontiersin.org/journals/chemistry/articles/10.3389/fchem.2024.1509407/full`

Designing DNA with LLMs

Let's now look at how LLMs can be used to design DNA sequences. These models are being used to design everything from genes, RNAs, and regulatory sequences to entire genomes used for the construction of synthetic organisms. For example, **CodonBERT** can design mRNAs for vaccines and optimize mRNAs for expression and translational stability, as well as analyzing mRNA properties; see Li et al., *CodonBERT: Large Language Models for mRNA design and optimization*, bioRxiv, Nov 2023 (`https://www.biorxiv.org/content/10.1101/2023.09.09.556981v2.full`). **Evo 2** (`https://github.com/ArcInstitute/evo2`) is a DNA foundation model trained on a huge number of genomes. It has 40 billion parameters, and simply by looking at so many genomes, it learned to predict the effects of pathogenic variants and understand features such as exon-intron boundaries. It can generate chromosome and genome-scale sequences containing natural features, including proteins that fold like real proteins; see Brixi et al., *Genome modeling and design across all* *domains of life with Evo 2*, bioRxiv, Feb 2025 – `https://www.biorxiv.org/content/10.11 01/2025.02.18.638918v1.full.pdf`. **DNABERT** is a bidirectional encoder that can predict promoters and other features in a genome; see Ji et al., *DNABERT: pre-trained Bidirectional Encoder Representations from Transformers model for DNA-language in genome*, Bioinformatics, Aug 2021 – `https://academic.oup.com/bioinformatics/article/37/15/2112/6128680`. DNABERT-2 built upon this work by training an even larger model; see Zhou et al., *DNABERT-2: Efficient Foundation Model and Benchmark for Multi-Species Genome*, arXiv, Mar 2024 – `https://arxiv.org/abs/2306.15006`. It is available through Hugging Face. In this recipe, we'll design genomic sequences and then use DNABERT-2 to evaluate and optimize those sequences, building our own small toy genome in the process.

By the end of this recipe, you'll know how to use LLMs to design DNA, including entire genomes. You'll understand how to analyze the resulting genomes and get a sense of how these techniques could be used to design powerful solutions in synthetic biology and biomedicine.

Getting ready

Let's install the packages that we need (some of these may already be installed from previous recipes):

```
! pip install transformers torch numpy biopython pandas matplotlib
seaborn plotly tqdm
```

The code for this recipe can be found in `Ch17/Ch17-3-genome-design.ipynb`.

How to do it...

Here are the steps to perform this recipe:

1. First, we import our libraries:

    ```python
    import torch
    from transformers import AutoTokenizer, AutoModel
    import numpy as np
    import pandas as pd
    import matplotlib.pyplot as plt
    import seaborn as sns
    import plotly.graph_objects as go
    import plotly.express as px
    from plotly.subplots import make_subplots
    from typing import List, Dict, Tuple
    import re
    import random
    import warnings
    import logging
    from tqdm import tqdm
    ```

```
from IPython.display import display, HTML, Markdown, clear_output
from io import StringIO
import sys
import time
```

We import the PyTorch library as before. From the `transformers` library, we will import `AutoTokenizer` and `AutoModel`. `AutoTokenizer` is a tool in the Hugging Face transformers library that helps you automatically choose the best tokenizer for your task. `AutoModel` helps you to automatically select and train various ML models for your project.

2. The next section of code will set up a few basic settings for our notebook and suppress some transformer warnings that we don't need:

```
plt.rcParams['figure.figsize'] = (12, 8)
plt.rcParams['font.size'] = 10
sns.set_style("whitegrid")
logging.getLogger("transformers").setLevel(logging.ERROR)
warnings.filterwarnings("ignore")
```

3. Next, we will define our genome generator class, which uses DNABERT. This is the heart of our recipe. Review the code for the `DNABERTGenomeGenerator` class in the notebook. This class has the following features:

I.   First, we define our `init()` function. Note that this function can take different arguments for running the algorithm. CUDA refers to the use of NVIDIA GPUs; however, this is no longer supported on modern Mac systems. **MPS** refers to **Metal Performance Shaders**, which is the type of GPU supported on modern M-series Macs. CPU is the default and safest option for running on your primary chip. You can play around with some of these options if you want to explore GPU acceleration, but you may run into some library compatibility issues; the code will run in a reasonable amount of time on the CPU option. This function next loads the DNABERT-2 model from Hugging Face. It then sets up our model and tokenizer. It defines some basic characteristics of core genome elements we will be using:

   • **Promoters**: These are core elements that sit in front of genes. Transcription factors bind to these regions to promote transcription of the gene into RNA. We will use a **TATA box** here, which is a well-known strong promoter; you can learn more about the TATA box here: Tang et al., *Promoter Architecture and Promoter Engineering in Saccharomyces cerevisiae*, Metabolites, Aug 2020 – `https://www.mdpi.com/2218-1989/10/8/320`

   • **Coding sequence (CDS)**: Including the **start codon** and **stop codons**.

   • **The terminator**: Typically, a GC-rich sequence at the end (3' region) of a gene that signals the end of transcription. Terminators can also influence mRNA stability, and hence gene expression; read more here: Ren et al., *Regulatory significance of terminator: A systematic approach for dissecting terminator-mediated enhancement of upstream mRNA stability,*

Synthetic and Systems Biotechnology, Mar 2025 – `https://www.sciencedirect.com/science/article/pii/S2405805X24001510`.

- **Intergenic regions**: In this case, the intergenic regions will just be a type of "filler"; we will not attempt to model **non-coding RNAs (ncRNAs)** – but this could be an interesting exercise to think about!

II.   The next function is an internal function called `_test_model_output()`. This function just performs a test with a simple short DNA sequence to see what types of outputs the DNABERTmodel is returning. This will be used in some later functions to help ensure compatibility.

III. Next is `_generate_reference_embeddings()`. This function will be used to train our model to see what a typical strong promoter or other genomic element might look like. Note that in this recipe, we'll be mixing some traditional genome design with the use of DNABERT. We'll design genome elements but then score them against learned examples using our model. In this function, we first define a few good examples for each common genomic element. For example, the promoters used might typically include the `lac`, or lactose operon, promoter, used in many experiments throughout history as a strong promoter; the `trp`, or tryptophan, promoter; and the `tac` promoter, which is a strong synthetic promoter. In this example, we simply provide three examples of strong synthetic promoters. In a real-world example, you would probably provide far more training data. After defining the genomic elements, we call `get_sequence_embedding()`, which is described next, to run the model and get representations. We then calculate the mean of our embedding and store this as an *ideal representation* of that element in our `references` dictionary.

Here is a figure explaining the embeddings:

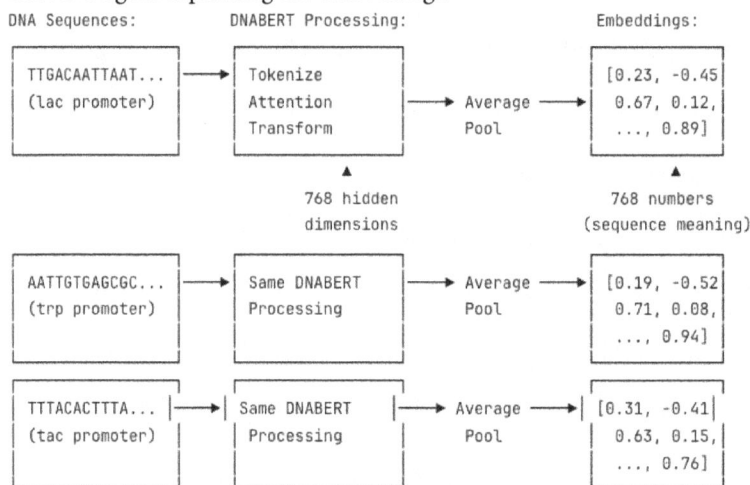

Figure 17.9 – Illustration of sequence embeddings

In the preceding figure, we see strong promoters fed into the DNABERT model. It represents each position in the sequence as a numeric score. For example, the first base may be a T with a score of 0.23, a weak positive score. Negative scores represent penalties. This overall score represents how "ideal" this promoter sequence is.

4. Now take a look at the `get_sequence_embedding()` function. This function converts DNA sequences into meaningful numerical representations for our model, as described in the preceding figure:

   I.   It first cleans the model by stripping out non-DNA characters. It then tokenizes the sequences. It turns off gradients to optimize for performance and then runs the sequence through the DNABERT model to get a representation, or hidden states from the model. These embeddings then represent a biological meaning of the sequence that is based on learnings from real genomes.

   II.   The next function is `preprocess_sequence()`. It is very simple and just strips our non-DNA characters.

   III. Next, we have `generate_random_sequence()`. This function simply generates a random nucleotide sequence of given average GC content.

   IV.   The `generate_promoter_sequence()` function is next. It will generate a random sequence and then insert the TATA box sequence partway into it.

   V.   The `generate_coding_sequence()` function will build a gene sequence. It begins with a start codon and then adds random codons in the middle, avoiding stop codons. It then attaches one of three stop codons at the end.

   VI.   The next function is `generate_terminator_sequence()`, which simply generates a random sequence of given GC content. Higher GC content implies potential **hairpin** formation, which adds to mRNA stability.

   VII. Next up is a critical function called `optimize_sequence_with_model()`. This function uses DNABERT to refine our genome and make it more realistic (or at least closer to what we have trained DNABERT to think is realistic!). It will take the sequence supplied and tweak it a bit by introducing mutations. It will then evaluate that sequence and compare it to the real biological sequences it was trained on. It will, in this way, attempt to create a sequence that is closer to optimal according to the model.

   VIII. The next function is `_score_sequence_with_model()`. This function will calculate a similarity score between the supplied sequence and the "ideal" model sequence. It first retrieves the reference embedding for that type of element.

   IX.   Up next, we have the `_mutate_sequence_guided()` function. This function will intelligently introduce mutations into our sequences so that they can be evaluated by the model. It first converts the sequence to a list for easier manipulation. It will introduce 1–3 mutations based on sequence length. The function chooses different strategies each time to provide different methods for exploring sequence space. You could extend this even further by specifying particular strategies for certain types of genomic elements.

X. The `generate_2kb_genome()` function is next, which is a core part of the functionality for this class. It will plan out the overall structure of our desired genome and then use the functions in this class to generate it.

XI. The next function in this class is `analyze_genome()`. It provides an overall summary analysis of the genome, including GC content and ORF finding. If you are not familiar with this term, we will define it ahead.

XII. The function to review next is `_analyze_element_quality`. This function evaluates elements using our DNABERT model scoring functions. It then calculates overall summary statistics on the scores.

XIII. The next function is `find_orfs()`. It will scan the genome for open reading frames.

XIV. Now look over the `display_genome_summary()` function. It will display an HTML summary of our genome with high-level characteristics.

That concludes our genome generator class!

5. Now we create a function to help us visualize our genome:

I. The `visualize_genome_notebook()` function will create a comprehensive visual overview of our genome. First, it builds a genome map using a horizontal bar chart.

II. Next, it creates a line graph of GC content. It then builds a pie chart of nucleotide composition, showing the percentages of A, C, G, and T, respectively, in our synthetic genome.

III. We then display the average quality scores from our model by element as a bar graph.

IV. The figure will be displayed and saved to the `Ch17-3-dnabert_genome_analysis_static.png` file in your working directory.

V. The next function helps us build an interactive genome viewer.

VI. This function builds an interactive HTML output with plots similar to our visualization function. It provides interactivity with the `text` and `hovertemplate` properties. The file will also be saved as `Ch17-3-dnabert_genome_analysis_interactive.html`.

6. The next function brings together all of our work to design and build a genome. Review `generate_and_analyze_genome()` in the notebook:

   • This function will run all of the key steps in our workflow. We'll run each of the steps individually later when we review our output.

7. The final function, `display_usage_instructions()`, will print out colorful instructions on how to use the notebook using HTML. This is a great example of how to provide instructions for your users.

8. This looks great! Let's now run our genome designer:

I.   The first step is to initialize the model:

```
generator = DNABERTGenomeGenerator()
```

This will give the following output:

```
🧬 Initializing DNABERT Genome Generator...
=======================================================
📱 Loading DNABERT model: zhihan1996/DNABERT-2-117M
    [abc] Loading tokenizer...
    🤖 Loading model...
    🔬 Testing model output format...
    🗔 Generating reference embeddings...
        📝 Computing promoter embeddings...
        🧬 Computing coding sequence embeddings...
        🔚 Computing terminator embeddings...
        🌐 Computing intergenic reference...
        ☑ Reference embeddings complete!

☑ DNABERT Model Ready!
Ready to generate AI-optimized genomes
```

II. This step loads our tokenizer and model and sets up the DNABERT model. Next, we will generate our 2Kb-long genome:

```
genome = generator.generate_2kb_genome()
```

Here is the output from this step:

```
🧬 Generating 2kb Genome with DNABERT Optimization
📋 Planned elements: 10 functional regions
=================================================================
🔧 [1/10] Generating promoter (80 bp)
    🔧 Optimizing promoter (initial score: 0.924)
    ☑ Optimization complete! Final score: 0.928 (+0.005, 1
improvements)
    📊 Progress: 4.0% (80/2000 bp)

🔧 [2/10] Generating coding_sequence (600 bp)
    Optimizing coding_sequence (initial score: 0.874)
    ☑ Optimization complete! Final score: 0.884 (+0.010, 1
improvements)
    📊 Progress: 34.0% (680/2000 bp)
```

🔧 [3/10] Generating intergenic (120 bp)
    🔧 Optimizing intergenic (initial score: 0.946)
    ☑ Optimization complete! Final score: 0.947 (+0.001, 1
improvements)
    Progress: 40.0% (800/2000 bp)

🔧 [4/10] Generating promoter (70 bp)
    🔧 Optimizing promoter (initial score: 0.904)
    ☑ Optimization complete! Final score: 0.913 (+0.009, 1
improvements)
    Progress: 43.5% (870/2000 bp)

🔧 [5/10] Generating coding_sequence (500 bp)
    Optimizing coding_sequence (initial score: 0.861)
    ☑ Optimization complete! Final score: 0.864 (+0.003, 1
improvements)
    Progress: 68.5% (1370/2000 bp)

🔧 [6/10] Generating terminator (50 bp)
    Optimizing terminator (initial score: 0.882)
    ☑ Optimization complete! Final score: 0.882 (+0.000, 1
improvements)
    Progress: 71.0% (1420/2000 bp)

🔧 [7/10] Generating intergenic (150 bp)
    Optimizing intergenic (initial score: 0.937)
    ☑ Optimization complete! Final score: 0.941 (+0.003, 1
improvements)
    Progress: 78.5% (1570/2000 bp)

🔧 [8/10] Generating promoter (60 bp)
    Optimizing promoter (initial score: 0.874)
    ☑ Optimization complete! Final score: 0.876 (+0.001, 1
improvements)
    Progress: 81.5% (1630/2000 bp)

🔧 [9/10] Generating coding_sequence (370 bp)
    Optimizing coding_sequence (initial score: 0.848)
    ☑ Optimization complete! Final score: 0.852 (+0.004, 1
improvements)
    📊 Progress: 100.0% (2000/2000 bp)

☑ Genome Generation Complete!
Generated 2000 bp genome with 9 functional elements

This step generates a series of elements, including promoters, genes, and terminators. We end up with nine elements in total.

III. Now we will analyze the genome we just created:

```
analysis = generator.analyze_genome(genome)
```

This gives us an analysis of the overall characteristics of the genome:

```
🔬 Analyzing Genome with DNABERT
📊 Computing basic statistics...
   Length: 2,000 bp
   GC Content: 23.6%
🗺 Computing GC content windows...
   Processing window 1/196
   Processing window 51/196
   Processing window 101/196
   Processing window 151/196
🔍 Finding Open Reading Frames...
   Found 3 ORFs
🌐 Computing DNABERT embedding...
⭐ Analyzing element quality with DNABERT...
   Analyzing 9 elements...
   Processing element 3/9
   Processing element 6/9
   Processing element 9/9
☑ Analysis complete!
```

This shows that we have an overall genome length of 2Kb. You may note that the GC content is quite low. This could be something we would tune depending on the application. We also report the number of **Open Reading Frame (ORF)** elements found. Recall that these are regions of the genome that begin with a start codon, meaning they code for a protein (https://en.wikipedia.org/wiki/Open_reading_frame).

IV. Now, let's look at our visualization:

```
static_file = visualize_genome_notebook(genome, analysis)
```

Here is the output:

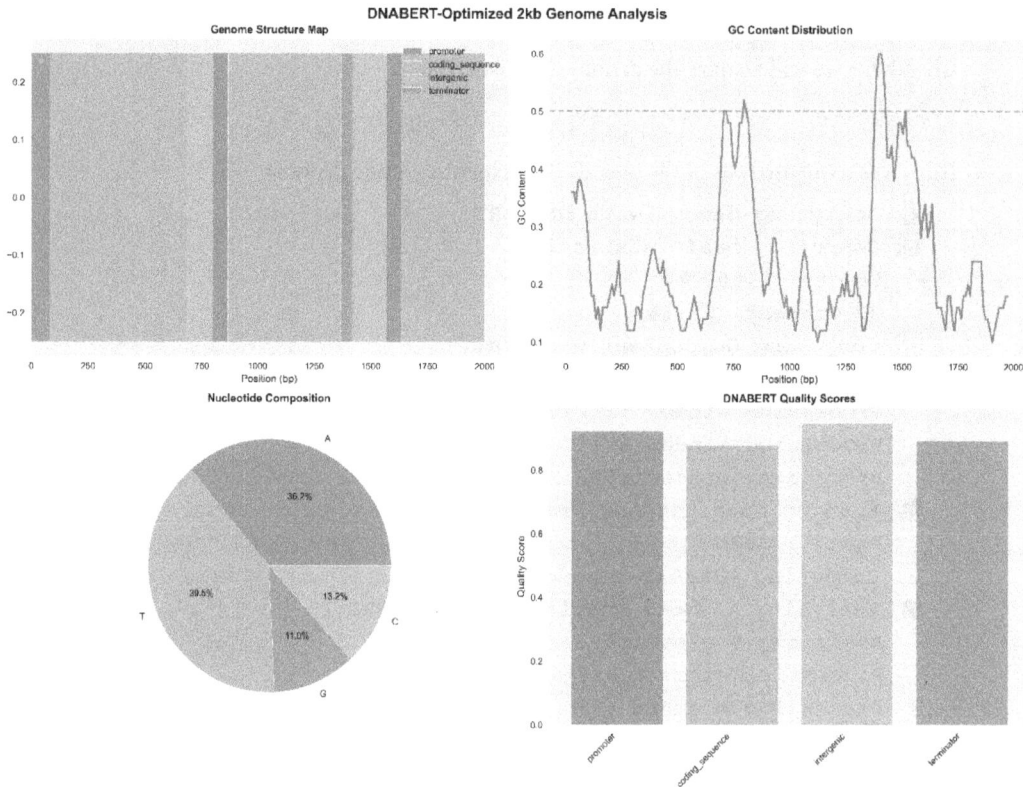

Figure 17.10 – Genome analysis overview

The preceding graph contains an overview of our genome.

V.   Finally, let's create our interactive analysis dashboard:

```
fig, html_file = create_interactive_genome_viewer_
notebook(genome, analysis)
```

This will give you an interactive analysis that looks like this:

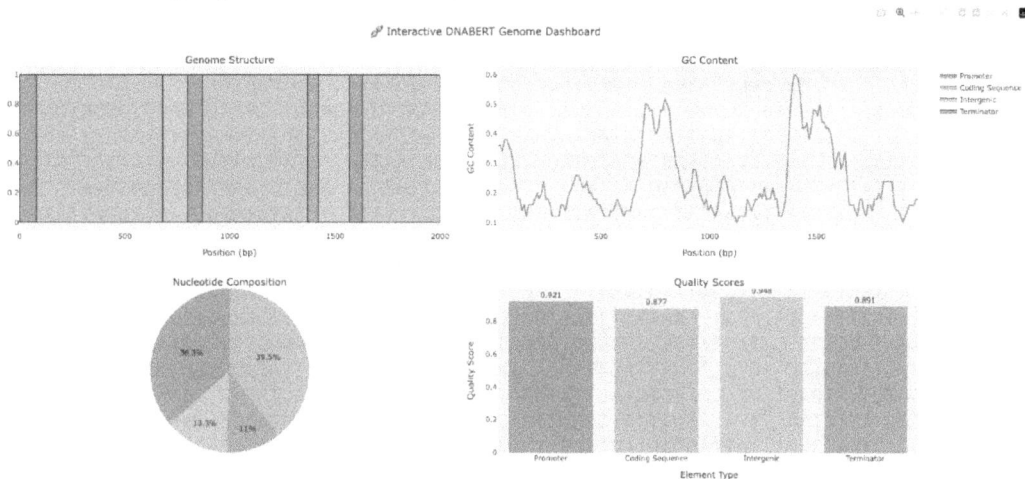

Figure 17.11 – Interactive genome analysis dashboard

You can go ahead and try out the dashboard! The top portion contains a high-level map of the genome with the main elements laid out. If you hover over a section, you will see its name and start and stop positions. The next graph is a line chart of the GC content in the genome. If you hover over it, you will see details on the total number and percentage for that base in the genome. On the bottom left, we have a pie chart of nucleotide composition across the genome. To the right of that, we have the element quality scores bar graph. This shows the average quality score by type of element. If you mouse over each bar, you will get the details.

There is one more line to run at the end of the notebook. This will run the entire workflow and also ensure that the generated genome is written out to a file. Run this cell in your notebook:

```
generator, genome, analysis, saved_files = generate_and_analyze_
genome()
```

Again, this will perform the entire genome generation workflow and save all the output files.

Let's review the output files created in the working directory.

The first file is the FASTA file for the genome we generated. It is called Ch17-3-dnabert_generated_genome.fasta. Go into your terminal and type the following:

```
less Ch17-3-dnabert_generated_genome.fasta
```

This is what we see:

```
>DNABERT_Generated_2kb_Genome
# Generated using DNABERT-optimized functional elements
# Promoters, coding sequences, terminators, and intergenic regions
CACTGGTGAGGATATATCGTGACGTAGAAGCTAGGCTTTCTAAAATTAATTATAAAAATTACATGTCACATGGCGTCATT
ATGATTACTAGTTTACATCTATTACATTACTATAATATAAATATTTTATTAATCAGTACTTTATTAATAATAATAAATTA
CGATGATTCAACTTATTTAGTATCAGTACTAGATGTACATACTAATGATAATCATATCATCAATACTAATGATATTATTA
TGGATATTGTATATTATGTAATTATCTATCTATATATGTTAAATATGTTATCAATAACTGTAGATAGTTCAATTATGTCA
TATTTATATTATATTTTAATTTTAATATTAATTTATTATATTGATTCATCATTAGTATATGTATACTACGTATATGATGA
TGATATCTTATTAATGAATTACCTGACTACTGTAAATATGAATATCTCAATGTTAATCTTAGTAAATCTAATTAATTATA
ATTATATGACTAATTCATTACTAGTACATATTTATCTAGTATTATTACTATCACAGATACTAATCGATATTTTATTACTA
AATATTATCATACATGATATTATCGATATATTAATAATGACTCTAATCAGTACTATAAATGATAATTTAATCATAAATGT
AATTAATGTAGATTTAAGTATAACTTTAATTGTAAATTGACCAAAATAGCCAGTATGACTCGCTGTCGTTAAATGGCAGC
GGTCTAGTAAATCCCGTTTCTTACTTTGACTTGCGCTGCCCTATATGTCTCAGAGAAACCGTGTCGGTAGAAATAAAAGA
CGTTGTCGCGCTACAGTACTTCTCTAACATTACATACAATTATAAATGGATCAAAACTTAGGGGTGATAGATGCATGATC
ATATAATGGATCTAATCAGTATTATATACTCAACTTATTCATATAGTATTTACAGTTACTATATTTCAAATATAAGTATA
AATAGTGTAAGTATTATAAATTTATATGATTATTACATTGATATTATGTTAGATGATAGTAGTCATAGTTTCTATATCTA
CGATATCAATAGTCATCATATAATTAGTAGTAGTACTATACATAGTGTAAATTTATCAGTAAATACTCATATAATGTATT
CAATTTACAATATAGTATATATTATTAGTTTATATATTAGTTAGTGTACATATAGTAATGACTAGTCATCTAATAGATTTA
TCAATAATTACTTCATTAATTATACATATAATAAATCATCTACTATATTTATATAGATTAATAATGATAATCATCACTGA
TAATTATTATGATTTAGATGATACTAATAATAATAATAGTTTACTAATCATTATATATGTAAATCTAAATAGTATGACTT
TAATCTAGATAAGCGGAGAGCACGTGCACTCGCATACAATGAGTCTTGTTCCCGGCTCGGCTCTGTATTAAAAGAAGAAG
CCCGCCAACTTCTGCGAAAGCAAGACATTGTTATACCCTCGGTCATTTATGCATCGGAGCTCGGTTATCTTCATTCTACC
ATACGTGATTAAAGTGTCGACCGACGTAATGTAAGAGTTGGAAGGTCGCTGAGACTAAGTGGTATTTGAGGTGCACTAAG
TATAAATTCTCAAGTTAAACGCTTCATCTAATGAATCATATGTATATCGTACTTATATATCATCTCAATTATTAATAG
TTATTACATTTTATATTATACGATTATTCTAATTGATTTAACTAATATAATAGATATTACTAGTTTAAGTAATTATAATA
TTAATGATACTTTAATTATAATAAATTACCATATCATCATTCTATTATTAACTTTATATATACTAAATGATAATTTAATA
ATTATCTACGTAAATACTTACCTAATGACTATCATCAATAGTTATATTTTATATATCCATTACAGTACTCATTACCATGT
ATATAATATGAGTAATATATACCTACATGATACTTACATTAGTATGATTTATATGATTGTAAGTGTAAATATAGATTAGG
Ch16-3-dnabert_generated_genome.fasta (END)
```

Figure 17.12 – Generated genome in FASTA format

This file simply provides the entire genome we designed in a FASTA format with the nucleotide sequence and header information.

The next file is the static version of our analysis graph, Ch17-3-dnabert_genome_analysis_static.png. We also provide the interactive HTML file Ch17-3-dnabert_genome_analysis_interactive.html. If you want to open it in your browser, go into your terminal and type the following:

```
open Ch17-3-dnabert_genome_analysis_interactive.html
```

This will open the interactive chart in your browser, and you can try it out.

That's it! Hopefully this recipe gave you a sense of how you can use ML to design and generate genomes.

There's more…

At this point in time, tools to generate entire genomes *de novo* are not widely available. For example, Evo 2 is not readily available as a standalone tool yet. So, for this recipe, we primarily used DNABERT to evaluate and optimize the genome, but not to generate everything from scratch. Soon we will see entire synthetic chromosomes and genomes being generated for important medical applications, food crop enhancement, industrial biofuel production, and more. Scientists have even embarked on an ambitious project to design an entire synthetic human genome: `https://gizmodo.com/ scientists-launch-wild-new-project-to-build-a-human-genome-from- scratch-2000620762`.

LLMs are being used in many areas of bioinformatics and biotechnology. For example, CRISPR-GPT helps design guide RNAs for gene editing experiments, as we covered in *Chapter 13, Genome Editing*; see Qu et al., *CRISPR-GPT: An LLM Agent for Automated Design of Gene-Editing Experiment*, bioRxiv, Apr 2024 – `https://www.biorxiv.org/content/10.1101/2024.04.25.591003v2`.

AlphaGenome is able to predict regulatory regions, gene expression, and many other features from genome inputs, and can predict the effect of variants in these regions as well; see Avsec et al., *AlphaGenome: advancing regulatory variant effect prediction with a unified DNA sequence model*, bioRxiv, Jun 2025 – `https://www.biorxiv.org/content/10.1101/2025.06.25.6 61532v1.full.pdf`.

Another key recent advance is **agentic AI**. These systems extend the autonomy of AI systems beyond just responding to prompts, so that they can plan, reason, and act in reactive loops; see Zhou et al., *Agentic Bioinformatics*, ResearchGate – `https://www.researchgate.net/ profile/Juexiao-Zhou/publication/389284860_Agentic_Bioinformatics/ links/67bd4a318311ce680c73a157/Agentic-Bioinformatics.pdf`.

AI tip

Let's explore building an AI agent for bioinformatics.

Prompt: Write an example of an AI agent that can search for genes and perform BLAST analysis. Make the code suitable for running in a Jupyter notebook. Have the code ignore SSL issues and handle BLAST timeouts. Provide a visualization of the BLAST result. Save the visualization as a PNG file as well.

You should see: Code to search NCBI for gene identifiers, perform BLAST, and parse and visualize the results. The code will likely provide outputs in the form of a CSV and FASTA file as well. The code should temporarily disable SSL verification; this is normally not safe for production but can be used for this exercise. Be sure to update the email address used to your own email address.

Here is an example of the results from our agent:

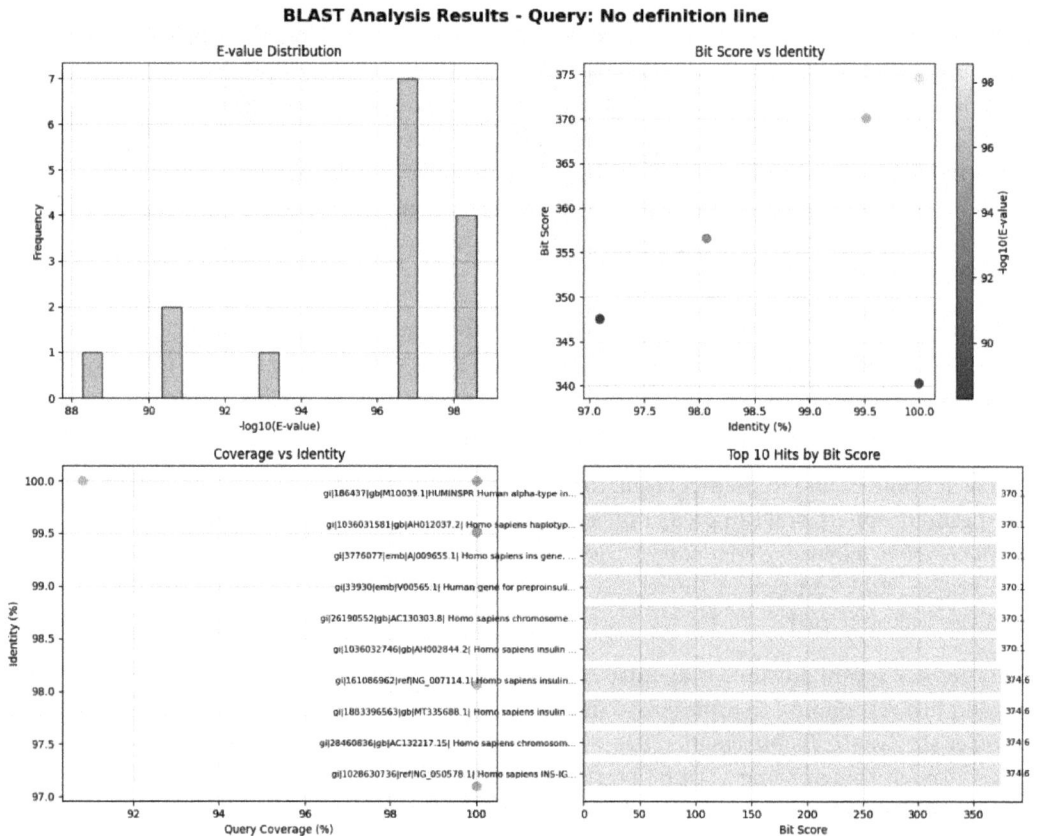

Figure 17.13 – BLAST results from bioinformatics agent

This code is available in the bonus Ch17-bonus-agent.ipynb notebook.

You've now seen how LLMs can be used for many powerful tasks, from designing proteins to genomes! I encourage you to deepen and strengthen your knowledge of these techniques and stay abreast of this fast-paced field!

OK, let's clean up and close down our conda environment:

```
conda deactivate
```

See also

- This article provides a great review of the impact of LLMs in genomic research: Li et al., *AI-Empowered Genome Decoding: Applications of Large Language Models in Genomics*, Frontiers of Digital Education, May 2025 – `https://link.springer.com/article/10.1007/s44366-025-0051-1`

- More discussion on Transformers can be found here: Zhang et al., *Applications of transformer-based language models in bioinformatics: a survey*, Bioinformatics Advances, Jan 2023 – `https://academic.oup.com/bioinformaticsadvances/article/3/1/vbad001/6984737`

- You may want to check out BioSeq-BLM: Li et al., *BioSeq-BLM: a platform for analyzing DNA, RNA and protein sequences based on biological language models*, Nucleic Acids Research, Dec 2021 – `https://academic.oup.com/nar/article/49/22/e129/6377401`

- Learn more about the use of DL in promoter analysis: Wang et al., *Accelerating promoter identification and design by deep learning*, Trends in Biotechnology, Jun 2025 – `https://www.cell.com/trends/biotechnology/abstract/S0167-7799(25)00174-X`

- Check out this interactive platform for developing ML pipelines in bioinformatics: Chen et al., *iLearnPlus: a comprehensive and automated machine-learning platform for nucleic acid and protein sequence analysis, prediction and visualization*, Nucleic Acids Research, Jun 2021 – `https://academic.oup.com/nar/article/49/10/e60/6154472`

- SpliceAI is another powerful DL tool for predicting splice sites in gene transcripts: Jaganathan et al., *Predicting Splicing from Primary Sequence with Deep Learning*, Cell, Jan 2019 – `https://www.cell.com/cell/fulltext/S0092-8674(18)31629-5`

- Learn more about the latest uses of Transformers in splice site prediction: Jonsson et al., *Transformers significantly improve splice site prediction*, Communications Biology, Dec 2024 – `https://www.nature.com/articles/s42003-024-07298-9`

- This tool uses DL to analyze nanopore data and three-dimensional genome structure: Ma et al., *DeepNanoHi-C: deep learning enables accurate single-cell nanopore long-read data analysis and 3D genome interpretation*, Nucleic Acids Research, Jul 2025 – `https://academic.oup.com/nar/article/53/13/gkaf640/8196083`

- **Hidden Markov Models** (**HMMs**) are another important ML tool in bioinformatics. This article provides an excellent review on the subject: Ma et al., *The Hidden Markov model and its applications in bioinformatics analysis*, Genes & Diseases, Jun 2025 – `https://www.sciencedirect.com/science/article/pii/S2352304225002181`

Get This Book's PDF Version and Exclusive Extras

Scan the QR code (or go to `packtpub.com/unlock`). Search for this book by name, confirm the edition, and then follow the steps on the page.

Note: Keep your invoice handy. Purchases made directly from Packt don't require an invoice.

18

Single-Cell Technology and Imaging

In this chapter, we will explore some of the most exciting new advances in fields where bioinformatics is being applied. We'll look at new capabilities, such as single-cell sequencing and droplet microfluidics, that are generating high-throughput datasets that demand big data approaches for analysis. We'll also learn about the exciting world of image analysis and how advanced microscopy technologies are enabling us to map the entire brain!

In this chapter, we will cover the following recipes:

- Building microfluidics devices
- Analyzing single-cell data
- Studying biological image data
- Mapping the brain

By the end of this chapter, you will have an understanding of how advanced technologies are being used to analyze biological systems at the cellular level, and how bioinformatics can be applied for analysis of these large datasets. You will also have learned the basics of image analysis using machine learning, and explored the exciting world of brain connectome mapping, in which scientists are finally beginning to understand the workings of intelligence.

Technical requirements

The code for this chapter can be found at `https://github.com/PacktPublishing/Bioinformatics-with-Python-Cookbook-Fourth-Edition/tree/main/Ch18`.

You will want to create a `Ch18` folder and set up your notebooks there.

Remember to activate your `conda` environment before beginning the recipes, like this:

```
conda activate bioinformatics_base
```

If you would like to set up a `conda` environment specific to this chapter, before activating `bioinformatics_base`, run the following:

```
conda create -n ch18-single-cell --clone bioinformatics_base
conda activate ch18-single-cell
```

You will be able to install the packages for the chapter as you go, or you can use the YAML file provided in the repository:

```
conda env update --file ch18-single-cell.yml
```

Building microfluidics devices

One of the core activities of biotechnology is screening. In order to test large numbers of cells, gene constructs, drugs, or proteins, scientists must test a huge number of possibilities. Historically, this has largely been done using **plate-based screening**. In this method, scientists use plates with 96, 384, or 1,536 wells arrayed in a rectangle (these are standard plate formats used in the field) to test small volumes of sample on a robotics platform.

Plate-based screening, also called **High-Throughput Screening** (HTS), suffers from numerous issues:

- By their nature, plates contain **bulk cells**; therefore, they aggregate the behavior of many different cells.

- Plates can suffer from dehydration and other artifacts.

- Plates must undergo careful normalization for row and column effects.

- Plates must be normalized against each other to control for batch effects.

- Plates can suffer from edge effects; see Mansoury et al., *The edge effect: A global problem. The trouble with culturing cells in 96-well plates*, Biochemistry and Biophysics Reports, Jul 2021 – `https://www.sciencedirect.com/science/article/pii/S2405580821000819`.

    HTS plates must be carefully designed to contain positive and negative controls in appropriate locations. Plates typically undergo normalization for artifacts, and then we calculate a **Z score** – `https://en.wikipedia.org/wiki/Standard_score`, which is the number of standard deviations a sample is above or below a control. We also calculate a Z prime, which is a measure of overall assay quality – `https://www.bmglabtech.com/en/blog/the-z-prime-value/`.

- Finally, the use of plate-based screening requires a large laboratory space with a significant number of large, expensive robots to run.

It is good to have a solid grounding in plate screening before we move on to microfluidics, as many of the same assay design and statistical concepts will apply, and it will give an appreciation of how microfluidic devices overcome many of the limitations of plates.

Microfluidics is a modern technology that addresses many of the concerns of plate-based screening and offers the potential for tremendous improvements in screening rates (throughput). It analyzes single cells, meaning the signal of individual cellular phenotypes is not obscured by bulk mixing. It also offers many other advantages, such as high levels of integration with multiple steps, including mixing and readout of results, in a single chip.

In this method, chips are designed using a variety of materials, and channels are etched into the chip. The channels are measured in **micrometers** (**um**), hence the name of the technology. Fluid flows down the channels through a variety of components, such as mixers, separators, and detection devices. This allows for fine-scale manipulation of microscopic objects and read-out of individual cellular events at a very high rate. Microfluidics setups and prototypes can actually be built fairly cheaply and then scaled up by working with manufacturing facilities. Microfluidics **lab-on-a-chip** solutions incorporate multiple lab functions shrunk down and integrated into a complex device. Compared to early expectations, these devices have not conquered the lab yet, but are likely poised for an explosion in the coming decade; see Elvira et al., *Materials and methods for droplet microfluidics device fabrication*, Lab on a Chip, Jan 2022 – `https://pubs.rsc.org/en/content/articlehtml/2022/lc/d1lc00836f`.

Here is a simplified overview of a microfluidics device:

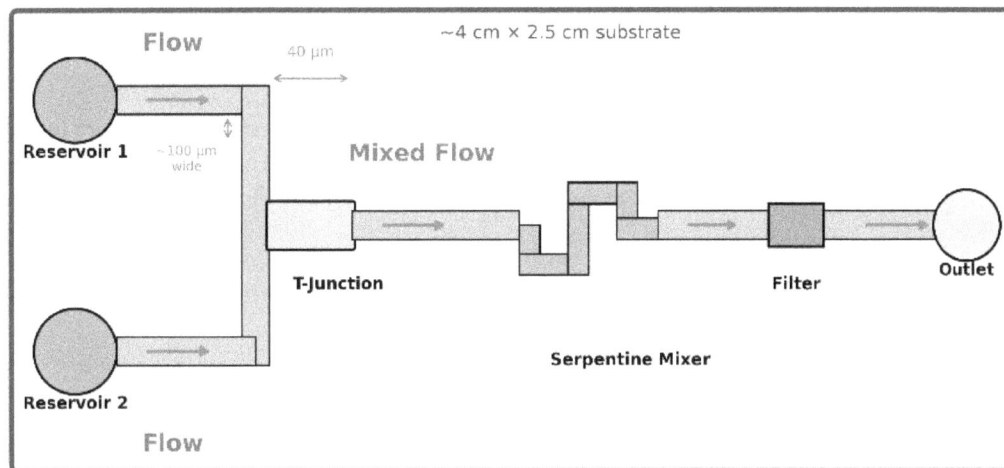

Figure 18.1 – A simple microfluidics device

One of the first applications of microfluidics was **Digital Droplet PCR** (**ddPCR**). In this technology, **polymerase chain reaction** amplifications are partitioned into many droplets, each acting as a reaction chamber to amplify a single molecule. It provides accurate and precise quantification of molecules, and can be used in applications such as gene expression or detection of food pathogens; see Hou et al., *Droplet-based digital PCR (ddPCR) and its applications*, Trends in Analytical Chemistry, Jan 2023 – `https://www.sciencedirect.com/science/article/pii/S0165993622003806`.

In this exercise, we will design a simple microfluidics device with a **T-junction** in which two channels come into a single junction and then flow down another channel. You'll learn the concepts of creating the design, storing it in a standard microfluidics file format, and analyzing the characteristics of your device.

Getting ready

Let's install the packages we need for this recipe:

```
! pip install shapely matplotlib numpy
```

How to do it...

Here are the steps to try this recipe:

1. First, we will import our libraries:

    ```
    import numpy as np
    import matplotlib.pyplot as plt
    import matplotlib.patches as patches
    from shapely.geometry import Polygon, Point
    from shapely.ops import unary_union
    import json
    import os
    ```

    In addition to using many of our standard libraries, we will use the `shapely` library in this example – https://shapely.readthedocs.io/en/stable/. `shapely` is a widely used geometry library for Python. We'll use it to help us draw out our microfluidics device.

2. Now we will define the dimensions of our microfluidics device:

    ```
    channel_width = 100
    channel_length = 2000
    inlet_length = 1000
    channel_depth = 50
    ```

    This device will have channels 100 um wide and up to 2,000 um (2 mm) long. The inlet length will be 1,000 um (1 mm) and the channels will be 50 um deep.

3. Next, we define our key class: `class MicrofluidicsDevice`. This class will manage the key aspects of defining our device and storing the geometry.

    This class will first initialize itself with the dimensions and other key parameters and define an `init` function to create an instance of a microfluidics device. Let's review the major functions of this class:

4. The `create_device()` function will set up the main outline of the device using the shapely `Polygon()` module – `https://shapely.readthedocs.io/en/stable/reference/shapely.Polygon.html`. Review the function in your notebook now. You will see we first define a list of tuples called `main_channel_coords`. This list defines a series of (x, y) coordinates (the tuples) which are the coordinates of the main channel. This is the standard way to define the vertices of a polygon in `shapely`. When we then call the `Polygon()` method, `shapely` will create the 2D geometry for the object. We will repeat this pattern several times for the inlets and T-junction elements (see steps 5 and 6 below). We will also make use of the `Point()` method –this method will be used to create circular objects representing **reservoirs** for fluid to flow in or out of our device.

5. We create a main channel that will act as our outlet channel. We define the coordinates of the bottom-left, bottom-right, top-right, and top-left corners. This will create a channel starting at *x=0* and extending to `channel_length`, vertically centered on the *y* axis, extending from minus half the `channel_width` value to positive half the `channel_width` value.

6. We define top and bottom inlet channels. We also define an area for the T-junction. Together, this will form our T-channel, with two inputs coming together to flow into an outlet. We define circular **reservoirs** – this is where you would put in your sample liquid (typically done with a syringe that can connect to the inlet) or remove your final sample from the outlet reservoir.

7. We use `shapely`'s `unary_union()` function to combine the geometric components into a single device. This function combines overlapping objects into a single shape, removing internal boundaries.

8. The `export_to_json()` function will write out our device specifications in JSON format. We first define a dictionary of key device parameters. We then loop over and convert the geometries of each component into coordinates. Finally, we open our output file and write out the information.

9. Another function, `export_to_dxf_format()`, will help us write out the specifications for our device in the popular **Drawing Exchange Format** (**DXF**). This format was created by Autodesk to share CAD drawings and can handle 2-D or 3-D formats. Review this function in the notebook now. The function first initializes an empty list `dxf_data`. Next, it loops over all the geometric items and extracts their coordinates. Note the use of the constant `LWPOLYLINE` – this is a lightweight polyline, a CAD entity for representing connected line segments. This information is stored in `dxf_data`. The function then opens an output file and prints out a human-readable file with the geometries. Note that this is not a true DXF file – to make one you would want to use `ezdxf(https://ezdxf.readthedocs.io/en/stable/)`. You can try this out on your own by using an AI prompt to upgrade this function to provide true DXF export. You may also want to check out a Python library for handling DXF designs – PyMicrofluidics - `https://github.com/guiwitz/PyMicrofluidics`.

10. Once you have reviewed the code for this class, run the cell in your notebook. This will define the class `MicrofluidicsDevice` for our use.

11. Great! Let's create an instance of our device:

```
device = MicrofluidicsDevice(
    channel_width, channel_length,
    inlet_length, channel_depth
)
```

This will use the class we've just defined to set up an instance of our `MicrofluidicsDevice` with the appropriate specifications.

12. Next, we'll build up an analyzer for our device – `class MicrofluidicsAnalyzer`. This class will help us calculate key metrics. Review the code for this class in your notebook.

   - The analyzer class takes in an instance of a device and initializes some key parameters. It calculates the cross-sectional area and the hydraulic diameter – a measurement that helps us treat non-circular pipes as circular to ease calculations. See `https://en.wikipedia.org/wiki/Hydraulic_diameter`.

13. We next define some key fluid properties. These include the density of water at 20 degrees Celsius (room temperature) and the **dynamic viscosity**, a measure of a fluid's resistance to force. The **kinematic viscosity** measures how a fluid flows in resistance to the force of gravity. The **diffusivity** measures the rate at which particles spread in a fluid – `https://www.sciencedirect.com/topics/physics-and-astronomy/diffusivity`.

14. The next section of the code will define several handy metric functions:

   - **Reynolds number**: An important factor in microfluidics design, it predicts whether fluid flow will be laminar (free-flowing) or turbulent –`https://www.simscale.com/docs/simwiki/numerics-background/what-is-the-reynolds-number/`.

   - **Peclet number**: This tells us whether particles in the device are primarily being carried by fluid flow or by random diffusion – `https://www.interfacefluidics.com/peclet-number-fluid-transport/`.

   - **Pressure drop**: The amount the pressure can drop across the length of a channel. We will use here the Reynolds number from the previous function and perform corrections based on the assumption that these are rectangular channels. We use the Hagen-Poiseuille equation – `https://sciencedemonstrations.fas.harvard.edu/presentations/poiseuilles-law`.

   - **Residence time**: The time needed for the volume in a chamber of the device to completely turn over

   - **Mixing length**: The distance fluids must travel along a channel for two fluid streams to become fully mixed

   - **Mixing efficiency**: The degree to which two different fluids become homogenous over time

15. The final function in this class generates a report for the analysis. It reports on key parameters of our device. It then simulates a variety of mixing velocities to enable efficiency analysis on the device. Review the code for `generate_analysis_report()` in the notebook.

16. OK, let's define and use our analyzer:

```
analyzer = MicrofluidicsAnalyzer(device)
test_velocities = [0.1, 0.5, 1.0, 2.0, 5.0]
mixing_data = analyzer.generate_analysis_report(test_velocities)
```

This code creates an instance of our `MicrofluidicsAnalyzer` class and passes in our device design. It then defines a range of test velocities and calls our reporting and analysis function. Testing different velocities is crucial for optimizing the performance of our device. Going too slowly will allow for better mixing but make for lower throughput experimentation. At higher throughput, we may experience poor mixing and extreme pressure. In a real-life scenario, we would want to optimize the performance of our application, whether it be chemical synthesis, cell sorting, or screening.

Nice work!

17. Now we are going to write code to visualize our device. Check out the `visualize_microfluidics_device()` function in the notebook:

    I.   This function will again use the `shapely` library to draw out our objects. We start by laying out our device with a title and set colors (you will see this in *Figure 18.3*, in the upper-left panel).

    II. Then, loop over the geometries and draw corresponding shapes to represent them.

    III. Add flow arrows to represent the direction of fluid flow.

    IV. Plot the cross-sectional profile (see *Figure 18.3*, the upper-right panel).

    V.   Build a visualization to summarize the mixing efficiency analysis (see *Figure 18.3*, the lower-left panel).

    VI. Create a chart showing the analysis of the design operating space (see *Figure 18.3*, the lower-right panel).

18. Let's run the visualization code:

```
visualize_microfluidics_device()
```

19. Awesome! Now we can export our device to the appropriate file formats:

```
json_file = device.export_to_json(
    'microfluidics_device.json')
dxf_file = device.export_to_dxf_format(
    'microfluidics_device_dxf.txt')
```

This will export files to our working directory in the JSON and DXF formats.

20. Finally, let's print a summary of what we've done. Review the *Print Summaries* section of the code.

Here is the summary of the microfluidics device design:

```
================================================================
                MICROFLUIDICS DEVICE ANALYSIS REPORT
================================================================


Device Specifications:
  Channel width: 100 µm
  Channel length: 2.0 mm
  Channel depth: 50 µm
  Hydraulic diameter: 66.7 µm
  Cross-sectional area: 5000.0 µm²

Fluid Properties (Water at 20°C):
  Density: 1000 kg/m³
  Dynamic viscosity: 1.0 mPa·s
  Diffusivity: 2.0 nm²/s

Flow Analysis for Different Velocities:
----------------------------------------------------------------
Velocity   Re      Pe      Pressure    Residence   Mixing
(mm/s)                     Drop (Pa)   Time (s)    Length(mm)
----------------------------------------------------------------
0.10       0.01    5       0.9         20.000      0.0
0.50       0.03    25      4.7         4.000       0.2
1.00       0.07    50      9.3         2.000       0.5
2.00       0.13    100     18.7        1.000       1.0
5.00       0.33    250     46.7        0.400       2.5
```

Figure 18.2 – Summary of microfluidics device design

Here is the visualization for our device:

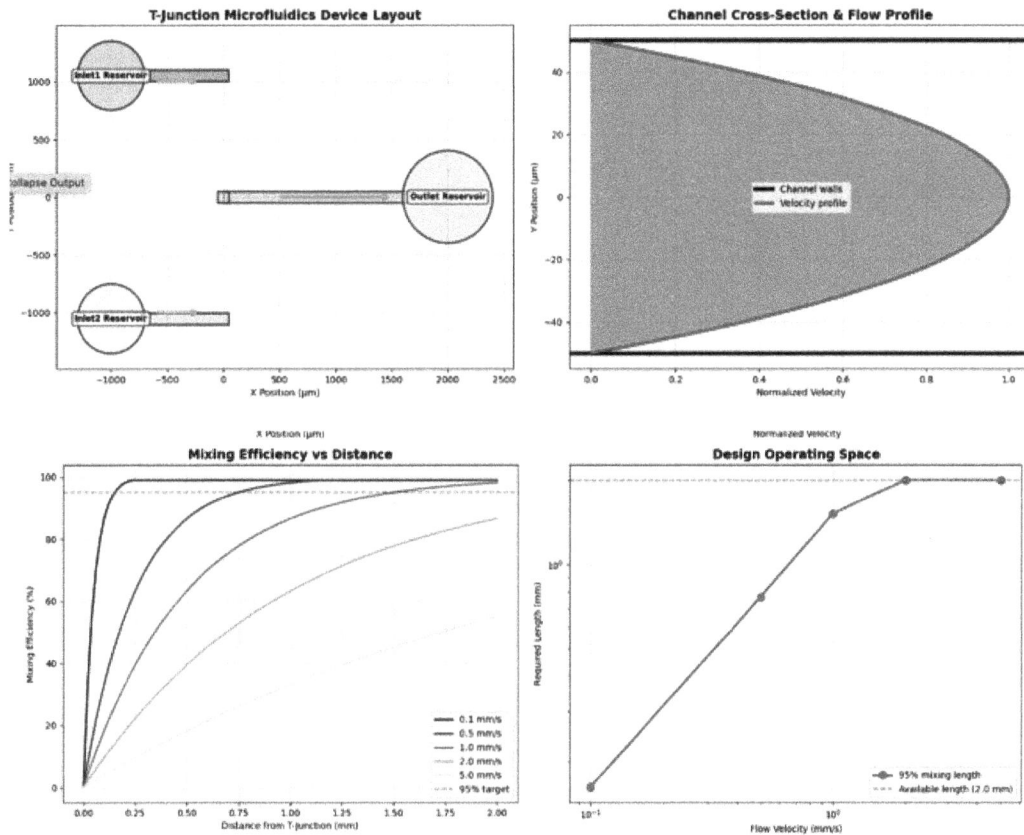

Figure 18.3 – Microfluidics device analysis

In the upper-left panel, we see an outline of our device. The fluid would be introduced in the upper and lower reservoirs (circles) and then flow into the channels, and then into the T-junction to go into the output channel and the final reservoir. (*Note: The T-junction isn't shown very well in the preceding figure, but it is where the two inlets come together with the outlet channel.*)

In the upper-right panel, we see the channel cross-section and flow profile. This shows a **parabolic velocity profile** in which the velocity of a fluid in a pipe is highest at the center, going to 0 at the walls (the thick lines at the top and bottom represent the walls of the channel) – https://www. sciencedirect.com/topics/engineering/velocity-profile.

In the bottom-left panel, we see an analysis of mixing efficiency. In this chart, we test different mixing velocities and look at how well the fluid is mixed as it moves down the channel. The dashed line near 100% represents our target of 95% mixing. As you can see, higher and middle velocities will reach effective mixing at some point down the channel. But lower velocities will never achieve effective mixing.

In the bottom-right panel, we have a **design operating space** chart. This helps us examine the trade-offs between velocity and mixing. If we want our device to work faster and process samples more quickly, we need to increase the velocity, but that will reduce the quality of mixing. At higher velocities, we will also need longer channels to get better mixing. Given a maximum device size of 2 mm (dashed line), we can only operate within the section of the curve under that line. Within that, we can choose options along the primary line that give us the best trade-off between device operation speed and mixing quality.

Finally, here is our summary output:

```
✓ Device exported to microfluidics_device.json
✓ File size: 16.8 KB
✓ DXF-format data exported to microfluidics_device_dxf.txt

File Export Summary:
--------------------------------------------
✓ JSON format: microfluidics_device.json
✓ DXF-compatible: microfluidics_device_dxf.txt
✓ Ready for CAD import and fabrication

Optimal Operating Conditions:
--------------------------------------------
• Flow velocity: 0.5-2.0 mm/s for good mixing
• Reynolds number: <1 (laminar flow confirmed)
• Mixing efficiency: >95% within channel length
• Pressure drop: <100 Pa (easily achievable)
• Residence time: 1-4 seconds for complete mixing

Alternative to GDSII:
------------------------------------
• JSON file contains all geometric data
• DXF-format file for CAD software import
• Both formats preserve design precision
• Compatible with most fabrication workflows
```

Figure 18.4 – Summary of microfluidics device design and analysis

Note that in this example, we used JSON and DXF outputs as an alternative to the more complex **GDSII** format, which is a standard file format for the exchange of **Integrated Circuit** (**IC**) designs – https://en.wikipedia.org/wiki/GDSII

This shows a basic example of how we can design a microfluidics device and analyze its properties *in silico*. You can see how one could design and simulate a much larger device and then have it fabricated for testing.

There's more…

Once you have a design, you can order prototypes for testing in your lab and eventually scale up for production. Alternatively, you can work with companies to develop assays on their microfluidics platforms. Here are some companies offering microfluidics fabrication or services:

- **Bio-Rad**: Offers ddPCR systems – `https://www.bio-rad.com/`

- **Bruker**: Offers the Beacon Quest optofluidic system – `https://brukercellularanalysis.com/products/instruments/beacon-quest-optofluidic-system`

- **Dolomite Microfluidics**: Emulsions and droplet encapsulation – `https://www.unchainedlabs.com/dolomite-microfluidics-systems/`

- **DropXcell**: Screens large numbers of antibody drug candidates – `https://www.dropxcell.com/`

- **Parallel Fluidics**: Can build a custom device in as little as three days – `http://parallelfluidics.com/`

- **Potomac Laser**: Offers a variety of microfluidics fabrication capabilities – `https://www.potomac-laser.com/application/microfluidics/`

- **uFluidix**: Offers custom microfabrication services – `https://www.ufluidix.com/`

Microfluidics is finding many uses throughout bioscience. Cells can be sorted, captured, and analyzed using their visual, fluorescent, mechanical, or electrical properties. The large datasets generated are perfect for machine learning analysis; see Jeon & Han, *Microfluidics with Machine Learning for Biophysical Characterization of Cells*, Annual Review of Analytical Chemistry, Vol 18, 2025 – `https://www.annualreviews.org/content/journals/10.1146/annurev-anchem-061622-025021`.

Microfluidics devices can be combined with machine learning to explore high-dimensional experimental spaces, acting as a type of "scientist on a chip"; see Damir et al., *Harnessing Synergies between Combinatorial Microfluidics and Machine Learning for Chemistry, Biology, and Fluidic Design*, Chemistry Methods, Jul 2025 – `https://chemistry-europe.onlinelibrary.wiley.com/doi/full/10.1002/cmtd.202500069`.

Microfluidics can be combined with high-complexity CRISPR libraries and single-cell analysis to systematically phenotype massive numbers of genomic alterations; Zheng et al., *Massively parallel in vivo Perturb-seq screening*, Nature Protocols, Feb 2025 – `https://www.nature.com/articles/s41596-024-01119-3`.

As you can see, microfluidics is poised to transform the biotechnology industry. By integrating entire lab functions onto chips, the speed at which data can be generated is increasing exponentially. Most importantly, this technology is powering the next generation of **single-cell analysis**, in which individual cells can be interrogated for their function. This capability is critical for unraveling the complexity of biology. Having a good grounding in the design and function of microfluidics devices will help you better understand the parameters of the experiments they can perform. Understanding these devices and the massive datasets they generate will be a key component of bioinformatics work in the future.

See also

- This review dives deeply into applications of droplet microfluidics: Breukers et al., *From specialization to broad adoption: Key trends in droplet microfluidic innovations enhancing accessibility to non-experts*, Biomicrofluidics, Mar 2025 – `https://pubs.aip.org/aip/bmf/article/19/2/021302/3338179`

- For a deeper review of plate normalization in HTS settings, see Mpindi et al., *Impact of normalization methods on high-throughput screening data with high hit rates and drug testing with dose-response data*, BioInformatics, Dec 2015 – https://academic.oup.com/bioinformatics/article/31/23/3815/208794

- Machine learning is impacting microfluidics design in numerous ways; see Lashkaripour et al., *Machine learning enables design automation of microfluidic flow-focusing droplet generation*, Nature Communications, Jan 2021 – `https://www.nature.com/articles/s41467-020-20284-z`

- For more information on the impact of machine learning on this field, see Park et al., *Machine Learning-Driven Innovations in Microfluidics*, Biosensors, Dec 2024 – `https://www.mdpi.com/2079-6374/14/12/613`

- Hybrid devices can be engineered that combine biological and electronic components on a microfluidic device; see Yazicigil et al., *Improving engineered biological systems with electronics and microfluidics*, Nature Biotechnology, Jun 2025 – `https://www.nature.com/articles/s41587-025-02709-6`

Analyzing single-cell data

In this recipe, we'll look at the exciting world of **single-cell analysis**. We've seen that microfluidics technology is poised to transform the throughput and accuracy with which we can perform biological experiments. This has powered the rise of single-cell technologies in which individual cells can be sorted, classified, and then handled for further analysis.

Before this technology came about, most experiments were performed using **bulk cell analysis**. This refers to the fact that large numbers of cells are mixed together and analyzed as a group, meaning that we are averaging out their properties. But individual cells each have different transcriptional programs, proteomic states, and metabolic profiles at any given time. To really understand biology, we need the ability to look at cells one at a time. This is where single-cell analysis comes in.

Here are some of the main types of single-cell technology:

- **Genomics**: Analyzes the DNA of the cell. This includes whole genome sequencing, chromatin accessibility, and methylation state. See Shao et al., *Advances in single-cell DNA sequencing enable insights into human somatic mosaicism*, Nature Reviews Genetics, Apr 2025 – `https://www.nature.com/articles/s41576-025-00832-3`.

- **Transcriptomics**: RNA sequencing to interrogate gene expression; see Duhan et al., *Single-cell transcriptomics: background, technologies, applications, and challenges*, Molecular Biology Reports, Apr 2024 – `https://link.springer.com/article/10.1007/s11033-024-09553-y`.

- **ChIP-seq**: Chromatin immunoprecipitation to identify DNA binding proteins and regulatory factors; see Grosselin et al., *High-throughput single-cell ChIP-seq identifies heterogeneity of chromatin states in breast cancer*, Nature Genetics, May 2019 – `https://www.nature.com/articles/s41588-019-0424-9`.

- **Proteomics**: Analyze the nature and level of protein expression; see Admad & Budnik, *A review of the current state of single-cell proteomics and future perspective*, Analytical and Bioanalytical Chemistry, Jun 2023 – `https://link.springer.com/article/10.1007/s00216-023-04759-8`.

- **Metabolomics**: Measures small molecules; Guo et al., *The limitless applications of single-cell metabolomics*, Current Opinion in Biotechnology, Oct 2021 – `https://www.sciencedirect.com/science/article/pii/S0958166921001269`.

- **Imaging**: Analyze cells using optical, fluorescent, X-ray, or spectroscopy modalities. Reconstruct tissues and organs; see Sun et al., *Deep Learning-Based Single-Cell Optical Image Studies*, Cytometry, Jan 2020 – `https://onlinelibrary.wiley.com/doi/full/10.1002/cyto.a.23973`.

- Many single-cell technologies combine multiple analysis modalities. For example, CITE-seq profiles both the transcriptome and proteome simultaneously; see Song et al., *Key Considerations on CITE-Seq for Single-Cell Multiomics*, Proteomics, Feb 2025 – `https://analyticalsciencejournals.onlinelibrary.wiley.com/doi/full/10.1002/pmic.202400011`.

There are many great tools available now to analyze and visualize single-cell data. Here are some of the major Python libraries used for single-cell analysis:

- **scanpy**: Popular tool for transcriptomic analysis – `https://scanpy.readthedocs.io/en/stable/`

- **anndata**: A key data structure library used heavily in single-cell analysis – `https://anndata.readthedocs.io/en/stable/`

- **scvi-tools**: Suite of tools to enable machine learning on single-cell data – `https://scvi-tools.org/`

- **SCeptre**: proteomics extension for `scanpy` – `https://github.com/bfurtwa/SCeptre`

- **SCMeTA**: Metabolomics analysis; see Pan et al., *SCMeTA: a pipeline for single-cell metabolic analysis data processing*, Bioinformatics, Sep 2024 – `https://academic.oup.com/bioinformatics/article/40/9/btae545/7750353`

- **Squidpy**: Spatial omics and visualization – `https://squidpy.readthedocs.io/en/stable/`

- **Deepcell-tf**: Biological image analysis package built on TensorFlow – `https://github.com/vanvalenlab/deepcell-tf`

- **Scbean**: Multi-omics data analysis; see Zhang et al., *Scbean: a python library for single-cell multi-omics data analysis*, Bioinformatics, Feb 2024 – `https://academic.oup.com/bioinformatics/article/40/2/btae053/7593744`

Another common use of single-cell analysis is to define **trajectories**. With single-cell analysis, we can find common developmental states that change over time, recognizable by their gene expression, proteomic, or other profiles. These states change in biological time, and the exact wall clock time is not important. We refer to this concept as **pseudotime**. Monocole is a classic tool for tracking single-cell trajectories – `https://cole-trapnell-lab.github.io/monocle-release/`. These tools can be used to understand the fate of cells as they develop in tissues and organs or as they undergo differentiation processes, respond to drug treatments, or experience mutations in a bioprocess. For a deeper review of trajectory analysis tools, read Saelens et al., *A comparison of single-cell trajectory inference methods*, Nature Biotechnology, Apr 2019 – `https://www.nature.com/articles/s41587-019-0071-9`.

In this recipe, we will use single-cell gene expression analysis to look at **Peripheral Blood Mononuclear (PBMC)** cells. PBMC cells are made up of the following key cell types:

- **T cells**: Involved in adaptive immunity

- **B cells**: Primary role is to produce antibodies for the immune system

- **NK cells**: Natural killer cells, focused on hunting down and killing cancer cells

- **Monocytes**: White blood cells

- For more information, see `https://www.ncbi.nlm.nih.gov/books/NBK500157/`.

We will use **Single-Cell Transcriptomic (RNA-Seq)** data that was obtained using 10x genomics Chromium scRNA sequencing technology – `https://www.10xgenomics.com/platforms/chromium`. This technique looks at cell state by studying gene expression. We will learn the core steps of data cleaning and normalization needed to effectively analyze this type of data. Then we'll perform clustering and visualization. The ultimate goal will be to find clusters of genes with related expression patterns that explain the types of cells seen and the roles they play biologically.

Getting ready

We will install the following tools:

```
! pip install scanpy igraph leidenalg
```

How to do it...

Here are the steps to try this recipe:

1. We begin by importing our libraries:

    ```
    import scanpy as sc
    import pandas as pd
    import numpy as np
    import matplotlib.pyplot as plt
    import seaborn as sns
    import warnings
    ```

    We will use the scanpy library for single-cell analysis.

2. The next section suppresses some warnings.

3. We will configure our scanpy settings. This section simply sets a verbosity level and some standard matplotlib parameters.

4. Next, we load the data for our PBMC3k dataset:

    ```
    print("Loading PBMC3k dataset...")
    adata = sc.datasets.pbmc3k()
    ```

    This loads the PBMC3k dataset. This built-in scanpy dataset consists of 3,000 PBMC cells that were obtained from a healthy donor and sequenced using 10x genomics technology – https://scanpy.readthedocs.io/en/stable/generated/scanpy.datasets.pbmc3k.html.

5. We will handle potential name duplications in our dataset.

6. Let's take a quick look at the shape of our data:

    ```
    print(f"Number of cells: {adata.n_obs}")
    print(f"Number of genes: {adata.n_vars}")
    print(f"Data shape: {adata.shape}")
    ```

    Here is what we see:

    ```
                Number of cells: 2700
                Number of genes: 32738
                Data shape: (2700, 32738)
    ```

    Figure 18.5 – Overview of the single-cell dataset

This dataset contains 2,700 cells with 32,738 genes being analyzed.

- We will next calculate some quality control metrics. This section of the code calls `calculate_qc_metrics()` from `scanpy` to generate key metrics for each cell, including the total expression counts for genes – `https://scanpy.readthedocs.io/en/stable/generated/scanpy.pp.calculate_qc_metrics.html`. We divide out all the metrics into some additional gene categories, as follows:

  - **Mitochondrial**: Used to filter out low-quality cell data. Dying cells tend to lose their cell membrane and hence their cytoplasmic mRNAs, while mitochondrial RNA will tend to stay intact.

  - **Ribosomal**: Indicates highly active cells due to high translation levels.

  - **Hemoglobin**: Used to detect contamination – red blood cells contain high levels of hemoglobin, but PMBC cells should not.

7. Let's visualize our QC metrics!

```
sc.pl.violin(
    adata,
    ['n_genes_by_counts', 'total_counts', 'pct_counts_mt'],
    jitter=0.4, multi_panel=True
)
sc.pl.scatter(adata, x='total_counts', y='pct_counts_mt')
sc.pl.scatter(adata, x='total_counts', y='n_genes_by_counts')
```

We will use the scanpy violin (`https://scanpy.readthedocs.io/en/latest/generated/scanpy.pl.violin.html`) and scatter (`https://scanpy.readthedocs.io/en/stable/generated/scanpy.pl.scatter.html`) functions. These will be used to understand the quality of our data.

Here is an example of a violin plot:

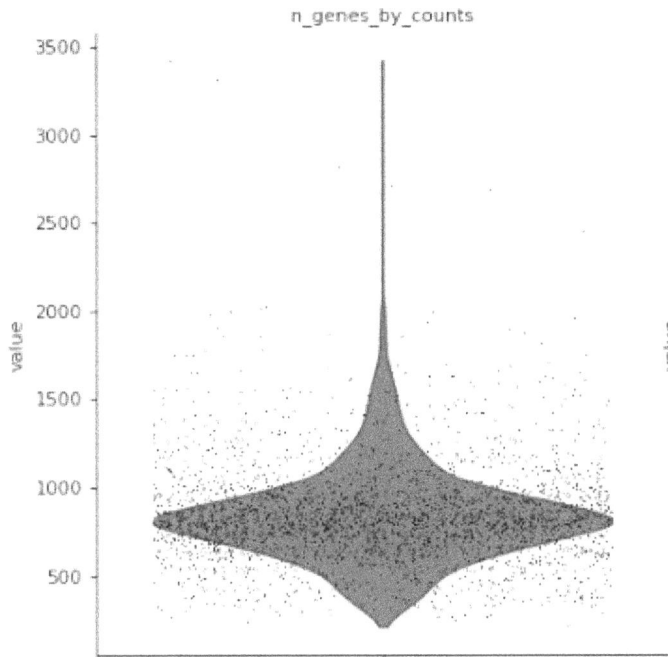

Figure 18.6 – Violin plot of number of genes detected per cell

Violin plots show the density of data for a distribution – https://en.wikipedia.org/wiki/Violin_plot. In this case, the *x* axis just contains one group, which is the number of genes detected in a cell. But you could have several groups displayed in the violin plot on the *x* axis. The *y* axis shows the actual count of genes detected in a cell. This plot shows us that most cells have around 750 or so genes detected (the bulk of the distribution), with tail ends of cells nearing the low hundreds or up to ~3,500.

8. We will now filter the cells and genes. Review the *FILTER CELLS AND GENES* section in the notebook.

    This code uses the scanpy filter_cells (https://scanpy.readthedocs.io/en/stable/generated/scanpy.pp.filter_cells.html) and filter_genes (https://scanpy.readthedocs.io/en/stable/generated/scanpy.pp.filter_genes.html) functions to clean our data. We will filter out any cell that has <200 genes expressed (this is biologically unrealistic). We will remove any gene that is only seen in fewer than three cells. We will also filter out any damaged cells based on our mitochondrial gene filter, that is, cells that may have been double-counted and cells with too few counts.

    Let's save a backup copy of our data before we normalize it and calculate some basic statistics. Review section *6. SAVE RAW DATA* in the notebook.

9. In this code, we save a backup copy of our data in the `adata.raw` variable. This is useful because in a couple of steps, we will scale and normalize our data, and we may want to have the original version to compare it to. We also calculate some basic statistics on the raw data, as follows:

```
Raw data statistics:
Raw data range: 1.00 to 419.00
Raw data mean: 2.79
Number of zeros: 0
```

Figure 18.7 – Raw data statistics

This tells us that we have a range of cell counts from 1 to 419, with the mean being 2.79. We don't see zeros because we are storing this data as a **sparse matrix** for efficiency – https://docs.scipy.org/doc/scipy/tutorial/sparse.html.

10. Next, we are going to find highly variable genes. Review section *7. FIND HIGHLY VARIABLE GENES* in the notebook:

   I.   First, we check for infinite or **Not-a-Number** (**NaN**) values that might throw off our calculations – https://en.wikipedia.org/wiki/NaN. This data will be cleaned and replaced with zeros as needed.

   II.   Next, we will look for highly variable genes. We will use the `scanpy highly_variable_genes` function (https://scanpy.readthedocs.io/en/stable/generated/scanpy.pp.highly_variable_genes.html) and implement the popular `Seurat` method. This method is excellent at standardizing the variance with respect to expression. It helps deal with technical noise and is more reproducible across experiments than other methods. The code also provides some backup methods using traditional variance estimation.

   III. We will keep only the highly variable genes for downstream analysis. This is done because they are the genes that contain the real biological signal we are looking for. They are the genes that will distinguish the different cell types within our population. This is in contrast to **housekeeping genes**, which are going to be highly expressed in all cells. These would be ribosomes involved in translation and other core components of cellular function that are going to be found in every cell – we want to avoid analyzing those.

11. We are now going to scale and normalize our data. Review section *8. NORMALIZATION AND SCALING* in the notebook:

   • We will use the `scanpy normalize_total` function so that every cell will have the same total gene count at the end. This helps us think about the difference between genes *within a cell* in more relative terms – https://scanpy.readthedocs.io/en/1.10.x/generated/scanpy.pp.normalize_total.html.

- Next, we **log-transform** the data. This is done to bring our data closer to a **normal distribution** – https://medium.com/@kyawsawhtoon/log-transformation-purpose-and-interpretation-9444b4b049c9. We will use the scanpy log1p function – https://scanpy.readthedocs.io/en/stable/generated/scanpy.pp.log1p.html.

12. We save the log-normalized data for later use in differential expression analysis.

13. We print out some checks to see whether there are still outliers after log normalization.

14. Finally, we scale the data using the scanpy scale method – https://scanpy.readthedocs.io/en/stable/generated/scanpy.pp.scale.html. This is a standard operation to give the overall distribution a variance of 1 and a mean of 0.

15. Now we will perform **PCA** (short for **principal components analysis**) on the data:

```
sc.tl.pca(adata, svd_solver='arpack')
sc.pl.pca_variance_ratio(adata, log=True, n_pcs=50)
```

This will perform PCA on our data to reveal the primary components (genes) that best explain the variance in our data. Recall that we covered PCA in *Chapter 4*, in the *Introducing scikit-learn with PCA* recipe.

16. Next, we compute the neighborhood graph:

```
sc.pp.neighbors(adata, n_neighbors=10, n_pcs=40)
```

This runs the scanpy neighbors function – https://scanpy.readthedocs.io/en/stable/api/generated/scanpy.pp.neighbors.html. It performs a nearest neighbors distance calculation to find the cells that are most similar to each other based on their expression. This will form the basis of the UMAP clustering we are about to create.

17. We will now perform UMAP clustering on the data:

```
sc.tl.umap(adata)
```

Recall that we covered this topic in *Chapter 4*, in the *Building a UMAP using Seaborn* recipe.

18. Clustering is performed next:

```
sc.tl.leiden(
    adata,
    resolution=0.5,
    flavor="igraph",
    n_iterations=2,
    directed=False,
    random_state=0
)
```

This uses **Leiden clustering**. It is a clustering method optimized to find modularity in networks – `https://en.wikipedia.org/wiki/Leiden_algorithm`. This is exactly what we want! We want to understand which gene modules most explain the different cell types in our dataset.

Let's visualize our UMAP!

```
sc.pl.umap(
    adata, color=['leiden'], legend_loc='on data',
    title='Leiden Clustering', frameon=False, save='.pdf'
)
sc.pl.umap(
    adata,
    color=['total_counts', 'n_genes_by_counts', 'pct_counts_
mt'],
    ncols=3
)
```

Here is the chart we see:

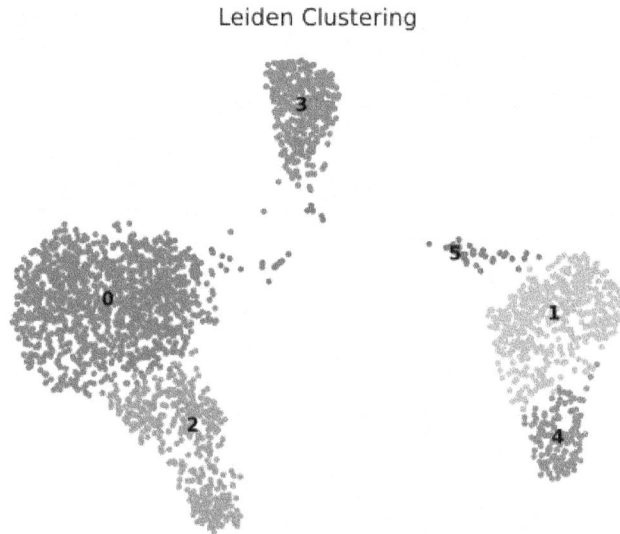

Figure 18.8 – UMAP visualization of single-cell data

We see here that the cells have neatly clustered into five groups (0–4). These clusters (hopefully) are going to represent the major types of cells found within PBMCs (e.g., B cells, T cells, etc.). Let's analyze the data more and find out!

> **Note**
>
> Clustering methods are not always deterministic, so if you don't get the exact same graph as in the preceding figure, do not be concerned.

19. The next section helps us find marker genes:

```
sc.tl.rank_genes_groups(
    adata, 'leiden', method='wilcoxon',
    use_raw=False, layer='log1p')
sc.pl.rank_genes_groups(adata, n_genes=5, sharey=False)
result = sc.get.rank_genes_groups_df(adata, group='0')
print(result.head(10))
```

This code uses the scanpy rank_genes_groups function to find the top genes that distinguish each cluster – https://scanpy.readthedocs.io/en/stable/generated/ scanpy.tl.rank_genes_groups.html. It uses the **Wilcoxon signed-ranked** statistical test to determine which genes are significant – https://www.statisticssolutions. com/free-resources/directory-of-statistical-analyses/how-to- conduct-the-wilcox-sign-test/.

20. Now we will plot specific genes. These are genes that are known to be informative of cell type in PMBC cells. Review the *15. PLOT SPECIFIC GENES* section in the notebook.

Here is an example of one of the charts we see:

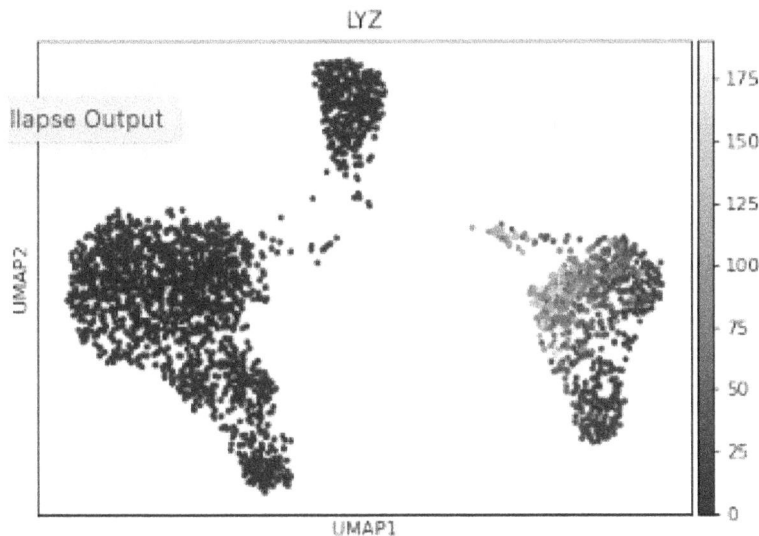

Figure 18.9 – UMAP clustering for LYZ gene

We see the UMAP clustering with the LYZ gene expression highlighted on it. It is considered a marker of monocytes. The LYZ gene makes lysozyme – https://www.genecards. org/cgi-bin/carddisp.pl?gene=LYZ.

Lysozyme is a key part of the immune system and is highly up-regulated in monocytes. We see that it is highly up-regulated in one of our clusters, which makes sense and suggests that our clustering is working well. For more background, read Schlachetzki et al., *A monocyte gene expression signature in the early clinical course of Parkinson's disease*, Nature, Jul 2018 – `https://www.nature.com/articles/s41598-018-28986-7`.

21. Now we will summarize our work. Review section *16. SAVE RESULTS* in the notebook.

    This code will make a `results` subdirectory in our working directory. We will write the h5ad file there. We will also export the cluster information and marker genes in CSV format. The h5ad file is a common format used by anndata for storing single-cell analysis data. For more on the h5ad file format, which is based on HDF5, check out `https://anndata.readthedocs.io/en/latest/fileformat-prose.html`.

22. The last section is optional and annotates cell types.

23. That's it! We have now performed an analysis of a single-cell dataset.

There's more...

Another key aspect of single-cell technology is the rise of **spatial biology**. In these techniques, the original positions of cells in tissues or organs are preserved and subjected to omics analyses. In this way, we can understand how expression profiles differ across different parts of a tissue and affect cell development and communication. These techniques may be applied in 2D tissue sections such as **Formalin-Fixed Paraffin-Embedded** (FFPE) samples, or even in 3D to analyze intact organs, organoids, or brain tissues. See Park et al., *Spatial omics technologies at multimodal and single cell/subcellular level*, Genome Biology, Dec 2022 – `https://link.springer.com/article/10.1186/s13059-022-02824-6`.

Spatial technologies can be used to construct **cell atlases**, which provide information on the location and types of cells in various settings, along with their associated omics data. For example, the **Allen Brain Cell Atlas** provides a wealth of multimodal cellular information on the mammalian brain – `https://portal.brain-map.org/atlases-and-data/bkp/abc-atlas`. The **Human Cell Atlas** maps every cell and tissue type in the human body – `https://www.humancellatlas.org/`.

Use of single-cell and spatial technology is exploding in bioscience. It is being used extensively in cancer research and will be used to guide treatment and therapy decisions; see Zhang et al., *Single-cell RNA sequencing in cancer research*, Journal of Experimental & Clinical Cancer Research, Mar 2021 – `https://link.springer.com/article/10.1186/s13046-021-01874-1`.

Single-cell technology is also massively expanding our understanding of the human brain, as we incorporate high-dimensional datasets into brain models; see Chen et al., *A brain cell atlas integrating single-cell transcriptomes across human brain regions*", Nature Medicine, Aug 2024 – `https://www.nature.com/articles/s41591-024-03150-z`.

The technology can also be used to perform massive experimentation and uncover biological circuits with high-throughput CRISPR screens; see Binan et al., *Simultaneous CRISPR screening and spatial transcriptomics reveal intracellular, intercellular, and functional transcriptional circuits*, Cell, Apr 2025 – `https://www.cell.com/cell/fulltext/S0092-8674(25)00197-7`.

The field is poised to experience continued growth and bring transformative breakthroughs to multiple areas of science, not to mention generate lots of big datasets for bioinformatics professionals!

Let's try one last example analysis before we leave.

AI tip

Prompt: Create a comprehensive single-cell proteomic analysis example using the SCeptre Python library with the following specifications: Import and use the SCeptre library with fallback handling; Generate realistic synthetic mass spectrometry proteomics data mimicking TMT multiplexed experiments; Implement a complete analysis pipeline following Sceptre's workflow methodology; Include extensive visualizations at each analysis step; Provide both synthetic data generation and real data usage examples; Include clustering and UMAP visualization

You should see: Complete code to implement a single-cell proteomic analysis using the SCeptre Python library, including methods to cluster, interpret, and visualize the results.

Note: If you have any trouble with SCeptre, you can try a fallback workflow using `scanpy` or another library.

See also

- For further review of single-cell technology, read Wen et al., *Single-cell technologies: From research to application*, The Innovation, Nov 2022 – `https://www.cell.com/the-innovation/fulltext/S2666-6758(22)00138-2`

- AnnSQL combines anndata and DuckDB to create a high-performance single-cell analysis system; see Pavan & Saunders, *AnnSQL: a Python SQL-based package for fast large-scale single-cell genomics analysis using minimal computational resources*, Bioinformatics Advances, May 2025 – `https://academic.oup.com/bioinformaticsadvances/article/5/1/vbaf105/8125003`

- rapids-singlecell provides GPU acceleration for single-cell analysis using anndata – `https://rapids-singlecell.readthedocs.io/en/latest/`

- Single-cell proteomics can be extended to analyze post-translational modifications and protein-protein interactions; see Mun et al., *Diversity of post-translational modifications and cell signaling revealed by single cell and single organelle mass spectrometry*, Communications Biology, Jul 2024 – `https://www.nature.com/articles/s42003-024-06579-7`

- Read about the use of Transformers in single-cell analysis: Szalata et al., *Transformers in single-cell omics: a review and new perspectives*, Nature Methods, Aug 2024 – `https://www.nature.com/articles/s41592-024-02353-z?fromPaywallRec=false`

- Longcell can accurately identify isoforms in different cells in specific tissue areas using single-cell combined with nanopore sequencing; see Fu et al., *Single cell and spatial alternative splicing analysis with Nanopore long read sequencing*, Nature Communications, Jul 2025 – `https://www.nature.com/articles/s41467-025-60902-2`

Studying biological image data

In this recipe, we will delve into the exciting world of image analysis. This is an area where AI has traditionally excelled, and it is now being applied with great success to biological imaging and analysis.

You have probably heard of machine learning being applied to recognize images of cats and have almost certainly heard about self-driving cars. Image recognition was one of the first big areas for AI, along with speech generation and recognition. These same tools are applied to image analysis of cells to determine their biological features.

Image analysis is useful in many areas of biology in its own right, including **digital pathology** to look at cancerous tissues, or imaging many individual cells in **flow cytometry**. We'll touch on those applications later in this recipe. Learning more about image analysis is also great preparation for the next recipe, where we'll learn how massive numbers of these images can be stitched together to build up a wiring diagram of the brain.

Let's make sure you have at least a basic knowledge of human cell structure. Review this diagram:

Human Cell Structure

A detailed view of a typical eukaryotic cell with major organelles

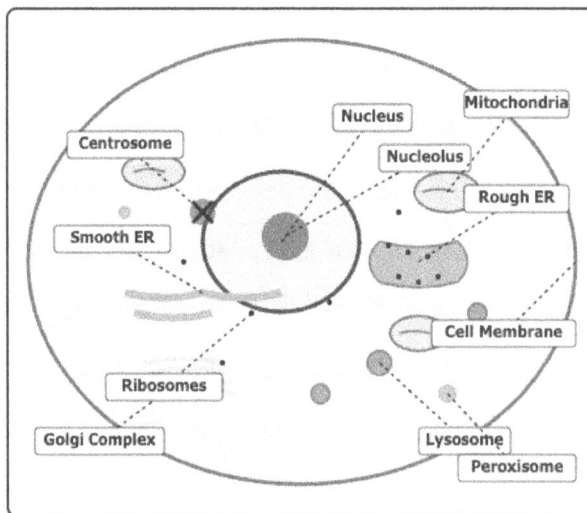

Figure 18.10 – Overview of human cell structure

This shows us a high-level overview of eukaryotic cell structure (prokaryotes, or bacteria, are, of course, simpler). Eukaryotes are organisms with cells that have a nucleus and other membrane-bound **organelles** inside their main membrane. Eukaryotic organisms include humans, animals, and plants. The cell is surrounded by a **cell membrane** filled with **cytoplasm**. Inside, we find numerous important cell structures:

- The **nucleus** houses the DNA, which is the instructions for making all cellular proteins.

- The **nucleolus** lies inside the nucleus and acts as a ribosome factory.

- The **centrosome** is involved in cell division.

- The **smooth ER** is a network of membranes in the cell that transports molecules through the cell; it is involved in lipid and steroid synthesis. It is called smooth because it does not contain ribosomes.

- The **rough ER** is called rough because it contains many ribosomes, which are busy synthesizing proteins. The rough ER is crucial for sorting, folding, and transporting membrane-bound proteins.

- **Ribosomes** are the protein factories of the cell, churning out proteins based on mRNA codes.

- The **Golgi complex** sorts, packages, and delivers proteins. It can build lipid-bound protein vesicles for transport inside or outside the cell.

- **Mitochondria** are the powerhouses of the cell. They create energy by performing cellular respiration, breaking down sugars to make **Adenosine Tri-Phosphate** (ATP), the primary *energy currency* of the cell.

- The **lysosome** is the cell's digestive system. It contains enzymes specialized in breaking down waste materials.

- **Peroxisomes** detoxify the cell and break down fatty acids; they generate hydrogen peroxide.

This gives you a basic idea of the structures we'll be looking for. Of course, different organisms, especially plants, have their own specialized cell structures, but the overall idea is the same. Amazingly, we are still understanding the cell and do not even know all of its core structures. Recently, scientists discovered a totally new organelle called the **hemifusome** – `https://www.livescience.com/health/scientists-discover-never-before-seen-part-of-human-cells-and-it-looks-like-a-snowman-wearing-a-scarf`.

For a deeper review of cell anatomy, see `https://www.youtube.com/watch?v=t5DvF5OVr1Y`.

In this recipe, we will build a cell image analyzer using the popular Python image analysis library scikit-image – `https://scikit-image.org/`. We'll construct a simplified synthetic cell with obvious features so that we can concentrate on the basics of image analysis. By the end of this recipe, you should have a good idea of the libraries and tools used in biological image analysis.

Getting ready

We will install scikit-image:

```
! pip install scikit-image
```

How to do it...

Here are the steps to try this recipe:

1. Let's import our libraries:

```
import numpy as np
import matplotlib.pyplot as plt
from skimage import measure, morphology, filters, segmentation
from scipy import ndimage
```

   We will use the scikit-image (`skimage`) library. From this, we will import a few different modules. The `measure` module will help us understand basic properties of our image, such as the area, perimeter, and center points – `https://scikit-image.org/docs/0.25.x/api/skimage.measure.html`. The `morphology` module helps us manipulate images through operations such as dilation or to remove small objects to clean up images – `https://scikit-image.org/docs/0.25.x/api/skimage.morphology.html`. The `filters` module is used for edge detection and noise reduction – `https://scikit-image.org/docs/0.25.x/api/skimage.filters.html`. The `segmentation` module helps us partition images into sub-regions – `https://scikit-image.org/docs/0.25.x/api/skimage.segmentation.html`. We will also use the `ndimage` module from `scipy`, which is useful for things such as texture analysis – `https://docs.scipy.org/doc/scipy/reference/ndimage.html`.

2. First, we will build a function that can create a synthetic cell image to analyze. Review the code for `create_cell_image()` in your notebook.

   This code will build a simple synthetic image of a cell with a few key features for us to identify. The features will be a cell membrane, a nucleus, a mitochondrion, and the endoplasmic reticulum. Although this is a highly oversimplified image, it will help you focus on the basics of image analysis and dealing with simple shapes that have obvious contrasting features. In the next recipe, *Mapping the brain*, we will use a real image.

We first initialize a blank image with zeros.

I.   Next, we create the cell membrane, which will form the outer boundary of our cell. We create a circle and then build a **mask** with true values inside the circle and false values outside the circle. We then set the pixels inside the circle to a gray value of 50 (this represents the cytoplasm).

II.   Then, we use a similar masking technique to create a bright boundary around the membrane.

III.   We then create a large central object that will represent the nucleus and include a nuclear membrane.

IV.   We add several mitochondria, which are small organelles that power the cell.

V.   We build an endoplasmic reticulum by creating a series of tubular structures and then connecting them with lines.

3. Next is our core function for cellular image analysis, `analyze_cell_image()`. Review this function in the notebook.

4. We first print a few basic image properties. Next, we perform **image segmentation** based on various thresholds. This is based on the idea that different components of the image will have different levels of contrast. For example, the background (outside of the cell) is expected to be very dark, membranes are very bright, mitochondria are medium-bright, and so on. We'll end up with area calculations for each component of the image. We will then perform a series of analyses, including morphological analysis, mitochondrial counting, edge detection, and texture analysis. Note that although we are using a synthetic image here, this type of analysis would be very typical of an **EM** (short for **electron microscopy**) image analysis. We perform the following types of analysis:For **morphological analysis**, we will first use our predefined nucleus mask, which looks like a large circular object as we would expect from a cell nucleus. We use the scikit-image `regionprops()` function to measure the properties of various regions in the image to see whether they match our nucleus mask. We take the first such match, assuming there is only one nucleus per cell.

5. We next perform **mitochondrial counting**. To do this, we use our predefined mitochondria component mask, based on the intensity thresholds we defined. We first clean up any small objects and then label our mitochondria and measure their region properties. We calculate the total mitochondrial area and the average size of a mitochondrion.

6. We next perform **edge detection**. In this technique, we use the **Sobel** filter to find sharp changes in the intensity gradient, which are likely to represent an edge – `https://homepages.inf.ed.ac.uk/rbf/HIPR2/sobel.htm`.

7. Then, we identify circular objects based on the edges, using a technique called the **Hough circle transform**, which looks for common points of intensity that are all at a similar distance from a defined center – in other words, something that looks like a circle. See `https://en.wikipedia.org/wiki/Circle_Hough_Transform`.

8. We next look for the biggest outer circle to help us identify the cell membrane. Edge pixels that cluster most frequently around a common distance from the center are likely the cell membrane.

9. We next perform **texture analysis**. This looks at how the intensity varies in different parts of the image. To do this, we make a small disk and move it around the image, calculating the standard deviation of intensity variation in each area. This allows us to identify areas with higher texture (rough) and lower texture (smooth). Rough areas may be indicative of the **endoplasmic reticulum**, for example.

10. Up next is a function to visualize our analysis. Review the `visualize_analysis()` function in the notebook.

    This code creates a six-panel visualization. We first set up a 2x3 grid of subplots.

    We first show the raw cell image. Next up is the result of our segmentation analysis, showing each of the major components revealed. We use a "hot" **colormap** (**cmap**), which will show strong edges (membranes) as bright lines.

    Next will be an image of our edge detection analysis.

    In the next two images, we highlight the nucleus and mitochondria, respectively.

    The final image shows our texture analysis using a "plasma" cmap, in which darker colors represent smoother textures and brighter colors represent rougher textures. Learn more about cmaps here: `https://matplotlib.org/stable/users/explain/colors/colormaps.html`.

11. OK, let's run our code! We will execute the functions in the `main()` code section:

    I.   We first create our synthetic cell image.

    II. Next, we run `analyze_cell_image()`.

    III. Finally, we perform our visualization and print a brief summary.

Here is our output:

```
Creating synthetic cell image...
=== CELL IMAGE ANALYSIS ===

Image dimensions: (200, 200)
Pixel value range: 0 - 255
Mean intensity: 53.91

1. COMPONENT SEGMENTATION
----------------------------------------
Background   : 14555 pixels (36.39%)
Cytoplasm    : 18762 pixels (46.91%)
ER           :   972 pixels ( 2.43%)
Mitochondria:    778 pixels ( 1.94%)
Nucleus      :  2894 pixels ( 7.23%)
Membranes    :  2039 pixels ( 5.10%)

2. MORPHOLOGICAL ANALYSIS
----------------------------------------
Nucleus area: 2894.0 pixels
Nucleus centroid: (95.4, 108.9)
Nucleus equivalent diameter: 60.70 pixels
Number of mitochondria detected: 4
Total mitochondrial area: 778.0 pixels
Average mitochondrial size: 194.50 pixels

3. MEMBRANE ANALYSIS
----------------------------------------
Total edge pixels detected: 3994
Estimated cell radius: 86.88 pixels

4. TEXTURE ANALYSIS
----------------------------------------
Average texture (local std): 21.95
Texture range: 0.00 - 120.60
High texture regions: 7998 pixels (20.0%)

Generating visualization...
```

Figure 18.11 – Overview of the image analysis

This tells us about the synthetic image we created and summarizes the results of our segmentation analysis. We can see that we correctly identified the major components of the cell. We then performed a morphological analysis and found the nucleus, its centroid, and its diameter. We identified several mitochondria.

We performed edge detection and found the cell membrane. Finally, we ran our texture analysis. Next, we see a visualization of our cell image in different states:

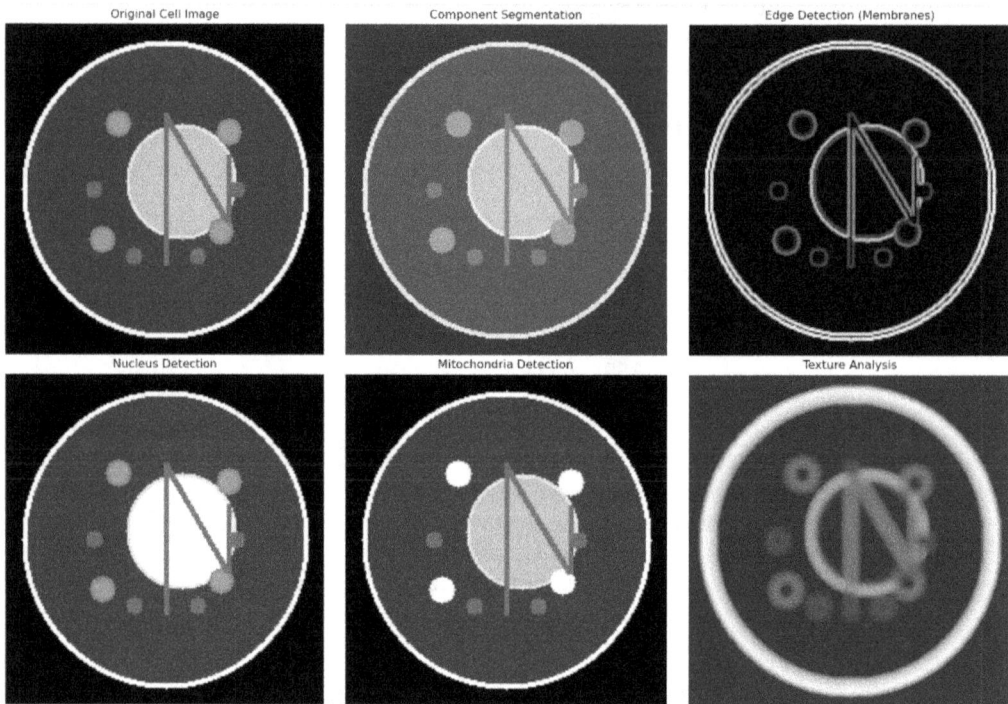

Figure 18.12 – Cell image analysis

The preceding figure shows various aspects of the cell highlighted as they are detected. You can see how the nucleus and key organelles are identified and then lit up to better understand what the algorithm detected.

There's more...

This was a simple introduction to cell image analysis to familiarize you with the basics. You can see how powerful tools such as scikit-image can be used to perform cellular analysis.

Cellular imaging technology is being used in many areas of science. In **digital pathology**, machine learning algorithms analyze tissue sections to identify and classify cancers; see Waqa et al., *Digital pathology and multimodal learning on oncology data*, BJR Artificial Intelligence, Sep 2024 – `https://academic.oup.com/bjrai/article/1/1/ubae014/7755042`.

Companies such as Pathomiq (`https://pathomiq.com/`) use machine learning to perform digital pathology, identifying cancer subtypes to guide appropriate treatment. Other companies in this space include PathAI (`https://www.pathai.com/`), Proscia (`https://proscia.com/`), and Paige.ai (`https://www.paige.ai/`).

Flow Cytometry is a technique for analyzing cells by moving them through a fluid stream and passing them under a laser or other devices to measure their characteristics. It can be combined with imaging to rapidly analyze cellular phenotypes; see Muffels et al, *Imaging flow cytometry-based cellular screening elucidates pathophysiology in individuals with Variants of Uncertain Significance*, Genome Medicine, Feb 2025 - `https://link.springer.com/article/10.1186/s13073-025-01433-9`

In **High-Content Imaging** (**HCI**), automated microscopy and image analysis is used to gather huge amounts of data. This can include cell morphology, protein expression, and other parameters. **Cell painting** is a type of HCI in which cells are stained with a variety of dyes that highlight various cellular features. Cells can be subjected to a variety of small molecule libraries or other perturbations, and machine learning algorithms can identify the patterns in the cell painting data that change and even make inferences on what biological pathways are being affected; see Stossi et al., *SPACe: an open-source, single-cell analysis of Cell Painting data*, Nature Communications, Nov 2024 – `https://www.nature.com/articles/s41467-024-54264-4`.

Companies such as ViQi (`https://www.viqiai.com/`) offer high-throughput image analysis solutions. Akoya Biosciences (`https://www.akoyabio.com/`) offers spatial phenotyping. Other companies in this space include Molecular Devices, Thermo Fisher Scientific, PerkinElmer, and Sartorius.

See also

- Perturb-tracing uses a CRISPR library combined with HCI to probe chromatin topology; see Cheng et al., *Perturb-tracing enables high-content screening of multi-scale 3D genome regulators*, Nature Methods, Apr 2025 – `https://www.nature.com/articles/s41592-025-02652-z`

- HCI and single-cell transcriptomics can be linked; see Tsuchida et al., *Opto-combinatorial indexing enables high-content transcriptomics by linking cell images and transcriptomes*, Lab Chip, Mar 2024 – `https://pubs.rsc.org/en/content/articlehtml/2024/lc/d3lc00866e`

- IACS goes beyond traditional approaches to provide high-content sorting of cells; see Ding et al., *Image-activated cell sorting*, Nature Reviews Bioengineering, Jul 2025 – `https://www.nature.com/articles/s44222-025-00334-1`

- Live-cell imaging looks at living (unfixed) cells while they are actively functioning; see Shroff et al., *Live-cell imaging powered by computation*, Nature Review MCB, Feb 2024 – `https://www.nature.com/articles/s41580-024-00702-6`

- **Hyperspectral imaging** (**HSI**) goes beyond normal wavelengths to capture cellular components such as proteins, lipids, and nucleic acids; see Liu et al., *Analysis of cellular response to drugs with a microfluidic single-cell platform based on hyperspectral imaging*, Analytica Chimica Acta, Feb 2024 – `https://www.sciencedirect.com/science/article/pii/S000326702301379X`

- ImagePy is another image processing library to check out – `https://github.com/Image-Py/imagepy`

Mapping the brain

In this recipe, we will learn about an exciting new area of science in which bioinformatics is playing a role, **connectome mapping**.

The connectome is defined as the map of all connections in a brain or a section of neural tissue. Using the image analysis techniques we discussed in the *Studying biological image data* recipe scientists can now combine high-throughput automated microscopy and machine learning to trace neurons and their connections in the brain. They then use this to reconstruct and study partial or whole connectomes of various organisms, including humans.

Let's review the structure of a neuron:

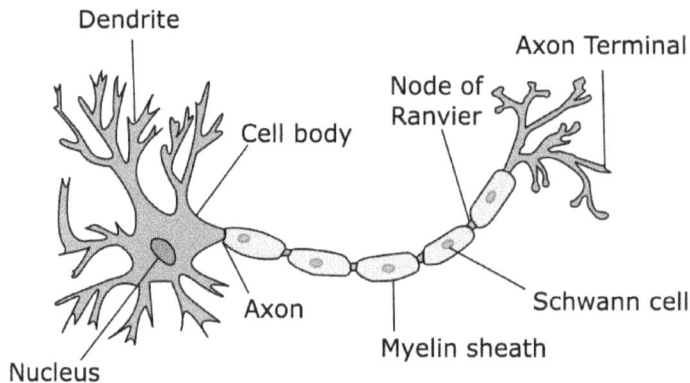

Figure 18.13 – Structure of a neuron

Source: Wikimedia Commons – https://commons.wikimedia.org/wiki/File:Neuron.svg (https://commons.wikimedia.org/wiki/Commons:GNU_Free_Documentation_License,_version_1.2)

Neurons contain a large cell body or **soma**, which contains the cell nucleus. Inputs come into the cell through **dendrites**, which are small projections full of **ion channels**. When two dendrites connect to each other, they form a **synapse**. At the synapse, small vesicles with neurotransmitters are released and travel across the **synaptic cleft** to interact with receptors on the surface of the next dendrite. These interactions may be **excitatory** or **inhibitory**.

If enough of the dendritic inputs to the neuron are excitatory and cross a **threshold**, the neuron will fire a pulse of electricity (referred to as an **action potential** or **spike**) down the **axon**. This is a long projection that connects one neuron to another (axons can project all the way from your brain down to your lower spinal cord!). The axons are sheathed in **myelin**, a fatty substance that insulates the axon and improves electrical conductivity. Periodically along the axon there are breaks in the myelin sheath called **nodes of Ranvier**; they are full of ion channels and promote jumping (acceleration) of the electrical signal. **Schwann cells** are a type of **glial cell**, which are support cells that help make the myelin sheath and support and repair neurons. For more information, read Holler et al., *Structure and function of a neocortical synapse*, Nature, Jan 2021 – https://www.nature.com/articles/s41586-020-03134-2.

You can see that the structures of a real neuron are analogous to the neural networks we learned about in *Chapter 17, Deep Learning and LLMs for Nucleic Acid and Protein Design*. Weighted inputs come into a neuron and are summed up to see whether they surpass a threshold. If they do, this results in an output (think of the output state as 1), and if not, there is no output (output state of 0).

Scientists are now able to fix brain tissue and section it using **automated microtomy** – a process in which a machine carefully makes thin slices of tissue. They can then stain the tissue in various ways and perform automated image analysis to find neurons and synaptic connections; see Kleinfeld et al., *Large-Scale Automated Histology in the Pursuit of Connectomes*, Journal of Neuroscience, Nov 2011 – `https://www.jneurosci.org/content/31/45/16125.full`.

In this example, we'll look at a cross-section of brain tissue and try to identify neuron bodies and dendrites that are close to each other, implying that they share a synaptic cleft and hence a connection. We'll use this approach to build up a putative connectome.

In this recipe, we will use a real brain image from the **Allen Brain Atlas** to explore connectome mapping - `https://brain-map.org/`. We'll learn how to identify synaptic connnections and build a representative network from the data. We'll be using Python imaging and network libraries to identify and represent our connectome!

Getting ready

Let's install the tools we need:

```
! pip install numpy matplotlib seaborn scipy scikit-image scikit-learn
networkx pandas
```

How to do it...

Here are the steps to try this recipe:

1. First, we will import the necessary libraries.

   We will use **PIL**, the **Python Image Library**. It will help us download a brain image and convert it to NumPy arrays for further analysis – `https://pillow.readthedocs.io/en/stable/reference/Image.html`.

   We will also use the `ndimage` and `networkx` libraries that we have used before.

2. The next section of code helps deal with making sure scikit-image (`skimage`) can be imported properly. The code attempts to import skimage and if this fails, it defines some simple alternative functions using scipy. Run this code in your notebook and then move on to the next cell.

3. Next up is an important function to perform peak detection in our image. Review the code for the `peak_local_maxima()` function in the notebook.

   This function will look for bright spots in the brain image that represent neuron bodies.

We use the `ndimage.maximum_filter()` method to find pixels brighter than any others in their neighborhood, which often indicate cell bodies – `https://docs.scipy.org/doc/scipy/reference/generated/scipy.ndimage.maximum_filter.html`.

We remove any peaks near the image border, as these are often artifacts.

4. Next, we have the heart of our connectome analyzer – the `NeuronConnectivityAnalyzer` class. Let's review what this class does:

We first initialize key data structures and set up our connection to the Allen Brain Atlas. We'll be using it to download our high-resolution brain image.

We define several IDs for high-resolution images as fallbacks in case our main download fails.

In the `download_neuron_image()` function, we define a known image ID and use the `requests` library to download it. As mentioned before, we will use the PIL imaging library to convert the image into a numerical NumPy array. We will also set some basic metadata about our image.

Next are some internal helper functions. `_crop_to_neuron_region()` trims an image down to focus on a region containing a neuron. `_create_synthetic_neurons()` is used as a fallback in case our download fails. `_enhance_image_contrast()` is used to provide additional contrast in the image if needed. `_add_enhanced_dendrite()` is used to make a clearly visible dendrite. We provide similar functions for enhanced axons and synaptic connections.

Now we come to a key function in our workflow, `preprocess_neuron_image()`. This function converts the image from **Red-Green-Blue (RGB)** to grayscale and enhances contrast using the NumPy `clip()` function. It applies smoothing to perform denoising using the `ndimage.gaussian_filter()`. It then applies some additional filtering to enhance neural structures within the image.

I.   The next section of the code detects neurons. The `detect_individual_neuron()` function will call sub-functions for analysis. It first uses **blob analysis** to find the neurons. In this technique, we look for regions that are similar to each other but different from their surroundings that are connected together – `https://scikit-image.org/docs/0.25.x/auto_examples/features_detection/plot_blob.html`.

II. If the blob analysis does not work, we fall back on using a peak detection approach.

III. We store the neurons in the `neurons` dictionary. We define a function for calculating the radius of each neuron. We store key properties for each neuron. We also provide some basic biological classification – large neurons tend to be motor neurons, medium-sized neurons may be "interneurons" that act as connections between two neurons, and small neurons may be involved in local computations.

IV. The next key function is `trace_neuron_connections()`. This function traces the connections between neurons. We create a NetworkX `Graph` object to store our connectome. We add each neuron in our dictionary as a **node** in the graph. We use **path tracing** to find connections between neurons. In this method, we move out from the neuron looking for bright areas that stand out from the background that may represent neural projections such as axons and dendrites. We continue until we find another neuron body, showing that the two neurons are connected.

V.   When we find a connection between two neurons, we call add_edge() to add this to our connectome graph. We also estimate a connection strength to indicate confidence in the connection. We print out the number of connections and a list of which neuron is connected to which.

VI.   The _trace_connection_path() helper function is used to look for an axonal path from one neuron to another. It starts with two neurons and defines their centroid points. We sample 20 points along the potential path between the 2 neurons. We look at 5x5 image neighborhoods for bright sections that stand out against the background – these might be the axon. If we find such areas as we go, we build up our confidence in the potential connection strength. Finally, we normalize by the number of times we needed to sample and the distance between the two neurons (greater distances imply less likelihood of connection). We return the putative connection strength between the two input neurons.

VII. After tracing the neuron connections, we will have built up a connectome graph with neurons as nodes and connections as edges. We next want to analyze and understand this network, using the analyze_neuron_network() function. This function will use the NetworkX module to perform various calculations on the connectome we have defined, as follows:

   i.   First, we use the density() function to calculate how well connected the graph is – https://networkx.org/documentation/stable/reference/generated/networkx.classes.function.density.html.

   ii.   We next compute the **average degree**. This represents the average number of connections per neuron.

   iii. We identify **hub neurons**, which are neurons that have maximum connectivity.

   iv.   We determine whether the graph is **fully connected** or **partially connected**. If the graph is fully connected, we calculate the **diameter** – the greatest distance you would ever need to travel to get from one node to another – https://networkx.org/documentation/stable/reference/algorithms/generated/networkx.algorithms.distance_measures.diameter.html.

   v.   For each neuron, we determine its type and measure its average size and size variability.

   vi.   Next is our function visualize_neuron_connectivity(). This code creates a six-panel image with the original brain image, detected neurons, neuron connections, a network representation of the connectome graph, a scatter plot of the neuron properties, and an analysis summary.

   vii. Once you have reviewed the NeuronConnectivityAnalyzer class in your notebook, go ahead and run the cell. This will define your class. We are now ready to use the code we have defined to run our connectome analysis!

5. Let's execute our analysis using the main() section:

    I.   We first check that skimage is installed.

    II.   Next, we initialize our NeuronConnectivityAnalyzer().

    III. We then find and download our high-resolution brain image.

    IV.   We run preprocessing on the brain image.

    V.   Then, we detect individual neurons.

    VI.   Next, we run connection tracing.

    VII. We then perform network analysis on our connectome graph.

    VIII. Finally, we run our visualization function and write out a brief summary.

    Amazing! We have now shown how we can analyze a brain image, identify neurons and synaptic connections, and represent it as a connectome graph.

6. Review the output.

    We see that our code has identified and downloaded a suitable brain image (this should only take a few seconds).

    We see a summary of image preprocessing and neuron detection. We have detected three neurons!

    We next trace connections between neurons. We find one connection, between neuron *1* and neuron *2*, with a connection strength of 0.43.

    We perform our network analysis – here are the results:

```
6. Analyzing neural network...
Analyzing neural network at cellular level...
  Cellular network analysis:
    n_neurons: 3
    n_connections: 1
    density: 0.3333333333333333
    avg_degree: 0.6666666666666666
    hub_neurons: [1, 2]
    connectivity: partially_connected
    avg_neuron_size: 421.0
    size_variability: 0.0
```

Figure 18.14 – Connectome network analysis

You can see that we have identified three neurons with one connection, giving a network density of 0.33. The average degree is 0.66. Neurons *1* and *2* are identified as hub neurons. This is a partially connected network. The average neuron size is 421.

Here is our visualization of the analysis:

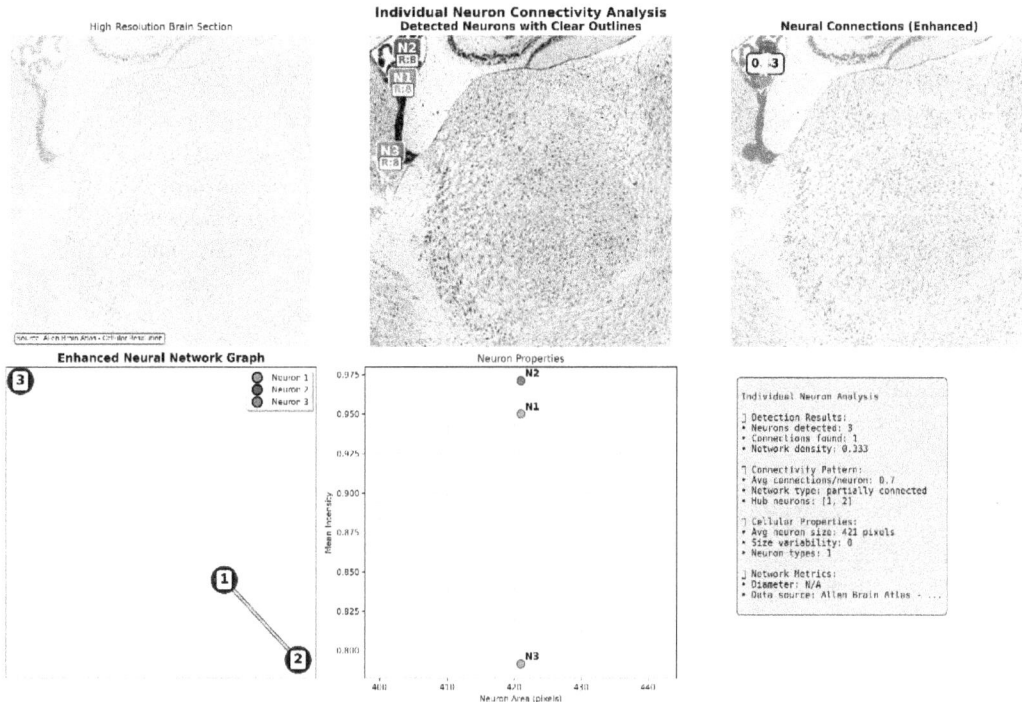

Figure 18.15 – Neural connectivity analysis

This figure shows the original brain slice in the upper left. You can see areas in the upper-left corner where two synapses have come to connect with each other. The next image shows neurons *1*, *2*, and *3* identified with an overlay. As you can see, neurons *1* and *2* lie close to each other in what looks like a synaptic connection. But neuron *3* is farther away by itself. In the third panel, we see the connections traced.

The fourth panel shows the connectivity graph, with neurons *1* and *2* connected and neuron *3* isolated. After that, we see the neuron properties analysis, with the average area of each cell on the *x* axis, and the brightness of the cell on the *y* axis. Finally, in the last panel, we have a text summary of the results.

This was, of course, a very simple example. You can easily imagine that bigger sections of brain tissue could be scanned, traced, and modeled into large-scale connectomes, given enough microscopy and computing horsepower, using this type of approach.

Imagine if we could take our brain connectome network and use it as the basis to build a neural network model like the ones we learned about in *Chapter 17, Deep Learning and LLMs for Nucleic Acid and Protein Design*!

AI tip

Prepare: Paste or upload your notebook for this recipe into your AI tool.

Prompt: Analyze the connectome analysis code I just uploaded.

Prompt: Build a connectome-to-neural network converter. I have biological connectome data (neurons, connections, network graph) from brain circuit analysis. Convert this into a PyTorch neural network that preserves the exact biological connectivity patterns and neuron properties. Generate complete working code for neuromorphic AI applications. Ensure that the new code can be called directly from the output of the connectome analysis code I just uploaded and provide me with instructions on how to run the neural network converter from the outputs of the connectome analysis code. Include code to generate a small test dataset. Include visualization of the network architecture, training results, and validation outputs.

You should see: Code to build and train a neural network using PyTorch, with a similar architecture to the one you discovered using your connectome analysis.

Note how we pass the example data from the *Mapping the brain* recipe into the converter:

```
analyzer, analysis = main()
converter, networks = convert_connectome_to_neural_
network(analyzer, analysis)
```

There's more...

We have now seen how to analyze brain images and build up a connectome. This exciting field is poised to bring transformative capabilities to our world. It will first transform the field of neuroscience and bring about amazing new treatments for mental disorders. It will also lead to more brain-like neurocomputing, driving the advancement of AI even faster. Ultimately, this field may lead to more human-like computers, AIs that can design chips and program themselves, household robots, and other new applications we may not even be able to imagine yet! Let's review some of the current advances happening in the field of connectome mapping:

- One of the core technologies for brain mapping remains EM. It is cumbersome and requires extensive fixation, but can reliably map large sections of the connectome and is advancing rapidly; see Schmidt et al., *RoboEM: automated 3D flight tracing for synaptic-resolution connectomics*, Nature Methods, Mar 2024 – https://www.nature.com/articles/s41592-024-02226-5.

- Connectome maps based on EM can serve as an excellent basis for registration of additional modalities. For example, the MICrONS consortium combined calcium imaging of ~75,000 neurons with an EM map of ~200,000 neurons to model a huge portion of the mouse visual cortex; see *Functional connectomics spanning multiple areas of mouse visual cortex*, Nature, Apr 2025 – https://www.nature.com/articles/s41586-025-08790-w.

- Scientists have now achieved the connectome mapping of a full brain, that of the fruit fly, *drosophila melanogaster*; see Dorkenwald et al., *Neuronal wiring diagram of an adult brain*, Nature, Oct 2024 – `https://www.nature.com/articles/s41586-024-07558-y`.

- New advances are pushing the field even further. **Expansion microscopy** uses hydrogels to expand brain tissue, giving optical microscopes access to resolution once only obtainable through EM. **Light-sheet microscopy** illuminates only a plane of the specimen at a time, providing excellent 3D reconstructions; see Pesce et al., *Expansion and Light-Sheet Microscopy for Nanoscale 3D Imaging*, Small Methods, Mar 2024 – `https://onlinelibrary.wiley.com/doi/full/10.1002/smtd.202301715`.

- Techniques such as Brainbow (`https://en.wikipedia.org/wiki/Brainbow`) are also expanding the connectome toolbox. In this method, transgenic animals are prepared by expressing combinatorial fluorescent barcode markers that assist with connection tracing. Similarly, sequencing technologies can be integrated to enhance connectome mapping; Chen et al., *Connectome-seq: High-throughput Mapping of Neuronal Connectivity at Single-Synapse Resolution via Barcode Sequencing*, bioRxiv, Feb 2025 – `https://www.biorxiv.org/content/10.1101/2025.02.13.638129v1.full`.

- Researchers around the globe are now rapidly building connectomes for numerous organisms. Companies such as E11 Bio (`https://e11.bio/`) aim to map entire mammalian brains. The Human Connectome project (`https://www.humanconnectome.org/`) seeks to map the human brain. The Developing Human Connectome Project seeks to understand the human connectome as it develops from birth to early childhood; see Ma et al., *The Developing Human Connectome Project: A fast deep learning-based pipeline for neonatal cortical surface reconstruction*, Medical Image Analysis, Feb 2025 – `https://www.sciencedirect.com/science/article/pii/S1361841524003190`.

- Real neurons are, of course, not as simple as we have presented here. The electrophysiology of neural spikes is such that they are not simply additive, but instead contain information encoded in complex patterns. For a great treatment on this topic, check out *Spikes: Exploring the Neural Code* by Fred Rieke – `https://www.amazon.com/Spikes-Exploring-Neural-Computational-Neuroscience/dp/0262681080/`.

- Likewise, dendritic computations are not simple summations but instead implement complex non-linear logic based on the branching structure of the dendrite and the nature and types of ion channels located within it; see Stingl et al., *A dendrite is a dendrite is a dendrite? Dendritic signal integration beyond the 'antenna' model*, European Journal of Physiology, Aug 2024 – `https://link.springer.com/article/10.1007/s00424-024-03004-0`.

- Inspired by real neurons, hardware engineers are building the next generation of **neuromorphic computing**, in which hardware emulates biology; see Kudithipudi et al., *Neuromorphic computing at scale*, Nature, Jan 2025 – `https://www.nature.com/articles/s41586-024-08253-8`.

- With further advances in connectome sophistication and hardware performance, scientists will begin to explore the simulation of consciousness in connectome models; see Klatzmann et al., *A mesoscale connectome-based model of conscious access in the macaque monkey*, bioRxiv, Oct 2024 – `https://www.biorxiv.org/content/10.1101/2022.02.20.481 230v4.abstract`.

This work will deepen our understanding of human consciousness and improve our AI models, leading to AI with more advanced traits. It may even give us the potential to model part or all of a human brain with its lifetime of experience in an AI platform, or even expand conscious computing beyond the limitations of the human mind. These possibilities are at present unknown, but with the exponential advances in technology and the skills you've learned, you can be part of the adventure!

OK, let's clean up our environment:

```
conda deactivate
```

See also

- Learn more about digital pathology by building a melanoma classifier in PyTorch – `https://www.youtube.com/playlist?list=PLWVKUEZ25V975TSDG6xdjIB-5chrrUP-S`

- Read more about scikit-image in Gouillart et al., *Analyzing microtomography data with Python and the scikit-image library*, Advanced Structural and Chemical Imaging, Dec 2016 – `https://link.springer.com/article/10.1186/s40679-016-0031-0`

- X-ray microscopy is another powerful tool for connectome analysis; see Hwu et al., *Q&A: Why use synchrotron x-ray tomography for multi-scale connectome mapping*, BMC Biology, Dec 2017 – `https://link.springer.com/article/10.1186/s12915-017-0461-8`

- High-field MRI is also revolutionizing connectome studies; see Cabalo et al., *Multimodal precision MRI of the individual human brain at ultra-high fields*, Scientific Data, Mar 2025 – `https://www.nature.com/articles/s41597-025-04863-7`

- PyTorch Connectomics provides a package for reconstructing connectomes; see Lin et al., *PyTorch Connectomics: A Scalable and Flexible Segmentation Framework for EM Connectomics*, arXiv, Dec 2021 – `https://arxiv.org/abs/2112.05754`

- Connectome Mapper is a Python connectome mapping pipeline; see Tourbier et al., *Connectome Mapper 3: A Flexible and Open-Source Pipeline Software for Multiscale Multimodal Human Connectome Mapping*, JOSS, Jun 2022 – `https://joss.theoj.org/papers/10.21105/joss.04248`

- Nilearn is another connectome analysis Python package you should check out – `https://nilearn.github.io/stable/index.html`

We covered a lot of new terms in this chapter! Here is a quick glossary for you to help solidify your learning.

Glossary

- **Bulk cell analysis**: Mixing together many individual cells to analyze them, typically with genomic, RNA, or protein profiling.

- **Connectome**: The complete map of neuronal connections in a brain or brain section.

- **CRISPR**: Standing for **Clustered Regularly Interspaced Short Palindromic Repeats**, this is a key genome editing technology derived from bacterial immune systems, in which guide RNAs bring a nuclease to a particular site in the genome to make modifications.

- **Electron Microscopy** (**EM**): A microscope that uses electrons, rather than light, to pass through a material. Since electrons have shorter wavelengths, they can reveal sub-nanometer features.

- **High Throughput Screening** (**HTS**): Uses arrays of many samples, typically in square plates, along with robotics to test and analyze many different conditions. Used in screening for drug development, strain engineering, or testing of culture conditions.

- **Microfluidics**: Devices containing channels on the micrometer scale. They can sort, mix, and manipulate cells and droplets.

- **Multi-omics**: Any technique that combines two or more layers of molecular analysis, including genomic (DNA), epigenomic (methylation), transcriptomic (RNA), proteomic (protein), or metabolomic (small molecule).

- **Organelle**: A small membrane-bound structure within a cell that compartmentalizes and carries out key functions (e.g., mitochondria).

- **Pseudotime**: A timeline of cell behavior that is based on a series of biological events as opposed to wall clock time.

- **Single-cell analysis**: Techniques in which individual cells are dissected from a tissue or sorted in a microfluidics device to allow for the analysis of a particular cell and its characteristics.

- **Spatial omics**: Performing single-cell analysis on cells that are from known locations in a tissue, for example, via fixation and laser capture dissection. Used to preserve and analyze information about gene or protein expression in particular locations within an organ or tissue.

Get This Book's PDF Version and Exclusive Extras

UNLOCK NOW

Scan the QR code (or go to `packtpub.com/unlock`). Search for this book by name, confirm the edition, and then follow the steps on the page.

Note: Keep your invoice handy. Purchases made directly from Packt don't require an invoice.

19
Unlock Your Exclusive Benefits

Your copy of this book includes the following exclusive benefit:

- ☁ Next-gen Packt Reader
- 📄 DRM-free PDF/ePub downloads

Follow the guide below to unlock them. The process takes only a few minutes and needs to be completed once.

Unlock this Book's Free Benefits in 3 Easy Steps

Step 1

Keep your purchase invoice ready for *Step 3*. If you have a physical copy, scan it using your phone and save it as a PDF, JPG, or PNG.

For more help on finding your invoice, visit `https://www.packtpub.com/unlock-benefits/help`.

> **Note**
> If you bought this book directly from Packt, no invoice is required. After *Step 2*, you can access your exclusive content right away.

Step 2

Scan the QR code or go to `packtpub.com/unlock`.

On the page that opens (similar to *Figure 19.1* on desktop), search for this book by name and select the correct edition.

<packt>   Q Search...   Subscription 🛒 👤

Explore Products   Best Sellers   New Releases   Books   Videos   Audiobooks   Learning Hub   Newsletter Hub   Free Learning

Discover and unlock your book's exclusive benefits

Bought a Packt book? Your purchase may come with free bonus benefits designed to maximise your learning. Discover and unlock them here

Discover Benefits   Sign Up/In   Upload Invoice

Need Help?

✦ 1. Discover your book's exclusive benefits   ∧

  Q Search by title or ISBN

  CONTINUE TO STEP 2

⚖ 2. Login or sign up for free   ∨

☁ 3. Upload your invoice and unlock   ∨

Figure 19.1: Packt unlock landing page on desktop

Step 3

After selecting your book, sign in to your Packt account or create one for free. Then upload your invoice (PDF, PNG, or JPG, up to 10 MB). Follow the on-screen instructions to finish the process.

Need help?

If you get stuck and need help, visit
`https://www.packtpub.com/unlock-benefits/help`
for a detailed FAQ on how to find your invoices and more. This QR code will take you to the help page.

> **Note**
>
> If you are still facing issues, reach out to `customercare@packt.com`.

Index

Symbols

A

‹packt›

packtpub.com

Subscribe to our online digital library for full access to over 7,000 books and videos, as well as industry leading tools to help you plan your personal development and advance your career. For more information, please visit our website.

Why subscribe?

- Spend less time learning and more time coding with practical eBooks and Videos from over 4,000 industry professionals
- Improve your learning with Skill Plans built especially for you
- Get a free eBook or video every month
- Fully searchable for easy access to vital information
- Copy and paste, print, and bookmark content

At www.packtpub.com, you can also read a collection of free technical articles, sign up for a range of free newsletters, and receive exclusive discounts and offers on Packt books and eBooks.

Other Books You May Enjoy

If you enjoyed this book, you may be interested in these other books by Packt:

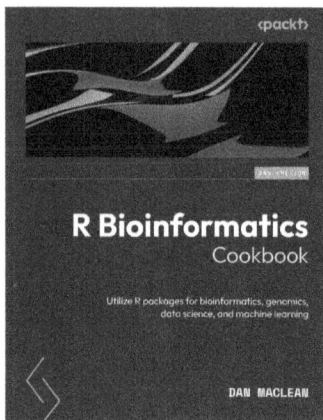

R Bioinformatics Cookbook

Dan MacLean

ISBN: 9781837634279

- Set up a working environment for bioinformatics analysis with R
- Import, clean, and organize bioinformatics data using tidyr
- Create publication-quality plots, reports, and presentations using ggplot2 and Quarto
- Analyze RNA-seq, ChIP-seq, genomics, and next-generation genetics with Bioconductor
- Search for genes and proteins by performing phylogenetics and gene annotation
- Apply ML techniques to bioinformatics data using mlr3
- Streamline programmatic work using iterators and functional tools in the base R and purrr packages
- Use ChatGPT to create, annotate, and debug code and workflows

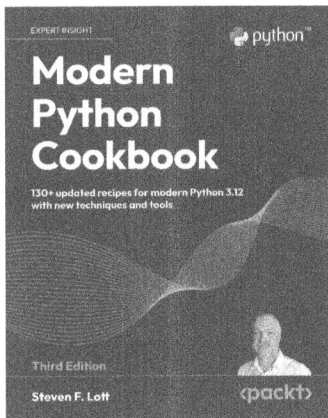

Modern Python Cookbook

Steven F. Lott

ISBN: 9781835466384

- Master core Python data structures, algorithms, and design patterns
- Implement object-oriented designs and functional programming features
- Use type matching and annotations to make more expressive programs
- Create useful data visualizations with Matplotlib and Pyplot
- Manage project dependencies and virtual environments effectively
- Follow best practices for code style and testing
- Create clear and trustworthy documentation for your projects

Packt is searching for authors like you

If you're interested in becoming an author for Packt, please visit `authors.packt.com` and apply today. We have worked with thousands of developers and tech professionals, just like you, to help them share their insight with the global tech community. You can make a general application, apply for a specific hot topic that we are recruiting an author for, or submit your own idea.

Share your thoughts

Now you've finished *Bioinformatics with Python Cookbook, Fourth Edition*, we'd love to hear your thoughts! Scan the QR code below to go straight to the Amazon review page for this book and share your feedback or leave a review on the site that you purchased it from.

`https://packt.link/r/183664275X`

Your review is important to us and the tech community and will help us make sure we're delivering excellent quality content.

www.ingramcontent.com/pod-product-compliance
Lightning Source LLC
Chambersburg PA
CBHW081212220326
41598CB00037B/6754